Mobile Web 2.0

Developing and Delivering Services to Mobile Devices

OTHER TELECOMMUNICATIONS BOOKS FROM AUERBACH

3D Television (3DTV) Technology, Systems, and Deployment: Rolling Out the Infrastructure for Next-Generation Entertainment
Daniel Minoli
ISBN: 978-1-4398-4066-5

Advances in Network Management
Jianguo Ding
ISBN: 978-1-4200-6452-0

Applied Aspects of Optical Communication and LIDAR
Nathan Blaunstein, Shlomi Arnon, Natan Kopeika, and Arkadi Zilberman
ISBN: 978-1-4200-9040-6

Bio-inspired Computing and Networking
Edited by Yang Xiao
ISBN: 978-1-4200-8032-2

Cloud Computing and Software Services: Theory and Techniques
Edited by Syed A. Ahson and Mohammad Ilyas
ISBN: 978-1-4398-0315-8

Cognitive Radio Networks: Architectures, Protocols, and Standards
Edited by Yan Zhang, Jun Zheng, and Hsiao-Hwa Chen
ISBN: 978-1-4200-7775-9

Evolved Cellular Network Planning and Optimization for UMTS and LTE
Edited by Lingyang Song and Jia Shen
ISBN: 978-1-4398-0649-4

Fundamentals of EMS, NMS and OSS/BSS
Jithesh Sathyan
ISBN: 978-1-4200-8573-0

HSDPA/HSUPA Handbook
Edited by Borko Furht and Syed A. Ahson
ISBN: 978-1-4200-7863-3

Integrated Inductors and Transformers: Characterization, Design and Modeling for RF and MM-Wave Applications
Egidio Ragonese, Angelo Scuderi, Tonio Biondi, and Giuseppe Palmisano
ISBN: 978-1-4200-8844-1

IP Communications and Services for NGN
Johnson I. Agbinya
ISBN: 978-1-4200-7090-3

Mobile Opportunistic Networks: Architectures, Protocols and Applications
Edited by Mieso K. Denko
ISBN: 978-1-4200-8812-0

Mobile Web 2.0: Developing and Delivering Services to Mobile Devices
Edited by Syed A. Ahson and Mohammad Ilyas
ISBN: 978-1-4398-0082-9

Orthogonal Frequency Division Multiple Access Fundamentals and Applications
Edited by Tao Jiang, Lingyang Song, and Yan Zhang
ISBN: 978-1-4200-8824-3

Overlay Networks: Toward Information Networking
Sasu Tarkoma
ISBN: 978-1-4398-1371-3

Product Release Planning: Methods, Tools and Applications
Guenther Ruhe
ISBN: 978-0-8493-2620-2

Revenue Assurance: Expert Opinions for Communications Providers
Eric Priezkalns
ISBN: 978-1-4398-5150-0

Security of Self-Organizing Networks: MANET, WSN, WMN, VANET
Edited by Al-Sakib Khan Pathan
ISBN: 978-1-4398-1919-7

Service Delivery Platforms: Developing and Deploying Converged Multimedia Services
Edited by Syed A. Ahson and Mohammad Ilyas
ISBN: 978-1-4398-0089-8

Transmission Techniques for Emergent Multicast and Broadcast Systems
Mario Marques da Silva, Americo Correia, Rui Dinis, Nuno Souto, and Joao Carlos Silva
ISBN: 978-1-4398-1593-9

Underwater Acoustic Sensor Networks
Edited by Yang Xiao
ISBN: 978-1-4200-6711-8

Wireless Sensor Networks: Principles and Practice
Fei Hu
ISBN: 978-1-4200-9215-8

AUERBACH PUBLICATIONS
www.auerbach-publications.com
To Order Call: 1-800-272-7737 • Fax: 1-800-374-3401
E-mail: orders@crcpress.com

Mobile Web 2.0

Developing and Delivering Services to Mobile Devices

Edited by
Syed A. Ahson · Mohammad Ilyas

CRC Press
Taylor & Francis Group
Boca Raton   London   New York

CRC Press is an imprint of the
Taylor & Francis Group, an **informa** business

AN AUERBACH BOOK

Auerbach Publications
Taylor & Francis Group
6000 Broken Sound Parkway NW, Suite 300
Boca Raton, FL 33487-2742

© 2011 by Taylor and Francis Group, LLC
Auerbach Publications is an imprint of Taylor & Francis Group, an Informa business

No claim to original U.S. Government works

International Standard Book Number: 978-1-4398-0082-9 (Hardback)

Library of Congress Cataloging-in-Publication Data

Mobile Web 2.0 : developing and delivering services to mobile devices / edited by Syed A. Ahson, Mohammad Ilyas.
    p. cm.
    Includes bibliographical references and index.
    ISBN 978-1-4398-0082-9 (hardcover : alk. paper)
    1. Mobile communication systems. 2. Web 2.0. 3. Web services. 4. Mobile computing--Programming. 5. Application software--Development. I. Ahson, Syed. II. Ilyas, Mohammad, 1953-

TK5103.2.M6327 2011
006.7'6--dc22   2010044375

Visit the Taylor & Francis Web site at
http://www.taylorandfrancis.com

and the Auerbach Web site at
http://www.auerbach-publications.com

Contents

Preface

Mobile devices today are beginning to reach a new level of sophistication. In terms of battery life and Internet connectivity, mobile devices have now reached satisfactory performance; moreover, the fact that they are portable is also an added advantage. The latest mobile devices have various new input mechanisms such as touch and voice inputs. In recent years, there have also been significant technological improvements in wireless networks, and this has led to a diffusion of more powerful mobile devices with increased hardware and software capabilities. The growth and penetration of mobile communication technologies, with up to 4.5 billion expected subscribers in 2012, has determined a scenario where users can access the Web directly from their mobile devices, a trend that is on the rise. Mobile devices, which, for reasons of portability, have become our companions in daily life, are likely to become the favorite platform of connecting, interacting, and sharing information with others. They are increasingly being used for functions other than making calls and texting, such as e-mailing, collecting and managing multimedia, playing games, browsing the Internet, and performing many other business and personal tasks.

Mobile phones are no longer a device for communication. The latest mobile phones cater to many broader categories such as entertainment, personal management, social activities, work, and health care. In the entertainment sector, mobile phones are used for gaming, music, video, and other interactive (e.g., match calculator)/Web applications. Personal management services offered by mobile phones range from organizers, schedulers, task lists, notes, reminders, and even pocket finance management software. Social activity services may include Facebook, Twitter, and other chat room applications. Many of these social activities can be performed over Wi-Fi/Bluetooth protocols rather than Hyper Text Transfer Protocol (HTTP), offering the user a cheaper option. Pocket Word, Pocket Excel, Pocket PowerPoint, finance and budgeting software, e-mailing suits, etc., provide greater support for work/business activities. Considering the importance of the Internet in people's daily lives, it is easy to see the Internet going mobile. The Internet will not be bound to personal computers (PCs); rather users can access it anywhere, anytime through mobile devices. Mobile devices can also be used for performing several tasks that were earlier possible only on a PC. People are increasingly using Internet services on the go as their mobile devices are now capable of offering access to Internet services.

The so-called *Mobile Web 2.0* originates from the conjunction of the Web 2.0 services and the proliferation of Web-enabled mobile devices. The term "Web 2.0" was coined by Tim O'Reilly in 2005 and is often associated with applications such as wikis, blogs, photo- and video-sharing sites, social-networking sites, Web-based communities, Web services, mashups, and folksonomies. All of these "Web 2.0" applications share some common aspects, including collaboration, interoperability, user-created information, and information sharing on the World Wide Web. The same trend is now spreading to the mobile Internet, for example, through moblog, various location-based mobile hypermedia services, and pervasive games. At the same time, advances in bandwidth and new mobile devices are promising access to information on the Web from anywhere, at anytime, and on any device, and thus allowing users to access and generate information even when mobile.

Making the Mobile Web 2.0 a reality is not only a matter of using highly capable devices and enhanced Web browsers. End users are expecting more from the new generation of the Web on the move. First of all, they want to use applications and portals adapted and optimized for every device. Another emerging requirement is the availability of rich user interfaces (AJAX, graphics, maps) supporting higher functionalities while optimizing traffic costs. Furthermore, context-awareness (e.g., enabling location-based services) is required to obtain the most relevant information or content depending on the situation or environment the user is in. In the last several years, the mobile applications field has been transformed in a dramatic way: From sophisticated new enabling hardware and human–computer interaction technologies, to attractive new devices, to powerful programming interfaces, and much-improved ubiquitous access to the Internet. These factors, together with a growing marketplace for mobile applications, the possibility of significant return on investment from application development, and an ever-growing need for always–on always available, have helped to make the development and distribution of mobile applications a lucrative prospect. In addition, application developers are seeking powerful, standards-based (and possibly open source) technologies that make the creation of advanced, high-quality mobile Web applications possible in time to market without too much investment or the necessity to hire specialized developers.

The conjunction of Web 2.0 and mobile Web accesses is leading to a new communication paradigm, where the mobile devices act not only as mere consumers of information, but also as complex carriers for receiving and providing information, and as platforms for novel services. We may assume that, in the near future, the demand for Web 2.0 services will be mainly from mobile devices. This assumption is confirmed by the current trend of popular Web 2.0 sites, such as MySpace and Facebook, that offer mobile access to users through specific applications preloaded on mobile devices. According to specialized studies, the most popular Web 2.0 sites are expected to have a mobile component within a few years. Mobile Web 2.0 represents both an opportunity to create novel services (typically related to user location) and an extension of Web 2.0 application to mobile devices. The management of user-generated content, of content personalization, of community and information sharing is much more challenging when dealing with devices with limited capabilities in terms of display, computational power, storage, and connectivity. Furthermore, novel services require support for real-time determination and communication of the user position. The choice of appropriate technological solutions that can effectively support Mobile Web 2.0 services will be a key element to determine its success.

This book provides technical information on all aspects of Mobile Web. The areas covered range from basic concepts to research grade material, including future directions. It captures the current state of Mobile Web and serves as a source of comprehensive reference material on this subject. It consists of 21 chapters authored by 51 experts from around the world. The targeted audience for this handbook include designers and/or planners for Mobile Web systems, researchers (faculty members and graduate students), and those who would like to learn about this field.

The book is expected to have the following specific salient features:

- To serve as a single comprehensive source of information and as reference material on Mobile Web
- To deal with an important and timely topic of emerging technology
- To present accurate, up-to-date information on a broad range of topics related to Mobile Web
- To present material authored by experts in the field
- To present information in an organized and well-structured manner

Although the book is not technically a textbook, it can certainly be used as one for graduate and research-oriented courses that deal with Mobile Web. Any comments from the readers will be highly appreciated. Many people have contributed to this handbook in their own unique ways. First and foremost, we would like to express our immense gratitude to the group of highly talented and skilled researchers who have contributed 21 chapters to this handbook. All of them have been extremely cooperative and professional. It has also been a pleasure to work with Rich O'Hanley and Jessica Vakili of CRC Press; we are extremely grateful to them for their support and professionalism. Our families have extended their unconditional love and support throughout this project and they deserve very special thanks.

Syed A. Ahson
Seattle, Washington

Mohammad Ilyas
Boca Raton, Florida

For MATLAB® and Simulink® product information, please contact

The MathWorks, Inc.
3 Apple Hill Drive
Natick, MA, 01760-2098 USA
Tel: 508-647-7000
Fax: 508-647-7001
E-mail: info@mathworks.com
Web: www.mathworks.com

Editors

Syed A. Ahson is a senior software design engineer with Microsoft, Redmond, Washington. As part of the Mobile Voice and Partner Services Group, he is currently working on new and exciting end-to-end mobile services and applications. Before joining Microsoft, he was a senior staff software engineer with Motorola, where he made significant contributions in creating several iDEN, CDMA, and GSM cellular phones. Syed has extensive experience with wireless data protocols, wireless data applications, and cellular telephony protocols. Prior to joining Motorola, he worked as a senior software design engineer with NetSpeak Corporation (now part of Net2Phone), a pioneer in VoIP telephony software.

Syed has authored more than 10 books and several research articles on emerging technologies such as Cloud Computing, Mobile Web 2.0, and Service Delivery Platforms. His recent books include *Cloud Computing and Software Services: Theory and Techniques* and *Mobile Web 2.0: Developing and Delivering Services to Mobile Phones.* He teaches computer engineering courses as adjunct faculty at Florida Atlantic University, Boca Raton, and has introduced a course on Smartphone technology and applications. Syed received his MS in computer engineering from Florida Atlantic University in July 1998. He received his BSc in electrical engineering from Aligarh Muslim University, India, in 1995.

Dr. Mohammad Ilyas is associate dean for research and industry relations and professor of computer science and engineering in the College of Engineering and Computer Science at Florida Atlantic University, Boca Raton. He is also currently serving as interim chair of the Department of Mechanical and Ocean Engineering. He received his BSc in electrical engineering from the University of Engineering and Technology, Lahore, Pakistan, in 1976. From March 1977 to September 1978, he worked for the Water and Power Development Authority, Pakistan. In 1978, he was awarded a scholarship for his graduate studies. He received his MS in electrical and electronic engineering from Shiraz University, Iran, in June 1980. In September 1980, he joined the doctoral program at Queen's University in Kingston, Ontario, Canada. He completed his PhD in 1983; his doctoral research was on switching and flow control techniques in computer communication networks. Since September

1983, he has been with the College of Engineering and Computer Science at Florida Atlantic University. From 1994 to 2000, he was chair of the Department of Computer Science and Engineering. From July 2004 to September 2005, he served as interim associate vice president for research and graduate studies. During the 1993–1994 academic year, he was on sabbatical leave with the Department of Computer Engineering, King Saud University, Riyadh, Saudi Arabia.

Dr. Ilyas has conducted successful research in various areas, including traffic management and congestion control in broadband/high-speed communication networks, traffic characterization, wireless communication networks, performance modeling, and simulation. He has published 1 book, 16 handbooks, and over 160 research articles. He has also supervised to completion 11 PhD dissertations and more than 38 MS theses, and has been a consultant to several national and international organizations. Dr. Ilyas is an active participant in several IEEE technical committees and activities. He is a senior member of IEEE and a member of ASEE.

Contributors

Harald Amelung
Coworking0711
Stuttgart, Germany

Daniel Aréchiga
Department of Sciences, Technologies
and Methodologies
Southern University Center
University of Guadalajara
Guadalajara, Mexico

Muhammad Asif
Department of Systems and Computer
Engineering
Carleton University
Ottawa, Ontario, Canada

Andy Brown
School of Computer Science
University of Manchester
Manchester, United Kingdom

Claudia Canali
Department of Information
Engineering
University of Modena and
ReggioEmilia
Modena, Italy

Michele Colajanni
Department of Information
Engineering
University of Modena and
ReggioEmilia
Modena, Italy

Cristian Rodríguez de la Cruz
Telefónica Investigación y Desarrollo
Valladolid, Spain

Francisco Javier Díaz
Laboratorio de Investigación en Nuevas
Tecnologías Informáticas
Facultad de Informática
Universidad Nacional de La Plata
La Plata, Argentina

Berna Erol
Ricoh Innovations
Menlo Park, California

Ben Falchuk
Telcordia Technologies, Inc.
Piscataway, New Jersey

José Manuel Cantera Fonseca
Telefónica Investigación y Desarrollo
Valladolid, Spain

Miguel Jiménez Gañán
School of Computer Science
Universidad Politécnica de Madrid
Madrid, Spain

Kaj Grønbæk
Center for Interactive Spaces
Department of Computer Science
Aarhus University
Århus, Denmark

Rafael Grote
Fraunhofer Institute for Open
    Communication Systems
    FOKUS
Berlin, Germany

Frank Allan Hansen
Center for Interactive Spaces
Department of Computer Science
Aarhus University
Århus, Denmark

Simon Harper
School of Computer Science
University of Manchester
Manchester, United Kingdom

Stephan Haslinger
Tisco GmbH
Vienna, Austria

Iiro Jantunen
Department of Communications
    and Networking
Aalto University
Espoo, Finland

Matthias Jarke
Information Systems and Databases
    Group
RWTH Aachen University
Aachen, Germany

and

Fraunhofer Institute for Applied
    Information Technology
Sankt Augustin, Germany

Caroline Jay
School of Computer Science
University of Manchester
Manchester, United Kingdom

Harald Kaaja
Nokia Research Center
Helsinki, Finland

Janne Pekko Kaasalainen
Nokia Head Office
Espoo, Finland

Eija Kaasinen
VTT Technical Research Centre
    of Finland
Tampere, Finland

Anne Kaikkonen
Nokia Corporation
Helsinki, Finland

Ibrahim Khalil
School of Computer Science & IT
RMIT University
Melbourne, Victoria, Australia

Riccardo Lancellotti
Department of Information
    Engineering
University of Modena
    and ReggioEmilia
Modena, Italy

David Linner
Department for Telecommunication
    Systems
Technische Universität Berlin
Berlin, Germany

Jiebo Luo
Kodak Research Laboratories
Rochester, New York

Shikharesh Majumdar
Department of Systems and Computer
    Engineering
Carleton University
Ottawa, Ontario, Canada

Matthias Müllenborn
OTICON
Smørum, Denmark

Christoph Ohl
Department of Computer Science
Friedrich-Schiller-University Jena
Jena, Germany

Ignacio Marín Prendes
R&D Department-Fundación CTIC
Gijón, Spain

Claudia Alejandra Queiruga
Laboratorio de Investigación en Nuevas
    Tecnologías Informáticas
Facultad de Informática
Universidad Nacional de La Plata
La Plata, Argentina

Sathish Rajasekhar
School of Computer Science & IT
RMIT University
Melbourne, Victoria, Australia

Jorge Horacio Rosso
Laboratorio de Investigación en Nuevas
    Tecnologías Informáticas
Facultad de Informática
Universidad Nacional de La Plata
La Plata, Argentina

Miguel Angel Santiago
Telefónica Investigación y Desarrollo
Madrid, Spain

Gabriele Schade
Department of Building Technologies
    and Computer Science
University of Applied Sciences Erfurt
Erfurt, Germany

Daniel Schall
Distributed Systems Group
Technical University of Vienna
Vienna, Austria

Vidya Setlur
Nokia Research Center
Palo Alto, California

Javier Sierra
Telefónica Investigación y Desarrollo
Madrid, Spain

Florian Skopik
Distributed Systems Group
Technical University of Vienna
Vienna, Austria

Satish Narayana Srirama
Distributed Systems Group
Institute of Computer Science
University of Tartu
Tartu, Estonia

and

Information Systems and Databases
  Group
RWTH Aachen University
Aachen, Germany

Fahim Sufi
Department of Health
Melbourne, Australia

Zahir Tari
School of Computer Science & IT
RMIT University
Melbourne, Victoria, Australia

Nicolas Tille
AARDEX Group
Sion, Switzerland

Martin Treiber
Distributed Systems Group
Technical University of Vienna
Vienna, Austria

Elina Vartiainen
Nokia Research Center
Helsinki, Finland

Jesús Vegas
Department of Computer Science
University of Valladolid
Valladolid, Spain

Juha Virtanen
GE Healthcare
Helsinki, Finland

Sascha Wagner
Department of Building Technologies
  and Computer Science
University of Applied Sciences Erfurt
Erfurt, Germany

Yeliz Yesilada
School of Computer Science
University of Manchester
Manchester, United Kingdom

Chapter 1

Designing for Mobility

Janne Pekko Kaasalainen

Contents

1.1 Introduction

The systems we, as designers and engineers, build each day increase in complexity as the technology advances and as our own needs develop. Increasingly, they start to be part of our daily lives in more or less visible roles. These systems are expected to do more for us, and at the same time help us concentrate on what we are doing and provide us a with a way to enjoy the time we have on our hands.

The life cycle of technology begins with an invention. In some cases, this invention makes it possible to do things that were not doable before. When this happens, the new opportunities that arise are infinitely better compared to the past. The user is empowered to do the impossible, even if it may not be usable nor even be a nice experience for them. Thus, a solution that allows people do what they could not have done before is a valid strategy when that something is valuable enough to justify the immaturity of the solution. It can, of course, come with the price of lost users who do not manage to put the invention to use.

When the technology matures, the attention starts to shift to softer values. It is no longer enough to provide solutions that only work in the technical sense of the word. Once the solution starts to have competition that also allows the user to do the same things, different and previously latent needs start to emerge. These solutions first need to be usable, so that people do not need to expend great effort in learning and continuing to use them. Once this is achieved, the race continues to provide a holistic experience to captivate the user and potential buyer.

It is at this stage that the questions about this holistic experience start to arise. The rules change the second time—the first time being the transition into being able to do what one wanted, and improving by doing it more easily. Now, the user starts to play an even more crucial role, as the application is no longer good enough unless it satisfies the emotional needs of the user as well, or at least satisfies those needs more than the competitors do. But can we really design experiences, and what does experience design really mean from a practical point of view?

1.2 Can We Design Experiences?

Before going deep into questions of how to design experiences, it could be questioned whether experience design is possible in the first place. This question is surely more complex than what it might appear at first. The simple answer could be that it is not. It could be said that user experience is ultimately a feeling that is affected by expectations and the person's history, opinions, and values, to name a few parameters (Roto 2006). These expectations could, however, be targeted with marketing, branding, and various other means.

The expectations are, hopefully, followed by actual usage. This too is a complex matter—not only can actual usage be divided into to the ability to learn and to aspects of continuous use, it can also be affected by the context where and when the system is being used. This is especially relevant for mobile systems.

Finally, there is the reflection period, when the user either consciously or unconsciously evaluates what the experience the product offered him or her really was. How does it feel a week or a month later? Do users long to use the system again or do they hate it from the bottom of their hearts? From these experiences, they form new expectations, whether toward the same product or toward a new one (Mäkelä and Fulton Suri 2001).

Thus, it could be argued that we cannot design holistic experiences.

It could also be argued that experience design is possible. There are movies, art, and exhibitions that seem to try to convey emotions, moods, and feelings. Rarely does a comedy not cause laughter, or a tragedy avoid sadness. We have companies that are widely appreciated for their delightful and seemingly exceptional products and offerings. I would be hard-pressed to argue that this is a mere coincidence. For if it is, how can some succeed in this time and time again?

So, perhaps we can design experiences.

It might not be impossible to include the exemplary cases of one view to the argumentation for the other. Thus, this chapter tries to explore the grey area in between these two opposing points of view, and to explain my attempts to approach the problem of designing for experiences.

These approaches are described in relation to Nokia Image Exchange, an experimental mobile image sharing service that was produced to study how a service could be tied to a mobile handset. I take a practical approach to the question, and try to set up basic assumptions that are applied into our software and service design.

1.2.1 User-Centered Design

A term that is often raised in design- and usability-related discussions is user-centered design. The Usability Professionals Association defines this as

> User-centered design (UCD) is an approach to design that grounds the process in information about the people who will use the product. UCD processes focus on users through the planning, design and development of a product. (UPA 2008)

Typically, this has been seen as an iterative, circular process. While exact procedures may change depending on the source, the process is typically started by researching the context in which the eventual product is to be used. This is followed by design and requirement specification, prototyping, and evaluations of the produced outcome. The whole procedure can be then started again to refine the product, and similar processes can be used within each iteration phase.

User-centered design methodology has many good and valuable concepts. Perhaps the most important lesson is to put the emphasis on the needs of the users for whom the systems are being designed, and on the value of iteration and testing with the target audience. As a philosophy it is neither right nor wrong, but it does not take away the responsibility from the design team to adapt it to their needs and practices, and not needlessly restrict them to one way of working.

The initial start of this iterative cycle remains one of the issues; however, much of the other criticism directed at user-centered design is not actually the fault of the design philosophy itself, but of rigorous adherence to it without independent thought. We also run into issues while designing systems that do not have a well established, existing context of use. Additionally, there is a marked difference in the aspiration to build great systems that stand out from the rest, not simply the wish to produce good systems or ones that aim at avoiding failure. While established methodologies seem to do well in raising the basic quality of the work, it can be questioned if they are the best solutions for creating outstanding work on their own.

However, this chapter is not meant to say that methodologies such as those presented by user-centered design are bad or obsolete. Many methodologies have their root in observing the practices that have worked on earlier projects. It is also worth noting that just having the right people is not enough to guarantee positive outcomes. For example, Apple had key people contributing to the success of iMac in-house from 1993 to 1997, but it was not before an organizational change that they were able to contribute to the success in the following years (Buxton 2008). Thus, it would seem preferable to have some understanding of working practices and methodologies together with having a team of skilled people.

Similar thoughts have also been raised in regard to usability testing, with reasoning close to what was presented above. Greenberg and Buxton write in their CHI 2008 paper "Usability evaluation considered harmful?":

> Usability evaluation, if wrongfully applied, can quash potentially valuable ideas early in the design process, incorrectly promote poor ideas, misdirect developers into solving minor vs. major problems, or ignore (or incorrectly suggest) how a design would be adopted and used in everyday practice. The curriculum stresses the teaching of evaluation methodologies as one of its major modules. This has certainly been taken up in practice, although in a somewhat limited manner. While there are many evaluation methods, the typical undergraduate HCI course stresses usability. (Greenberg and Buxton 2008)

In conclusion, there seems to be more to design than following any set of methodologies or utilizing specific tools, and simply relying on standard practices is no guarantee of producing meaningful results.

1.2.2 Usability and User Experience

The basis for human–computer interaction (HCI) lies greatly on the concept of usability. Luckily, HCI has been a rather well-established field, and the definition of usability is an ISO standard (ISO 9241-11):

> [Usability refers to] the extent to which a product can be used by specified users to achieve specified goals with effectiveness, efficiency and satisfaction in a specified context of use.

However, it has also been noted that usability, while containing the term "satisfaction," has not traditionally been concerned with pleasure and joy (Hassenzahl 2001). This may have been caused by the history of system design and the emphasis on industrial applications. Traditionally, usability has been functional and goal-centric (Hassenzahl et al. 2001). Mark Hassenzahl has argued (Hassenzahl 2003) that the appeal of a system consists of pragmatic and hedonistic values. An interactive system is often trying to fulfill a pragmatic need, such as (in an extreme case) a hammer that is used to put nails into walls. Besides these pragmatic values, however, there are hedonistic needs as well, including self-expression, social status, and enjoyment.

Due to these issues, the term "user experience" has been both proposed and used, even if without a precise definition. It tries to better encompass factors such as pleasure and satisfaction, and to understand what makes certain systems more appealing than others. The following is a list of some suggested definitions:

> A result of motivated action in a certain context. (Mäkelä and Fulton Suri 2001)

> A consequence of a user's internal state (predispositions, expectations, needs, motivation, mood, etc.), the characteristics of the designed system (e.g. complexity, purpose, usability, functionality, etc.) and the context (or the environment) within which the interaction occurs (e.g. organizational/social setting, meaningfulness of the activity, if the use is voluntary, etc.). (Hassenzahl and Tractinsky 2006)

> All the aspects of how people use an interactive product: the way it feels in their hands, how well they understand how it works, how they feel about it while they're using it, how well it serves their purposes, and how well it fits into the entire context in which they are using it. (Alben 1996)

In regard to terminology, this chapter uses the definition for user experience proposed by Virpi Roto (Roto 2006). In her doctoral dissertation, she argues that the term "user experience" would be narrowed down to mean the interaction between a person and a machine. This view is also supported by the following definition:

> Every aspect of the user's interaction with a product, service, or company that makes up the user's perceptions of the whole. User experience

> design as a discipline is concerned with all the elements that together make up that interface, including layout, visual design, text, brand, sound, and interaction. UE (user experience) works to coordinate these elements to allow for the best possible interaction by users. (UPA 2007)

However, experiences—including user experience—consist of previous experiences and expectations that the user has toward the system he or she is going to use. The user has a motivation to use the new system, and he or she undertakes an action by using it in a context (to be understood rather vaguely as "on lunch break," e.g., or "finding commuting routes"). Motivation, action, and context form the present experience at the time of the use. The present experience then moulds future experiences and expectations (Mäkelä and Fulton Suri 2001).

An issue with the definitions above is that their abstraction level is relatively high. They are also insufficient to offer designers more practical help to do their work, even if they can be used to provide a mental model that itself can be used as an aid.

Thus, the term "experience" is used throughout this chapter to cover aspects that are beyond interface and device design, and should be considered as a separate term from "user experience." These aspects include but are not limited to marketing, brands, social interaction, as well as the eventual departure from using the product. Perhaps the largest difference, according to Roto, is the difference of user activity; user experience is limited to the domain where the user himself or herself acts with the system. The term "experience" allows the user be a passive participant (Roto 2006). It is this passivity that allows more methods to shape the expectations toward the system. A traditional movie would cause an "experience," whereas an interactive movie would be considered to create a "user experience."

It has also been suggested that we ought not to ask if we can design experiences, but instead ask if we can design for experiences. The difference is subtle, but philosophically meaningful; designing for experiences does not design experiences themselves, but gives them opportunities to emerge (Hassenzahl et al. 2008). This terminology would allow the experience to be personal and subjective to emphasize that it includes the user's personal aspects, such as his or her personal history and previous experiences.

This difference, however, seems to make little or no practical difference to the practical design work itself. It seems unlikely that practical design work would be much different if a solution for a problem would be thought of as "I hope this will create a fun moment" versus "I do this in the hope that it will be fun." Not much experience is needed to notice that the feelings and the emotions of users cannot be universally guaranteed, and it seems unlikely that many designers would imagine this were the case. What is likely to be meant in both cases lies among the lines "I hope that this feature will be found fun by a sufficiently large portion of our targeted audience." To accomplish this, better, practical, understanding of the intended target audience is needed.

The very nature of user experience covers most if not all activities that relate to the particular system in question. For practical reasons, not all of these activities are regarded as experience design because of what it is possible to manage, and to retain some level of understanding of what is meant by being an "experience designer." Figure 1.1 lists a few, broad-level topics that are aspects that contribute directly to the experience design. It should be noted that the classification in this image is meant to illustrate that the field is broad, and should not be taken as a definition of the field (Figure 1.1).

Understanding	Communities
	Context research
	Market analysis
	Psychology, physiology
	Trends
	User research
Creating	Audio/sound
	Concept design
	Graphic design
	Haptics, gestures
	Industrial design
	Interaction design
	Narration
	User interface design
	Input devices
Evaluation	Experiential, feelings
	Psychology, physiology
	Usability
Supporting	Methodology
	Metrics
	Models
	Processes
	Tools

Figure 1.1　List of some broad-level topics contributing directly to the experience design.

1.3 Brief to Interaction Design

Interaction design is often understood as a creation of dialogue between a product, service, or system (Kolko 2007). It is often utilized to reduce user dissatisfaction and to increase productivity and satisfaction. Interaction design also has common aspects with user-centered design.

In practice, in interaction design, one needs to balance between multiple stakeholders, including but not limited to the business, users, and legal issues. After all, the system needs to be maintainable to remain functional, this often being a prerequisite of being profitable. The systems being designed should not break laws and ethical codes. The system should also bring value to its users or it would be unlikely to meet its other goals. While the underlying decision does not necessarily rest on interaction designers, the design deliverables need to achieve a balance.

However, interaction can be understood as a broader concept in some cases. In the case of Nokia Image Exchange, it was not only a question of a person using a system, but people mediating with each other through the system. Thus, interaction design took place between people, and the system's role was to facilitate this interaction. The difference is slight, but vital. If the starting point is to deal with a single person and a system, there is the danger of scoping the problem space in a manner that concentrates on answering how the system can be used best instead of answering what is the best system.

1.3.1 Emotional Communication

Only about 45% of the feelings and attitude in messages shared between people occur through speech (Mizutani 2006). Actions, such as shaking hands, deliver vast amounts of information via gestures such as the firmness of the shake and whether the partner is smiling or being serious. In fact, Mehrabian proposed a theory that in each face-to-face communication situation that deals with emotional messages, 7% of the message comes from the words themselves, 38% comes from the tone of the words, and 55% comes from the body language of the speaker (Mehrabian 1981).

If the above-mentioned messages differ, Mehrabian states that the most powerful ones dominate the message and rule how it is interpreted. Thus, to create a strong message, each element needs to support the other. It should be emphasized that Mehrabian only stated the relative importance to apply when the messages dealt with feelings and attitudes. The rule he proposed was not meant to be generalized for any form of communication.

However, if it is so that in emotional communication only 7% of the message is delivered by words, it puts communication via technical means in an odd situation. Via a voice call, we can still hear the tone of the speaker's voice, but when that is taken away, the means to express emotion are relatively thin. Perhaps this has influenced the usage of *emoticons* (often better known as smileys) in text-based mediums such as Internet Relay Chat (IRC) and instant messaging (IM) applications like Microsoft Messenger.

Whether or not the birth of emoticons was caused by limitations of the text-based medium, the phenomena is documented in the context of instant messenger software. It appears that emoticons are used to make sure that messages meant to be humorous are not taken seriously. Emoticons do have a set of problems, such as the duration and scope of the emotion, and thus methods to expand the emotional communication in those mediums have been attempted (Sánchez et al. 2005). Some methods have even been offered to tackle transferring the gestural messages within text (Adesemowo and Tucker 2005).

Emotional factors also lead to the subject of massively multiplayer online games (MMOs). These types of games rely heavily on other users and the interactions between people to make the player feel more immersed into the world. The game itself can at times be mastered relatively quickly, and offers little more than what is available on stand-alone games often played on computers and game consoles. The real hook is to provide a social context for the players to stay with the game and its subscription even after the initial objectives have been reached.

Some have, however, stated that the importance of these social networks is over-stated, although obtained results have not indicated that the social aspects do not matter. Instead, games such as "World of Warcraft" benefit from merging games to the community activity, offering players both, at least to some extent (Ducheneaut et al. 2006). Others have also noted that players with heavy interactions with other gamers tend to play the games longer (Chen and Lei 2006).

Each of the examples emphasizes the importance of human factors, and hints that the emotional bond between individuals would affect their feelings for a device or an application, either directly or indirectly, by providing them the connection they could not get otherwise. In the case of MMOs, choosing a new one would require the player to rebuild his or her social network again (unless, of course, the existing network could move with him or her).

Thus, we are left with an interesting question: "How can we facilitate people to create a bond with either the software itself or with other people reached through this software?" Michihito Mizutani proposes a "French bulldog theory" in his thesis:

> Suppose that you have a French bulldog at home. He is really lovely, but none of your friends understand that. One day while walking your dog in a park, you come across another French bulldog. You excitedly start a conversation with its owner even if you have never met them before.

The idea he presents is not too surprising, and has been quite well known among dog owners for some time. Another example he mentions is a "Sole Bag" designed by Naoto Fukusawa. Sole Bag is a shopping bag with a shoe bottom fitted to it. The shoe bottom supports the bag when it is placed onto the ground. Again, the actual product value is experienced only by the user and easily omitted by the casual observer.

The situations he describes come from the object facilitating the discussion and the creation of a new social contact. The individuals have something in common, a peer group of sorts. A product can also be a tie to a specific social group an individual wants to belong to or to be identified by. Such trends can be seen in fashion, where clothes are a method to express oneself.

In fact, a similar phenomenon is not uncommon in the software and digital world either. There are people who are acknowledged supporters of various applications, such as operating systems. The common interest has caused people to create communities (such as Mac and Linux supporters) and made people feel strongly for their chosen software.

1.3.2 Theory of Broken Perfection

While the messages that pass between people are undoubtedly important to create an emotional contact, there are other aspects as well that affect how we perceive things. If we consider the "French bulldog" theory from the perspective of the initial owner of the dog, the social needs that are fulfilled by owning such a dog are not likely to be the ruling factors (though they can be, on occasion) when you choose which kind of a dog you would like to have. After you have made the decision to get the dog, the emotional connection to other dog owners comes naturally. But why would one get a French bulldog in the first place?

Some objects appeal to people because they stand out from the rest. This is particularly notable in regard to the fashion industry, but is not limited to it. Some software have also created an avid audience by (or, one could argue, despite) providing unique interfaces that stand out from the norm. Blender could be considered an application that differs greatly from the dominating 3D software line-up (such as Autodesk Maya, Autodesk XSI, and SideEffects Houdini). Another example is Pixologic ZBrush, which too introduces a new interface style for sculpting 3D surfaces (in contrast to Autodesk Mudbox and Nevercenter's Silo, for instance).

If one designs what one wants to in detail, there is a good chance one might end up with something that one does not like in the long run (assuming that the designer is the user as well). Knowing every little detail and being able to predict the future behavior of an object can take part of the fun out of it. In fact, aleatory filmmaking has been experimented on by directors such as Fred Camper with the movie "SN" and Barry Salts with "Permutations." These avant-garde films took the chance of randomness to the very experience of their movies.

For example, Cindy Crawford has a mole next to her mouth. That mole breaks the symmetry or "perfection," and makes her face more interesting. A similar oddity is seen in the placement of the Saabs (Swedish automobile manufacturer) keys, which were located next to the hand brake and not in the steering wheel like most other car manufacturers have decided to do. Mac OS X introduced the small-window closing, minimizing, and resizing icons that are on the "wrong" side of the window (on the left, while on many other window managers they are on the right). All these little quirks make things different and, at times, more attractive. It could even be thought that such personalities together create the very soul of a product.

Together with the appeal to a certain set of people comes the risk of alienating a large (or even largest) group of users. There is no denying that a good percentage of the population is rather conservative, and feels intimidated when things do not work exactly as they used to. In general, people have a tendency to resist change or a preference for familiar systems (Butler 1996). Such behavior, however, should not be taken too strictly. When cars emerged, people were rather suspicious toward them as well. For example, there were strict speed limits and a person was required to walk in front of the car with a red flag to warn others of the vehicle approaching them. If we do not challenge people to think differently, little progress can be made. Still, when walking one's own road too much, one needs to remember that there is a fine line and possible penalty to pay.

To emphasize, this does not mean that usability should be neglected, nor does it mean that usability has no value. On the contrary, usability has immense value for the end user, and this theory only proposes that usability is not the single aspect end users care about. Quirks should not contradict usability and traditions too much; a little might be allowed, but if they destroy the users' ability to use the device or program, there is little benefit from being different for the mere sake of it.

1.3.3 Emotional Value Chain

In the previously noted "French bulldog" theory, the communication is in fact occurring between people. The object that makes them bond and starts the conversations merely presents their likings, and signals they may have something in common. The emotional value, the message exchange with the other person, can be seen from transmitting with the help of the facilitating objects.

Similarly, if we receive a photograph that portrays a beloved family member, the significance is not in the physical photograph. In this case, the value of the image is the emotional contact we have with the persons in it. Again, the object merely transfers the emotions, possibly over distance and time.

1.4 Paradox of Choice

In light of market segmentation, it might seem appealing to design products that offer features for people who are looking for different aspects in software and services. It is, in fact, possible to offer people options to customize even individual software or to offer multiple software to choose from.

In some cases, choice has even been considered a virtue in itself. Offering a choice is also an easy solution for a design problem. Instead of making a decision, it may be quite easy to implement multiple alternative ways to do the same thing and offer a toggle that users can change to alter the behavior. But this approach has the danger of leading to needlessly complicated interfaces—and worse, causing user anxiety (Schwartz 2005). Barry Schwartz lists four reasons for this in his TED talk: regret and anticipated regret from the choice, the opportunity costs of what other good things the choice cost us, escalated expectations since we have so much to choose from, and finally the possible self-blame from making the wrong choice. The following excerpt demonstrates issues with trade-offs, which are essentially also choices (Schwartz 2003):

> Participants were told that Car A costs $25,000 and ranks high in safety (8 on a 10-point scale). Car B ranks 6 on the safety scale. Participants were then asked how much Car B would have to cost to be as attractive as Car A. Answering this question required making a trade-off, in this case, between safety and price. It required asking how much each extra unit of safety was worth. If someone were to say, for example, that Car B was only worth $10,000, they would clearly be placing great value on the extra safety afforded by Car A. If instead they were to say that Car B was worth $22,000, they would be placing much less value on the extra safety afforded by Car A. Participants performed this task with little apparent difficulty. A little while later, though, they were confronted with a second task. They were presented with a choice between Car A, safety rating 8 and price $25,000, and Car B, safety rating 6, and the price they had previously said made the two cars equally attractive. How did they choose between two equivalent alternatives?

> Since the alternatives were equivalent, you might expect that about half the people would choose the safer, more expensive car and half would choose the less safe, cheaper car. But that is not what the researchers found. Most participants chose the safer, more expensive car. When forced to choose, most people refused to trade safety for price. They acted as if the importance of safety to their decision was so great that price was essentially irrelevant...

> Even though their decision was purely hypothetical, participants experienced substantial negative emotion when choosing between Cars A and B. And if the experimental procedure gave them the opportunity, they refused to make the decision at all. So the researchers concluded that being forced to confront trade-offs in making decisions makes people unhappy and indecisive.

> Confronting any trade-off, it seems, is incredibly unsettling. And as the available alternatives increase, the extent to which choices will require trade-offs will increase as well.

It thus follows that users should not be forced to make choices any more than what is necessary. In practice, this means assuming defaults that hopefully work for the majority of the target audience and, in case that the audience is mostly undecided, offer an option only if absolutely needed. It also means that if there are multiple ways to achieve a given functionality, in many cases the design might be better off making a decision, even if it was not optimal for everybody.

1.5 Role of the Computer

One of the fundamental philosophies for designing Nokia Image Exchange was to see a computer as a tool that is built to help us. In that role, our thought was that computers should adapt to our needs instead of us adapting to them. This could be debated, and different views have been offered by researchers such as Dougas Engebart (Moggridge 2006). Our stand was more closely aligned with Raskin's (Raskin 2000).

If the computer is there to help us and make doing things easier for us, we need to consider a few scenarios. First, one needs to deal with the easiness of knowing what to do with the system and the system's learnability. In the long run, though, other factors such as efficiency need to be taken into consideration as well. At times, even these two factors may not be fully orthogonal, and thus compromises are needed.

Part of making life easier for the user leads to some other aspects that need balancing. Some of these are crucial, such as privacy and trust between the user and the system. There are also more easily implementable philosophical drivers, such as the idea that the system should not prompt the user for information if it already has a chance to know it. An example of this concept is our user identification based on phone number—an aspect that the system should already know without the user ever typing it in.

1.6 User Experience as a Design Driver

As mentioned earlier, user experience is a concept separate from ease of use or usability. While a system can be easy to use, it does not guarantee a positive experience for the user. For one, a system with a single button in it can be extremely easy and intuitive to use for people, but unless it fills the expectations of what it should do, it can turn out to be a failure.

Similarly, we can create interfaces that present users with multiple simple questions to fulfill their task. However, splitting a complex task into small, simple parts may make a system easy to use, but can also make it inefficient. In extreme cases, it may make the system unsuitable for actual use despite being simple.

We tried to study how far we could go in making users feel pleased and happy to use our system. As such, we could not rely on ease of use alone. Furthermore, this chapter does not deal much with usability, but with concepts dealing with the subjective and hedonistic aspects in the design process.

1.6.1 Laziness and Personal Time

One of the key concepts that seems to be easily forgotten in the hands of enthusiastic developers is that the actual end users may not consider the application being developed to be the most important thing in their lives. In fact, this is unlikely to be the case, especially when dealing in the consumer domain. Thus a conflict arises. On one hand, the developers are keen to invent functionality and features that seem to make the user's life easier, but come at the cost of learning. On the other hand, users may not know what the application does when they start to use it, and have limited motivation to see the benefits it might offer.

In general, if we look at the offerings from a consumer's point of view, the world is full of offerings. Naturally, some of these are more relevant than others, depending on the user's needs and the context he or she is in. Some of these needs are basic, such as the need of food and shelter, whereas some deal with more abstract wishes. However, all of the needs can be fulfilled in multiple ways and often by multiple vendors. The user needs to choose which offering he or she will take.

In this context, the offering will be an application. Regardless of the actual monetary price, users will always make an investment when they take an application into use. Economical value is the most direct example of this, but even if the piece of software is free, there will be at least an investment of time. This time consumption consists of several factors, such as

1. Finding out about the product and obtaining it
2. Taking the product into use and learning to use it
3. Using the product
4. Moving on from the product

It is often impossible to learn and study each and every alternative available before making a decision on which offering to choose. Furthermore, as people try to compare more alternatives, the time for each individual alternative decreases. As the number of options increases, the time to give each one a fair chance diminishes.

This chapter is not about marketing, and thus the question of finding the product lies outside the scope of this writing and much of my personal work as well. This does not make these aspects any less important, but speaking of these matters will be left to the marketing experts.

The second aspect after finding out about a product is the ability to put the product into use and learn how to use it, which is of great importance as well. Once the product is found, the user needs to put it into use to determine if it will fulfill his or her expectations. All obstacles users face will increase the pay-off expectations that the application needs to deliver for the exchange of the effort the user needs to spend to learn to use it. This naturally varies, depending on the application and the user. The requirements and needs for enjoyment and fun are quite different between, for example, media applications and banking software. In any case, if the user has given the application time and effort in learning how to use it, this trade-off should be acceptable in the continued use as well.

I've tried to avoid using concepts such as usability, as that is not the only factor (or, in some cases, even the main factor) that the user is looking for. While usability is often an integral part of the balance between time spent and the returns, other factors such as enjoyment can be equally important. These factors depend on the application or other product in question. As an example, a movie can be viewed during a 2 h long train ride. The same situation arises when the user decides how he or she would like to spend his or her time and whether the movie in question meets his or her expectations.

An important conclusion is that the reactions and the lack of attention toward an application is not necessarily a result of users' ignorance, but a failing with the application in offering appealing benefits that justify the investment of time and effort the user has given to your work.

In a similar vein, the user should not be forced to make choices that are irrelevant to him to her. This is even worse if he or she does not understand what he or she is asked, which is often the case when taking something new into use. However, it remains debatable what these critical questions are. My personal opinion is that the user should not initially be asked any questions that do not deal with fundamental aspects of the system or to the user's own well-being. A bad example about the opposite can be found from the initial start-up process that many mobile phones put the user through, as they ask the user to specify time and date instead of defaulting to using network time (time and date that the mobile phone can retrieve from the mobile network). Such behavior does not cause user harm, and the cases where the result is correct ought to outweigh those where it is not. On the other hand, the example application (Nokia Image Exchange) is forced to ask permission from the user to use network connections, as making any assumption could lead to direct monetary harm. As it was the only question asked from the user and because network integration is an important part of the whole concept, the question was not considered to be too obtrusive.

1.6.2 Responsiveness and Speed

A crucial aspect of creating positive user experiences seems to be speed and responsiveness of the user interface. Naturally, being fast does not guarantee positive results. However mundane and uninteresting speed optimization might be for the development, it does seem to be one of the most important showstoppers if it is found to be inadequate.

Such factors are emphasized in a mobile environment where the user's attention span is short. Typical usage situations are short waiting periods, such as waiting for the bus to arrive or killing time while waiting for a friend on a street corner. Often, the surroundings also demand concentration and create additional cognitive load, as happens while walking along the street. Not only does the user need to pay attention to the mobile device, but also to the traffic that surrounds him or her. Thus, it is easy to see why speed plays an important role.

However, the system does not need to be extremely fast; it only needs to be fast enough to not make the user wait. This includes both the actual performance as well as latency. In short, the system should perform so well that the user does not need to wait unless it is absolutely unavoidable (Tognazzini 2008). Once this limit is reached, benefits to user experience start to diminish.

1.6.3 Perception Equals Reality

Reality is a curious beast, and often overrated when it comes to offering experiences. Perhaps one of the most obvious examples, filmmaking, is based solely on creating an illusion that is immersive but not real. This is especially the case with computer animation, where everything visible is created for the purpose of the production. Furthermore, in many cases, what does not show in the final image can be omitted. To create an illusion, houses can have only the front walls, the rain can come from sprinklers, and altering object positions in depth can give a false impression of size and place.

Aside from the artificial illusions, there are many other occasions where reality gives way to perception. Even everyday concepts such as colors are result of perception, and not absolute truths. This can be observed by altering the background of a solid rectangle and noticing how the perceived color changes with different color combinations.

Illusions are actually what computers are based on, only we do not call it "magic" or "illusion" but "abstractions" (Dourish 2004). Computer systems operate on electrical signals, which in turn consist of electrons moving in conducting materials. However, abstraction after another has been built to hide the physics of the machine and been replaced with various libraries and blocks of code. Finally, the user is presented with an interface that is yet another abstraction. Thus, what truly happens inside electronic gadgets is a mystery to most of us.

In a similar fashion, abstraction levels can be increased to hide technical implementation to the extent that interfaces just seem to work. Applications can, and in my personal opinion should, guess what the user would do next and prepare for it. Furthermore, it is possible to take advantage of user behavior to do things that otherwise might seem impossible given the current technology. I'll describe two scenarios that are used by the Nokia Image Exchange imaging application that relate to the aspects of responsiveness and behavioral factors.

The first of these two examples deals with the speed of the Nokia Image Exchange S60 interface. Due to technical issues in early prototypes, the speed of rendering text was not sufficiently fast enough to smoothly resize and rotate the text. This was needed to support rotation of the screen between landscape and portrait modes. Such action created a delay that forced the user to wait before he or she could see the comments in the new aspect ratio, but this was masked by the means of simple animation. Comments were made to fade out just before the image would be rotated and fade back in after the rotation was done. This actually took more time than simply placing them onto the image as quickly as possible, but the constant motion and the perception that the application was doing something made the situation feel more pleasant. Similar tricks are used elsewhere, such as in Apple's iPhone. When taking an image with the built-in camera, the iPhone needs a moment to save the sensor data into the memory and create the actual image file. This moment is masked by an animated shutter that hides the resulting delay. Situations such as these are not, in fact, much different from the reasoning behind loading icons and progress bars, but concentrate more on the hedonistic aspects instead of the utilitarian ones.

The second example takes advantage of behavioral aspects and typical usage patterns. One of the driving use cases for the imaging service was to allow users to easily move and view their images on their personal computers. In practice, this operation involves transferring data from their cameras to their computers, and is often, but not always, done manually with memory card readers, over Bluetooth or by uploading to image sharing sites such as Flickr. However, mobile cameras are often used to capture surprising events of everyday life, and are mobile by nature. In such circumstances, the photographer is not likely to immediately view his or her new images on any computer, even less on his or her own. Thus, it is possible to detect newly taken images and start transferring them to an image service or to a PC without user intervention. When the user eventually sits before a computer screen and browses the images with a Web browser, he or she can already see the images he or she has taken. In reality, transferring large images via limited cellular connections takes time, but in the above scenario the perception is that there is no wasted time and effort.

While the idea of creating illusions sounds alluring, the problem with abstractions comes when the illusions break for one reason or another (Dourish 2004). The software can have programming errors or, in the case of networking applications, the network itself may be unusable or unavailable for unknown reasons. In fact, all sorts of reasons can cause the user to run into error situations. The more the number of layers, abstractions, and illusions, the higher the chance that some of them fail.

The errors themselves are not too big an issue. The real issue is that a carefully crafted illusion may have disconnected the user from the events that happens underneath the hood of the application. As a result, the reason for the error may be incomprehensible and may not make any sense. Some of these problems may be alleviated by careful explanations of the situation and ways to recover from it. Furthermore, the error situations may not be solely technical, but also social. For example, it would be possible to create a communication system that would combine the usage of phone calls, e-mail, SMS, MMS, and IM. However, each of these has its own characteristics when it comes to how it is being used. An SMS is instant, but typically less urgent than a direct phone call. An e-mail is assumed to be delivered with delay, and can be more formal than an IM message.

For reasons such as these, there are occasions when it may be better to present the system state and functionality to the user without artificial abstractions. Such situations are likely to be heavily dependant on the exact system being built, and thus generalizations seem difficult to state.

1.6.4 On Physical Interaction and Simplicity

It is the people who will eventually use consumer services that we design and implement. As the system is ultimately a collection of logic rules that happen inside electronic devices, there will also be an interface between the machine and the person using it. However, interaction with this logic is a wider concept that what it might first seem to be.

Systems and products are not used in vacuum. In this case particularly, we can even generalize that the meaningful interaction actually happens between the people themselves or, at the very least, between a person and the surrounding world. This happens since the images that are taken are taken for a purpose. This purpose can be a selfish act, where the image is taken and utilized by the same person. In other cases, the image is possibly shared between people, and thus the interaction occurs between them. Other examples also exist, but for the purposes of this chapter none of these scenarios take away the role of a human actor who acts in a physical world together with a physical world. In fact, even simplistic activities such as seeing can be tied to the act of doing. Further still, being and experiencing is integral part of even abstract activities such as thinking (Heidegger 1927).

Following this thought, if the experiences are caused and observed in relation to the physical world and if our ultimate motives of interaction do not deal with electronic devices, the interfaces should focus on aiding users to fulfill their goals. In the scope of this writing, these goals are social more often than dealing with the user and the physical world. For example, a user needs study we organized in Tokyo showed that, for many people interviewed, the motivation was not to capture beautiful images, but show others what they are doing, how their surroundings feel, and how they are thinking about their dear ones. Interaction design, thus, is about facilitating communication between users and, in some cases, the surrounding world itself. Before venturing onward, let us consider the hierarchy of interactions. The *Oxford American Dictionary* (accessed May 22, 2008) defines the verb "interact" as:

> [to] act in such a way as to have an effect on another; act reciprocally: all the stages in the process interact | the user interacts directly with the library.

Interacting, by definition, is not limited to human-to-human actions. I would argue that the reasons for interaction often come from social motivations. These motivations are carried out by interacting with the physical world and ultimately coming down to pressing buttons to make digital devices do things that we wish from them. In the case of digital artifacts, the human-to-machine interface is a needed abstraction to allow us to deliver messages contained in these artifacts to other people. Thus, and especially in the case of our imaging service, we had

1. Social motives that are carried out by…
2. Digital artifacts that are manipulated by…
3. Human-to-machine interfaces that consist of…
4. Physical interfaces with which the user controls possible…
5. Software interfaces of the device

The exact order of items remains arguable, and the separation between physical device and the interface it contains is becoming hazy in some sectors (Buxton 2008). Interestingly though, the hierarchy described above also lends itself to highlight the concept of direct manipulation. In principle, it ought to happen that the higher in the hierarchy we can move the cognitive load of the user and the fewer abstractions there are between the user and his or her intent, the easier and more pleasant the system ought to feel. This naturally assumes positive outcomes that are not always guaranteed even in social environments.

To argue further for direct manipulation, let us have a look at the history of computers and computing as described by Paul Dourish (Dourish 2004). Computers started as mechanical devices that helped people compute. Due to various reasons, the mechanical implementation changed to electronic signaling, which allowed greater focus on the logic of the computing instead of the mechanical implementation aspects. A level of abstraction was taken off from the shoulders of the designers of such machines. Later on, the designs were adapted to use customizable wiring and subsequently started to utilize punch cards (cards that had holes in them that described what the machine was to do). This time, it was the electrical implementation that was taken off from the user's shoulders. The interaction further focused on the user and the task that he or she tried to accomplish.

The punch cards turned into command lines, which in turn developed into the graphical interfaces that are prevalent today. Interestingly, the development toward direct manipulation was also present, even to the extent of light pens that, however, did not gain traction. Instead of writing what the machine should do, joysticks allowed one to move the cursor on these graphical displays to select what should be done. Mice then turned this moving of a cursor into pointing. While the difference is subtle, it is meaningful; a moving cursor forces one to utilize an interface abstraction, whereas pointing deals directly with the object in question.

In the last few years, touch interfaces have been becoming increasingly common, especially after the commercialization of multi-touch. While neither of these technologies is particularly new, their acceptance and utilization in the consumer space is. Each can also be seen as a continuum toward directly manipulating digital artifacts. Basic touch technology accomplishes this simply by removing the task of operating a mouse and by multi-touch, allowing multiple pointing devices to be used (multiple fingers, for one).

But it is not only computers that have gone toward interaction models where the user deals more directly with the content. Radios have gained automatic tuners instead of finding the radio stations manually, and car manufacturers are including increasingly sophisticated technologies such as automatic gearboxes and traction control systems to help drivers concentrate on driving. The argument can also be used to advocate perceived simplicity, as well as certain parts of ubiquitous and tangible computing practices. Simplicity itself can often be seen as a by-product of good interaction design.

Nokia Image Exchange had to operate in an existing environment and utilize existing platform and devices on top of which it was implemented. In practice, this meant that the physical devices were mostly fixed and could not be tampered with. However, existing physical features could be taken advantage of when they did exist. For example, Nokia N95 multimedia computer included an accelerometer that was used to rotate the image according to the orientation in which the device was held. This effectively meant that users did not need to operate both physical and software interfaces and could concentrate on the physical.

Furthermore, the interaction paradigm focused on utilizing the joystick as an extension of the user's finger where possible. The joystick directions moved the focus from one element to another on the selected object, and pressing the joystick initiated features on top of the selection. Not only did this behavior simulate touching to some extent, it also allowed context-sensitive menus to be drawn based on what the user had selected. This in turn decreased the visual clutter on the interface, and was one of the carrying themes on user interface design.

However, this solution was still not as good as direct physical touching, but the latter was simply not possible with the given devices. The devices themselves were fixed, given the market situation and expected size of the audience.

1.6.5 Delivering More Than Promised

It may seem self-evident, but good experiences are triggered by positive events. Thus, it is imperative that the positive events occur in the first place. Better still, experiencing positive surprises can be hoped to be more memorable and thus offer greater emotional impact. In practice, this can be accomplished by providing positive reactions over time.

These kinds of aims easily conflict with the most obvious marketing aims to some extent. To let people know about your product, service, or exhibition, it needs to be made interesting to the audience one way or another. An easy solution to this is to market why it is helpful or desirable to users. However, if all information about the product is revealed beforehand, surprises no longer occur.

In the case of Nokia Image Exchange, we deliberately left some of the features we considered fancy without a mention and avoided giving any direct hints on finding them. The most notable of these was automatic screen rotation that utilized an accelerometer in the targeted devices. When viewing images, the accelerometer animated the image rotation so that the image itself was always kept in closest 90-degree orientation (the top of the image pointing up). When the user physically rotated the phone, the image remained in the orientation and was fit to the screen if scaling was needed. As a result, we heard many joyful stories about positive surprises this feature had given to our test users.

1.6.6 What Should a System Do?

As easy as it may sound, deciding what a system should do is not always self-evident before the system is built. It is well known that users often use systems differently than what the designers have planned. However, even in this case, the audience does fulfill some of their desires. To maximize user satisfaction, we might do better if we could know and optimize for the actual needs for the majority of users. While it may be difficult or even impossible to know the distributions and exact needs of a diverse audience due to the complexities of data gathering, it would be hard to justify not even trying to do so.

Somewhat simple methods do exist to help understand users better. Perhaps the easiest method is to review previous research work that describes the experiences of others that have worked in similar fields. Other methods are often utilized in design research as well. Furthermore, it is entirely possible to find representative users for interviewing and observing their current behavior to understand them better. It does not harm the team to start actively participating in the activities they are designing for. It should be noted, however, that even interviews and observations might not reveal future or latent needs.

1.6.7 Sketching Interaction

Sketching has multiple benefits when used in design. By sketching, one is forced to think about what one is doing. Furthermore, by the definition of sketching, what one creates is not the final outcome, and thus the sketch itself can provoke new ideas. Finally, sketching is cheap, and can easily be thrown away if found unsuitable (Buxton 2007).

Unfortunately, sketching interactive systems is not quite as easy as working with static drawings. By definition, interactive systems change over time, by the system's users. In other than the simplest cases, the interaction forms a non-linear structure of actions in which users somehow navigate.

Further complexities often arise from technical limitations. When considering the user experience of the final product, it seems fair to say that it is dependent on the implementation. Thus, the design aspects and the technology together form an embodiment. This causes issues in approximating what the final system should feel like. While it is possible to create simple demonstrations of the ideal case, these can hinder the final outcome by being too finished and inflexible to adapt to changes that almost necessarily arise in the production phase of the system. Despite this, a holistic view would be preferable from the beginning.

Acknowledging these imperfections, our interaction sketching was done in stages. An instrumental tool in the process was an interface diagram, which tried to map how the system would be capable of supporting the previously mentioned scenarios and would ensure that the resulting navigation hierarchy remained logical within the system (Figure 1.2).

Figure 1.2 shows a version of the mobile interface diagram between iterations. It approaches the interface synthesis in a manner that might not be the most commonly used. The lines that connect the various screen mock-ups describe actions that trigger the transitions. The mock-ups come at different levels, depending on the problems that they present. Some of the notes simply indicated missing screens and what functionality the screens should contain, whereas others were pixel-perfect visualizations of the screens. These were occasionally needed to ensure that the information can fit to the screen and that the screens remain easy to

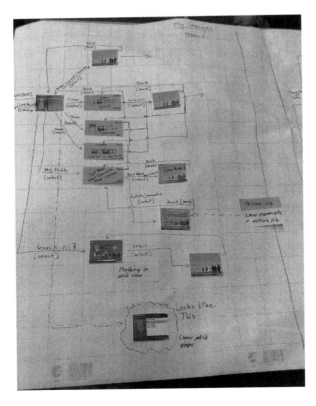

Figure 1.2   Version of the mobile interface diagram between iterations.

understand. This is in contrast to the rather common tradition of producing the interface diagrams as simple wireframes that leave the graphical visualization more uncertain.

Furthermore, some of the issues can be resolved at multiple levels. For example, visualizations can emphasize affordances that are mandated by the hardware but do not flow as well with the visual presentation as they would if the hardware could be redesigned.

Interface diagrams are not able to solve all the issues with experience sketching. By its very nature, the diagram is not interactive and it does not even try to explain what happens between the screens. Diagrams are not very efficient in modeling time.

We produced interactive prototypes to quickly estimate the feel of the software on the computer screen. These were not complete, and only concentrated on specific questions or mandated usage flows. This was mostly done to save time, as it was clear that the actual implementation would demand changes. The interaction allowed, however, better feedback from expert evaluations that were conducted with usability and design professionals to see if there would be obvious faults in the proposed interface.

1.6.8 Lowering the Barriers of Usage

If designing how something should work is not easy, then designing how something can work is even harder. While there are existing practices that deal with many aspects on how Web services work, few take advantage of mobile-specific matters. Some of these advantages deal directly with the issues outlined previously in regard to laziness.

To lower the effort needed to try out and test the imaging service, it would need to be easy to put into use. Putting a service into use usually starts with a registration process where the user creates credentials for himself or herself and fills in basic information such as his or her e-mail address that can be utilized, for example, to recover lost passwords. Other typical information usually includes at least the username and password that can be utilized to log into the service. The password is typically filled in twice to make sure it is not mistyped. Additional techniques such as CAPTCHA can be used to make sure that a computer program is not filling in the information for spamming purposes.

It ought to be needless to mention that filling out this information with a mobile keypad or touch-screen is a time-consuming process. This is an even less delightful process given that the scenarios in which the mobile phone is used are often mobile. Time is possibly fragmented, and the person using the phone is likely not going to have long moments of time to begin with.

One of the research questions was to study how far one can go to provide pleasant user experiences, and the typical solutions were not found satisfactory. In the Western world where mobile devices are in many cases personal devices, the process is also against the basic principles mentioned in Section 1.5. Filling in credentials is, eventually, redundant information that a personal device should know already. Of course, part of the blame lies on the handset makers themselves for not utilizing these aspects already.

In our case, it turned out to be possible to identify and establish a user account automatically, based on the information that can be derived from the communication network. Eventually, the needed input from the user was reduced to a single question (accompanied by explanatory text) at the initial start-up of the application: "Do you want to allow network connections?" This question was unavoidable for the time being, as it would lead to issues of cost. At the time of writing, there are no standard protocols available to detect whether or not network traffic generates costs to the end user. However, functionality that has been added later on has introduced a few more dialogs to the installation process.

A user account can, however, be created automatically if these network connections are allowed. The default user name is set to the user's phone number to allow easy access to the Web service, and it can later be changed to the user's liking. A default password is delivered to the mobile handset via SMS, and is set to be readable from the S60 client interface for as long as it remains unchanged. The main benefit of the system is that the service becomes usable via confirmation screens that require minimal user input. The registration system and its aspects dealing with user experience are further analyzed in an IMSA paper (Vartiainen et al. 2008). Further benefits of tying the user account to the user's phone number include the possibility to let the user change to new devices and have his or her account follow without any extra configuration steps. Phone numbers also benefit in scenarios where the user wants to share content with friends that are using the system.

Accessing the Web interface does prompt for more information at the initial login, but in no way demands it to be filled in. The user can continue the usage with the mobile-created credentials alone, and the functionality that depends on additional information such as e-mail simply remains disabled. Furthermore, the information is prompted only at the initial login, after which the user needs to specifically go to set it under Settings. It was hoped that this would annoy the user as little as possible, yet still indicate that the functionality exists.

To address the continued usage, it was also necessary to ease image management as much as possible. As one of the basic design drivers was the assumption of pervasive Internet connectivity, the mobile client also implemented automatic uploading and downloading of images to and from the user's private account. When new images were put on the phone via the file system or by taking new photographs, the client detected them and uploaded the images to the server in the background, so as not to disturb the user from what he or she was doing. Similarly, if images were uploaded to the server via a Web browser, they were transferred to the user's mobile phone automatically to create an illusion of a central storage space.

1.6.9 *Issues of Joy*

A large part of designing for experiences is to make sure that there is as little as possible that makes the experience negative, but concentrating mainly on minimizing these issues easily neglects the positive aspects. Taking this thought to the extreme, it might even lead to a situation where the system has nothing wrong with it, but it is still lacking the qualities that would give its users gratification.

Earlier studies as well as the ones conducted for this design work indicated several possibilities for enjoyment. Images themselves can be emotionally loaded, mediating feelings from one person to another. They are often shared with people close to each other and thus having an already existing emotional bond. Sharing can happen either physically in the same place or via transferring the images to the recipients. A user can be the one sharing the image or receiving it.

To address this, the first phase of the implementation tried to make publishing images as easy as possible. Sharing was implemented later on, and is now available in the published versions of S60 and S40 clients. However, the earlier tests simulated the sharing scenarios to some extent, due to the limited number of users within the system. The actual sharing was implemented later, and it took the advantage of the user's existing social network in the form of the address book. Instead of creating new contacts, the system maps existing contacts to those in the user's phone book and allows direct sharing to the ones that are already users. For the rest, other methods of sharing existed. The Web interface allows sharing links to images, while the mobile application connects to both Short Message Service (SMS) and Multimedia Messaging Service (MMS) features provided by the platform.

Additionally, the user's joy can come from passing time while, for example, commuting. Images that are viewed in such situations do not necessarily need to be one's own, even if that seems to be a common habit with the interviewed people. An alternative can be the exploration of other interesting images.

This led to the need to optimize image browsing as much as possible, as well as to concentrate on the image content itself. Images, by default, were shown in full screen. The access to one's own images was made as fast as possible, while still providing visible options for more complex features. Image browsing itself was made almost as fast as possible.

Finally, it is not unfathomable that some part of the pleasure comes from the interface itself. While it was unlikely that this would be the major driver for this system, various niceties were implemented to delight the user or at least make sure that the interface would not get in the way of his or her browsing the images.

Examples of this in Nokia Image Exchange are the transitions between the views and the persistence of images. For example, the main menu shows the latest taken or viewed image in the background, with semi-transparent menus layered on top of it. Some browsing speed is sacrificed to make the images swipe in and off the screen, making the system more fluid. Images were zoomed to move between single images and image grids. Automatic screen rotation on devices that had accelerometers was also an instance where the joy was combined with functionality.

1.6.10 *Issues of Trust*

Elemental aspect of making an acceptable system is to make it reasonably trustworthy. The need of trust depends naturally on the service itself; for example, the needs for banks are quite different from instant messengers. This becomes more understandable if we consider the nature of interaction people perform with these entities. Banks, as an example, deal with money and personal savings that have direct (and possibly dire) consequences to people's lives if anything goes wrong. In the case of instant messengers, the biggest threats are about the usage with a possibility of eavesdropping. Major losses are likely to be related to finding a new service, as very little personal data is stored in instant messaging systems. Imaging services such as the one presented here lie somewhere in between. While it holds personal data that is potentially very private and certainly personal, the loss of data is not likely to be as devastating as losing one's bank account.

Costs are a factor in the current mobile ecosystem. At the time of writing, mobile data plans are not usually guaranteed to be sold together with the service contract. Furthermore, data is often charged by the amount of traffic, which leads to the urge to minimize the data traffic and its cost. The design was based on the assumption of flat-fee data plans, and thus the system presented here provided minimal support for more fine-grained monitoring of data traffic. Some corner cases are notable, however. While roaming, the application does stop from making data connections, as roaming charges for data can be prohibiting. It is to be noted that this behavior is by design, and thus efforts should be made to communicate the behavior clearly to new users. If this were omitted, the lack of trust in the system would be a real threat to the usage.

Another concern of trust comes from the automatic uploading of images to the service. Nokia Image Exchange tries to keep the user's image collection in synchronization with the server at all times by uploading all taken images to the user's private account. It is understandable that the acceptance relies heavily on trust, and that not all people will be willing to give their data to external parties. For example, such worries are quickly raised in companies whose employees take images to document their sketches from whiteboards. Again, this behavior is by design, and needs to be communicated clearly to avoid loss of trust.

Being a research prototype also introduced an aspect that could not be downplayed—the imaging service was, from the start, planned to eventually go live and public. However, it was a prototype with limited resources, and could give no guarantees of existence for a long period of time. Even technical failures could be damaging to the service, which operated on a relatively low budget.

Nokia Image Exchange's two-way connection to personal computers was designed to allow both uploading and downloading images to and from the service.

While the uploading itself was a crude means to allow people to have images from sources other than their mobile to be transferred to the service and to their handsets, downloading was implemented to make it easy to get one's images from the service if the project would run into issues.

The Web interface contains options to download all images belonging to the given user in a zip package or to limit the downloading to the images that have not yet been downloaded. These options were a balance of implementation effort, ethical demands, as well as usability issues.

Some ethical issues were clear. It was simply unimaginable to not offer a way for users to retrieve their images if the system was to be shut down. This was even more critical due to the technicalities that operated under the cover. Namely, the S60 client stored original images onto the Web service and was only required to have scaled-down versions of the images on it for quick viewing. Were the service to go down, the originals could be lost for good. This was simply not acceptable from the team's ethical point of view.

On the other hand, the full implementation of the image packaging and downloading could easily lead to a complex user interface that offered very little benefit for the user if all went well, and was taxing compared to resources available. There was need for a simple solution that would still guarantee safety. Furthermore, this solution should not be overly demanding for the servers that might have thousands of users.

Finally, the downloading system could not be taxing for users if the need for it should arise, or if users would start to use it for their own purposes.

Our solution was to create a page that offered users two options to create a zip package of the images they had on their user accounts. The first, and default, option was to only package and download images that they had not yet downloaded. The second option was to create a package of all the images. The last workaround allowed users to simply save the images one by one while viewing them. Together, it was hoped that these options would be sufficient for users, given the prevailing restrictions.

1.6.11 Compromises on Mobile Client

Preliminary designs and concepts are rarely perfect, as is the case with many other methods of formalization as well. For one, documentation is very rarely unambiguous. The reasons for these imperfections are many, but often the mere complexity of the project at hand is so vast that fully understanding it from all aspects becomes extremely difficult if not impossible. Thus, it essentially comes down to the fact that people need to deal with imperfect plans for whatever they are doing. When, not if, surprises occur, compromises are often needed to adjust the plans to what is feasible.

As an example, the S60 client offered a menu entry "Latest" which allowed browsing the latest images in the service. While it would have been possible to do the updating in the background, quick calculations can be used to demonstrate the issues:

P = Number of users in the system
I = Number of published images, average per user per time unit
T = Total number of image transfers

$$T = (P * I) * (P + 1)$$

This basically tells us that the average number of images is sent to every person within the system. If we now imagine that the system has 10,000 users, each publishing an image per day, the total number of transfers would be of this scale:

$$T = 10,000 * 1 * (10,001) = 100,010,000$$

Being optimistic, 10,000 users is a pessimistic figure, as the hopes for user count are much higher. Similarly, however, a published image per day per user is an optimistic estimate. Finally, the calculation did not take into account traffic generated by other activities. Due to this reason, the estimation should be treated as an approximation that at best should give an idea about scale, but not exact figures. Nonetheless, we can see that a simple functionality that would keep the public images up to date all the time soon starts to push technical boundaries. The number of image transfers grows exponentially with the user count, and with a mere user base of 10,000 the transfers would already hit 100 million per day. One hundred thousand users would already result in 10 billion image transfers per day.

In this light, the compromise of making the user wait for a brief moment while the newly published images were downloaded was a necessity. This was aided by the fact that the Internet, for one, has taught people to wait. Further optimization was also done to make the wait time as short as possible. The sizes of the downloaded images are reduced at the server end, and they are downloaded in batches of multiple images to lessen the impact of network latency if they were to be transferred separately. As a result, a modern 3G mobile connection could theoretically be able to support downloading and viewing 15 images per second, albeit with a moderate hit on the wait time the user needs to tolerate at the beginning of the browsing session. In practice, though, such numbers are not likely to be reached in most situations; in others, the delay can even be much longer. Finally, the already downloaded images were cached onto the phone to lessen the need of bandwidth and, much later on, make it possible to introduce offline features such as access to a vast image collection even when there is no network coverage at all.

Other discussed option was "virtually latest," where the images would not be required to be the absolute latest, but instead a collection that was update at specific intervals. This, however, could result in situations where new content would not be seen even if it were available. As one of the major functions of the application was identified to be killing time, this trade-off was not acceptable.

In contrast, however, the dynamics do change when people are sharing images with their friends. The number of image transfers goes down dramatically when approximately following the pattern below:

P = Number of users in the system
F = Average number of friends per user
I = Number of published images, average per user per time unit
T = Total number of image transfers

$$T = (P * F) * 1$$

Whereas the last calculation showed exponential growth based on the number of users, this pattern is actually linear. Thus, for 10,000 users each having eight friends on average, the number of image transfers is

$$T = 10,000 * 8 * 1 = 80,000$$

This is obviously a lot less than a hundred million transactions mentioned in the previous example. Previous disclaimers to the formula apply to this one as well, but the growth of the transactions as user count increases becomes evident.

This made it feasible to share images, as well as the accompanying metadata, instantly to users friends as well as to receive new images automatically from these friends. While not yet implemented, the team needed to develop underlying technology for purposes such as this one. The exact technical functionality is beyond the scope of this chapter, but in principle it allows clients to retrieve images in a few seconds after they are made available to the user and to have the latest information from one's friends pre-loaded to allow immediate access.

1.7 Conclusion

In summary, I have tried to outline my view that developing and designing services deal with many more aspects than only the technology. Human factors ought to be considered throughout the development process to increase chances of a successful service, and should be assisted by technology where feasible. Technology and the human interface are also embodied, as neither exists on its own.

1.7.1 Usability and User Experience

Some examples on how we have dealt with the hard questions of usability and user experience have been outlined in context of Nokia Image Exchange, an experimental imaging client that, at the time of writing this, is freely available for downloading. I would encourage people to try it out to see and form their own opinions on how interfaces can or should work in mobile context. I find it likely that there will be no status quo in experience design, and new ideas will be prototyped and tested together with new technologies and infrastructural changes that provide new possibilities for developers and users. It is our job to improve upon what we have here and now.

References

Adesemowo, A. and Tucker, W. 2005. Instant messaging on handhelds: An affective gesture approach. *Proceedings of the South African Institute for Computer Scientists and Information Technologists*, White River, South Africa, July 2005.

Alben, L. 1996. Quality of experience: Defining the criteria for effective interaction design. *Interactions*, 3(3): 11–15.

Butler, A. January 1996. Usability engineering turns 10. *Interactions*, 3(1): 58–75.

Buxton, B. 2007. *Sketching User Experiences*. Morgan Kaufmann: San Francisco, CA.

Buxton, B. 2008. The design eco-system, Keynote. *IxDA Interaction 08*, SCAD, Savanna, GA, February 10, 2008.

Chen, K. and Lei, C. 2006. Network game design: Hints and implications of player interaction. *Proceedings of ACM Netgames 2006*, Singapore, October 2006. ACM Press: New York.

Dourish, P. 2004. *Where the Action Is*. Massachusetts Institute of Technology: Cambridge, MA.

Ducheneaut, N., Yee, N., Nickell, E., and Moore, R. 2006. Alone together? Exploring the social dynamics of the massively multiplayer online games. *Proceedings of the SIGCHI Conference on Human Factors in Computing Systems*, Montreal, Quebec, Canada, pp. 407–416. ACM Press: New York.

Greenberg, S. and Buxton, B. 2008. Usability evaluation considered harmful (some of the time). *Proceedings of the 26th Annual SIGCHI Conference on Human factors in Computing systems (CHI 2008)*, Florence, Italy, pp. 111–120. ACM: New York.

Hassenzahl, M. 2001. The effect of perceived hedonic quality on product appealingness. *International Journal of Human-Computer Interaction*, 13: 481–499.

Hassenzahl, M. 2003. The thing and I: Understanding the relationship between user and product. In M. Blythe, C. Overbeeke, A.F. Monk, and P.C. Wright, (eds.), *Funology: From Usability to Enjoyment*, pp. 31–42, Kluwer Academic Publishers: Boston, MA.

Hassenzahl, M. and Tractinsky, N. 2006. User experience—A research agenda. *Behaviour & Information Technology*, 25(2): 91–97.

Hassenzahl, M., Beu, A., and Burmester, M. January/February 2001. Engineering joy. *IEEE Software*, 18: 70–76.

Hassenzahl, M., Law, E., Roto, V., Vermeeren, A., and Kort, J. 2008. Towards a shared definition of user experience. *Conference on Human Factors in Computing Systems*, Florence, Italy, pp. 2395–2398. ACM: New York.

Heidegger, M. 1927. *Being and Time*, English translation 1962. Harper & Row: New York.

Kolko, J. 2007. *Thoughts on Interaction Design*. Brown Bear LLC: Savannah, GA.

Mäkelä, A. and Fulton Suri, J. 2001. Supporting users' creativity: Design to induce pleasurable experiences. *Proceedings of the International Conference on Affective Human Factors Design,* Singapore, pp. 387–394. ASEAN/Academic Press: London, U.K.

Mehrabian, A. 1981. *Silent Messages: Implicit Communication of Emotions and Attitudes.* Wadsworth: Belmont, CA.

Mizutani, M. 2006. *Emotional Communication.* Media Lab, University of Art and Design Helsinki, Finland.

Moggridge, B. 2006. *Designing Interaction.* MIT Press: Cambridge, MA.

Raskin, J. 2000. *The Human Interface.* Addison-Wesley Professional: Boston, MA.

Roto, V. 2006. Web browsing on mobile phones—Characteristics of user experience. PhD dissertation, Helsinki University of Technology, Espoo, Finland.

Sánchez, J, Kirschning, I, Palacio, J., and Ostróvskaya, Y. 2005. Towards mood-oriented interfaces for synchronous interaction. *Proceedings of the 2005 Latin American Conference on Human-Computer Interaction,* Cuernavaca, Mexico, October 2005. ACM Press: New York.

Schwartz, B. 2003. *The Paradox of Choice: Why More Is Less.* Ecco: New York.

Schwartz, B. 2005. The paradox of choice. TED, http://www.ted.com/index.php/talks/barry_schwartz_on_the_paradox_of_c hoice.html (accessed October 5, 2008).

Tognazzini, B. 2008. First principles of interaction design. AskTog, http://www.asktog.com/basics/firstPrinciples.html (accessed May 16, 2008).

UPA (Usability Professionals' Association). 2007. Usability body of knowledge. Usability Professionals' Association, http://www.usabilitybok.org/glossary (accessed April 1, 2007).

UPA (Usability Professionals' Association). 2008. What is user-centered design. Usability Professionals' Association, http://www.usabilityprofessionals.org/usability_resources/about_usability/what_is_ucd.html (accessed May 15, 2008).

Vartiainen, E., Kaasalainen, J., and Strandell, T. 2008. Designing user experience for a mobile imaging application. *Proceedings of the IADIS Interfaces and Human Computer Interaction (IHCI 2008),* Amsterdam, the Netherlands.

Chapter 2

Designing Mobile User Interfaces for Internet Services

Elina Vartiainen

Contents

2.1 Introduction

Considering the importance of the Internet in people's daily lives, it is easy to see the Internet going mobile. The Internet will not be bound to personal computers (PCs), but mobile devices will provide users access to the Internet anywhere, anytime. Mobile devices can be used for performing many tasks that were earlier possible only on a PC. People are increasingly using Internet services on the go, as their mobile devices are always with them and capable of offering access to Internet services.

However, as the technological sophistication of a mobile device has grown, user interfaces of mobile applications are becoming more and more complex. Compared to a PC, a mobile device has a limited set of resources in terms of input and output capabilities, processing power, connectivity, and memory. This means that the user interface design cannot be directly transferred to a mobile device from a PC. The mobile context can also be totally different from the one where a PC is used. When the user is using a mobile device, he or she might be on the move and have only a limited and possibly fragmented time to spend on a task (Oulasvirta et al., 2005). The fragmented nature of mobile environment needs to be considered in interaction design for mobile applications.

Today, mobile devices are starting to reach maturity in terms of new technologies. The level of battery performance and Internet connectivity are satisfactory, and mobile devices are small in size and lightweight. The latest mobile devices have various new input mechanisms such as touch and voice input. Hence, as mobile devices have become mass-market products, more emphasis needs to be put on user experience of mobile devices and applications: aesthetics, usability, utility, and emotional aspects need careful consideration (Hiltunen et al., 2002; Moore, 2002; Jones and Marsden, 2006).

This chapter defines design implications for mobile user interfaces when a user is accessing Internet services on a mobile device. It covers the challenges and limitations of a mobile device, and outlines the design implications that enable a positive user experience of Internet services. User interfaces should be designed for mobile use, should be simple and enjoyable to use, and should hide the unnecessary technical details from the user.

2.2 Mobile User Interfaces for Internet Services

Internet services are mainly used via a Web browser on a PC. This is a convenient way of accessing the services, as they all can be accessed via the same application: the Web browser. On a mobile device, the situation becomes more complicated. The mobile device has a limited set of resources compared to a PC. Thus, it is vital to consider the special characteristics of the device when designing mobile user interfaces for Internet services instead of copying user interface style from a PC (Hiltunen et al., 2002; Moore, 2002; Jones and Marsden, 2006; Roto, 2006).

When a user uses Internet services or browses Web pages on a mobile device, the following limitations need to be considered in the user interface design:

- *Input and output capabilities.* A mobile device may not have a pointing tool or a full keyboard. Also, the screen-size is relatively small.
- *Device resources.* Compared to a PC, a mobile device has limited resources in terms of disk space, processing power, and connection speed.
- *Cost.* Downloading content via a cellular connection may be very expensive for a user when he or she does not have a flat-fee agreement for mobile data traffic.
- *Mobile context.* In the mobile context, the user might be on the move and have only a limited and possibly fragmented time to spend on a task; therefore, the user interface of a mobile application should be intuitive and easy to use. In addition, the mobile context introduces its own set of use cases: on-the-go lookup and entry of information and quick communication.
- *Usage patterns on PC.* It is important to consider how to design a unique user interface for a mobile device without displacing the deep-seated usage patterns that people have found useful on a PC.

Currently, the use of Internet services can be supported in two ways on a mobile device (Kaikkonen, 2009): first, on a mobile Web browser that is optimized for mobile use and takes into consideration the limitations of a mobile device while enabling access to various available Internet services; second, on a mobile client application that is fully integrated with one Internet service, and optimized for the requirements and functionalities of the Web service and mobile use. The mobile Web browser is not optimized for a specific Web page or service, but rather is a common platform for all Web content. Furthermore, it has to copy many elements of the interaction paradigm from the PC, as most Web sites are currently designed for desktop Web browsers. However, next-generation mobile devices overcome some of the earlier limitations and enable development of mobile client applications that are fully integrated to Internet services, offering a positive user experience. These mobile client applications represent an alternative way to support Internet services on a mobile device in the future. In addition to a mobile Web browser and a mobile client application, mobile widgets have recently gained a lot of interest in the mobile industry, as they present a new intermediate solution between the two approaches.

The following sections describe the earlier research on mobile Web browsers and mobile client applications for Internet services, and also introduce the current situation of mobile widgets.

2.2.1 Mobile Web Browsers

A lot of research has been conducted on how to support Web page viewing on mobile devices, as many of the Web pages are originally designed for PC screens. There has not been a clear answer on whether the user interface style on a mobile

Web browser should be consistent or inconsistent with a PC Web browser and in which circumstances (Ketola et al., 2006). History has shown that with the Wireless Access Protocol (WAP), having a separate mobile Internet is not a viable solution. People want to access all the content in the World Wide Web on their mobile devices as well, even though the usability of WAP services is better (Kaasinen, 2005).

Several methods for viewing Web pages on mobile devices apply the overview plus the detail method, where an overview is used to display the whole Web page and a detailed view shows a close-up of a part of the Web page (Björk et al., 1999; Buyukkokten et al., 2000; Milic-Frayling and Sommerer, 2002; Wobbrock et al., 2002; Lam and Baudisch, 2005). Generally, the overview plus the detail method requires a pointing device, a touch screen, or a Personal Digital Assistant (PDA) screen. Commercial Web browsers on touch devices, such as Apple's iPhone and Google's Android phone, utilize the overview plus the detail method and let users zoom between the views.

In addition to the overview plus the detail method, one approach is to simply eliminate some of the content without offering any possibility to view the page in its original form (Trevor et al., 2001; de Bruijn et al., 2002; Gupta et al., 2003; Yang and Wang, 2003). By using this method, the layout and the content of a Web page are modified for good, and a user is not able to view the Web page as he would on a PC.

Many commercial mobile Web browsers use Narrow Layout as a visualization method. Narrow Layout reformats a Web page into one column that fits the width of a mobile device display. This way, the need for horizontal scrolling is eliminated and the user will see all the content just by scrolling down. Narrow Layout, however, has several drawbacks (Roto and Kaikkonen, 2003):

- It often destroys the intended logical grouping of content, leading to situations where users cannot recognize even familiar pages.
- It hinders users from realizing that they have proceeded to a new page after selecting a link, because the first screen of the new page may look exactly the same as that of the previous page.
- It forces Web pages into a one-dimensional layout, which may break pages that rely on a two-dimensional layout, such as timetables and maps.
- It is not compatible with dynamic Web content, where client-side scripting is used to modify the document.

To fix the issues of Narrow Layout, mobile Web browsers using the method also include functionality to show the Web page in its original layout, as on a PC. This, however, introduces the usage of modes in the user interface, causing interaction to work differently in different views, which in turn is difficult for the user to comprehend.

Rolling back to previously visited Web pages is one of the main activities for a Web browser, as more than half of Web page visits are to pages previously visited by the user (Tauscher and Greenberg, 1997). However, this area has not been investigated much in the mobile context. Instead, many studies have been focusing on desktop Web browsers and their functionality for Web history (Ayers and Stasko, 1995; Hightower et al., 1998; Milic-Frayling et al., 2004). The results have shown that a Back menu of visited pages is more efficient than individual Back button presses for distant navigation tasks (Cockburn et al., 2002). Thumbnails of Web pages, in particular, can help users to identify the correct page in the history (Cockburn and Greenberg, 1999). Commercial Web browsers for touch-enabled mobile devices have introduced graphical solutions for the Web history, where thumbnails are utilized in the visualization (e.g., The Iris Browser (2009)).

Another important use case for a Web browser is multiple windows management. People encounter this functionality through opening a new window on purpose or as pop-ups (Hawkey and Inkpen, 2005). There has not been much research in this area, not even related to a desktop Web browser. What has been found out is that people frequently move between windows on PCs when they browse the Web, but they do not necessarily remember which window the wanted page is in. Solutions designed for PCs are not applicable to mobile Web browsers, as they rely on having a big screen (Brown and Shillner, 1996; Kandogan and Shneiderman, 1997). Both Google's Android phone and Apple's iPhone visualize multiple windows in a separate view, where each window is presented as a thumbnail indicating the currently open Web page inside a certain window. As iPhone and Google's Android phone are touch-enabled device, the tab view is designed to work with touch input.

2.2.2 Mobile Client Applications for Internet Services

A mobile client application is a separate application installed in a mobile device and is directly connected to an Internet service. It is implemented to run in the mobile device environment, enabling access to the device resources but requiring an advanced device and development and deployment of the software (Ryan and Gonsalves, 2005). A mobile client application is capable of optimizing, for example, network use and provides sophisticated interaction styles. Mobile client applications can offer a more immediate experience, as they directly connect to a corresponding Internet service and do not fully rely on the request/response paradigm inherent in Web browsers and sites. Thus, it might be difficult for the user to distinguish what exists on the mobile device and on the Internet service (Ketola and Röykkee, 2001). Furthermore, mobile client applications can be used offline, and the information can be synchronized with the Internet service once the connection is reestablished.

Mobile image sharing has been an important topic in the research literature covering the use of Internet services on mobile devices. The focus has been on how people share images (Van House et al., 2005; Kirk et al., 2006) and how to improve the image sharing process (Counts and Fellheimer, 2004; Ahern et al., 2005; Shneiderman et al., 2006). A pleasant user experience has also been an important aspect of the design in these studies. In the Flipper project, one of the design goals was to provide a minimal set of features but to maintain focus on photo content (Counts and Fellheimer, 2004), while the Zurfer project aimed at enabling simple and easy access to users' own photos and their contacts' photos (Naaman et al., 2008). The design also endeavored to be intuitive and have playful interaction with the content.

Nowadays, the state-of-the-art mobile applications for mobile photo sharing (e.g., Facebook, 2009; Kodak EasyShare Gallery, 2009; Pictavision, 2009; Radar, 2009; Share Online, 2009; ShoZu, 2009; Yahoo! Go, 2009) are essentially upload tools for specific Internet services. The applications are usually add-ons to existing gallery applications offering functionalities for separately uploading and download-ing images and the data linked to them. However, the image gallery application and the user's image collection are not fully integrated and synchronized with the service. Furthermore, the upload tool applications require account creation and configuration of settings before they can be used.

2.2.3 Mobile Widgets

Mobile widgets offer an intermediate way to access a specific Internet service on a mobile device compared to a mobile Web browser or client application. Mobile widgets are neither independent applications nor traditional mobile or Web sites, but they run inside a widget engine that may determine the rules for capabilities, appearance, and interaction of mobile widgets. However, people might perceive mobile widgets as separate applications.

A widget engine can be implemented in different ways (Boström et al., 2008; Mihalic, 2008). The most common solution is to use the Web browser as a wid-get engine and implement mobile widgets with Web technologies such as HTML, JavaScript, and Asynchronous JavaScript And XML (AJAX). Such solutions are cur-rently provided by Apple iPhone, Nokia, and Opera widgets. Another approach is to implement the widget engine as a proprietary system, where mobile widgets are developed with specific tools and languages. In this case, the widget engine is often written in Java, as shown by the examples of Nokia WidSets, Plusmo, and Yahoo! Go.

As mobile client applications, mobile widgets allow easy and quick access to Internet services. However, there are substantial issues with mobile widgets. The standardization of mobile widgets is still work-in-progress (Caceres, 2008) and each widget engine offers different capabilities and features, which makes it dif-ficult for developers to design and implement widgets (Mihalic, 2008). In addition, this may introduce problems to users as they cannot use the same mobile widgets with different widget engines. Widget engines also entail different user interface and interaction styles, which forces users to adopt their behavior each time they use another platform.

Mobile widgets also have many limitations compared to mobile client applica-tions. The development of mobile widgets should be as simple as possible, to ensure a wide developer base and selection of widgets. Widget engine APIs, however, tend to be quite constricted compared to platform APIs, and do not offer an extensive set of tools for designing user interfaces within a mobile widget. As an exception, the API for Apple's iPhone widgets provides a wide range of methods to access platform components and create rich interaction methods. iPhone widgets are based on Web technologies, and the iPhone platform provides touch as an input method which fits better to the Web interaction style than a 5-way navigation control.

2.3 Minimap and Image Exchange

This chapter introduces two approaches to support the usage of Internet services on a mobile device: Minimap and Image Exchange. Minimap is a mobile Web browser, while Image Exchange is a photo sharing Internet service including a mobile client application. The solutions in the user interface design are justified for both cases through the results of user experience evaluations.

2.3.1 Mobile Web Browser: Minimap

The Minimap mobile Web browser currently presents the most common approach to support the use of Internet services on mobile devices (Roto et al., 2006; Vartiainen et al., 2007; Vartiainen et al., 2008a). We considered the limitations of mobile devices in the user interface design and developed a solution that would enable a compelling user experience. Minimap and its user experience were evaluated in several user studies during the development process.

2.3.1.1 User Interface Design

The first task in the user interface design was to tackle the limitation of a small screen. This was particularly important when defining how to visualize Web pages, as they are mainly designed for desktop screens. In addition, we needed to outline a good design for the visualization of a Web history and multiple windows, as they are important use cases on a PC but have many challenges when the metaphors transferred to a mobile device.

For Web page visualization, we designed a method that scales down the layout of a Web page to fit more content to the screen (Figure 2.1). The method modified the size of the text relative to the rest of the Web page contents and limited the maximum width of the text paragraphs to the width of the screen of the mobile device. Hence, the text paragraphs were at most as wide as the screen, and the need for horizontal movement was eliminated while reading text. The Web page was still navigated by scrolling, as today's mobile devices are capable of performing scrolling efficiently (Jones and Marsden, 2006). As for a navigational aid, we rendered an overview of the Web page overlay transparently on top of the Web page view. In the overview, the user could see his location within the Web page when scrolling.

Figure 2.1   The user interface of the Minimap Web browser including the overview.

For the user's browsing history and windows, we designed a separate view, where the user can move within the Web history and between browser windows with a navigation key (Figure 2.2). The view consisted of graphical representation of the Web history of the current window and the other open windows, where the Web pages are visualized as thumbnails. The Web history is aligned on the horizontal line and the currently open browser windows were displayed on a vertical line. Each thumbnail in the Web history presented a Web page the user had visited during the browsing session in the same browser window, while in the window list a thumbnail image illustrated a Web page that was currently opened in the window. We named this view for the Web history and multiple windows as "Rolling History."

Both views were optimized for a 5-way navigation key, a joystick. In the view for Web page visualization, the vertical and horizontal movements of the joystick were reserved for scrolling the page, while the select action activated a link. Rolling History used the horizontal movement for navigating in the Web history of the current window, while the vertical movement was to switch between browser windows. A user selected a page in Rolling History with a joystick select action. The views did not necessitate a pointing device or a touch screen for input. Moreover, the views did not require a zoom key to work, as most mobile devices do not provide a dedicated key for zooming.

In their user interfaces, the Web page view and Rolling History did not include modes enabling interaction mechanisms to be always analogous in one view. This is essential in the mobile context, as the user might be interrupted and have difficulties

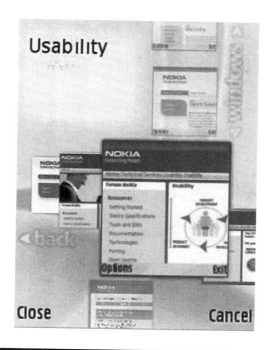

Figure 2.2   Rolling History is a user interface for navigating in the Web history and between multiple windows.

in remembering which mode was currently active. Using modes in the user interface design introduces a frequent source of errors and frustration to users, and should be avoided if possible (Nielsen, 1994). Rolling History also employed animations in the user interface to give users a visual aid. Rolling History was a relatively new concept for users, so we needed to make it more intuitive and usable on the go.

The Web page visualization method of Minimap used an algorithm for the hyperlink selection: when a user scrolled within the Web page, the link selection moved accordingly. The user could select any link that was visible in the browser view. The link selection aimed at being as natural as possible to the user, who is accustomed to using a mouse cursor for hyperlink selections on a PC screen. In this case, it was beneficial to copy a usage pattern from a PC, even though other user interface paradigms might be more effective. Rolling History also demonstrated how to design a unique user interface for a mobile device without displacing the deep-seated usage patterns that people have found useful on a PC. It used the same paradigm for multiple windows as on a PC, even though windows did not really exist on a S60 user interface style. However, people need multiple windows on a PC in several situations, and having the functionality on the mobile device improves the user experience. Nonetheless, the user interface can be optimized for a mobile device, although the paradigm would remain.

Rolling History also contributed to overcoming the limitation of restricted mobile device resources. Getting a previous Web page back on the screen on a mobile device may take a while due to the bandwidth and processing power, and may also consume battery and memory. Furthermore, some Web pages require non-cacheable content from the Web, so downloading content via a cellular connection while back-stepping may even add to browsing expenses. Rolling History enables the user to select only the page he or she desires, as the page thumbnails are easily recognizable. Furthermore, the user does not need to load all Web pages along the way if he or she needs to step multiple pages back in the Web history.

2.3.1.2 User Experience Evaluations

We organized several rounds of user studies to evaluate the user experience of different functionalities of Minimap, as our ambition was to develop user interfaces in an iterative fashion. After each study, we went through the issues discovered by the study and further developed the solutions and their user interfaces. I will go through the most essential user study results of Minimap in this chapter.

Minimap's method for Web page visualization was evaluated in two user studies: first, in a laboratory test with 8 subjects; second, in a field study with 20 participants. In the field study, Minimap was compared to a commercial Web browser that used Narrow Layout as a method for Web page visualization. The participants used Minimap and the other browser for 8 days each. Evaluation data was collected through questionnaires, task feedback, diaries, and logs. We also gathered qualitative data in focus groups, where the participants discussed Minimap and the study. Comparing two applications enabled the participants to have a reference point to their evaluation: it is easier to compare two options than to evaluate a single solution.

The user experience was measured by asking the participants' preference between the two browsers. We asked them to evaluate which browser they would prefer if they needed to browse Web pages on a mobile phone. The results shown in Figure 2.3 clearly demonstrate a preference for Minimap, as 12 users out of 20 strongly preferred it.

Rolling History was evaluated in two laboratory studies. The results of the first study indicated that the initial version of the solution was not intuitive enough, as participants without technical background or previous experience on mobile Web browsing were confused with pages and windows shown in the same view. We analyzed the problem in the graphical design. After improving the solution, we conducted another user study, where the solution was compared to a state-of-the-art tab approach. Participants conducted a set of tasks, including all the basic interaction that users experience when handling pages in the Web history and multiple windows. After completing the tasks, the participants gave feedback by filling out a questionnaire. The user experience was again evaluated according to the preference,

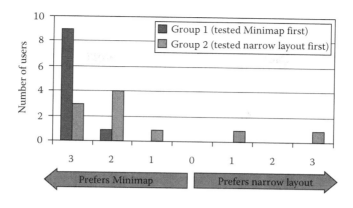

Figure 2.3 **Eighteen out of twenty users preferred Minimap Web browser after a field study.**

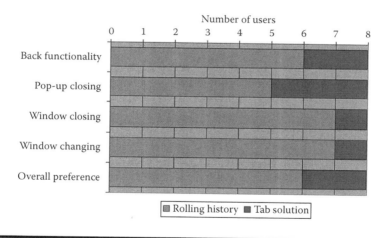

Figure 2.4 **Rolling History was preferred by most users across all functions.**

and the results showed that most participants preferred our solution across all functions, as seen in Figure 2.4.

2.3.2 Mobile Client Application: Image Exchange

Image Exchange exemplifies a future solution for supporting the usage of Internet services on a mobile device consisting of a mobile client application and a corresponding Internet service (Vartiainen et al., 2008b,c; Vartiainen, 2009). The mobile client application was fully integrated to the corresponding Internet service: the application was always connected to, and the user's image collection was up to date with, the service. The mobile client application provided a way for users to share and interact with photos in real-time on the go.

As before, we took into account the limitations of a mobile device in the user interface design for the mobile application of Image Exchange. However, this time we were capable of fully exploiting device resources and rich user interface possibilities, contrary to Minimap. Hence, we developed a solution that aimed at offering a better user experience of Internet services on a mobile device than a Web browser. The user experience of Image Exchange was evaluated in two user studies.

2.3.2.1 User Interface Design

The mobile client application of Image Exchange was developed for advanced mobile devices equipped with an always-on, flat-fee data connection, longer battery lifetime, and enough CPU power to run fast and rich user interfaces. In addition, the application was designed exclusively for mobile use, as Internet services on a PC are commonly used on a Web browser. Therefore, the user interface design could exploit the opportunity for unique mobile-specific solutions. We still needed to consider the limited input and output methods and the mobile context in the design process, as those limitations were to remain.

The user interface of Image Exchange mobile client application was designed by identifying the most essential use cases for mobile photo sharing. The use case definitions were formed and prioritized through user interviews and evaluations of other mobile photo sharing applications. Finally, they were utilized to define the requirements for the mobile client application. The aim was to find a simplistic and pleasant user interface design that would offer a positive user experience.

The main views of the Image Exchange mobile user interface are presented in Figure 2.5. When the application was launched, the main menu was displayed on the screen and the latest captured image was shown full-screen in the background. A user could access images through different categorizations in the main menu by using the 5-way navigation key. Thus, the need for text input was eliminated. All selections in the main menu item led to the image browsing view, which optimized the screen area and displayed a specific image full-screen. The left and right arrow keys were used to flip through the images in the image browsing view.

The image menu was used to show functions that were available for the selected image. We decided to create our own menu style instead of using the option menu that is offered by the platform in many mobile devices. The reasoning for this was that usually the options menu does not separate the functionality and the settings of the application, but has all of them presented in the same list. We wanted to make a clear separation between the functionality that is available for a certain image versus the application-wide settings. As we were developing an application

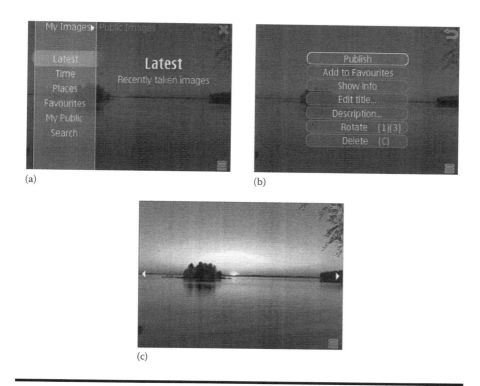

Figure 2.5 The user interface of Image Exchange: (a) the main menu, (b) the image menu, and (c) the image browsing view.

that would directly run on top of the platform, we could take a full advantage of the rich user interface styles.

As the mobile context requires that people's tasks on their mobile devices are quick and simple to complete, we needed to make the interaction between the mobile client application and the Internet service easy and fluent:

■ To facilitate the registration process to the service, we decided to use the identification number of the device as an initial user name for the Internet service, to minimize effort. A user account was created to the service without any user input except asking for permission to use the network connection. By removing the need for any input, we lowered the entry barrier for the user to start using the application.

■ To enable people to share their images right on the spot, Image Exchange transferred captured images transparently without requiring any user input. As a result, a user could trust that his image collection was always present on the device and in the service.

■ To provide a user a way to add additional data to the images, Image Exchange enabled him or her to modify the title, description, and comments through the image menu with a couple of clicks. Whenever the user decided to change a title or description or add a comment either on the mobile device or in the service, the changes also appeared immediately at the other end.

2.3.2.2 Benefits in Contrast to Mobile Web Browsers

When a mobile Web browser is used for accessing Internet services, it requires at least some amount of text input for registration and possibly for logging in. Typing text on a mobile device can be a laborious task, especially when a task such as a registration or login requires text entry without any typographical errors. For Image Exchange, we were able to simplify and facilitate the registration process, as the client application could access the device resources and make use of the platform data. Similarly, when a Web browser is used for uploading content to an Internet service, it usually requires filling out a form by inputting text. In Image Exchange, we transferred the captured mobile images transparently without requiring any user interaction. As a result, the user's image content was automatically up to date with the service, without any effort from him or her.

When using the Web browser to access content on an Internet service, the user might need to navigate through many Web pages and links until he can access the latest content. In addition, the Web browser may need to send many requests to the service and wait for the responses until it can show the content to the user. Therefore, the content browsing on a Web browser might not be fluent and immediate from the user's perspective. When the user started the Image Exchange application, he or she could access the latest online photos by two clicks. This led the user to an image browsing view, where he or she could browse online images with the left and right arrow keys. The browsing between images occurred immediately without any waiting periods, as the application fetched the latest online images published and cached them to the device memory. The service would only send screen-size thumbnails of the images to make the transfer fast and to save network bandwidth. Still, the whole screen area was used for showing the image.

Modifying content on an Internet service with a Web browser might also require many link selections, filling out a form, submitting the form, and waiting for a response. The image Exchange application used a dialog activated directly with a mouse click to enable modifying of image content, such as titles, descriptions, or comments. The title, description, and comments of an image were up to date with the corresponding Internet service, without any need to specifically submit or update the information.

2.3.2.3 User Experience Evaluations

Image Exchange was evaluated in two field studies. First, we launched the mobile application and the service to a closed group of participants and conducted a small-scale user study to find out if the concept was useful and fun in practice. The results showed that the integration of the mobile application and the service was appreciated because of the automatic synchronization of data and how the whole concept worked seamlessly. The user interface and the user experience of the mobile application were positively rated by the participants.

To evaluate Image Exchange more thoroughly, we conducted a field study of 2 weeks to compare Image Exchange with a state-of-the-art gallery application combined with an add-on tool for photo sharing. The earlier studies have shown that a mobile client application offers a superior way to interact with an Internet service compared to its Web-based counterpart (Ryan and Gonsalves, 2005). That is why we decided to compare Image Exchange to a state-of-the-art mobile client application, Gallery, instead of a mobile Web browser. In addition, Image Exchange could be used as an image browser application to view the user's own image collection on his mobile device, so the Gallery application presented a good point of comparison in that extent as well. The Gallery application was not fully integrated to an Internet service, but used an upload tool for transferring and updating the content. We wanted to prove that the full integration of a mobile client application and an Internet service is crucial to enable a compelling user experience.

The field study included two groups each containing five participants. Both groups used the Image Exchange application and the Gallery application for 7 days each. One group started with Image Exchange, and the other with the Gallery application and the upload tool. The focus of the study was on the overall user experience of implemented features and how that would affect the social activity within the group during the testing period. We collected data through task feedback, questionnaires, logs, and focus group discussions.

Figure 2.6 shows the preference of the participants after using both applications for 7 days each. We used a 7-point scale, 3 meaning strong preference for either application and 0 meaning no preference. Of the total of 10 participants, 7 preferred Image Exchange very strongly and 8 participants preferred it in total (Figure 2.6).

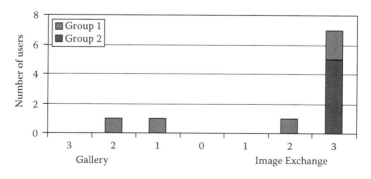

Figure 2.6   Seven out of ten participants preferred Image Exchange very strongly and eight participants preferred it in total.

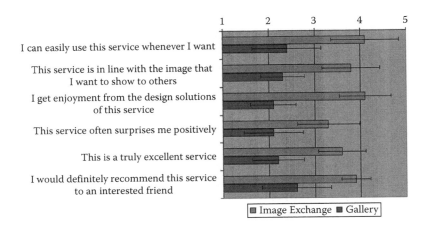

Figure 2.7   The Image Exchange mobile application scored significantly better in five out of six user experience evaluation questions. The results include standard deviation.

The Image Exchange application was especially appreciated because of the user experience (Figure 2.7). The results showed that the Image Exchange application scored better in five out of six questions. The ease of use and the design solutions of the application, in particular, attracted the participants, and they listed the ease of use and simplicity as the key design solutions of Image Exchange. The participants enjoyed the use so much that they would even recommend it to an interested friend. Furthermore, the participants considered the Image Exchange application and service to be rather excellent and to support the image that they want to show to the others. The participants explained that they highly appreciated the visual looks of the Image Exchange application, describing it as "stylish," "modern," and "beautiful." The application also managed to surprise the participants positively from time to time, and they commented that Image Exchange was fun to use.

The user experience evaluation also explained why Image Exchange was strongly preferred by the participants compared to the Gallery application. The participants took pleasure in using Image Exchange even though they also got things done with the Gallery application. Image Exchange enabled the participants to enjoyably interact with each other in real time and on the go, while the Gallery application still required the participants to take care of many tasks (e.g., uploading an image, synchronizing the image data) before they could concentrate on the actual communication.

The results of the social activity within the participant groups are shown in Figure 2.8. The social activity was measured through the number of published images and comments during the test periods. The results revealed that the participants were more socially active when using Image Exchange, which indicated that a better user experience encourages users to use Internet services more actively.

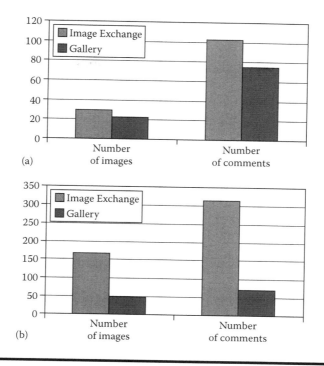

Figure 2.8 **The participants of the study were more socially active when using the Image Exchange mobile application. (a) The results of social activity of Group 1. (b) The results of social activity of Group 2.**

Moreover, the social activity was not dependent on the order in which participants used the applications and services; both groups used Image Exchange more frequently.

2.4 Conclusions

The Internet is going mobile. Mobile devices have become more technologically sophisticated, and they can be used to access the Internet anytime, anywhere. Many Internet services are particularly relevant for people on the go, and people find it meaningful to access them via their mobile devices. However, the user experience of Internet services on a mobile device still requires improvement. Mobile devices have many enablers to offer, but these have not yet been fully utilized in supporting Internet services.

As stated earlier, Internet services can be supported by two approaches on a mobile device: on a mobile Web browser and on a mobile client application. It has also been proven by related work that a mobile client application has a superior performance over its Web-based counterpart: Ryan and Gonsalves (2005) discovered that a mobile client application can optimize network usage and use client-side processing more extensively instead of requesting data from the server, while Weiss (2002) noted that mobile client applications can take advantage of the rich user interface features of the mobile device without the limitations of mobile device Web browsers.

The Minimap Web browser presents a solution, where a user interface paradigm is transferred from a PC to a mobile device in a way that preserves the familiar desktop usage patterns but offers unique solutions in the mobile user interface design. The user experience evaluations show that the designed solutions to surmount the limitations of a mobile device were successful and support the usage of Internet services on a mobile device.

The benefits of Minimap can be summarized as follows:

- It is optimized for a small-screen mobile device.
- It does not require a pointing device or a touch screen but makes good use of 5-way navigation key.
- It does not present modes in the user interface.
- It utilizes the familiar desktop usage patters.
- It presents unique solutions for new usage behavior that the mobile context introduces.
- It introduces unique user interface solutions to compensate the limited resources of a mobile device.

If we look at the use of a specific Internet service on a mobile device, even Minimap is not fully optimized for the mobile context. It copies the interaction style from a PC, where it tries to simulate the navigation with a hyperlink selection algorithm. Thus, it is not capable of fully utilizing the rich user interface features or device resources (Weiss, 2002; Ryan and Gonsalves, 2005). Moreover, a mobile Web browser is technology-wise tied to the request/response paradigm of the Web, which prevents it from providing an immediate experience to the user when he is interacting with an Internet service. For example, when a registration or a login to an Internet service is done on a mobile Web browser, it demands a considerable amount of user input. As text input is laborious and mobile context is full of interruptions, completing such a test might be too difficult for a user and prevent him or her from using Internet services on a mobile device. Thus, the usage of a specific Internet service requires an optimized solution to be able to successfully offer a pleasant user experience.

Image Exchange demonstrates a solution in which the mobile usage of an Internet service is supported via a fully service-integrated mobile client application. The application is optimized for the requirements of the service, and it strives to take full advantage of mobile device resources and capabilities. The user interface design of the application is unique as it is developed purely for the mobile use and the corresponding Internet service.

The main benefit of a fully service-integrated mobile client application compared to a mobile Web browser is its capability to provide a more pleasant user experience for the usage of Internet service. There are several reasons for this:

- The mobile user interface can be optimized for the needs of the Internet service: Navigation in the user interface becomes simpler, and the service can send an update to the client without a separate request from the user.
- The mobile user interface can directly utilize device capabilities: Technical details are hidden from the user, and smart defaults can be offered to remove the need to modify complex settings. Many functions can also be automated, and a need for user input decreases. Furthermore, the user interface is able to offer an immediate experience and becomes more aesthetically pleasing and fast, as it can benefit from the rich user interface features. Finally, the mobile client application can also be used offline.
- The full integration with the corresponding Internet service improves the user experience: The content is always up to date between the mobile client application and the Internet service. There is no need for manual updating, thus fitting better to the mobile context of use.

In practice, developing a mobile client application for each Internet service is not feasible. It requires a lot of design and implementation work, especially if it is aimed at functioning on many platforms. Hence, in cases such as the deployment of a new Internet service, it may be more practical to concentrate on making the user experience of the Web site of the Internet service as pleasing as possible, and to let users access it with their mobile Web browser.

As a result, the following design implications drawn from the conclusions of Minimap and Image Exchange are intended as guidelines on how to design mobile interfaces for Internet services:

Optimize content. In the mobile Web browser, the Web content should be optimized to match the output capabilities of a mobile device. It should at least have a visualization method that modifies the content for the small screen. The mobile client application can optimize the user interface for the needs of a specific Internet service, including its content.

Utilize desktop and mobile usage patterns. The mobile Web browser should utilize the familiar desktop usage patterns, but also introduce novel solutions for mobile use. The mobile client application can be uniquely designed for a mobile device and mobile use.

Fully exploit device capabilities. The user interface of the mobile Web browser should offer a pleasant user experience via a good use of the 5-way navigation key, a small-screen, and a consistent interaction. The mobile client application can utilize the rich user interface features that a mobile device has to offer, including enhanced graphics and immediate feedback.

Compensate for device resources. The mobile Web browser should introduce unique user interface solutions to compensate for the limited mobile device resources. The mobile client application has an opportunity to fully utilize device capabilities to hide technical details from the user, offer smart defaults, automate functions, and enable offline usage.

Update content. The mobile Web browser is able to offer manual updating as well as automatic downloading of Internet content. It should optimize the updating in such a way that it is useful to the user. The mobile client application can be fully integrated with the corresponding Internet service, and can provide access to up-to-date content without requiring manual updating.

2.5 Future Evolution

The design implications for mobile interfaces of Internet services were defined through two studies: Minimap and Image Exchange. However, Image Exchange exemplifies only one type of mobile client application, and is not able to provide general guidelines. Every Internet service is a unique combination of features, interaction, and content, and needs to be considered separately. More work needs to be conducted, and the design implications should be tested with alternative mobile client applications in the right context to form generalizable design implications for mobile client applications. Our design process in both studies included multiple rounds of user needs studies, evaluations, and refinements. It seemed to fit well for developing mobile user interfaces for Internet services, and can be recommended for future studies as well.

Additionally, it might be beneficial in the future to explore more deeply how two approaches, a mobile Web browser and a mobile client application, compare with each other. By designing and evaluating mobile client applications of different Internet services, the design guidelines could be expanded to answer questions such as when it is beneficial to develop a mobile client application for an Internet service and when the support via a mobile Web browser is enough.

Another topic for future research is to explore new enablers for mobile client applications. Today, Web technologies such as JavaScript and HTML enable the creation of Web applications that may appear as separate applications in a user interface of an operating system, but actually run on top of the Web browser and its engine. The benefit of these applications is that they are platform-independent, and can be updated and maintained easily without software distribution or installation. The client-side processing of these applications enable immediate experience that does not require the Web page to reload after submitting information.

In addition, mobile devices are also starting to support Flash as a method for creating rich Web applications. Many Internet services nowadays offer a mobile Web site to support mobile usage. People can access the Web site through their mobile Web browser. Furthermore, there are recent developments in Web technologies related to user interface graphics that are able to offer new tools for mobile user interface design for Internet services, including SVG, Java applets, and XForms.

In the future, all of these methods have a chance to serve as powerful methods for implementing a mobile client application to support the usage of Internet services on a mobile device in addition to client applications that run directly on top the mobile platform. In particular, as mobile devices are increasingly equipped with new input mechanisms, such as touch, it provides an opportunity to create a positive user experience through more lightweight mechanisms (e.g., Web applications) than through implementing a mobile client application. Moreover, the methods that are capable of solving the issue with interoperability of different platforms have a change to succeed in large-scale. Internet services will continue to be an important part of people's online lives, and mobile devices have a good chance to become the prime method to access them.

References

Ahern, S., King, S., and Davis, M. Mmm2: Mobile media metadata for photo sharing. In *Proceedings of the 13th Annual ACM International Conference on Multimedia* (*MULTIMEDIA'05*), Singapore, 2005, pp. 790–791, ACM, New York.

Ayers, E. Z. and Stasko, J. T. Using graphic history in browsing the world wide web. In *Fourth International World Wide Web Conference*, Boston, MA, 1995, pp. 11–14.

Björk, S., Holmquist, L. E., Redström, J., Bretan, I., Danielsson, R., Karlgren, J., and Franzen, K. West: A web browser for small terminals. In *Proceedings of the 12th Annual ACM Symposium on User Interface Software and Technology* (*UIST'99*), Ashville, NC, 1999, pp. 187–196, ACM, New York.

Boström, F., Nurmi, P., Floréen, P., Liu, T., Oikarinen, T.-K., Vetek, A., and Boda, P. Capricorn—An intelligent user interface for mobile widgets. In *Proceedings of the 10th International Conference on Human Computer Interaction with Mobile Devices and Services* (*MobileHCI'08*), Amsterdam, the Netherlands, 2008, pp. 327–330, ACM, New York.

Brown, M. H. and Shillner, R. A. The deckscape web browser. In *Conference Companion on Human Factors in Computing Systems* (*CHI'96*), Vancouver, Canada, 1996, pp. 418–419, ACM, New York.

de Bruijn, O., Spence, R., and Chong, M. Y. RSVP browser: Web browsing on small screen devices. *Personal and Ubiquitous Computing*, 6(4):245–252, 2002.

Buyukkokten, O., Garcia-Molina, H., Paepcke, A., and Winograd, T. Power browser: Efficient web browsing for PDAs. In *Proceedings of the SIGCHI Conference on Human Factors in Computing Systems* (*CHI'00*), Hague, the Netherlands, 2000, pp. 430–437, ACM, New York.

Caceres, M. Widgets 1.0: Packaging and configuration. World Wide Web Consortium Working Draft WD-widgets-20081222, W3C, Cambridge, MA, December 2008.

Cockburn, A. and Greenberg, S. Issues of page representation and organization in web browser's revisitation tools. *Australian Journal of Information Systems*, 7:120–127, 1999.

Cockburn, A., McKenzie, B., and Smith, M. J. Pushing back: Evaluating a new behaviour for the back and forward buttons in web browsers. *International Journal Human Computer Studies*, 57(5):397–414, 2002.

Counts, S. and Fellheimer, E. Supporting social presence through lightweight photo sharing on and off the desktop. In *Proceedings of the SIGCHI Conference on Human Factors in Computing Systems (CHI'04)*, Vienna, Austria, 2004, pp. 599–606, 2004, ACM, New York.

Facebook. http://www.facebook.com/ (accessed September 2009).

Gupta, S., Kaiser, G., Neistadt, D., and Grimm, P. Dom-based content extraction of html documents. In *Proceedings of the 12th International Conference on World Wide Web (WWW'03)*, Budapest, Hungary, 2003, pp. 207–214, ACM, New York.

Hawkey, K. and Inkpen, K. Web browsing today: The impact of changing contexts on user activity. In *CHI '05 Extended Abstracts on Human Factors in Computing Systems (CHI'05)*, Portland, OR, 2005, pp. 1443–1446, ACM, New York.

Hightower, R. R., Ring, L. T., Helfman, J. I., Bederson, B. B., and Hollan, J. D. Graphical multiscale web histories: A study of padprints. In *Proceedings of the Ninth ACM Conference on Hypertext and Hypermedia: Links, Objects, Time and Space Structure in Hypermedia Systems (HYPERTEXT'98)*, Pittsburgh, PA, 1998, pp. 58–65, 1998, ACM, New York.

Hiltunen, M., Laukka, M., and Luomala, J. *Mobile User Experience*. Cromland, Bethlehem, PA, 2002.

Jones, M. and Marsden, G. *Mobile Interaction Design*. John Wiley & Sons, Chichester, U.K., February 2006.

Kaasinen, E. User acceptance of mobile services—Value, ease of use, trust and ease of adoption. PhD thesis, Tampere University of Technology, Tampere, Finland, 2005.

Kaikkonen, A. Mobile internet: Past, present, and the future. *International Journal of Mobile Human Computer Interaction*, 1(3):29–45, 2009.

Kandogan, E. and Shneiderman, B. Elastic windows: A hierarchical multiwindow world-wide web browser. In *Proceedings of the 10th Annual ACM Symposium on User Interface Software and Technology (UIST'97)*, Banff, Canada, 1997, pp. 169–177, ACM, New York.

Ketola, P. and Röykkee, M. Three facets of usability in mobile handsets. In *Proceedings of the CHI 2001, Workshop, Mobile Communications: Understanding Users, Adoption & Design Sunday and Monday*, Seattle, WA, 2001, ACM, New York.

Ketola, P., Hjelmeroos, H., and Räihä, K.-J. Coping with consistency under multiple design constraints: The case of the Nokia 9000 www browser. *Personal and Ubiquitous Computing*, 4(2–3):86–95, 2006.

Kirk, D., Sellen, A., Rother, C., and Wood, K. Understanding photowork. In *Proceedings of the SIGCHI Conference on Human Factors in Computing Systems (CHI'06)*, Montreal, Canada, 2006, pp. 761–770, ACM, New York.

Kodak EasyShare Gallery. http://www.kodakgallery.com/ (accessed September 2009).

Lam, H. and Baudisch, P. Summary thumbnails: Readable overviews for small screen web browsers. In *Proceedings of the SIGCHI Conference on Human Factors in Computing Systems (CHI'05)*, Portland, OR, 2005, pp. 681–690, ACM, New York.

Mihalic, K. Widgetization of mobile internet experience. In *The Second International Workshop on Mobile Internet User Experience (MIUX'08), in Conjunction with MobileHCI'08 Conference*, Amsterdam, the Netherlands, 2008.

Milic-Frayling, N. and Sommerer, R. Smartview: Flexible viewing of web page contents. In *Poster Proceedings of the Eleventh International World Wide Web Conference*, Honolulu, HI, May 2002.

Milic-Frayling, N., Jones, R., Rodden, K., Smyth, G., Blackwell, A., and Sommerer, R. Smartback: Supporting users in back navigation. In *Proceedings of the 13th international conference on World Wide Web (WWW'04)*, New York, 2004, pp. 63–71, ACM, New York.

Moore, G. *Crossing the Chasm*. HarperBusiness, New York, 2002.

Naaman, M., Nair, R., and Kaplun, V. Photos on the go: A mobile application case study. In *Proceedings of the 26th Annual SIGCHI Conference on Human Factors in Computing Systems* (*CHI'08*), Florence, Italy, 2008, pp. 1739–1748, ACM, New York.

Nielsen, J. *Usability Engineering*. Morgan Kaufmann Publishers, San Francisco, CA, 1994.

Oulasvirta, A., Tamminen, S., Roto, V., and Kuorelahti, J. Interaction in 4-second bursts: The fragmented nature of attentional resources in mobile HCI. In *Proceedings of the SIGCHI Conference on Human Factors in Computing Systems* (*CHI'05*), Portland, OR, 2005, pp. 919–928, ACM, New York.

Pictavision. http://www.pictavision.com/ (accessed September 2009).

Radar. http://radar.net/ (accessed September 2009).

Roto, V. Web browsing on mobile phones—Characteristics of user experience. PhD thesis, Helsinki University of Technology, Espoo, Finland, 2006.

Roto, V. and Kaikkonen, A. Perception of narrow web pages on a mobile phone. In *Human Factors in Telecommunications*, Berlin, Germany, 2003.

Roto, V., Popescu, A., Koivisto, A., and Vartiainen, E. Minimap: A web page visualization method for mobile phones. In *Proceedings of Human Factors in Computing Systems Conference* (CHI'06), Montreal, Canada, 2006, pp. 35–44.

Ryan, C. and Gonsalves, A. The effect of context and application type on mobile usability: An empirical study. In *Proceedings of the 28th Australasian conference on Computer Science* (*ACSC'05*), New Castle, Australia, 2005, pp. 115–124, Australian Computer Society, Darlinghurst, Australia.

Share Online. http://www.nokia.com/betalabs/shareonline/ (accessed September 2009).

Shneiderman, B., Bederson, B. B., and Drucker, S. M. Find that photo!: Interface strategies to annotate, browse, and share. *Communications of the ACM*, 49(4):69–71, 2006.

ShoZu. http://www.shozu.com/ (accessed September 2009).

Tauscher, L. and Greenberg, S. Revisitation patterns in world wide web navigation. In *Proceedings of the SIGCHI Conference on Human Factors in Computing Systems* (*CHI'97*), Los Angeles, CA, 1997, pp. 399–406, ACM, New York.

The Iris browser. http://www.irisbrowser.com/ (Accessed September 2009).

Trevor, J., Hilbert, D. M., Schilit, B. N., and Koh, T. K. From desktop to phonetop: A UI for web interaction on very small devices. In *Proceedings of the 14th Annual ACM Symposium on User Interface Software and Technology* (*UIST'01*), Orlando, FL, 2001, pp. 121–130, ACM, New York.

Van House, N., Davis, M., Ames, M., Finn, M., and Viswanathan, V. The uses of personal networked digital imaging: An empirical study of camera phone photos and sharing. In *CHI'05 Extended Abstracts on Human Factors in Computing Systems* (*CHI'05*), Portland, OR, 2005, pp. 1853–1856, ACM, New York.

Vartiainen, E. Improving the user experience of a mobile photo gallery via supporting social interaction. *International Journal of Mobile Human Computer Interaction* (*IJMHCI*), 1(4):42–57, 2009.

Vartiainen, E., Roto, V., and Popescu, A. Auto-update: A concept for automatic downloading of Web content to a mobile device. In *Proceedings of the Fourth International Conference on Mobile Technology, Applications, and Systems and the First International Symposium on Computer Human Interaction in Mobile Technology*, Singapore, 2007, pp. 683–689.

Vartiainen, E., Roto, V., and Kaasalainen, J. Graphical history list with multi-window support on a mobile Web browser. In *Proceedings of the Third International Conference on Internet and Web Applications and Services* (*ICIW'08*), Athens, Greece, 2008a, pp. 121–129.

Vartiainen, E., Kaasalainen, J., and Strandell, T. Designing user experience for a mobile imaging application. In *Proceedings of IADIS International Conference Interfaces and Human Computer Interaction*, Amsterdam, the Netherlands, 2008b, pp. 197–204.

Vartiainen, E., Strandell, T., and Kaasalainen, J. Fully service-integrated mobile application for photo-sharing. In *Proceedings of the 12th IASTED International Conference on Internet and Multimedia Systems and Applications* (*IMSA 2008*), Kailua-Kona, HI, 2008c, pp. 38–43.

Weiss, S. *Handheld Usability*. John Wiley & Sons, New York, 2002.

Wobbrock, J. O., Forlizzi, J., Hudson, S. E., and Myers, B. A. WebThumb: Interaction techniques for small-screen browsers. In *Proceedings of the 15th Annual ACM Symposium on User Interface Software and Technology* (*UIST'02*), Paris, France, 2002, pp. 205–208, ACM, New York.

Yahoo! Go. http://mobile.yahoo.com/go/ (accessed September 2009).

Yang, C. C. and Wang, F. L. Fractal summarization for mobile devices to access large documents on the web. In *Proceedings of the 12th International Conference on World Wide Web* (*WWW'03*), Budapest, Hungary, 2003, pp. 215–224, ACM, New York.

Chapter 3

Mobile Usability

Harald Amelung, Christoph Ohl,
Gabriele Schade, and Sascha Wagner

Contents

3.1 Usability—The Human Factor

Simplicity is not always the best. But the best is always simple.

Heinrich Tessenow, architect

In recent years, many companies have realized that there is more to a successful product than the product being rich in features. As software becomes more complex, it is important to narrow the focus to the expected user, and not on the technical system. The relationship between the product and its users is addressed by usability. Quite often, usability is defined as a quality attribute that assesses how easy user interfaces are to use (cf. Nielsen, 2003). However, usability is more than just the quality of the user interface; it is all about building products that people can and will use. Usability describes the entire field of research that is concerned with the interaction of technical systems where users are performing tasks, such as a mobile Web site or application. Jakob Nielsen, a guru on the subject, defines the term as follows:

> Usability is the measure of the quality of the user experience when interacting with something-whether a Website, a traditional software application, or any other device the user can operate in some way or another. (Nielsen, 1997)

Usability is also an umbrella term that can have different meanings depending on the context (cf. Spencer, 2004). On the one hand, usability is an attribute of the quality of a system; on the other, it is a process or set of techniques used during the project. The second aspect is also known as usability engineering.

3.1.1 Usability Engineering

Usability engineering is a user-centered process that aims at gaining a high level of effectiveness, efficiency, and satisfaction for products (cf. Hix and Hartson, 1993). The goal is to study the human–computer interaction and cognitive behavior of a person to make the product more usable. Usability engineering is not a one-shot affair where the user interface is fixed up before the product is released. Rather, it is a set of activities that ideally take place throughout the life cycle of the product (cf. Nielsen, 1993, p. 71). The main idea behind usability engineering is the urge to reduce the amount of complexity so that people can concentrate on their tasks better, rather than on the product they are using to perform their tasks. Usability engineering is a flexible process that can be applied at any stage of the development life cycle. Depending on the development stage of a product, different techniques can be applied. Typical usability engineering activities are shown in Figure 3.1. The results of such a usability engineering process are products that meet the customers' needs.

3.1.2 Reasons for Usability

If you have not already incorporated usability considerations into your design process, you might wonder why you should indeed try it. Many experts point out that a usability-oriented software development process provides significant added

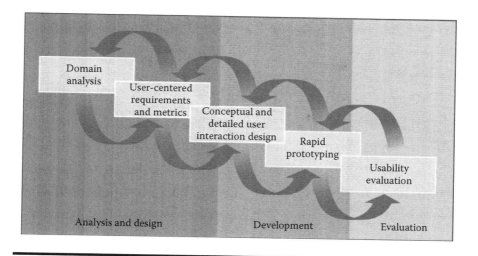

Figure 3.1   The usability engineering process. (From Gabbard, J.L. et al., Usability engineering for complex interactive systems development, in *Proceedings of Human Systems Integration Symposium*, Vienna, VA, pp. 1–13, 2003. With permission.)

value to the product. Further on, you will be given reasons why it is advisable to pay more attention to usability, which are based on the statements of Hinderberger (2003) and Himmelweiss and Ball (2002).

Attribute of the quality

 Products only serve a purpose if they are easy to use. Usability is the capability of the software to be understood, learned, and used, and can be verified by usability testing. It is safe to say that usability is a characteristic for the quality of a product.

Saving of costs

 Project managers usually worry that initiating a user-centered design process and executing appropriate usability testing will require intolerable amounts of time and money. This is a typical and misleading notion. The cost spent by focusing on the user pays off, and it actually saves both time and money. Usability supports a development process by placing the user at the focal point. Early user feedback helps to fix potential problems faster and more economically. According to Landauer (1996), a correction of any error will cost six times more if it is done after the implementation stage and ten times more if this occurs in the additional maintenance stage, in comparison to the costs during the development. A further positive effect of usability is the reduction of costs incurred in documentation, training, and support at a later time.

Certainly, it can be stated that users probably need less support for a product that complies with their mental conceptions and abilities.

Saving of time

In the early stages of product development, many decisions must be made which have a strong influence on time management. Usability engineering is helpful in uncovering the real needs and tasks of users, in addition to prioritizing features that have to be developed. Features that are not significant to users can be disregarded or developed at a later time. The knowledge about user needs determines the best decision. In the end, you get further cost and time savings by usability testing, e.g., if you detect design problems earlier in the development process.

Increase in sales

More and more people desire products or services that are intuitive and easy to use. Usability is helpful in drawing a line between your products and those of your competitors. For example, if two products are basically similar, the product with better usability will most likely be considered superior. Usability may therefore be considered to be a competitive advantage. Many enterprises have recognized this promotional aspect, and have gone on to successfully integrate usability in their advertising strategy. Since user satisfaction is one of the most important tasks of usability, long-term customer loyalty can be achieved by doing this. Users develop confidence and trust in the product, become loyal in due course, and they may even recommend your product to others.

Encourage ideas

Usability testing identifies problems, but it can also generate new ideas and point out alternatives. In many cases, the best advice and inspirations for product improvements come from the comments of the actual potential users. Testing with early prototypes and real users makes it easy to find a balance between different concepts. All in all, it can be said that usability testing indeed encourages the development of innovative products.

Joy of use and user satisfaction

Joy of use can be characterized as the positive, subjective feeling a user experiences while using a product (Reeps, 2004). Many products do not appear to be designed to make the life easier for us. Technology should facilitate our everyday life instead of making it more complicated. The primary goal of usability is not only to reduce these technical barriers but also to provide a good joy of use (cf. Nielsen, 2002).

3.2 Usability Design for Mobile User Interfaces

Success in mobile interaction design requires that one understands and considers the mobile situation. Simply transferring a full-sized desktop service to mobile devices almost always results in a suboptimal mobile experience (cf. Ballard, 2007, p. 70). The first step is to determine the context in which people will be using the service. Being mobile often means that the user's location, physical, and social context may change with time (cf. Ballard, 2007, p. 10). Furthermore, the contexts vary, and are more difficult to predict and figure out. In contrast to stationary desktop services, users of mobile devices are additionally influenced by external factors. The mobile user initially has all the sources of interruption from the real world, just like those of a desktop user. He or she additionally inherits new distractions resulting from the mobility factor. The user of a mobile device still has to be in a situation that allows him or her to focus on other things even as he or she uses the device (cf. Oulasvirta, 2005). This makes it necessary to provide the presented information with greater accessibility and with ease. For instance, pedestrians must stop to type a short text message on their mobile phones. Typing by itself requires a high level of attention. In this situation, the pedestrians must remain cautious by ensuring that they keep track of their direction while avoiding accidents, such as running into other people and not being hit by moving vehicles, as they work on their mobile devices.

3.2.1 The Mobile Context

There are a variety of models that clarify the mobile context. An interesting model, which will be introduced in a short while, comes from Savio and Braiterman (2004). The model is shown in Figure 3.2 and consists of several layers. The external influences are composed of culture (such as economics, religion, etiquette, law, social structures), environment (such as sound, light, space, privacy, distractions, other people), and the current activity (walking, driving, eating). A user who performs an activity has certain goals, distributes his or her attention, and attends to subtasks. The capabilities of the mobile device depend on the manufacturer's contract obligations, the connecting options, and the actual technology. The interrelation between the user and the device is addressed by the interface.

We note that the handling of mobile devices is strongly influenced by the current situation, and that designing for mobiles is more complex than designing for desktops.

3.2.2 Mobile Design Recommendations

For the designer, there are some resources for design recommendations available, called style guide or design patterns. Most of these references focus on a particular

Figure 3.2   The mobile context: Savio and Braiterman model. (From Savio, N. and Braiterman, J., Design sketch: The context of mobile interaction, Giant Ant, San Francisco, CA, Version: 2007, http://www.giantant.com/publications/mobile_context_model.pdf, 2007. With permission.)

platform and a particular set of devices. According to Ballard, the carrier and manufacturer style guides and design patterns, e.g., Microsoft Windows Mobile "UI Guidelines" (Windows, 2009) or "Nokia Mobile Design Patterns" (Nokia, 2009), are good points of references for developing in a limited environment (cf. Ballard, 2007, p. 92). Nonetheless, there is no universal solution that magically fits into any situation or context. It is always good to take a look at these recommendations, but it is more important to consider the context of the application in the design process.

The summary below shows some examples of universal design recommendations, which are partially derived from results of research and indicate how design can prevent usability problems.

In order to support quick user-driven monitoring, a concise yet powerful representation of changes should be offered (cf. Oulasvirta, 2005). It is also recommended that modality-targeted feedback needs to be given. For example, on one hand the receipt of a new short message on mobile phones is indicated on the display, and on the other hand is also clarified by an acoustic signal or the vibration of the mobile device.

Furthermore, the end user should have the temporal control and delayed reactions that do not cause serious consequences (cf. Oulasvirta, 2005). Moreover, considerable automation or elimination of unimportant tasks is expected.

In this context, we highlight one of John Maeda's Laws of Simplicity,

> Simplicity is about subtracting the obvious and adding the meaningful. (Maeda, 2006)

This shows that adding the wrong features may mean more cognitive competition.

3.2.2.1 Designing for Small Screens

Mobile devices are typically lightweight and with an obviously much smaller screen than that of stationary computers (cf. Weiss, 2002, p. 5). Even if there are mobile devices that have the ability to display a typical Web site in a reasonable way by intelligent zoom functions, the mobile Web pages should have their own page layout (cf. Panzirsch, 2008, p. 89 f.). Unfortunately, there is no standard screen size because of the wide range of different mobile devices (cf. Bieh, 2008, p. 68; Weiss, 2002, p. 90). Moreover, many mobile devices do not have a comfortable size to provide a typical Web layout. The end user needs to get the information easily and quickly enough by an appropriate representation. Apart from that, a logical and sophisticated menu and navigation concept also needs to be offered to users. The combination of these two goals is a difficult task for the designer (cf. Bay, 2006, p. 50).

3.2.2.2 Effective Usage of Limited Space

Since handling of mobile devices is a technical barrier for some people, data input effort has to be reduced effectively. According to Cordes, mobile devices are often used to view information, but they are unsuitable for entering large amounts of data (cf. Cordes, 2007, p. 32 f.). The aim of mobile Web sites is not to present as much content as typical Web sites do, but rather to provide a minimalistic design. This involves minimal use of long content pages and scrolling, doing away with all forms of gimmick such as fancy background images, with an aim of supporting good usability (cf. Bieh, 2008, p. 68 ff.). There is no room for irrelevant elements that do not contribute to navigation or provide a comfortable handling. The limited space available should be used effectively. Try to place as much information as possible in the limited space without causing information overload and disorientation. The use of icons on small screens is justifiably very popular in this context, since they save on space. They have the ability to deliver information fast and clearly. The most important characteristic of icons is that they are highly recognizable in comparison to words (cf. Wieser, 2004, p. 30).

3.2.2.3 Providing a Good Typography

With the presentation of text, it is necessary to use legible fonts, which do not appear too big for the small screens. Typographically, the main task of the designer is to assure optimal legibility regardless of the medium (cf. Cordes, 2007, p. 45; Zwick et al., 2006, p. 118). According to Zwick, the most significant factor for legibility of mobile devices is the ambient illumination (cf. Zwick et al., 2006, p. 118). While the brightness contrast of 30% is high enough for large screens, small displays need at least 50% for effective legibility in all situations. The smaller the device or the font, the higher the degree of brightness required. Other important factors for legibility include font, font size, and line space.

3.2.2.4 Choosing Appropriate Color

Color is well suited as a means of design. Its implementation should be done sparingly and in a well-conceived manner. Information could be lost because the display does not produce sufficient color depth to represent the desired colors (Bieh, 2008, p. 69 f.). All substantial information should thus be opened with a few colors. In order to use color most effectively, it is necessary to consider some of its aspects. Colors can activate emotions and be interpreted in different ways depending on cultural background. The color effect of individual colors and in combination with others should be taken note of and used distinctively. Color is especially suited for the distinction of functional areas (Stapelkamp, 2007, p. 78 ff.).

3.2.3 Further Reading

More information about designing mobile user interfaces can be found in Zwick et al. (2006), Jones and Marsden (2006), Ballard (2007), and Saffer (2006).

3.3 Usability Evaluation for Mobile User Interfaces

Evaluation methods may be classified by two means:

1. Formative and summative methods
   Formative evaluation techniques take place if the usability evaluation is a consistent part of the whole design process. This method enables early

identification of usability problems. Summative methods, in comparison, check the usability of a system only in conclusion to its design. This allows a better insight into the overall quality of the system (cf. Sarodnick and Brau, 2006, p. 20).

2. Empirical or analytical methods

   Empirical methods are based on direct user input, and include interviews, observations, or questionnaires. Analytical methods do not take note of user input. Instead, experts evaluate a system by the stipulated special guidelines (cf. Sarodnick and Brau, 2006, p. 113).

The best solution is always a combination of different methods. Analytical methods are better suited for the beginning of the design process and may be supported by an empirical evaluation later (cf. Sarodnick and Brau, 2006, p. 113 f.). A minimum of project realization (e.g., first paper-prototypes) is the prerequisite in any case.

3.3.1 Usability Evaluation Techniques for Mobile User Interfaces

There is little scientific backing on the relevance and importance of specific evaluation techniques in measuring the usability of mobile user interfaces. Nevertheless, all known methods from "classical" usability evaluation may be perceived as relevant for the analysis of mobile user interfaces. A literature study delivers some examples of comparative tests of usability evaluation methods for mobile user interfaces as well as for some new approaches, with the intention of finding the best-suited one.

Those experiments may be divided into four groups:

1. Comparison of expert-based evaluation versus user-based usability tests
2. Comparison of user-based usability tests in a laboratory but with a different touch to reality
3. Comparison of laboratory-based usability test with outside field tests
4. Tests of new laboratory-based methods that try to reproduce mobile context

Direct observation, questionnaires, and interviews of test persons in a usability lab or field test are the most widespread techniques for the testing of mobile gadgets and applications. However, there are a lot of challenges, and it is essential to provide some fundamental thoughts in terms of the research approaches, selection of technical equipment, and data collection. The most important distinguishing feature is the application's mobile context, which cannot be fully reproduced in a test environment (whether in a laboratory or in a field test) without loss of data integrity of the test results. Ideally, the experiment tries to take many of the possible situations into consideration, but it will never be possible to simulate a realistic and complete image of the real usage context. Thus, a usability test in the laboratory might disregard certain aspects and focus only on some parts of the application. The situation in a field test differs from this since there is a more realistic test environment, but with the risk of losing the aspects of test control (cf. Waterson et al., 2002, p. 1; Zhang and Adipat 2005, p. 2 f.).

A good balance is the reproduction of mobile situations in a usability lab. Specific usage situations can be simulated almost realistically (cf. Kjeldskov et al., 2004a). Beside, there have been tests to simulate mobility, e.g., with test persons who had to walk on a treadmill or follow a painted pathway through the laboratory (cf. Beck et al., 2003; Kjeldskov and Stage, 2004). There have also been experiments in driving simulators with the view of reproducing the distributed attention, while other tests let the test persons play an active game (cf. Beck et al., 2003; Kjeldskov and Stage, 2004).

The experiments described above come to the conclusion that a test person who is seated in one place can tackle usability problems. However, when the test person is mobile and is simultaneously focusing on different things, such as multi-tasking, some problems would occur more often. This is a consequence of the additional external factors.

Usability tests might be accomplished with real mobile devices as well as with emulators or prototypes (cf. Panzirsch, 2008). The latter are not suited for field tests, though. Emulators running on a normal desktop computer allow the collection and analysis of performance data, such as the number of clicks or the time taken to complete a task. However, the mobile context is completely disregarded, such as latency of loading or answering times, or the difficult handling of data input on a mobile device. Therefore, the recommended practice is to test layout or menu structures only during the development process of a mobile application using emulators (cf. Zhang and Adipat, 2005, p. 14).

Using real mobile devices when testing mobile applications has a greater fundamental influence on identifying usability problems than using desktop computers for the test of computer applications. Testing an application on the real-end devices enables the collection of realistic information. However, there are difficulties in the collection of adequate and sufficient details of the user's behavior as the application is used (cf. Zhang and Adipat, 2005, p. 15). Another problem lies in the lack of an applicable video recording technique. Cameras are still too large for the small screens of mobile devices, and they cannot be mounted directly by the test person. If cameras are used to record the display of the mobile device, it is almost impossible for the test person to freely move around in or outside the room. However, if the recording technique is mounted to the test person who carries stuff around, he or she has to wear a heavy backpack or its equivalent, which is an additional burden. A test of different miniature cameras in combination with different mounting techniques showed to some extent unsatisfying results, both in the quality of the recorded videos as well as in handling the construction (cf. Wagner, 2008).

3.3.2 Which Usability Evaluation Technique Is the Best?

There is no final answer to this question, which seeks to obtain an answer for the method that is best suited for the usability evaluation of mobile user interfaces. In fact, different experiments produce results that differ and do not offer final clarification. While some studies show that laboratory usability tests deliver sufficient results (cf. Holtz Betiol and Abreu Cybis, 2005; Kaikkonen et al., 2005; Kjeldskov et al., 2004a), other experiments have shown that, with more complex and extensive field tests, more usability problems come about (cf. Duh et al., 2006; Nielsen et al., 2006). Also, expert-based evaluation methods and the new idea of "rapid reflection" (Kjeldskov et al., 2004b) deliver significant results (cf. Kjeldskov et al., 2005).

Prototypes play a special role in the evaluation of different stages of the development of mobile applications (cf. Kempken and Heinsen, 2003, p. 260 ff.). Simulators, paper prototypes, or functional prototypes allow the involvement of end users early in the development process and let them evaluate the interfaces. This is a quick and cheap way of testing, but the prototypes have to be as detailed and accurate as possible.

Finally, it can be said that expert-based evaluation methods are good for the early product-development stages. However, end users need to be included in the evaluation during the concept phase by different measuring methods as well as simple user tests. When functional prototypes exist, their usability can be tested in a laboratory setting. The final application should be tested during the introduction phase with field tests since the real usage-context sometimes discovers other problems that have not been uncovered before. Questionnaires and interviews may be combined with these usability tests; users have practical knowledge of the application and its user interface and have an emotional impression, which is a good prerequisite for answering the questions. In the long run, data may be collected and evaluated through log files, or by newer approaches such as weblogs or recurring online questionnaires, and also via mobile devices.

References

Ballard, B. (2007). *Designing the Mobile User Experience*. John Wiley & Sons, Chichester, U.K.

Bay, S. (2006). Komplexe Menüs und Kleine Displays: Experimentelle Untersuchugen zur Evaluation und Optimierung der Dialoggestaltung an Mobiltelefonen.dissertation.de.

Beck, E. T., Christiansen, M. K., Kjeldskov, J., Kolbe, N., and Stage, J. (2003). Experimental evaluation of techniques for usability testing of mobile systems in a laboratory setting. In *Proceedings of OZCHI 2003*, Brisbane, Australia, CHISIG, pp. 106–115.

Bieh, M. (2008). *Mobiles Webdesign: Webseiten für mobile Endgeräte*. Galileo Press, Bonn, Germany.

Cordes, B. (2007). *Design und Usability im mobilen Zeitalter. Potentiale von hybriden Netzen*. VDM Verlag Dr. Müller, Saarbrücken, Germany.

Duh, H. B., Tan, G. C. B., and Chen, V. H. (2006). Usability evaluation for mobile device: A comparison of laboratory and field tests. In *Proceedings of the Eighth Conference on Human–Computer Interaction with Mobile Devices and Services (MobileHCI'06)*, Helsinki, Finland, pp. 181–186. ACM, New York.

Gabbard, J. L., Hix, D., Swan II, E. S., Livingston, M. A., Höllerer, T. H., Julier, S. J., Brown, D., and Baillot, Y. (2003). Usability engineering for complex interactive systems development. In *Proceedings of Human Systems Integration Symposium*, Vienna, VA, pp. 1–13.

Himmelweiss, A. and Ball, G. (2002). Ten good reasons for usability. Version: 2002. http://www.contentmanager.net/magazine/article_204_ten_good_reasons_for_usability.html

Hinderberger, R. (2003). Usability als Investition. In S. Heinsen and P. Vogt (Hrsg.), *Usability praktisch umsetzen—Handbuch für Software, Web, Mobile Devices und andere interaktive Produkte*. Hanser, München, Germany, pp. 24–41.

Hix, D. and Hartson, H. R. (1993). *Developing User Interfaces: Ensuring Usability Through Product and Process*, John Wiley & Sons, New York.

Holtz Betiol, A. and de Abreu Cybis, W. (2005). Usability testing of mobile devices: A comparison of three approaches. http://www.se.auckland.ac.nz/courses/SOFTENG350/lectures/L5%20Usability%20Mobile%20Devices.pdf

Jones, M. and Marsden, G. (2006). *Mobile Interaction Design*. John Wiley & Sons, New York.

Kaikkonen, A., Kekäläinen, A., Cankar, M., Kallio, T., and Kankainen, A. (2005). Usability testing of mobile applications: A comparison between laboratory and field testing. Version: 2005. http://www.upassoc.org/upa_publications/jus/2005_november/mobile.pdf

Kempken, A. and Heinsen, S. (2003). Usability für mobile Anwendungen. In S. Heinsen and P. Vogt (Hrsg.), *Usability praktisch umsetzen—Handbuch für Software, Web, Mobile Devices und andere interaktive Produkte*. Hanser, München, Germany, pp. 249–265.

Kjeldskov, J. and Stage, J. (2004). New techniques for usability evaluation of mobile systems. *International Journal of Human–Computer Studies* (*IJHCS*), 60, 599–620.

Kjeldskov, J., Skov, M. B., Als, B. S., and Hoegh, R. T. (2004a). Is it worth the hassle? Exploring the added value of evaluating the usability of context-aware mobile systems in the field. In *Proceedings of the Sixth International Mobile HCI 2004 Conference*, Glasgow, Scotland, pp. 61–73. Springer, Berlin, Germany.

Kjeldskov, J., Skov, M. B., and Stage, J. (2004b). Instant data analysis: Evaluating usability in a day. In *Proceedings of NordiCHI 2004*, Tampere, Finland, pp. 233–240. ACM, New York.

Kjeldskov, J., Graham, C., Pedell, S., Vetere, F., Howard, S., Balbo, S., and Davies, J. (2005). Evaluating the usability of a mobile guide: The influence of location, participants and resources. *Behaviour and Information Technology*, 24, 51–65.

Landauer, T. K. (1996). *The Trouble with Computers: Usefulness, Usability, and Productivity*. The MIT Press, Cambridge, MA.

Maeda, J. (2006). *The Laws of Simplicity* (*Simplicity: Design, Technology, Business, Life*). The MIT Press, Cambridge, MA.

Nielsen, J. (1993). *Usability Engineering*. AP Professional, Boston, MA.

Nielsen, J. (1997). Usability testing. In G. Salvendy (ed.), *Handbook of Human Factors and Ergonomics*, 2nd edn. John Wiley & Sons, New York, pp. 1543–1568.

Nielsen, J. (2002). Jakob Nielsen's Alertbox: User empowerment and the fun factor. Version: 2002. http://www.useit.com/alertbox/20020707.html

Nielsen, J. (2003). Jakob Nielsen's Alertbox: Usability 101: Introduction to usability. Version: 2003. http://www.useit.com/alertbox/20030825.html

Nielsen, C. M., Overgaard, M., Pedersen, M. B., Stage, J., and Stenild, S. (2006). It's worth the hassle!: The added value of evaluating the usability of mobile systems in the field. In *Proceedings of the Fourth Nordic Conference on Human-Computer Interaction* (*NordiCHI'06*), Oslo, Norway, pp. 272–280. ACM, New York.

Nokia (2009). Mobile design patterns. Version: 2009. http://wiki.forum.nokia.com/index. php/Category:Mobile_Design_Patterns

Oulasvirta, A. (2005). The fragmentation of attention in mobile interaction, and what to do with it. *Interactions*. 12(6), 16–18.

Panzirsch, M. (2008). Web-usability und usability von mobilen Endgeräten. Master thesis, Fachhochschule Erfurt, University of Applied Science, Erfurt, Germany.

Reeps, I. (2004). Joy-of-use—eine neue Qualität für interaktive Produkte. Master thesis, University of Konstanz, Constance, Germany.

Saffer, D. (2006). *Designing for Interaction*: *Creating Smart Applications and Clever Devices*. Peachpit Press, Berkeley, CA.

Sarodnick, F. and Brau, H. (2006). Methoden der Usability Evaluation—Wissenschaftliche Grundlagen und praktische Anwendung. In E. Bamberg, G. Mohr, and M. Rummel (Hrsg.), *Praxis der Arbeits- und Organisationspsychologie*. Huber, Bern, Switzerland.

Savio, N. and Braiterman, J. (2007). Design sketch: The context of mobile interaction. Giant Ant, San Francisco, CA. Version: 2007. http://www.giantant.com/publications/mobile_context_model.pdf

Spencer, D. (2004). What is usability? KM column. Step Two Designs Pty Ltd, Windsor, Australia. Version: 2004. http://www.steptwo.com.au/files/kmc_whatisusability.pdf

Stapelkamp, T. (2007). Screen- and Interfacedesign. *Gestaltung und Usability für Hard- und Software*, m. CD-ROM, Springer, Berlin.

Wagner, S. (2008). Konzeption, Aufbau und Test eines Usability-Messplatzes für mobile Endgeräte. Fachhochschule Erfurt, University of Applied Science, Erfurt, Germany.

Waterson, S., Landay, J. A., and Matthews, T. (2002). In the lab and out in the wild: Remote web usability testing for mobile devices. http://citeseer.ist.psu.edu/613508.html

Weiss, S. (2002). *Handheld Usability*. John Wiley & Sons, New York.

Weisser, V. (2004). Usability versus Design -ein Widerspruch? Theorie und Praxis der Gestaltung von Websites. In U. Rautenberg and V. Titel (eds.), *Alles Buch. Studien der Erlanger Buchwissenschaft* VI, 2nd edn. Universität Erlangen–Nürnberg, Erlangen.

Windows (2009). Windows Mobile Developer Center: Design guidelines. Version: 2009. http://msdn.microsoft.com/en-us/library/bb158602.aspx

Zhang, D. and Adipat, B. (2005). Challenges, methodologies and issues in the usability testing of mobile applications. http://userpages.umbc.edu/~zhangd/Papers/IJHCI1.pdf

Zwick, C., Schmitz, B., and Kuhl, K. (2006). *Designing for Small Screens*. AVA Publishing SA, Lausanne, Switzerland.

Chapter 4

The Blind Leading the Blind: Web Accessibility Research Leading Mobile Web Usability

Andy Brown, Yeliz Yesilada, Caroline Jay, and Simon Harper

Contents

4.1 Introduction

The World Wide Web (Web) is characterized by a set of innovative and rapidly changing technologies. While the power of the Web is due in no small part to its ability to facilitate these rapid changes, a Web constantly in flux introduces problems and complexities for both users and slower-moving technologies such as hardware devices, programming paradigms, and operating systems. Nevertheless, the Web is repurposing these slower-moving technologies by becoming mobile, by becoming the platform, and by reinventing the programming paradigm. In all cases, the Web updates old tools, techniques, and technologies placing network connectivity at their heart, to build a world of remote procedure calls, representational state transfer (RESTful) interfaces, and Web-centric application programming interfaces (API). Indeed, the rush of technical innovation has happened so quickly that terms have been proposed to differentiate the original Web from its genuses; Web 2.0 is one such term.

While everyone agrees that we are in the Web 2.0* age, there is still a lack of understanding as to what "Web 2.0" actually means, implies, or requires; the term can mean radically different things to different people. For instance, many people talk about Web sites degrading gracefully, but the addition of asynchronous JavaScript and XML† (AJAX) scripting to Web pages can render them completely unusable to anyone browsing with JavaScript turned off, using a slightly older browser, or using a browser as part of a mobile device. In the rush to implement these new ideas, insufficient regard has been given to enabling all users to access, and interact with, the information. In particular, people with visual impairments can struggle to understand even what a page is offering, let alone use it effectively, and those with motor impairments find it difficult to input information.

Web accessibility (see Section 4.2) aims to help people with disabilities to perceive, understand, navigate, interact with, and contribute to the Web. There are millions of people who have disabilities that affect their use of the Web. Currently, most Web sites have accessibility barriers that make it difficult or impossible for many people with disabilities to use the sites. Web accessibility depends on several different components of Web development and interaction working together, including software, developers, and content. The World Wide Web Consortium (W3C) and the Web Accessibility Initiative (WAI)‡ recognize these difficulties and provide guidelines for each of these interdependent components. There are also other organizations that have produced guidelines, but, the WAI guidelines are more complete and cover the key points of all the others.

---

* A phrase coined by O'Reilly Media, referring to a supposed second generation of Internet-based services.
† Extensible Markup Language, http://www.w3.org/XML/
‡ See http://www.w3.org/WAI/

There are, however, no homogeneous set of guidelines that designers can easily follow. Disabled people typically use assistive technologies* to access Web pages in alternative forms such as audio or Braille, or to ease data input, and research has shown that many of the difficulties encountered by these users are also encountered by people using mobile devices to access the Web. Indeed, the link between mobility and impairment is suggested by Sears and Young (2003) who define a new type of impairment (see Section 4.3.1) in which an able-bodied user's behavior is impaired by both the characteristics of a device and the environment in which it is used. This behavioral change is defined as a situationally induced impairment and is often associated with small devices such as mobile phones or personal digital assistants (PDAs) used in a mobile setting or constrained environment.

Web technologies have become key enablers for access to content through desktop and notebook computing platforms. These same technologies have the potential to play the same role for content access from mobile devices too. However, mobile Web access suffers from interoperability and usability problems that make the Web difficult to use for many people (see Section 4.3). Current work in mobile Web accessibility focuses in two main areas: (1) developing a set of technical "best practices" and associated materials in support of the development of Web sites that can be easily viewed and interacted with on mobile devices, and (2) identifying device information required for content adaptation, which includes the development of services that provide device descriptions in support of Web-enabled applications. In practice, there is a tremendous commonality between the mobile Web and Web accessibility, and this is why the lessons learned creating Web accessibility are important for the mobile Web. However, little is understood of the interplay, which will take place between the mobile Web and Web 2.0 domains (see Section 4.5). Indeed, accessibility specialists are only just beginning to find solutions for making Web 2.0 content, in the form of AJAX, accessible. The main focus of this accessibility effort is directed toward Accessible Rich Internet Applications (ARIA and now WAI-ARIA), which defines new ways for functionality to be provided to assistive technology, and thereby tries to make these updates visible in "live regions," in addition suggesting how updates are handled (see Section 4.5.2).

Given the overlap between the needs of disabled users and mobile users, we examine this relationship in the context of Web 2.0 and conclude (see Section 4.6) that the emerging solutions for Web 2.0 accessibility, such as WAI-ARIA, can also benefit those people wishing to access the same sites using mobile devices. The aspects of WAI-ARIA, that enable visually disabled users to access Web pages may also enable mobile users to be alerted to off-screen updates, trivial and nontrivial information updates, and interaction conformation.

---

* Assistive technology refers to hardware and software designed to facilitate the use of computers by people with disabilities.

4.2 Accessibility

The Web is a primary source of information, employment, and entertainment for people with disabilities (European Union Policy Survey 2005), and its importance cannot be underestimated. Ensuring that it is accessible is not always straightforward; however, both the technology used to access the Web and the underlying design of a site can have a huge impact on how open it is to someone with a disability.

Input devices and displays can significantly affect not only Web access, but computer access more generally. As the standard keyboard, mouse, and monitor setup is not always appropriate for, or indeed physically usable by, people with certain disabilities, assistive technologies provide an alternative means of access. There is no "catch-all" solution to providing accessibility to Web-based resources for disabled users. Instead, each disability requires a different set of base access technologies and the flexibility for those technologies to be personalized by the end user.

Motor impaired users may be able to read a standard display, but can find it difficult to operate a standard keyboard or mouse (Trewin 2008). A number of specially developed input devices can help users with varying types and levels of disability, including keyboards with specially programmed shortcuts and sticky keys, software (on screen) keyboards operated using a joystick, switch or mouse, pointing devices attached to a movable limb, or the head and speech recognition or eye tracking input software.

Profoundly blind users are unable to see a display or operate a mouse, so generally use a keyboard to provide input to a computer, and receive audio output through a screen reader, which "speaks" the contents of the display (Barreto 2008). Visually impaired users who have some vision may use a screen magnifier, with or without a screen reader, to enlarge small parts of the display.

Although deaf users initially appear to be able to access a computer, and therefore the Web, using a standard setup, this is not necessarily the case (Cavender and Ladner 2008). The primary language of many hearing-impaired users is not English but sign language, and textual rendering is therefore not necessarily accessible or appropriate (Huenerfauth 2005). Users with cognitive impairments (termed "learning disability" in the United Kingdom and "intellectual disability" in Australia, Europe, and Canada), may also find some content difficult to understand (Lewis 2008).

It is not only users with hearing or cognitive impairments who find that it is aspects of the Web content itself, rather than the equipment used to display or interact with it, which renders it inaccessible. The serial nature of audio output means that the visual interaction model, and XHTML structure created to support this model on the Web, must be moved into an audio interaction paradigm (World Health Organization 2004) so that visually impaired users can have access to it. At a basic level, screen readers perform this translation by reading the page from top-left to bottom-right. As the complexity of Web pages has increased, screen readers and Web browsers for visually impaired users have been created to access the deeper document structure, by directly examining the XHTML or the document object model (DOM). Examining the precise linguistic meaning of the text allows more complex meanings (associated with style, color, etc.) to be derived, and enables users to move between different HTML elements on the page, such as links or headings. This technology is still largely unable to comprehend the meaning of the underlying structure; however, so much of the implicit information provided by layout and appearance is lost.

People with disabilities cannot necessarily be divided into neat groups—many will have a combination of needs that must be met in order to ensure they can effectively access the Web. One section of the population in particular often exhibit unique combinations of disabilities: "seniors" (Czaja 2006). Many members of the ageing population have multiple disabilities that are not necessarily severe, but in combination can still cause decreased Web accessibility (Hawthorn 2000, Kurniawan 2008). Although assistive technologies can be useful to the older user, there is sometimes a perceived stigma in using such devices due to a reluctance to admit to a limitation or disability. As such, the assistive technology developer must achieve Web accessibility along with the additional challenge of camouflaging of the access technology itself.

4.3 Disabled or Mobile: Same Barriers

There are millions of people whose disabilities affect their use of the Web in ways like those described above, and although an accessible Web means unprecedented access to information for these people, their access remains limited. Likewise, accessing Web pages from mobile devices (e.g., PDAs and other portable gadgets) is becoming increasingly popular. But here too, difficulties arise, this time due to device limitations, such as the small screen size, low bandwidth, and different operating modalities.

The problems that mobile users face when interacting with the mobile Web are similar in many ways to the problems that disabled users face when interacting using a traditional desktop computer. For example, hearing-impaired users have a hard time accessing multimedia content if the content does not have captions, while mobile users will miss auditory information if they have the sound turned off (e.g., in public places such as trains or hotel lobbies) or if they are in noisy places. The similarities and commonalities are many (more examples are given below) and the challenges for designing for the mobile Web are similar to those of designing for the accessible Web (Leventhal 2006, Trewin 2006), but despite this, there has been very limited work on systematically comparing these two domains. Unfortunately, previous work has only looked at accessibility problems from a single user group perspective such as visually impaired (Asakawa 2005), motor-impaired (Mankoff et al. 2002), or mobile users (Brewster 2002). If designers want to create a page that is accessible by both mobile and disabled users, they have to follow a number of different guidelines and validation tools,* which means it will be time consuming and costly.

---

* Web accessibility evaluation tools: http://www.w3.org/WAI/ER/tools/

If we understand the overlaps and integrate research into the accessible and mobile Webs, we can develop a common infrastructure where both users and developers could benefit. However, to compare these two domains, we need to consider all the key factors in the human–Web Interaction. These include *input*, which refers to various approaches and devices used to deliver information to the Web; *content*, which is the information that forms Web pages (the code and markup); and *output*, which refers to the rendering of the Web content by a particular user agent. In the following sections, we discuss each of these factors and pose the question "do disabled desktop users and mobile users experience similar interaction barriers?"

4.3.1 Input Impairment

Regarding input, existing studies show that there are two common problem domains, target acquisition and typing input. A dexterity impairment can have high impact on these with regard to the accuracy, completion times, and error rates of clicking and pointing to a target. For example, because of arthritis (an inflammation of the joints) one might be unable to make the necessary movements. Similarly, spasms (sudden, involuntary contractions of a muscle) can cause unwanted diversions or mouse clicks during pointing movements. People with Parkinson's disease click more slowly, giving even greater opportunity for slippage (Trewin 2006). Similarly, a situationally induced impairment, such as a change in lighting (Barnard et al. 2005) or movement (Lin et al. 2007), can also have high impact on accuracy, completion times and error rates of typing and clicking and pointing to a target. Bradykinesia (a slowness in the execution of movement) will cause typing rates in a physical keyboard to be greatly reduced (Sears and Young 2003). Similarly, Brewster (2002) show that when a mobile device is used in a more realistic situation (whilst walking outside), usability is significantly reduced (with increased workload and less data entered) than when used in a laboratory.

Different techniques and devices are available for text input such as keypad, voice recognition system, eye tracking, head tracking, etc.; however, the keyboard is still the most widely used. Either because of an impairment or a situationally induced impairment (Sears and Young 2003), the task of inputting text can easily become challenging. The common barriers experienced range from very well-defined problems, such as pressing a key unintentionally, to very generic problems, such as not being able to use a keyboard. Detailed information about common input errors can be found in Chen et al. (2008). For example, *bounce error* occurs when a user unintentionally presses a key more than once, producing unwanted copies of the intended key. This problem is observed for both motor impaired and ageing users, and is mainly due to a user's finger twitching when releasing a key (Trewin and Pain 1999). Small device users mainly experience this problem when they use a multi-tap input system where the user presses each key one or more times to specify the desired letter (James and Reischel 2001, MacKenzie et al. 2001). A multi-tap method works by cycling through letters on a key with each successive press; however, this causes problems when two letters on the same key are entered consecutively (Silfverberg et al. 2000). Another common input error is *transposition error*, which refers to the situation where two characters are typed in reverse order (Trewin and Pain 1999). This problem is observed with motor impaired, ageing, and mobile users (MacKenzie and Soukoreff 2002, Trewin and Pain 1999).

The second major task when considering input is target acquisition: pointing and selecting. The two main reasons for these problems are the difficulty of positioning the device cursor within a confined area, and the challenge of accurately executing a click (selection). *Pointing** and dragging* with a mouse is difficult for motor impaired and ageing users due to their limited hand movement and control. Trewin and Pain (1999) suggest that motor impaired users have problems pointing at small on-screen objects using a mouse, and that the smaller the object is, the harder it is to pinpoint (Hwang et al. 2004). Pointing accuracy also affects mobile users who rely on the touch-screen and stylus for input. Brewster (2002) illustrates that as the on-screen button becomes smaller, the subjective workload of mobile users increases, and the overall performance decreases. Pointing is also a problem for visually impaired users who cannot see clearly (Jacko et al. 2003). *Clicking error* refers to the situation where a user slightly moves the mouse while performing a clicking task. This may cause the cursor to move out of scope of the target object[†] and thus generate a clicking error. Studies (Trewin et al. 2006) suggest that clicking error affects motor-impaired users as well as ageing users (Chaparro et al. 1999, Moffatt and McGrenere 2007, Smith et al. 1999). Additional studies (Brewster 2002) find that "slip-off" error is also experienced by mobile touch screen users and similar work (Jacko et al. 2003) shows that clicking error also affects visually impaired users. Finally, another generic common problem is the *inability to use a mouse*. Severely motor-impaired users find mouse use difficult (Sears and Young 2003) as do some visually impaired users with low vision (Jacko et al. 2003). In addition, mobile users cannot use the mouse due to device size limitations (Greenstein 1997).

4.3.2 Output Impairment

Output refers to the rendering of the Web content through user agents. User agents include "any software that retrieves and renders Web content for users, such as Web browsers, media players, plug-ins and assistive technologies" (Gunderson and Jacobs 1999). People encounter output problems when their disabilities hinder the access to Web content of certain formats. On the other hand, it is the user agent that delivers Web content to end users, which means a badly designed user agent can result in output problems even if the Web content is created accessible for

---

* Pointing is also referred as "target acquisition," "area pointing," "mouse pointing," etc., which is the action of acquiring on-screen targets with the mouse cursor or with a pen/stylus (Wobbrock and Gajos 2007).

† Also referred to as accidental clicks (Trewin et al. 2006) or drifting errors (Moffatt and McGrenere 2007).

disabled users. Furthermore, disabilities and limitations of the device can also affect the installation and configuration of the user agents with which people access the Web. Small device users and disabled users can experience similar output problems. The User Agent Accessibility Guidelines Working Group (UAWG) from WAI has published *User Agent Accessibility Guidelines 1.0* (UAAG 1.0) (Jacobs et al. 2002). UAAG 1.0 provides guidelines for designing user agents that lower barriers to Web accessibility for people with disabilities.

When we look at the accessibility problems experienced by disabled and small screen device users, we can see that there are also common barriers regarding output. For example, people with visual impairments have limited access to information that is visually presented, thus making it difficult to perceive or use. Similarly, mobile users' vision is limited by the size of the screen and lighting conditions (Barnard et al. 2007, Duchnicky and Kolers 1983, Jones et al. 1999, Reisel and Shneiderman 1987). Therefore, both of these user groups can experience barriers in accessing and using the user agent. Another common barrier example is having limited access to structured elements in Web pages. Lazar et al. (2007) conducted a survey on what frustrates screen reader users on the Web with 100 blind users; one of the top causes of user frustration was the confusion caused by page layout. Screen reader users sometimes have difficulties locating themselves within a page, especially when tables are used to display texts or other objects in proximity. Similarly, it is difficult to fit large tables into small displays on mobile phones or PDAs due to the restriction posed by the screen size. To reduce the width of a table, either the Web content developer or the user agent must reduce the number of columns by splitting a table into multiple, narrower tables. However, this will increase the use of vertical scrolling, and a user's ability to compare information in one table will be hindered by the complications of reading multiple tables (Watters et al. 2003).

Finally, both mobile users and disabled users experience an output barrier when there is a lack of confirmation. Zajicek and Hall (2000) have conducted experiments with older visually impaired Web users using a voice Web browser. After a user made input, the browser provided speech dialogues that reassured a user that the interaction was going well. Results showed that such confirmatory messages increased a user's confidence in the interaction and aided the construction of conceptual models of the Web pages. The trade-off, however, was between increasing confidence and adding to the length of the interaction. Confirmatory messages demanded yet more attention from the user and could be irritating to confident users (Zajicek 2004). Another problem that sighted aged Web users experience is that they do not know what is clickable and what is not (Chadwick-Dias et al. 2003, Coyne and Nielsen 2002). They need specific confirmation, such as the pointer changes to a pointing finger over a link, to indicate that an item is clickable. Morell (2002) suggested that additional feedback should be provided to aged users for confirmation after they click a clickable item. Small device users experience similar problem: Chen et al. (2008) showed that when accessing the Web from a PDA, mobile users can easily get confused if not enough feedback was provided after an input. For example, a mobile user would keep clicking a button on a Web page several times if no feedback was given after the first click. They suggested that immediate feedback should be provided for user actions, such as page redirecting, and widget loading. The function of providing feedback should be configurable: for aged and inexperienced Web users, more feedback should be provided for basic actions, for example, clicking on a link; and for experienced Web users, feedback should only be provided for advanced actions, for example, form submission.

4.3.3 Content Impairment

As with input and output, existing research suggests that when people without disabilities access Web *content* with mobiles they experience situationally induced impairments and they face similar barriers to when people with disabilities access the Web (Sears and Young 2003, Trewin 2006, Wobbrock 2006). To illustrate the common barriers to content between mobile and disabled users, we give some examples below under the following four principles (Chuter and Yesilada 2008, Yesilada et al. 2008):

1. *Perceivable*: Information and user interface components must be present-able to users in ways they can perceive. For example, if information is con-veyed solely with color, then users who are blind or color blind can miss or misunderstand information (Coyne and Nielsen 2001, Disability Rights Commission (DRC) 2004). Small device users might also experience similar problems—many screens have a limited color palette and color difference is not presented, or the device may be used in poor lighting (e.g., outdoors) where colors are not as clearly perceived (Barnard et al. 2007, Duchnicky and Kolers 1983, Jones et al. 1999). Due to the decline in vision, older users usually have problems with visual acuity, distinguishing colors of similar hue, contrast discrimination, and a reduction in the efficacy of parafoveal vision (Kurniawan 2008, Newell et al. 2006).

   Another common example of a perception barrier is when non-text objects in a page have no text alternative. A user who is blind, or using a browser or assistive technology that does not support the object (Earl and Leventhal 1999, Takagi et al. 2004), cannot perceive this type of content so will be unable to access the information it contains. A similar problem can face mobile users, who may turn off images to reduce download times and costs, or whose user agents have limited support for non-text objects. Equally, some mobile user agents shrink images in size to fit the device's screen; this can make them meaningless.

2. *Operable*: User interface components and navigation must be operable. An example of a barrier to operability is when a page does not have a title or if the title is inappropriate. In this case, disabled, aging, and mobile users, all of

whom have difficulty scanning a page, cannot get an overview of the content. Additionally, blind users typically use a screen reader feature to get a list of the currently open windows, by window title (Disability Rights Commission (DRC) 2004). Therefore, if the page title is long, inappropriate, or missing, the user cannot perceive the content. With mobile devices, the page title is truncated to fit the narrow viewport (Rabin and McCathieNevile 2005).

Another common barrier to operability is that faced when link labels are not descriptive. In this situation, the user cannot determine whether or not to follow a link because the link label does not contain sufficient information. Screen reader users often access a list of the links on a page, which are presented without context—if a link label is not descriptive, the user will be unable to determine the purpose of that link (Penev and Wong 2008). Older users have also been shown to be hindered by nondescriptive hyperlinks, such as "click here" (Caldwell et al. 2008, Dalal et al. 2000), and to benefit from expanding the link text to describe the result of clicking (Sayago et al. 2009).

3. *Understandable*: Information and the operation of the user interface must be understandable. Examples of the types of problems that can be faced by users here are those caused when content spawns new windows without warning. In this situation, the user can become disoriented among windows, is unable to use the "back" button as expected (Bailey et al. 2005, Chadwick-Dias et al. 2003), or can close a window, not realizing it is the last in stack, accidentally closing the browser instance. While potentially difficult for all users, these problems are exaggerated for users with low vision, or restricted field of vision, or cognitive disabilities, who do not realize the active window is new (Coyne and Nielsen 2001, Craven and Brophy 2003), and for those using small-screen devices, where multiple stacked windows hide each other. Blinking, moving, scrolling, or auto-updating content can also be difficult to read or comprehend or interact with, particularly for older users and people with reading or learning disabilities or cognitive limitations (Ellis and Kurniawan 2000, Groff et al. 1999, Lewis 2008). The reduced viewport size or poor ambient lighting can make it difficult to see this type of content on mobile devices.

4. *Robust*: Content must be robust enough that it can be interpreted reliably by a wide variety of user agents, including assistive technologies. An example of the problems that may arise when content is insufficiently robust is what happens when content has invalid and unsupported markup. While such content can sometimes be rendered correctly by standard browsers, assistive technologies (Edwards 2008) and older mobile browsers (Siek et al. 2004) may be unable to display it, causing problems for both disabled and mobile users. Similarly, if a user's assistive technology or browser does not support scripting, or the facility is turned off (Rabin and McCathieNevile 2005) (e.g., to enhance security), content that requires scripting will not be accessible.

4.4 Disabled or Mobile: Same Solutions

At present, most solutions are designed only to solve problems for a single user domain. However, with an understanding of the common problems discussed in the previous section, it is possible to migrate solutions from one domain to another to benefit all users. Some solutions are common between mobile and disabled users and some solutions are disjoint; however, there are some solutions that are not common and can potentially be migrated between these user groups. For example, regarding input, small-device users share the following solutions with motor impaired users: soft keyboard (Hinckley 2007), joystick, voice, prediction facility, tablet, touch screen, and multimodal interface (Chau et al. 2006, Felzer and Nordmann 2006, Mankoff et al. 2002, Silfverberg et al. 2000, Wobbrock et al. 2004). Similarly, voice interaction is also used on small devices for speech dialing or editing text messages (Karpov et al. 2006) and used by motor-impaired (Oliveira Neto et al. 2009) and visually impaired users (Manaris and Harkreader 1998) as a substitute for both the keyboard and mouse. A text entry facility, which "predicts" the words a user is entering by looking for the most relevant key combination in its internal dictionary (Minneman 1986) has also been adopted to speed up input. Multimodal interfaces, which combine a number of modalities such as head movement and speech for motor-impaired users (Malkewitz 1998) and handwriting and speech for mobile users (Serrano et al. 2006), have also been suggested as possible input solutions, and are gaining popularity.

Some solutions exist for disabled users, but not for small device users. For example, regarding input, the following solutions exist for motor-impaired users but not for small device users: One-handed keyboards (Matias et al. 1996), trackballs (Wobbrock and Myers 2006), eye tracking (Majaranta and Räihä 2002), and switch interfaces (Mankoff et al. 2002), as well as predefined texts or graphical icons (Majaranta and Räihä 2002). Similarly, regarding output, non-speech output and tactile output are used to help visually impaired desktop users and aged desktop users to access the Web (Ramloll et al. 2001). These solutions may also be useful to small device users who access multimedia content in a noisy environment or small device users who cannot access heavy Web content due to traffic cost or connection quality.

Finally, some solutions exist for small device users but not for disabled users. For instance, regarding input, the following solutions exist for small device users but not for motor-impaired users: chording (where a few keys are used in different combinations to represent a large number of characters), auditory feedback, haptic feedback, and handwriting input. As chording requires fewer keys than a standard keyboard, a chording keypad can be used with just one hand and thus may be useful to small device users who usually need to type with one hand (Lyons et al. 2004) only. Some motor-impaired users have difficulty in pressing keys simultaneously, so they may not be able to use chording keyboards. However, for motor-impaired users who lose one hand but have fine control over the other, chording keyboards may improve their typing performance.

4.5 Web 2.0

So far, we have discussed the similarities in Web access between mobile users and disabled users in the context of the "traditional" Web. The Web is evolving, however, and "Web 2.0" is widely available on mobile devices (Holmquist 2007) ("Mobile 2.0" (Holmquist 2007) or "Mobile Web 2.0" (Jaokar and Fish 2006)). Even though the benefits of Web 2.0 in mobile settings is indisputable, there are some techno-logical limitations: while some mobile devices, especially smart phones, are quite powerful, devices can still be a technological barrier to using Web 2.0 on mobile devices. User agents, critical in supporting mobile Web 2.0, are also still limited. This means that barriers to effective Web use are changing for both disabled users (Gibson 2007) and mobile users.

We have seen that users of mobile devices encounter barriers to Web access that are the same as, or similar to, the barriers that people with a range of disabilities face when using traditional desktop equipment. In Section 4.3, these barriers were introduced with reference to input, content, and output, and issues were identified in each area. So while the similarities in barriers, and overlapping solutions (both actual and potential), are valid for the "traditional" Web, how are they different when the content is Web 2.0?

4.5.1 Web 2.0 Accessibility Challenges

Web 2.0 is a mixed bag, with different meanings to different people. Often it is characterized by social networking, and user-generated content; to others it is more technological—the use of technologies such as AJAX, allowing Web pages to appear and behave more like an application than a document. Which of these aspects of Web 2.0 pose accessibility problems, what are the characteristics of these problems, and what is being done about them?

While noting the potential for social exclusion if sites are inaccessible, social networking is not inherently problematic for either disabled or mobile users. That is not to say that current implementations are easy to use: indeed they suffer the same accessibility barriers as many other traditional Web sites (e.g., registration problems (Meiselwitz and Lazar 2009)). No, the problems with social networking sites arise from the way in which they typically incorporate the other major aspects of Web 2.0: user-generated content and dynamically updating content.

As with social networking, of which it is an integral part, user-generated content need not necessarily create accessibility problems. In reality, however, it often appears to lead to pages that are busy, cluttered, and disorganized, and can thus be difficult to understand. Users inexperienced in Web development will often generate content with sole consideration for how it looks, an approach that can lead to poor use of semantic tags from HTML. For example, head-ings might be differentiated from the rest of the text only by color or font size, instead of using the HTML heading tags. When these pages are encountered in a non-visual medium, such as through a screen reader, the differentiation is lost and understanding the text becomes more challenging. Even many of the more sophisticated users generating this type of content will be unaware of recommendations such as *Web Content Accessibility Guidelines 2* (WCAG 2); indeed, many will probably not even consider that their content will be accessed using any means other than those they employ themselves.

Authoring tools, both stand-alone and those built into pages, have an important role to play here, in terms of both their input and their output. User-generated content involves users at both ends of the process, so true accessibility means that not only must the output, the resulting Web page, be accessible, but also that the tools used for generating this content must be accessible (Power and Petrie 2007). Since content generation requires much more input than traditional "Web surfing," it is here that the input barriers become particularly problematic.

Perhaps the greatest challenges of Web 2.0 for users of assistive technologies arise from the technological advances that enable content to update dynamically. Traditionally, Web pages did not change once they had been served. In this context, they have many of the characteristics of a document. Despite the potential for multimedia, with sound and video components, the page acted as a page: it could be read (or watched, or listened to), and not (except in a very limited sense) interacted with. Web pages did not contain controls and did not change over time: they were static. The rise of scripting technologies, and particularly AJAX, has changed this. Pages are often now dynamic and interactive—the whole model of a page has changed quite dramatically. Not only do readers have the challenge of navigating the Web, and understanding and relating different sections of a page, but now they may also need to understand whether the page has changed, which parts have changed, and whether the changes are of interest.

A typical use for AJAX is in Web-based e-mail access. These are often designed so that the page appears like a standard desktop mail client, with a list of folders, a list of messages in the currently selected folder, and the contents of the currently selected message. The list of messages is a list of links, but selecting one does not take the user to a new page; instead it dynamically updates the region of the page showing the current message. The list of folders behaves in a similar manner. Consequently, users are able to read many different e-mails, all without actually moving to a new page. Making changes, then perceiving and understanding what has changed is critical to the usability of the system.

While these Web applications are seen as being appealing to users, in fact they can pose multiple problems to some people. Returning to the four principles of WCAG, for applications to be accessible they must be perceivable, operable, understandable, and robust (see Section 4.3.2). AJAX Web applications can fail on all counts. Since HTML has limited native controls, more sophisticated ones, such as sliders and trees, must be generated using graphics and scripting; these are known generically as *widgets*. When the function of a widget is only implicit in its appearance, screen reader users cannot recognize it (the control may appear simply as a list of links): the widget is not perceivable. Furthermore, keyboard access to these widgets is usually lacking, meaning that only mouse users are able to control them: the widget is not operable. A different barrier faces users of sites with "hotkeys" enabled, since using these can require screen readers to be put into an unfamiliar and difficult-to-use mode (Hailpern et al. 2009). Even if a user manages to operate a control, it can be difficult or impossible to discover its effect—while sighted users can scan the whole page to quickly identify changes, users with a restricted visual or audio view of the page must search around to find it (screen readers do not typically notify users of dynamic updates (Brown and Jay 2008, Hailpern et al. 2009)). The widget operation is not understandable. Finally, the use of scripting in widgets necessarily means that the Web application is not usable by people whose access technology does not support this: the widget is not robust.

4.5.2 Research

Currently, the most significant attempt to tackle the accessibility of Web 2.0 widgets is WAI-ARIA: the ARIA (Gibson 2007) from the W3C's WAI. The essence of the ARIA solution is to add semantic metadata to a page, and through doing so can address the accessibility problems outlined above.* First, WAI-ARIA extends the HTML tabindex, which allows controls to be focusable, allowing users to perceive and operate controls as they move around the page. Second, it introduces the role attribute to define the function of a widget, for example, role = "treeitem," allowing the information implicit in the visual construction to be made explicit to both the user and their assistive technology. Third, WAI-ARIA has tags to hold any states or properties of a widget; these may be queried to allow the user to understand the current state of a control (e.g., a slider is at position 30 out of 200). Note that ARIA tags are not expected to be hand-coded into static HTML, but set and modified as part of the scripting. Finally, ARIA provides a mechanism for noting live regions—those parts of a page that may be updated. These can have further attributes to denote what types of update to announce (e.g., just additions), how important it is to announce the update, and how much of a region should be read when part is updated.

While the information the ARIA attributes and properties provide can help, it cannot be used in a naïve manner. Thiessen and Chen (2007) explored the difficulties in scaling ARIA for highly active pages; they found using ARIA effectively became problematic when updates occur frequently. Understanding how to use ARIA tags, both from the point of view of the developer (which settings to choose) and from that of the user-agent and assistive technologies (how to use the tags to help determine exactly when and how to present an update), is difficult, and we believe that a good understanding of how sighted users interact with dynamic updates, and why, is essential.

---

* The WAI has a primer on ARIA: http://www.w3.org/TR/wai-aria-primer/

At the University of Manchester, we are undertaking such a project—SASWAT: Single Structured Accessibility Stream for Web 2.0 Access Technologies.* Eye-tracking studies with sighted users have given us an understanding of how attention is allocated to different types of dynamic update, and the model resulting from these studies has been used to propose techniques for audio interaction with these pages. This research showed that while most updates that happen automatically are ignored, those triggered by user actions are usually attended to, at least briefly (Jay and Brown 2008). This latter group could be split further, with those updates that were explicitly requested (e.g., by clicking a control) viewed by 98% of participants, and those that were simply initiated by user action (e.g., typing in an input box) viewed by 82%.

Borodin et al. (2008) have also been investigating the problem of presenting updates to screen reader users, but have concentrated more on the ways of detecting changes than how they should be presented. In particular, their system, known as Dynamo, takes the approach of dealing with all changes, both within the page (i.e., dynamic updates) and between pages (i.e., moving to a new page) in the same way. Users were notified that a change had taken place by a short non-speech sound, and could move to the new content with a command. They found that participants in their evaluation liked the ability to jump to new content when pages dynamically updated.

Unifying these two types of change could be advantageous for pages using older technologies. For example, where a page needs to be refreshed to show new information, comparing the two could allow the user's focus (e.g., screen reader focus, or the position of the view port on a small-screen device or magnification system) to remain roughly constant, eliminating the need to navigate past the top of the page to reach the new content.

4.5.3 Crossover

In Section 4.3, we saw how mobile users share many of the problems that different disabled users face when Web browsing. We have also seen (Section 4.5.1) some of the particular challenges that Web 2.0 is posing disabled users; which of these are likely to pose problems to mobile users?

Probably the most obvious similarity is the difficulty screen reader users have in identifying that an update has had an effect, and finding which region of the page it has changed. The small screen can mean that only a small region of the page is in view at any one time, so it may be necessary to use the control, then move around the page trying to find the new region. Spotting such changes when the region is visible is usually easy, since the attention is attracted toward such events (Carmi and Itti 2006), but if not visible, it is necessary to use a combination of memory and context to discover it.

Keyboard access is another issue that may also affect mobile users. Even using a stylus on a touch screen, high-resolution screens displaying large portions of a page will end up with small controls that make clicking errors more likely, particularly

---

* http://hcw.cs.manchester.ac.uk/research/saswat/

if the user is actually moving. Interactions initiated by hovering the cursor over a field prove difficult for mobile and disabled users alike. Mobile and screen reader users often have no pointer to hover, while people with motor impairments can find holding a mouse in one position difficult (Hwang et al. 2004). Indeed, research for the SASWAT project showed that even sighted users have difficulties with this type of interaction (Jay and Brown 2008).

So we see that Web 2.0 poses difficulties not only for disabled users, but also for people wishing to access it via mobile devices (while noting that there is overlap between these groups). This apparent bad news is potentially, however, beneficial to all. As we saw for the traditional Web, when two groups face similar barriers, solutions may also be transferred. So those technologies currently being developed to help disabled users may be leveraged by mobile systems, and vice versa.

The information provided by WAI-ARIA markup, and the models produced by research such as SASWAT can be used, independently or together, to improve handling of dynamic updates on small-screen devices. Changing the position of the view to show updates can help users understand how an interface works, and to use it more efficiently, but it is not appropriate for all types of update. If WAI-ARIA were to be used to help mobile users, we run into the possibility of a virtuous circle: technology developed for disabled users helps mobile users, which accelerates the uptake of ARIA amongst developers, thereby benefiting disabled users. The main obstacle to this, that mobile browsers do not currently support ARIA, demonstrates how little the overlap between these communities has been recognized thus far.

The same process might be seen for the problems of data input. If content is designed to be accessible to assistive technology users (e.g., by following WCAG 2.0), Widgets become perceivable and operable without a mouse, so it should also be usable by mobile users. Here, we see benefits for both groups: if developers reject interfaces, which are unusable for mobile users, the results should be better for users of assistive technologies. Additionally, the increasing importance of data input might speed migration of those techniques mentioned in Section 4.4. This could benefit disabled groups as increased demand leads to further development and lower costs. The very ability to use the Web while mobile can cause problems, with usage of mobile devices when driving associated with an increased risk of accidents. This, along with the other difficulties highlighted above is prompting research and development from device manufacturers and software developers into simplifying interfaces so that a minimum of attention is required. There is potential here for improvements, both in hardware and software, which may also benefit aging or disabled users.

While concrete technologies can be used to aid users, improving the quality of the content is less easy. One way of achieving accessible content is to give users authoring tools that ensure, or at least facilitate, accessible content generation. Another is to give users (and, indeed, authoring tool developers) guidelines as to what constitutes good practice. Even though W3C recognizes barriers experienced by both user groups and provides such guidelines, these are published as two independent resources—WCAG (Caldwell et al. 2008) and the Mobile Web Best Practices (MWBP) (Rabin and McCathieNevile 2005). However, identifying the overlaps would provide benefits to both groups. Allowing pages to be developed and evaluated for both accessibility and mobile Web support together would mean that designers do not need to follow two separate methodologies independently, saving time and reducing the costs.

4.6 Conclusions

We have seen that people using mobile devices to access the Web suffer limitations that reduce the ease with which they can use sites, or even make it impossible. These may result from the limitations of the device or from the circumstances and environment in which it is used. We have also seen that the barriers they face are often remarkably similar to the barriers faced by other groups, such as disabled users or older people, and that, although Web 2.0 is changing the barriers, many commonalities remain.

Sadly, the majority of research and development concentrates on one group or another. The commonalities are not just in the barriers, however—the solutions can also be transferred. Technologies, techniques, and guidelines designed to benefit one set of users can often also be applied to help a different set. This has been the case for the traditional Web, and remains the case for Web 2.0. Developments such as WAI-ARIA, designed for disabled users, and visually disabled users in particular, have the potential to help mobile users too.

Importantly, we believe that recognizing the relationships between the barriers faced by different users has the potential to deliver benefits far beyond a few improvements in assistive technologies or mobile user agents. As the Web transforms into Web 2.0 and the world fully embraces mobile technologies, it is possible that these changes will have a transformative effect on content too. Thus far, businesses and organizations have often paid only lip-service to accessibility, wrongly believing that it is an unnecessary cost, rather than a motivator for really considering the usability of their site. It is hard to believe, however, that the same attitude will be taken with mobile accessibility—this is a growing group, and one that is perceived as a wealthy group of customers worth attracting. Furthermore, while very few developers will have explored their sites with screen readers or other assistive technologies, it is likely that the majority will have used mobile devices. In this way, any shortcomings become quickly apparent to those best placed to fix them. We believe that a growing consciousness of the issues that surround accessing the Web using devices other than the desktop computer can only benefit all users.

References

Asakawa, C. What's the web like if you can't see it? In *W4A'05*, Chiba, Japan, pp. 1–8, 2005, ACM Press, New York. ISBN 1-59593-036-1.

Bailey, S., Barrett, S., and Guilford, S. Older users' interaction with websites. In *Workshop on HCI and the Older Population, British HCI*, Edinburgh, U.K., 2005.

Barnard, L., Yi, J.S., Jacko, J.A., and Sears, A. An empirical comparison of use-in-motion evaluation scenarios for mobile computing devices. *International Journal of Human Computer Studies*, 62(4):487–520, 2005. ISSN 1071-5819. doi: http://dx.doi.org/10.1016/j.ijhcs.2004.12.002.

Barnard, L., Yi, J.S., Jacko, J.A., and Sears, A. Capturing the effects of context on human performance in mobile computing systems. *Personal and Ubiquitous Computing*, 11(2):81–96, 2007. ISSN 1617-4909. doi: http://dx.doi.org/10.1007/s00779-006-0063-x.

Barreto, A. Visual impairments. In S. Harper and Y. Yesilada (eds.), *Web Accessibility: A Foundation for Research*, Springer, London, U.K., 2008.

Borodin, Y., Bigham, J.P., Raman, R., and Ramakrishnan, I.V. What's new?—Making web page updates accessible. In *Proceedings of the 10th International ACM SIG ACCESS Conference on Computers and Accessibility (Assets'08)*, Halifax, Canada, pp. 145–152, 2008, ACM, New York. ISBN 978-1-59593-976-0.

Brewster, S. Overcoming the lack of screen space on mobile computers. *Personal and Ubiquitous Computing*, 6:188–205, 2002.

Brown, A. and Jay, C. A review of assistive technologies: Can users access dynamically updating information? Technical Report, University of Manchester, Manchester, U.K., 2008. http://hcw-eprints.cs.man.ac.uk/70/

Caldwell, B., Cooper, M., Reid, L.G., and Vanderheiden, G. *Web Content Accessibility Guidelines 2.0* (WCAG 2.0), W3C, 2008. http://www.w3.org/TR/WCAG20/

Carmi, R. and Itti, L. Visual causes versus correlates of attention selection in dynamic scenes. *Vision Research*, 46:4333–4345, 2006.

Cavender, A. and Ladner, R.E. Hearing impairments. In S. Harper and Y. Yesilada (eds.), *Web Accessibility: A Foundation for Research*, Springer, London, U.K., 2008.

Chadwick-Dias, A., McNulty, M., and Tullis, T. Web usability and age: How design changes can improve performance. In *Proceedings of the 2003 Conference on Universal Usability (CUU'03)*, Vancouver, Canada, pp. 30–37, 2003, ACM, New York. ISBN 1-58113-701-X. doi: http://doi.acm.org/10.1145/957205.957212.

Chaparro, A., Bohan, M., Fernandez, J., Choi, S.D, and Kattel, B. The impact of age on computer input device use: Psychological sciences. *International Journal of Industrial Ergonomics*, 24(5):503–513, 1999.

Chau, D.H., Wobbrock, J., Myers, B., and Rothrock, B. Integrating isometric joysticks into mobile phones for text entry. In *Extended Abstracts on Human Factors in Computing Systems (CHI'06)*, Montréal, Canada, pp. 640–645, 2006.

Chen, T., Yesilada, Y., and Harper, S. Investigating the typing and pointing errors of mobile users. Technical Report 51, School of Computer Science, University of Manchester, Manchester, U.K., 2008.

Chuter, A. and Yesilada, Y. Relationship between mobile web best practices (MWBP) and web content accessibility guidelines (WCAG), W3C, 2008. http://www.w3.org/TR/mwbp-wcag/

Coyne, K.P. and Nielsen, J. Beyond ALT text: Making the web easy to use for users with disabilities. Nielson Norman Group, Fremont, CA, 2001.

Coyne, K.P. and Nielsen, J. Web usability for senior citizens: 46 design guidelines based on usability studies with people age 65 and older. Nielsen Norman Group Report, Fremont, CA, 2002.

Craven, J. and Brophy, P. Non-visual access to the digital library: The use of digital library interfaces by blind and visually impaired people, 2003. Library and Information Commission Research Report 145, Centre for Research in Library and Information Management, Manchester, U.K.

Czaja, S.J. Technology and older adults: Designing for accessibility and usability. In *Proceedings of the Eighth International ACM SIGACCESS Conference on Computers and Accessibility (Assets'06)*, Portland, OR, pp. 1, 2006, ACM, New York. ISBN 1-59593-290-9. doi: http://doi.acm.org/10.1145/1168987.1168988.

Dalal, N.P., Quible, Z., and Wyatt, K. Cognitive design of home pages: An experimental study of comprehension on the World Wide Web. *Information Processing and Management*, 36(4):607–621, 2000. doi: http://dx.doi.org/10.1016/S0306-4573(99)00071-0. http://www.ingentaconnect.com/content/els/03064573/2000/00000036/00000004/art00071

Disability Rights Commission (DRC). *The Web: Access and Inclusion for Disabled People*, The Stationary Office, London, U.K., 2004.

Duchnicky, R.L. and Kolers, P.A. Subscribed content readability of text scrolled on visual display terminals as a function of window size. *The Journal of the Human Factors and Ergonomics Society*, 25(6), 683–692, 1983.

Earl, C. and Leventhal, J. A survey of windows screen reader users: Recent improvements in accessibility. *Journal of Visual Impairment and Blindness*, 93(3), 174–177, 1999.

Edwards, A.D.N. Assistive technologies. In S. Harper and Y. Yesilada (eds.), *Web Accessibility: A Foundation for Research, Human–Computer Interaction Series*, 1st edn., Springer, London, U.K., pp. 142–162, September 2008, Chapter 10. ISBN 978-1-84800-049-0. doi: http://dx.doi.org/10.1007/978-1-84800-050-6_10.

Ellis, R.D. and Kurniawan, S.H. Increasing the usability of online information for older users: A case study in participatory design. *International Journal of Human–Computer Interaction*, 12:263–276, 2000.

European Union Policy Survey. E-accessibility of public sector services in the European union. European union policy survey, U.K. Government Cabinet Office (RNIB and AbilityNet and DCU and RNID and Socitm), November 2005. http://www.cabinet-office.gov.uk/e˜government/resources/eaccessibility/index.asp

Felzer, T. and Nordmann, R. Alternative text entry using different input methods. In *Proceedings of the Eighth International ACM SIGACCESS Conference on Computers and Accessibility*, Portland, OR, pp. 10–17, 2006.

Gibson, B. Enabling an accessible web 2.0. In (*W4A'07*): *Proceedings of the 2007 International Cross-Disciplinary Conference on Web Accessibility (W4A)*, Banff, Canada, pp. 1–6, 2007, ACM, New York, 2007. doi: http://doi.acm.org/10.1145/ 1243441.1243442.

Greenstein, J.S. Pointing devices. In M.G. Helander, T.K. Landauer, and P.V. Prabhu (eds.), *Handbook of Human-Computer Interaction*, 2nd edn., North-Holland, Amsterdam, the Netherlands, 1997, pp. 1317–1348.

Groff, L., Liao, C., Chaparro, B., and Chaparro, A. Exploring how the elderly use the web. *Usability News*, 1(2), 1999. Software Usability Research Laboratory (SURL), Wichita State University, Wichita, KS, http://www. surl.org/usabilitynews/12/elderly.asp

Gunderson, J. and Jacobs, I. *User Agent Accessibility Guidelines 1.0*, W3C, 1999. http://www.w3.org/TR/WAI-USERAGENT/

Hailpern, J., Guarino-Reid, L., Boardman, R., and Annam, S. Web 2.0: Blind to an accessible new world. In *Proceedings of the 18th International Conference on World Wide Web* (*WWW'09*), Madrid, Spain, pp. 821–830, 2009, ACM, New York. ISBN 978-1-60558-487-4. doi: http://doi.acm.org/10.1145/1526709.1526820.

Hawthorn, D. Possible implications of aging for interface designers. *Interacting with Computers*, 12(5):507–528, 2000. http://www.sciencedirect.com/science/article/B6V0D-3YKKDNJ-5/2/9e0ae58d2d9be3ed8302fa25fc4a482f

Hinckley, K. Input technologies and techniques. In A. Sears and J. Jacko (eds.) (reversal of first edn.). *The Human–Computer Interaction Handbook*, 2nd edn., Lawrence Erlbaum Associates, Mahwah, NJ, Chapter 7, pp. 161–176, 2007.

Holmquist, L.E. Mobile 2.0. *Interactions*, 14(2):46–47, 2007. ISSN 1072-5520. doi: http://doi.acm.org/10.1145/1229863.1229891

Huenerfauth, M. Representing coordination and non-coordination in an American sign language animation. In *Proceedings of the Seventh International ACM SIGACCESS Conference on Computers and Accessibility* (*Assets'05*), Baltimore, MD, pp. 44–51, 2005, ACM, New York. ISBN 1-59593-159-7. doi: http://doi.acm.org/10.1145/1090785.1090796

Hwang, F., Keates, S., Langdon, P., and Clarkson, J. Mouse movements of motion-impaired users: A submovement analysis. In *Proceedings of the Sixth International ACM SIGACCESS Conference on Computers and Accessibility* (*Assets'04*), Atlanta, GA, pp. 102–109, 2004, ACM, New York. ISBN 1-58113-911-X. doi: http://doi.acm.org/10.1145/1028630.1028649

Jacko, J., Vitense, H., and Scott, I. Perceptual impairments and computing technologies. In J.A. Jacko and A. Sears (eds.), *The Human–Computer Interaction Handbook: Fundamentals, Evolving Technologies and Emerging Applications*, Lawrence Erlbaum Associates, Mahwah, NJ, pp. 504–522, 2003.

Jacobs, I., Gunderson, J., and Hansen, E. *User Agent Accessibility Guidelines 1.0* (UAAG), W3C, 2002. http://www.w3.org/TR/WCAG20/

James, C.L. and Reischel, K.M. Text input for mobile devices: Comparing model prediction to actual performance. In *Proceedings of the SIGCHI Conference on Human Factors in Computing Systems* (*CHI'01*), Seattle, WA, pp. 365–371, 2001, ACM Press, New York. ISBN 1-58113-327-8. doi: http://doi.acm.org/10.1145/365024.365300

Jaokar, A. and Fish, T. *Mobile Web 2.0*, Future Text Publications, London, U.K., 2006.

Jay, C. and Brown, A. User review document: Results of initial sighted and visually disabled user investigations. Technical Report, University of Manchester, Manchester, U.K., 2008. http://hcw-eprints.cs.man.ac.uk/49/

Jones, M., Marsden, G., Mohd-Nasir, N., Boone, K., and Buchanan, G. Improving Web interaction on small displays. *Computer Networks*, 31(11–16):1129–1137, 1999. citeseer.nj.nec.com/jones99improving.html

Karpov, E., Kiss, I., Leppänen, J., and Olsen, J. Short message dictation on symbian series 60 mobile phones. In *Proceedings of the Eighth International Conference on Multimodal Interfaces*, Banff, Canada, pp. 126–127, 2006.

Kurniawan, S.H. Ageing. In S. Harper and Y. Yesilada (eds.), *Web Accessibility: A Foundation for Research*, Springer, London, U.K., 2008.

Lazar, J., Allen, A., Kleinman, J., and Malarkey, C. What frustrates screen reader users on the web: A study of 100 blind users. *International Journal of Human Computer Interaction*, 22(3):247–269, 2007.

Leventhal, A. Structure benefits all. In *W4A: Proceedings of the 2006 International Cross-Disciplinary Workshop on Web Accessibility (W4A)*, Edinburgh, U.K., pp. 33–37, 2006, ACM Press, New York, ISBN 1-59593-281-X. doi: http://doi.acm.org/10.1145/1133219.1133226

Lewis, C. Cognitive and learning impairments. In S. Harper and Y. Yesilada (eds.), *Web Accessibility: A Foundation for Research*, Springer, London, U.K., 2008.

Lin, M., Goldman, R., Price, K.J., Sears, A., and Jacko, J. How do people tap when walking? An empirical investigation of nomadic data entry. *International Journal Human Computer Studies*, 65(9):759–769, 2007. ISSN 1071-5819. doi: http://dx.doi.org/10.1016/j.ijhcs.2007.04.001

Lyons, K., Starner, T., Plaisted, D., Fusia, J., Lyons, A., Drew, A., and Looney, E.W. Twiddler typing: One-handed chording text entry for mobile phones. In *Proceedings of the SIGCHI Conference on Human Factors in Computing Systems*, Vienna, Austria, pp. 671–678, 2004.

MacKenzie, I.S. and William Soukoreff, R. Text entry for mobile computing: Models and methods, theory and practice. *Human-Computer interaction*, 17:147–198, 2002.

MacKenzie, I.S., Kober, H., Smith, D., Jones, T., and Skepner, E. Letterwise: Prefix-based disambiguation for mobile text input. In *Proceedings of the 14th Annual ACM Symposium on User Interface Software and Technology (UIST'01)*, Orlando, FL, pp. 111–120, 2001, ACM Press, New York. ISBN 1-58113-438-X. doi: http://doi.acm.org/10.1145/502348.502365

Majaranta, P. and Räihä, K.-J. Twenty years of eye typing: Systems and design issues. In *Proceedings of the 2002 Symposium on Eye Tracking Research and Applications*, New Orleans, LA, pp. 15–22, 2002.

Malkewitz, R. Head pointing and speech control as a hands-free interface to desktop computing. In *Proceedings of the Third International ACM Conference on Assistive Technologies*, Marina del Rey, CA, pp. 182–188, 1998.

Manaris, B. and Harkreader, A. Suitekeys: A speech understanding interface for the motor-control challenged. In *Proceedings of the Third International ACM SIGCAPH Conference on Assistive Technologies*, Marina del Rey, CA, pp. 108–115, 1998.

Mankoff, J., Dey, A., Batra, U., and Moore, M. Web accessibility for low bandwidth input. In *Proceedings of the Fifth International ACM Conference on Assistive Technologies*, Edinburgh, U.K., pp. 17–24, 2002.

Matias, E., MacKenzie, I.S., and Buxton, W. One-handed touch typing on a qwerty keyboard. *Human–Computer Interaction*, 11:1–27, 1996.

Meiselwitz, G. and Lazar, J. Accessibility of registration mechanisms in social networking sites. In A.A. Ozok and P. Zaphiris (eds.), *Online Communities and Social Computing, Lecture Notes in Computer Science*, Vol. 5621, Springer, Berlin/Heidelberg, Germany, pp. 82–90, 2009.

Minneman, L. Keyboard optimization technique to improve output rate of disabled individuals. In *RESNA Ninth Annual Conference*, Minneapolis, MN, pp. 402–404, 1986.

Moffatt, K.A. and McGrenere, J. Slipping and drifting: Using older users to uncover pen-based target acquisition difficulties. In *Proceedings of the Ninth International ACM SIGACCESS Conference on Computers and Accessibility (Assets'07)*, Tempe, AZ, pp. 11–18, 2007. ISBN 978-1-59593-573-1. doi: http://doi.acm.org/10.1145/1296843.1296848

Morell, R.W. *OlderAdults, Health Information and the World Wide Web*, Lawrence Erlbaum Associates, Mahwah, NJ, 2002.

Newell, A.F., Dickinson, A., Smith, M.J., and Gregor, P. Designing a portal for older users: A case study of an industrial/academic collaboration. *ACM Transactions on Computer-Human Interactions*, 13(3):347–375, 2006. ISSN 1073-0516. doi: http://doi.acm.org/10.1145/1183456.1183459

Oliveira Neto, F.G., Fechine, J.M., and Pereira, R.R.G. Matraca: A tool to provide support for people with impaired vision when using the computer for simple tasks. In *Proceedings of the 2009 ACM Symposium on Applied Computing*, Honolulu, HI, pp. 158–159, 2009.

Penev, A. and Wong, R.K. Grouping hyperlinks for improved voice/mobile accessibility. In *Proceedings of the 2008 International Cross-Disciplinary Conference on Web Accessibility (W4A'08)*, Beijing, China, pp. 50–53, 2008, ACM, New York. ISBN 978-1-60558-153-8. doi: http://doi.acm.org/10.1145/1368044.1368055

Power, C. and Petrie, H. Accessibility in non-professional web authoring tools: A missed web 2.0 opportunity? In *Proceedings of the 2007 International Cross-Disciplinary Conference on Web Accessibility (W4A'07)*, Banff, Canada, pp. 116–119, 2007, ACM Press, New York, ISBN 1-59593-590-X. doi: http://doi.acm.org/10.1145/1243441.1243468

Rabin, J. and McCathieNevile, C. Mobile web best practices 1.0, W3C, 2005. http://www.w3.org/TR/mobile-bp/

Ramloll, R., Brewster, S., Yu, W., and Riedel, B. Using non-speech sounds to improve access to 2D tabular numerical information for visually impaired users. In *IHM-HCI2001*, Lille, France, 2001.

Reisel, J.F. and Shneiderman, B. Is bigger better? the effects of display size on program reading. In *Proceedings of the Second International Conference on Human–Computer Interaction*, Honolulu, HI, pp. 113–122, 1987.

Sayago, S., Camacho, L., and Blat, J. Evaluation of techniques defined in WCAG 2.0 with older people. In *(W4A'09): Proceedings of the 2009 International Cross-Disciplinary Conference on Web Accessibility (W4A)*, Madrid, Spain, pp. 79–82, 2009, ACM, New York. ISBN 978-1-60558-561-1. doi: http://doi.acm.org/10.1145/1535654.0397

Sears, A. and Young, M. Physical disabilities and computing technologies: An analysis of impairments. In J.A. Jacko and A. Sears (eds.), *The Human–Computer Interaction Handbook: Fundamentals, Evolving Technologies and Emerging Applications*, Lawrence Erlbaum Associates, Mahwah, NJ, pp. 482–503, 2003. ISBN 0-8058-3838-4.

Serrano, M., Nigay, L., Demumieux, R., Descos, J., and Losquin, P. Multimodal interaction on mobile phones: Development and evaluation using ACICARE. In *Proceedings of the Eighth Conference on Human–Computer Interaction with Mobile Devices and Services (MobileHCI'06)*, Helsinki, Finland, pp. 129–136, 2006.

Siek, K.A., Khalil, A., Liu, Y., Edmonds, N., and Connelly, K.H. A comparative study of web language support for mobile web browsers. In *WWW@10: The Dream and the Reality*, Terre Haute, IN, 2004.

Silfverberg, M., MacKenzie, I.S., and Korhonen, P. Predicting text entry speed on mobile phones. In *Proceedings of the SIGCHI Conference on Human Factors in Computing Systems*, Seattle, WA, pp. 9–16, 2000.

Smith, M., Sharit, J., and Czaja, S. Aging, motor control, and the performance of computer mouse tasks. *Human Factors*, 40(3):389–396, 1999.

Takagi, H., Asakawa, C., Fukuda, K., and Maeda, J. Accessibility designer: Visualizing usability for the blind. In *ASSETS'04*, Atlanta, GA, pp. 177–184, 2004.

Thiessen, P. and Chen, C. Ajax live regions: Chat as a case example. In *W4A'07: Proceedings of the 2007 International Cross-Disciplinary Conference on Web Accessibility (W4A)*, Banff, Canada, pp. 7–14, 2007, ACM, New York. ISBN 1-59593-590-X. doi: http://doi.acm.org/10.1145/1243441.1243450.

Trewin, S. Physical usability and the mobile web. In *Proceedings of the 2006 International Cross-Disciplinary Workshop on Web Accessibility (W4A)*, Edinburgh, U.K., pp. 109–112, 2006, ACM Press, New York. ISBN 1-59593-281-X. doi: http://doi.acm.org/10.1145/1133219.1133239.

Trewin, S. Physical impairment. In S. Harper and Y. Yesilada (eds.), *Web Accessibility: A Foundation for Research*, Springer, London, U.K., 2008.

Trewin, S. and Pain, H. Keyboard and mouse errors due to motor disabilities. *International Journal of Human Computer Studies*, 50:109–144, 1999.

Trewin, S., Keates, S., and Moffatt, K. Developing steady clicks: A method of cursor assistance for people with motor impairments. In *Proceedings of the Eighth International ACM SIGACCESS Conference on Computers and Accessibility (Assets'06)*, Portland, OR, pp. 26–33, 2006. ISBN 1-59593-290-9. doi: http://doi.acm.org/10.1145/1168987.1168993

Watters, C., Duffy, J., and Duffy, K. Using large tables on small display devices. *International Journal of Human-Computer Studies*, 58(1):21–37, 2003. ISSN 1071-5819. doi: http://dx.doi.org/10.1016/S1071-5819(02)00124-6

Wobbrock, J.O. The future of mobile device research in HCI. In *CHI'06 Workshop*, Montréal, Canada, 2006.

Wobbrock, J.O. and Gajos, K.Z. A comparison of area pointing and goal crossing for people with and without motor impairments. In *Proceedings of the Ninth International ACM SIGACCESS Conference on Computers and Accessibility (Assets'07)*, Tempe, AZ, pp. 3–10, 2007. ISBN 978-1-59593-573-1. doi: http://doi.acm.org/10.1145/1296843.1296847

Wobbrock, J. and Myers, B. Trackball text entry for people with motor impairments. In *Proceedings of CHI'06*, Montréal, Canada, pp. 479–488, 2006.

Wobbrock J.O. and Myers, B.A., Aung, H.H., and LoPresti, E.F. Text entry from power wheelchairs: Edgewrite for joysticks and touchpads. In *Proceedings of the Sixth International ACM SIGACCESS Conference on Computers and Accessibility*, Atlanta, GA, pp. 110–117, 2004.

World Health Organization. Magnitude and causes of visual impairment, World Health Organization, 2004. http://www.who.int/mediacentre/factsheets/fs282/en/

Yesilada, Y., Chuter, A., and Henry, S.L. Shared web experiences: Barriers common to mobile device users and people with disabilities, W3C, 2008. http://www.w3.org/WAI/mobile/experiences

Zajicek, M. Successful and available: Interface design exemplars for older users. *Interacting with Computers*, 16(3), 2004.

Zajicek, M. and Hall, S. Solutions for elderly visually impaired people using the internet. In *HCI 2000*, Sunderland, U.K., pp. 299–307, 2000.

Chapter 5

Technological Solutions to Support Mobile Web 2.0 Services

Claudia Canali, Michele Colajanni,
and Riccardo Lancellotti

Contents

The widespread diffusion and technological improvements of wireless networks and portable devices are facilitating mobile access to the Web and Web 2.0 services. The emerging Mobile Web 2.0 scenario still requires appropriate solutions to guarantee user interactions that are comparable with present levels of services. In this chapter, we classify the most important services for Mobile Web 2.0 and we identify the key functions that are required to support each category of Mobile Web 2.0 services. We discuss some possible technological solutions to implement these functions at the client and at the server level, and we identify some research issues that are still open.

5.1 Introduction

The so-called Mobile Web 2.0 originates from the conjunction of the Web 2.0 services and the proliferation of Web-enabled mobile devices.

The term Web 2.0, first introduced in 2004–2005 [20], indicates an evolution of the World Wide Web that aims to facilitate interactive information sharing, interoperability, user-centered design, and collaboration among users. Although it is difficult to precisely define Web 2.0, its novel essence refers to a user-centric environment that is characterized by two predominant features [13]:

■ Users may actively create and upload contents in many forms and have prominent profile pages, including heterogeneous information.
■ Users may belong to a sort of virtual community determined by social interactions. For example, users may form connections among each other via explicit links to users who are denoted as "friends" or through "membership" groups of heterogeneous nature.

In the last years, we have also observed significant technological improvements in wireless networks and the diffusion of more powerful mobile devices with increased hardware and software capabilities. The growth and penetration of mobile communication technologies, with an expected number of global mobile phone subscribers reaching up to 4.5 billion in 2012 [19], has determined a scenario where users can access the Web directly from their mobile devices, and this trend is increasing. Mobile devices, which for portability reasons are companions in the user's daily life, are likely to become the favorite platform to connect, interact, and share content with other people.

The conjunction of Web 2.0 and mobile Web accesses is leading to a new communication paradigm, where mobile devices act not only as mere consumers of information, but also as complex carriers for getting and providing information, and as platforms for novel services [28].

We may expect that in the near future the demand for Web 2.0 services will mainly come from mobile devices [4,22]. This expectation is confirmed by the current trend of popular Web 2.0 sites, such as MySpace and Facebook, which offer mobile access to the users through specific applications preloaded on mobile devices. According to specialized studies, the most popular Web 2.0 sites are expected to have a mobile component within a few years [4].

Mobile Web 2.0 represents both an opportunity for creating novel services (typically related to user location) and an extension of Web 2.0 applications to mobile devices. The management of user-generated content, of content personalization, of community and information sharing is much more challenging in a context characterized by devices with limited capabilities in terms of display, computational power, storage, and connectivity. Furthermore, novel services require support for real-time determination and communication of the user position. The choice of appropriate technological solutions that can effectively support Mobile Web 2.0 services will be a key element to determine its success [17].

In this chapter, we propose a classification of Mobile Web 2.0 services and we evidence some key functions that are required for their support. We identify the main requirements for the implementation of each function. Finally, we discuss possible technological solutions for functions implementation at the client and server level, and identify some open issues.

This chapter is organized as follows. Section 5.2 proposes a classification of the emerging services for Mobile Web 2.0. Section 5.3 identifies the key functions required to support the Mobile Web 2.0 services. Section 5.4 describes some technological solutions to implement the supporting functions and identify possible open issues. Section 5.5 concludes with some final remarks.

5.2 Mobile Web 2.0 Services

Mobile Web 2.0 includes a wide range of heterogeneous and complex services. In this section, we propose a two-level classification of these services based on what we consider their predominant feature and other more specific characteristics. In the taxonomy shown in Figure 5.1, at the first level we have:

- *Sharing services* that are characterized by the publication of contents to be shared with other users
- *Social services* that refer to the management of social relationships among the users
- *Location services* that tailor information and contents on the basis of the user location

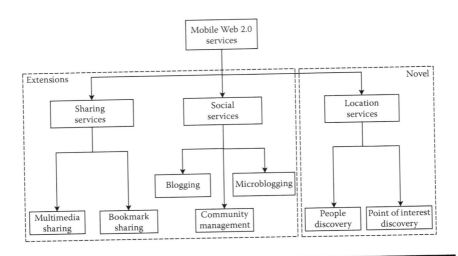

Figure 5.1   Taxonomy of Mobile Web 2.0 services.

As indicated by the dotted boxes in the figure, the sharing and social service classes represent extensions of existing Web 2.0 to the mobile scenario, while the location services represent a completely novel class of services, that exploits information on the user mobility.

It is worth noting that the service classes shown in Figure 5.1 as well as their subclasses, described below, are not completely disjointed categories.

5.2.1 Sharing Services

Sharing services offer the users the capability to store, organize, search, and manage heterogeneous contents. These contents may be rated, commented, tagged, and shared with specified users or groups that can usually visualize the stored resources chronologically, by category, rating or tags, or via a search engine.

The subclass of multimedia sharing considers management services related to multimedia resources, such as photos or videos. These resources are typically generated by the users that exploit the sharing service to upload and publish their own contents. Popular examples of Web portals offering a multimedia sharing service include Flickr, Zooomr, YouTube, Mocospace, and Mobimii.

The class of bookmark sharing services allows users to manage a common collection of Web page bookmarks. In this case, the shared contents are publicly available links to Web pages that users consider as interesting resources and want to share with other users. Many bookmarking services provide Web feeds for their lists of bookmarks, so that subscribers may become aware of new bookmarks as they are shared and tagged by other users. Among the most popular portals that offer bookmark sharing services we cite Del.icio.us, Reddit, Digg and its recently developed mobile version called Dgm8.

5.2.2 Social Services

The management of user relationships is the main feature of the social services that allow users to create social connections based on common interests, hobbies, or experiences, and to actively interact with each other.

The services belonging to the *Community management* subclass allow registered users to maintain a list of contact details of people they know. Their key feature is the possibility to create and update a personal profile including information such as user preferences and his list of contacts. These contacts may be used in different ways depending on the purpose of the service, which may range from the creation of a personal network of business and professional contacts (e.g., LinkedIn), to the management of social events (e.g., Meetup), and up to the connection with old and new friends (e.g., Facebook, MySpace, Friendster).

The *Blogging* services allow a user to create and manage a blog, that is, a sort of personal online journal, possibly focused on a specific topic of interest. Blogs are usually created and managed by an individual or a limited group of people, namely *author(s)*, through regular entries of heterogeneous content, including text, images, and links to other resources related to the main topic, such as other blogs, Web pages, or multimedia contents. A blog is not a simple online journal, because the large majority of them allow external comments on the entries. The final effect is the creation of a discussion forum that engages readers and builds a social community around a person or a topic. Other related services may also include blogrolls (i.e., links to other blogs that the author reads) to indicate social relationships to other bloggers. Among the most popular portals that allow users to manage their own blog we cite BlogSpot, LiveJournal, Wordpress, and Splinder.

In *Microblogging* services, the communication is characterized by very short message exchanges among the users. Although this class of services originates from the blogging category, there are important differences between microblogging and traditional blogs: (1) the size of the exchanged messages is significantly smaller, (2) the purpose of microblogging is to capture and communicate instantaneous thoughts or feeling of the users, and (3) the recipient of the communication may differ from that of traditional blogs because microblogging allows authors to interact with a group of selected friends. Twitter, Jaiku, Plurk, Folkstr, GUSHUP, and Mobikade are examples of portals providing microblogging services.

5.2.3 Location Services

The ability to continuously trace user position represents one of the most innovative features in the context of Mobile Web 2.0, which emphasizes the important role of mobile devices in accessing the Web [26,31].

The knowledge of the user current location may be exploited in several ways to offer value added services. One of the most popular uses concerns *people discovery*, that basically aims to locate user friends; significant examples of this service may be found in Loopt, Brightkite, and Buddy Beacon applications. Usually these services, also called "friend finder" applications, plot the position of the user and his friends on a map; the geographical location of the users is uploaded to the system by means of a positioning system installed on the user mobile devices.

Another class of location services, that we call *points of interest (POIs) discovery*, exploits geographical information to locate POIs, such as events, restaurants, museums, and any kind of attractions that may be useful or interesting for a user. These services offer the users a list of nearby POIs selected on the basis of their personal preferences and specifications. POIs are collected by exploiting collaborative recommendations from other users that may add a new POI by uploading its geographical location, possibly determined through a GPS positioning system installed on the mobile device. Users may also upload short descriptions, comments, tags, and images or videos depicting the place. POI discovery services are provided by POIfriends, Socialight, and Mobnotes portals.

5.3 Functions to Support Mobile Web 2.0 Services

The previous section has pointed out that Mobile Web 2.0 includes complex and heterogeneous services, some totally new, others as extensions of existing Web 2.0 services. In this section, we identify some key functions that are at the basis of Mobile Web 2.0 services by separating functions that are required to extend Web 2.0 services to a mobile context from functions that are specifically related to the novel class of location services.

Among the functions required to extend Web 2.0 services to a mobile scenario, we identify

- Information input
- Large file upload
- Personalization
- Fruition of multimedia content

Other functions that are related to the possibility of localizing the mobile device are

- Computation of device location
- Geo-referenced information management

5.3.1 Functions Description

We give a brief description for each of the above functions to support Mobile Web 2.0 services.

The *information input* function refers to the communication of small size data, typically in a text format, from the users to the service through a mobile device.

Inserting comments in a blog or in a forum, tagging a resource, assigning ratings, updating personal information or simply adding a new entry in a microblogging service represent typical examples of information input. These operations occur very frequently in the Mobile Web 2.0 scenario, where users are not only consumers but also providers of information that actively interact with the services.

The *large file upload* function shares several common traits with the information input function; however, the focus is mainly related to multimedia resources, instead of textual information. The upload of user-generated large multimedia files (e.g., images, audio, and video resources) is a characteristic feature of Mobile Web 2.0 services that has been inherited by Web 2.0, and is becoming increasingly popular thanks to the diffusion of mobile devices equipped with built-in cameras.

The *personalization* function aims to tailor contents to the user preferences and needs [18]. Information about the users is collected by the services and may be exploited to offer personalized content in several ways, for example, through customized layout, information filtering, recommendation systems, subscription to specific channels or news feeds, and specification of lists of contacts.

Another feature of Mobile Web 2.0 that comes from Web 2.0 concerns the *fruition of multimedia contents*, which refers to the high demand for multimedia resources, such as images, video, and audio. Besides multimedia sharing services, also blogs, community management or POI discovery involve a considerable exchange of multimedia contents among mobile users.

The last two functions refer to the most innovative feature of Mobile Web 2.0 services, that is, the capability of identifying the current location of a mobile device. The geographical information may be exploited by the services to locate a user or a suggested POI. The *computation of device location* function is required to identify the geographical position of the mobile device. Once computed, the location data should be communicated to the service and stored as a geo-referenced data. The operations related to the storage and management of this data are accomplished by the *geo-referenced information management* function. It is worth to recall that a location data may refer both to a POI and to a user. If referred to a user, the current location may represent a volatile data that is subject to frequent updates.

5.3.2 Services and Functions

Each Mobile Web 2.0 service requires the support of at least one of the above functions. In Table 5.1, we map the functions and the service classes: each function is considered mandatory (*Yes*), not required (*No*) or optional (*Maybe*) for a specific service class. We do not include in the table the two functions related to the localization of mobile devices because their mapping on the services is quite straightforward and may easily be summarized as follows: computation of device location and geo-referenced information management are mandatory only for location services, but optional for all other services, where geo-referenced data may represent an additional information about users and uploaded contents.

Table 5.1 Functions to Support Mobile Web 2.0 Services

Mobile Web 2.0 Service	Functions			
	Input	Upload	Personalization	Multimedia Fruition
Multimedia sharing	Yes	Yes	Yes	Yes
Bookmark sharing	Yes	No	Maybe	No
Community management	Yes	No	Yes	Maybe
Blogging	Yes	No	Maybe	Maybe
Microblogging	Yes	Maybe	Yes	Maybe
People discovery	Yes	No	Yes	No
POI discovery	Yes	Maybe	Yes	Maybe

We observe that some input function is required by all the services through one or more operations, such as posting comments, adding ratings, or tagging resources. The upload function is mandatory for multimedia sharing services, where the users upload self-generated large size files and where pictures taken on-the-fly may be added to enrich the short posts of the provided POI description; the upload function is optional in microblogging and POI discovery services. The personalization function is optional for bookmark sharing and blogging services, where it may be exploited to filter contents and subscribe to news feeds, while is mandatory for all the other services that strongly rely on personal information maintained in the user profiles.

Although multimedia fruition does not represent a characterizing feature, it is mandatory in multimedia sharing services, while it is optional for community management, blogging, microblogging, and POI discovery services, where the presence of multimedia contents is possible.

5.3.3 Mobile Device Limitations

Supporting the previously described functions in a mobile scenario is not trivial due to the limited capabilities of mobile devices in terms of connection, CPU/storage capabilities, display size, and interface usability.

The bandwidth available to mobile devices has been greatly increased with the diffusion of 3-G wireless networks, and further enhancements are expected with the advent of 4-G technologies [11]. However, the wireless network connection remains relatively unstable and heterogeneous, because it is affected by coverage issues.

The computational power and storage capacities have experienced significant improvements in mobile devices; however, they remain significantly lower than those of a laptop/desktop that is typically used to access Web 2.0 services. The computational power and storage capacity may introduce problems when the mobile device must handle complex or multimedia-rich services, preventing the fruition and the storage of some kinds of resource formats [12]. Furthermore, computationally expensive functions consume a significant amount of energy, thus affecting the lifetime of batteries.

The display is characterized by minor improvements because of an intrinsic constraint of portability of the mobile devices. Even if the trend is toward devices with at least 3 inches screens and resolutions of 480 × 320 pixels [32], the limited display size determines very unpleasant user navigation while accessing services designed for desktop computers. Finally, the device interfaces, which force users to insert input data through tiny keypads or small on-screen keyboards, are hardly satisfactory for the users who actively interact with the service.

Even if the technological evolution has substantially improved the scenario of the mobile devices population [28,32], the above limitations may hinder the deployment of the key functions required for Mobile Web 2.0 services. In Section 5.4, we describe some technological solutions that may be exploited to implement the functions to support Mobile Web 2.0 services, anticipating that the implementation of some functions presents open issues that deserve further research efforts.

5.4 Technological Solutions for Function Implementation

The main functions to support Mobile Web 2.0 services (information input, large file upload, personalization, fruition of multimedia contents, and location-related functions) may be implemented through different technological solutions. Each solution may follow a *client-side* or *server-side* approach depending on where the functions are implemented. The best approach is not an absolute choice, but it strongly depends on the specific context of each service, as discussed in the rest of this section. We can anticipate that solutions following a server-side approach may rely on more powerful platforms and usually do not present severe issues. However, in the Mobile Web 2.0 context many functions necessarily have to be implemented on the client-side, even if this means coping with the limitations of the mobile devices.

5.4.1 Information Input

When the information input is carried out through a mobile device, the presence of intermittent or low bandwidth connections, as well as the small display and the peculiar input methods of these devices, introduce novel challenges that are not present in Web 2.0. For example, network connection quality can lead to poor navigation experience because tasks such as page refresh may be very slow with an unreliable and limited wireless connection. Furthermore, typing input information represents a cumbersome task in most mobile devices because the user must rely on a stylus or must interact with small on-screen keyboards. These limitations may result in an unsatisfactory experience that could move away users from these services. Allowing a comfortable navigation and supporting information input through a mobile device is a key issue in the context of Mobile Web 2.0, because this function is a fundamental element for all the offered services.

The problem of managing user input is usually addressed by client-side solutions and requires operating on two distinct elements: first, the communications between mobile devices and servers should be optimized to reduce the need for (synchronous) data exchange; second, the interface of the service should be redesigned to allow comfortable user navigation and input operations in a mobile device.

To cope with the first problem, the typical solution is to adopt asynchronous communications between the client and the server. This solution allows the browser on the mobile device to communicate with the server in the background in an asynchronous way with respect to the user interactions and without interfering with the current state of the page. The responses from the server are handled asynchronously by the browser that updates the Web pages without having to keep the user attention frozen. This benefit is particularly valuable in a mobile context, where the interaction with the service occurs while the user is on the run and cannot stay continuously focused on the device. By sending requests just for the required data that typically represent a small portion of the information managed by the whole Web page, asynchronous communications between the client and the server allows to greatly reduce the amount of reloading and data transferred. Furthermore, the solution allows improving the user input speed and reduces the display processing requirements.

Typical examples of technologies supporting this approach are Asynchronous JavaScript and XML (AJAX) and Flex. AJAX is considered one of the most important enabling technologies for implementing interactive services. Actually, the term AJAX indicates a mixture of several technologies that integrate Web page presentation, interactive data exchange between client and server, client-side scripts, and asynchronous updates of server response [24]. Flex is an evolution of the widespread Flash technology commonly used to create animation, advertisements, and to integrate video into Web pages. Flex objects offer functionalities that resemble the AJAX approach, allowing the deployment of services that exploit asynchronous communication with the server to improve the user experience in the Mobile Web 2.0 scenario.

A limitation for asynchronous communications at the Web level in this scenario is that not all mobile Web browsers support the required technologies with adequate performance. While the most recent mobile devices satisfy this requirement, older devices with less computational power and memory may not be suitable to support asynchronous communication with the server. An increasingly popular alternative is to redesign the interface at the client level, without need to use Web browsers. In this case, the service is provided directly by an application installed in the mobile device, and interaction with the server is managed asynchronously, possibly using the same Web Application Programming Interface (API) that are already available for interaction through the Web browser. This approach, typically relying on the Java platform for mobile devices or through some device specific Software Development Kit (SDK) (as in the case of the iPhone Objective-C API), has the potential to reduce the computational and memory requirements that may hinder the popularity of a service accessed through a Web mobile browser. However, this approach requires a significant amount of effort for the development of the application. Furthermore, the client applications usually need frequent updates that in most cases have to be done manually.

The second critical problem is the difficulty for users to interact with user interfaces and type data for input in a mobile device. The need to rely on stylus or the small size of on-screen keyboards in mobile devices suggest that, even if asynchronous communication can improve the user experience, input operations remain a key issue for the diffusion of Mobile Web 2.0 services. The problem of supporting seamless input from user remains an open issue to address, although some possible research direction seems more mature and promising than others.

A first solution may be to redesign the user interface to avoid input whenever possible. The typical approach is to rely on fill-in forms that are pre-compiled based on default settings that may be personalized for each user. In this way, the user does not have to compile forms, but can simply choose between one or more options.

The alternative approach is to exploit client-side technologies to simplify the input operation by defining novel interfaces for human and computer interaction. To this aim, speech recognition and user gestures recognition (based on accelerometers) are gaining popularity in mobile devices [28]. The main drawbacks of this solution are the difficulty to integrate user interaction in a Web-based interface and the need to adapt these systems to mobile devices.

Alternative solutions may be too computationally expensive for current mobile devices. For example, speech recognition with a large vocabulary may be computationally unfeasible on CPU-power constrained devices. In a similar way, access to accelerometer for gesture recognition needs to be tailored to every specific device characteristic, thus hindering the adoption of general solutions. We expect that this area of alternative methods for human–computer interface will receive significant attention from researchers and industries, with the goal of simplifying the task of information input from the user thus enabling the development of even more sophisticated and interactive services.

5.4.2 Large File Upload

As for information input, the limitations of mobile devices may represent an issue that hinders the possibility for users to directly upload their self-generated contents, such as pictures and videos.

The user interface of a mobile device is often inadequate to support large file uploads to Mobile Web 2.0 services. For example, uploading a picture from a mobile device requires the off-line creation of the resource with the built-in camera, the temporary storage of the multimedia data in the device file system, and finally to seek the saved file from among the directories for the upload operation. While this behavior is acceptable when working on a PC or a laptop, it becomes unacceptable for the users of a Mobile Web 2.0 service. A user interface that simplifies the upload of a large file and especially multimedia content (for example, through a click-and-upload feature) will play a key role for the success of Mobile Web 2.0.

The issues related to the upload of large files are typically addressed at the client-side. A solution is to exploit specialized clients that provide direct and easy upload of user-generated contents without the need to rely on Web-based upload forms or on Multimedia Messaging Service (MMS). For example, specialized client applications may allow pictures or videos to be taken directly through the device built-in camera and uploaded with a single click. This support has the potential to strongly affect the success of Mobile Web 2.0 services. A clear confirmation of this claim can be found in the recent announcement by Google: daily YouTube uploads directly from mobile devices have increased 400% in 6 days after the release of the last model of iPhone [30] that provides users with an easy interface to upload videos to the YouTube portal.

Another critical issue for uploading large files from a mobile device is related to the quality of the wireless connection. The upload time may increase to an unacceptable level, and, in the case of disconnections, upload may fail. The exchange of large amount of data may also reduce the battery lifetime and limit the possibility for the user to interact with the service. The problem seems to be even more critical if we consider the current trend of installing high resolution camera (up to several Mpixel per image) in mobile devices. To overcome this limitation, it is possible to carry out some content adaptation before the upload. For example, images can be cropped or scaled directly on the mobile device to reduce the amount of data transferred through the network.

However, the actual effectiveness of these solutions remains an open problem. Indeed, there is a trade-off involving computational power, network connection, and battery power. On one hand, content adaptation on the device requires a significant amount of computational power that may not be available on every device. On the other hand, transferring the high resolution resources without any adaptation to the server consumes network resources that in mobile devices may be scarce as well. Furthermore, both CPU-intensive operation and wireless data transfer have a significant impact on the mobile device batteries. The research for solutions that can address this trade-off, for example, combining client-side and server-side adaptation, represents an open issue that is likely to receive growing amount of attention in the next years.

5.4.3 Personalization

The personalization function requires an initial phase of user data collection, possibly from different sources. Then, the gathered information has to be stored in a user *profile* and maintained for subsequent use. The collection and the management of the user information are typically accomplished through a server-side approach. The user information may be obtained basically from two sources:

■ Explicitly communicated by the user
■ Implicitly acquired from the user behavior

In the case of explicit communication by the user, personal information are provided through apposite fill-in forms to add/edit user preferences; this communication may occur when the user registers himself for the access to a service or may be filled/modified later.

When implicitly acquired, the user information is typically inferred through the analysis of the user behavior, for example, through data mining operations on a Web site log files or on sets of Really Simple Syndication (RSS) feeds the user subscribed to [14]. Collaborative filtering techniques may also be exploited for grouping users based on similar preferences or click history [15]: missing information about a user may be integrated by considering the corresponding information in the profile of other users belonging to the same group. These techniques for implicit user profiling represent server-side solutions and are usually carried out off-line because they involve time-consuming operations, such as data mining.

The user profiles are usually maintained in database(s) on the server infrastructures. We should consider that the infrastructures to support Mobile Web 2.0 services typically consist of distributed systems with multiple servers, such as Content Delivery Networks [5]. Solutions for replicating data storages on a distributed infrastructure have been widely studied in the context of databases [21].

The simplest solution to manage database replication in distributed Web environments is based on a centralized master copy and replicated secondary copies. In case of updates, data are modified on the master copy and the changes are then propagated to the secondary copies. However, the access patterns for the user profiles present a unique feature that may help the management in case of replication. Specifically, each user typically interacts with only one server; hence the profile of a given user is accessed by one server for the whole duration of a user session. This access pattern has a significant impact on consistency and replication policies. Indeed, the whole dataset of user profiles can be partitioned and distributed over the servers depending on the user access patterns. Since no replication is needed, consistency issues are limited to guarantee that the user profiles on the servers are consistent with the data of the master copy.

User migration among multiple servers, however, may occur between consecutive sessions. Therefore, the user profile data should migrate following the user. The support for this behavior is not explicitly optimized in most replication strategies for back-end databases. An example of proposal to handle profile migration is Tuxedo [25], a generic data caching framework that supports user mobility by allowing data to follow the user.

It is worth to note that a unique opportunity offered by Mobile Web 2.0 concerns the use of Subscriber Identity Module (SIM), removable cards for personalization purposes. SIM cards have always been used to store data such as international number of the mobile user, billing information, security authentication and ciphering data, subscriber address books, etc. However, the advent of Mobile Web 2.0 has created the opportunity to exploit SIM cards as a place to store user authentication data for personalization purposes. This solution follows the philosophy of a "unified login" that allows users to log in to many Mobile Web 2.0 services using just one account, that is maintained by simply moving the SIM card from device to device.

The possibility to exploit a sort of unified login is particularly important to provide users with customized information coming from different Mobile Web 2.0 services. A typical behavior of Mobile Web 2.0 users, indeed, is to subscribe to several services and provide different information to each of them. For example, for each service a user may specify a list of contacts; to communicate or share contents with all his contacts, the user should separately access all the subscribed services. *Mash-up* technologies are increasingly being used to provide users with updated information coming from different subscribed services without the need of separately accessing all of them. The term mash-up indicates an approach that allows easy and fast service integration of data and functionalities from two or more external sources by using publicly available APIs. Thanks to this content aggregation, users may have available on a single page information coming from different services and updated in real time. Furthermore, mash-up solutions also allow integrating personal profile information maintained by different services without duplication of the information itself, thus simplifying update operations. Architecturally, the content aggregation usually takes place on the client-side, by exploiting the Web browser of the mobile device to combine and reformat the data retrieved from multiple services.

5.4.4 Fruition of Multimedia Content

The poor connections of wireless networks and the reduced hardware capabilities of mobile devices may determine critical issues for the fruition of multimedia contents such as (1) the low and unreliable network bandwidth may cause long latency while downloading multimedia resources, (2) the computational power may be insufficient to decode and render high quality multimedia resources and (3) the small display size may not support high resolution formats. These limitations give rise to the need for adapting multimedia contents to match the capabilities of mobile devices and network connections.

Content adaptation may involve a wide range of heterogeneous transformations that are applied to the original contents to generate adapted versions suitable to be consumed by mobile device [9]. The basic idea behind content adaptation is that mobile users often do not need a best-quality experience when consuming multimedia resources, but rather a good-enough quality and acceptable latency to convey the needed information [34].

The adaptation is typically applied to multimedia resources with the main goal of reducing their size. Size reduction helps to decrease downloading time and storage requirements, and may also reduce, depending on the type of adaptation, the computational demand to render the resource on the mobile device. A large and heterogeneous set of resource attributes can be considered for each type of multimedia resource to perform the adaptation [9]. For example, image adaptation typically includes scaling, cropping, or compressing the image; audio resources are adapted by reducing the bit rate [16]; common transformations for video resources are frame size and color depth reduction [16].

The adaptation of multimedia contents may be performed at the client-side or at the server-side.

A critical aspect of content adaptation is the high computational cost of the transformations, especially when applied to large-sized multimedia resources. The computational cost of content adaptation may easily exacerbate the capabilities of the supporting infrastructures [7]; for this reason, determining the platform where content adaptation should be carried out represents a strategic choice.

In a client-side approach, contents are adapted directly on the mobile device. The advantage of this solution is that the adaptation may generate a resource version that perfectly matches the device limitations, thanks to the exact knowledge that the device has of its capabilities. However, this approach is not always feasible or convenient due to the device limitations. The reduced storage, computational power, and battery energy may prevent performing locally the expensive adaptation tasks. Furthermore, a client-side solution does not address the issue of poor connections: since the multimedia resources have to be entirely downloaded on the mobile device, long latency may be experienced while transmitting the content over the wireless connection. Hence, we may observe that, even if mobile devices are becoming more powerful platforms with medium-large connections, their limitations still prevent relying only on client-side adaptations. The technological evolution, however, allows the mobile devices to consume larger size and better quality resources with respect to the past and, if necessary, to carry out locally some final adjustments on multimedia contents.

The server-side approach represents a more feasible solution where the content adaptation is carried out on the server infrastructure. In this case, there are two main alternatives about *when* multimedia resources should be adapted: *on-the-fly* and *off-line* adaptation.

If on-the-fly adaptation is applied, the server infrastructure generates an adapted resource version for the specific mobile device at the moment of the request. However, the high computational costs of adaptation may hinder the effectiveness of this solution in the Mobile Web 2.0 scenario. Solutions based on on-the-fly adaptation have been proposed in the past [27], usually integrated with caching strategies at the intermediary level [3,10]. However, this approach was feasible in a context characterized by a limited amount of available multimedia resources and a small fraction of requests coming from mobile devices and, consequently, requiring adaptation. On the other hand, an on-the-fly approach may lead to excessive computational costs for the server platforms in the Mobile Web 2.0 scenario, even if coupled with caching strategies.

The off-line approach consists in pre-generating multiple adapted versions of multimedia contents that are maintained on the server infrastructure or cached at an intermediary level and then delivered to the user when requested. Relying completely on off-line adaptation means to pre-generate adapted versions of all multimedia resources for any class of device/connection, thus avoiding the expensive cost of on-the-fly adaptation. Furthermore, the use of layered encoding technologies for the off-line generation of adapted versions allows achieving significant advantages. Layered encoding allows generating only one adapted version of the content from which it is possible to obtain a suitable version for any mobile device. Basically, this approach generates a base layer and one or more enhanced layers to achieve the desired resolution of the multimedia content. Layers may be added or dropped depending on the requirements of the mobile device. A popular technology for layered encoding is Scalable Video Codecs (SVC) [23], where the enhanced layers may add temporal and/or spatial quality to the base layer. The use of SVC provides important benefits from the computational and storage points of view for systems adopting off-line adaptation solutions for supporting Mobile Web 2.0 services. The original multimedia content, indeed, has to be encoded only once, and the result is a scalable adapted version from which representations with lower quality can be obtained by discarding parts of the data. This solution avoids the need of storing multiple versions of the same multimedia content to satisfy any possible combination of requirements of mobile devices and wireless networks, thus simplifying even the server-side approach to the content adaptation.

To this aim, we should consider that the technological evolution of the mobile devices may have positive consequences for the off-line approach, because resources will not need to be tailored exactly for every type of client device as it happened until now. For example, while the first generation of devices ranged from monochrome to full-color capabilities, modern devices can display at least 16-bit color images; hence previous adaptations from color to B/W videos are now useless. Thanks to the technological improvements, different devices are now able to consume the same version of a multimedia resource, thus reducing the number of adapted versions that must be generated for every original resource. On the other hand, the presence of user-generated content in the Mobile Web 2.0 scenario is causing an explosion of multimedia contents in terms of quantity and heterogeneity [1,8]. We should also consider that the working set of accessed multimedia resources in Mobile Web 2.0 is highly volatile. Indeed, the resources are characterized by a short life span, because they typically concern real-world events or hot topics for which user interest rapidly subsides. For these reasons, a pure off-line solution may be not feasible or convenient due to the excessive waste of storage and computational power caused by pre-generating and maintaining adapted versions for every multimedia resource.

A hybrid solution that combine on-the-fly and off-line content adaptation may represent a better choice to support Mobile Web 2.0 services. A possible solution consists in applying off-line adaptation only to a limited set of the most popular resources, while adapting on-the-fly the remaining resources [6]. The rationale behind this approach originates from the popularity of multimedia resources in Mobile Web 2.0, that follows a Zipf-like distribution [8,33]. This means that pre-generation of adapted versions for a limited fraction of popular resources allows a system for Mobile Web 2.0 services to satisfy a high number of user requests. For the remaining requests, adaptation may be applied on-the-fly without overcoming the capabilities of server infrastructures. However, identifying the most popular resources represents an open challenge especially in the context of Mobile Web 2.0, whose workload is characterized by high volatility, short resource life span, and sudden popularity peaks.

5.4.5 Location-Related Functions

The possibility to geographically locate a mobile device and exploit this information to enrich the user experience is one of the most innovative features of Mobile Web 2.0. Two main functions are needed to support this feature: computation of device location and geo-referenced information management.

5.4.5.1 Computation of Device Location

Several solutions may be exploited for positioning purposes, that is, determining the geographical location of a mobile device. Positioning is usually performed by following a client-side approach, where the computation of the location is carried out on the device, then communicated to the service.

The most popular positioning technology is the Global Positioning System (GPS), whose wide adoption is due to the large diffusion of mobile devices equipped with GPS receivers that provide reliable three-dimensional location (latitude, longitude, and altitude). However, relying on the GPS technology for positioning may present two main drawbacks: the long time taken by the mobile device during the start-up phase to look for available satellites (between 45 and 90s on average), and the consequent considerable cost of computational and battery power.

To overcome these limitations, the assisted GPS (A-GPS) has been introduced in the last few years. A-GPS is a carrier network dependent system that can improve the initial performance of a GPS satellite-based positioning system. Basically, the A-GPS uses an assistance server that communicates to the mobile device information on the available satellites to accelerate the signal acquisition.

In specific conditions, such as in indoor environments, an alternative technique to provide positioning is to exploit cellular or Wi-Fi triangulation, based on the device distance from cell towers or Wi-Fi access points. It is worth to note that GPS- and triangulation-based technologies may also be employed together to improve positioning accuracy [31].

Although mobile devices have become more powerful, they do not always have the computational power necessary to compute their current location through GPS or triangulation techniques. We should also consider that the computational cost of the operation depends on the accuracy required by the specific service, and on the frequency/speed of the users' movements, that may cause frequent recomputations of the exact location. In this case, the computation may be executed on the server side. The server infrastructure receives from the mobile devices GPS- and/or cell-based information and calculates the location and transmits the result to the devices.

5.4.5.2 Geo-Referenced Information Management

The device location represents a geo-referenced data that has to be stored and managed on the server-side. The presence of geo-referenced data requires the use of technological solutions that allow operations on spatial data. For example, the system should be able to find all the POIs that are close to a given user location or to identify the shortest path among two given locations. This requirement may be addressed by specialized Geographic Information Systems (GIS) or by database systems that support the storage and management of spatial data.

However, this requirement does not represent an open issue, due to the wide diffusion of GIS technologies and database systems with spatial support (e.g., MySQL, PostgreSQL, Oracle, Microsoft SQL Server).

Another important characteristic for management purposes is the potentially dynamic nature of the device location data that may change frequently due to user movements. Traditional approaches for data replication are not suitable to store and maintain potentially dynamic user location due to consistency issues. For this reason, a commonly adopted solution is to maintain the user location at the application server level just for the duration of the current user session. On the other hand, location data referring to POIs may be stored in databases due to their more stable nature.

A last consideration about the management of geo-referenced information is that the current location is usually considered by the users as a sensitive data [29]. Hence, Mobile Web 2.0 services should provide users with appropriate mechanisms to control the disclosure of their location. This issue is typically addressed on the server-side. A first solution consists in allowing the user to edit a list of authorized contacts that may access his location. The user must have the possibility of modifying/updating the authorization list that is maintained in the user profile on the server infrastructure, at any moment. This is particularly important in the context of people discovery services, because the users' movements are continuously tracked to communicate their presence to nearby contacts even when they are not actively interacting with the application. A more sophisticated approach to preserve the privacy of the user consists in revealing the user position with different accuracies, depending on the specific location and/or on the recipient of the information [2].

5.5 Conclusions

The popularity of Web 2.0 services, coupled with the diffusion of increasingly powerful Web-enabled mobile devices, has led to the advent of Mobile Web 2.0. This emerging scenario includes very complex and heterogeneous services: some services are totally new, based on the notion of user location, while others are extensions of existing Web 2.0 services to a mobile context. The deployment of new and extended services may be hindered by the limited capabilities of mobile devices in terms of display, computational power, storage, and connectivity. Hence, the choice of appropriate technological solutions is a key element to effectively support Mobile Web 2.0 services.

We classify the emerging Mobile Web 2.0 services and we identify some key functions required for their support, discussing possible technological solutions for the implementation of each function. We show that existing technological solutions are sufficient to implement most functions; hence the problem is a correct integration and capacity design. On the other hand, the implementation of other functions represents a challenge that deserves further research efforts, as in the case of providing comfortable interfaces for user input and large file upload.

References

1. Berg Insight AB. Mobile Internet 2.0. Research Report, May 2007.
2. E. Bertino. Privacy-preserving techniques for location-based services. *SIGSPATIAL Special*, 1(2):2–3, 2009.
3. S. Buchholz and T. Buchholz. Replica placement in adaptive content distribution networks. In *Proceedings of the 2004 ACM Symposium on Applied Computing (SAC'04)*, pp. 1705–1710, Nicosia, Cyprus, March 2004.
4. BusinessWeek. Social networking goes mobile. May 2006.
5. C. Canali, V. Cardellini, M. Colajanni, and R. Lancellotti. Content delivery and management. In *Content Delivery Networks*, R. Buyya, M. Pathan, and A. Vakali (eds.), pp. 105–126. Springer, Berlin, Germany, 2008.
6. C. Canali, M. Colajanni, and R. Lancellotti. Resource management strategies for mobile web-based services. In *Proceedings of 4th IEEE International Conference on Wireless and Mobile Computing (WIMOB'08)*, pp. 172–177, Avignon, France, October 2008.
7. C. Canali, M. Colajanni, and R. Lancellotti. Performance evolution of mobile-web based services. *IEEE Internet Computing*, 13(2):60–68, March/April 2009.
8. M. Cha, H. Kwak, P. Rodriguez, Y.-Y. Ahn, and S. Moon. I tube, you tube, everybody tubes: Analyzing the world's largest user generated content video system. In *Proceedings of the 7th ACM SIGCOMM Conference on Internet Measurement (IMC'07)*, pp. 1–14, San Diego, CA, October 2007.
9. S. Chandra. Content adaptation and transcoding. In *Practical Handbook of Internet Computing*, M. P. Singh (ed.), pp. 1–144, Chapman & Hall/CRC Press, Baton Rouge, LA, 2004.
10. C.-Y. Chang and M.-S. Chen. On exploring aggregate effect for efficient cache replacement in transcoding proxies. *IEEE Transactions on Parallel and Distributed Systems*, 14:611–624, June 2003.

11. H.-H. Chen, M. Guizani, and W. Mohr. Evolution toward 4G wireless networking. *IEEE Network*, 21(1):4–5, January/February 2007.

12. Y.-K. Chen and S.-Y. Kung. Trends and challenges with system-on-chip technology for multimedia system design. In *Proceedings of Emerging Information Technology Conference*, p. 4, Santa Clara, CA, March 2005.

13. G. Cormode and B. Krishnamurthy. Key differences between Web 1.0 and Web 2.0. *First Monday*, 13(6), June 2008.

14. M. Eiriniaki and M. Vazirgiannis. Web mining for web personalization. *ACM Transaction on Internet Technology*, 3(1):1–27, 2003.

15. S. Flesca, S. Greco, A. Tagarelli, and E. Zumpano. Mining user preferences, page content and usage to personalize website navigation. *World Wide Web*, 8(3):317–345, August 2005.

16. L. Guo, E. Tan, S. Chen, Z. Xiao, O. Spatscheck, and X. Zhang. Delving into Internet streaming media delivery: A quality and resource utilization perspective. In *Proceedings of the 6th ACM SIGCOMM Conference on Internet Measurement (IMC'06)*, pp. 217–230, ACM, New York, 2006.

17. A. Jaokar and T. Fish. *Mobile Web 2.0: The Innovator's Guide to Developing and Marketing Next Generation Wireless/Mobile Applications*. Futuretext, London, U.K., 2006.

18. T.-P. Liang, H.-J. Lai, and Y.-C. Ku. Personalized content recommendation and user satisfaction: Theoretical synthesis and empirical findings. *Journal of Management Information Systems*, 23(3):45–70, 2007.

19. Market Intelligence Center. Global mobile phone subscribers forecasted to reach 4.5 billion by 2012. Press Release, 2008.

20. T. O'Reilly. What Is Web 2.0. Design patterns and business models for the next generation of software. O'Reilly Media, March 2005.

21. M. Pati ~no-Mart′inez, R. Jim′enez-Peris, B. Kemme, and G. Alonso. Consistent database replication at the middleware level. *ACM Transactions on Computer Systems*, 23(4):1–49, 2005.

22. Tekrati Research Report. 950 million users will access social networking sites via mobile devices. February 2008.

23. H. Schwarz, D. Marpe, and T. Wiegand. Overview of the scalable video coding extension of the H.264/AVC standard. *IEEE Transactions on Circuits and Systems for Video Technology*, 17(9):1103–1120, 2007.

24. N. Serrano and J. P. Aroztegi. Ajax frameworks in interactive web apps. *IEEE Software*, 24(5):12–14, September–October 2007.

25. W. Shi, K. Shah, Y. Mao, and V. Chaudhary. Tuxedo: A peer-to-peer caching system. In *Proceedings of the International Conference on Parallel and Distributed Processing Techniques and Applications (PDPTA'03)*, pp. 981–987, Las Vegas, NV, June 2003.

26. W. Shu, M. Jungwon, and K. Y. Byung. Location based services for mobiles: Technologies and standards. In *Proceedings of the IEEE International Conference on Communication (ICC'08)*, Beijing, China, May 2008.

27. G. Singh. Guest editor's introduction: Content repurposing. *IEEE Multimedia*, 11(1):20–21, March 2004.

28. S. R. Subramanya and B. K. Yi. Enhancing the user experience in mobile phones. *Computer*, 40(12):114–117, 2007.

29. Y. Sun, T. L. Porta, and P. Kermani. A flexible privacy-enhanced location-based services system framework and practice. *IEEE Transactions on Mobile Computing*, 8(3):304–321, March 2009.

30. TechCrunch Press Release. YouTube Mobile Uploads Up 400% Since iPhone 3GS Launch. June 2009.
31. S. J. Vaughan-Nichols. Will mobile computing's future be location, location, location? *IEEE Computer Magazine*, 42(22):14–17, 2009.
32. M. Walker, R. Turnbull, and N. Sim. Future mobile devices? An overview of emerging device trends, and the impact on future converged services. *BT Technology Journal*, 25(2):120 – 125, April 2007.
33. T. Yamakami. A Zipf-like distribution of popularity and hits in the mobile web pages with short life time. In *Proceedings of the International Conference on Parallel and Distributed Computing, Applications and Technologies (PDCAT'06)*, pp. 240–243, Taipei, Taiwan, December 2006.
34. D. Zhang. Web content adaptation for mobile handheld devices. *Communications of the ACM*, 50(2):75–79, 2007.

Chapter 6

Mobile Internet Services or Services for Multiplatform Use

Anne Kaikkonen

Contents

6.1 Introduction

In recent years, Internet on mobiles has become an increasingly interesting subject. During the past 10 years, the number of mobile Internet users has increased all over the world. Japan and South Korea have been leading the development, but lately the number of users in Europe and North America has increased. Many things have changed from the initial stages: mobile devices are very different, mobile networks are faster, and billing models have changed. The situation has evolved considerably since the mid-1990s, when personal digital assistants (PDAs) and Nokia communicators provided access to Internet. Technology and business infrastructure build framework for the service adaptation: they can act both as enablers or as barriers. Based on technology, business model, service content, and ease of use, users evaluate if the new technology adds value to their life.

During the past years, I have investigated the question: "How to design and create mobile Internet services that people can use and want to use?" My research has focused on human factors and usability, but I have lately been forced to admit that the amount of knowledge on the subject is not enough. The knowledge and insights must be used in different decision-making situations. Good mobile service creation requires that all three factors: technology, business infrastructure, and users be taken into consideration with the right balance when making decisions. When using knowledge on users in decision making, it is important to understand that different phases of the service development cycle require different kinds of information on users. This is important to realize, as in service development, the teams are multidisciplinary and may have different perceptions on the critical information. Decision making in mobile service development can be seen as a part of risk management: To decrease the risk in decision making, it is crucial to know what information is needed and when, what is the reliable information in different situations, and who can provide that information. Information is already available about user interfaces and interaction design for mobile applications and services. Material on user-centric processes and methods that are suitable in the mobile service development is also available. The challenge is to utilize the available information in decision making.

In this chapter, I will discuss mobile Internet, what it is, what the alternatives are, and how the people use these different alternatives. I will also handle what is Web 2.0 and how the social aspects are linked with mobile devices. In the last section, I handle the mobile service development process and give examples of decisions that companies need to make when creating mobile Internet services.

6.2 What Is Mobile Internet?

The Internet on mobiles can be described in many different ways. In some markets the Internet use on portable computers (laptops) is considered as Internet on mobiles. This is however excluded from my definition: when talking about Internet on mobiles, I discuss specifically about Internet access with mobile phones.

In 2008, four studies were published on mobile Internet use. I use these studies as examples to illustrate how differently the Internet on mobiles can be defined.

Cui and Roto (2008) studied mobile Web usage and their definition of the Web on mobiles is about viewing Web pages with mobile browsers; this covers both mobile-tailored and full Web content. Hinman et al. (2008) compared mobile phone and computer Web use in a computer deprivation study. In this study, the use of the mobile Web is mainly related to full Web site use on mobiles. For Taylor et al. (2008) the mobile Web is mostly about relevant, mobile-tailored content, and services.

The fourth definition of the mobile Web is a combination all three of the previous approaches: Kaikkonen (2008, 2009) defines the mobile Web as an access to the Internet via a mobile device. This not only includes full Web and mobile-tailored content on the browser, but also the cases when mobile applications or widgets interact with online content. This approach can rather be described as Internet access on mobiles than the "mobile Internet." The different alternatives for using and accessing the Web on mobiles today can be seen in Figure 6.1. Web access from mobiles can be divided to browser-accessed and client-accessed. The difference is very clear from the user's perspective. For browser-accessed approaches, there are

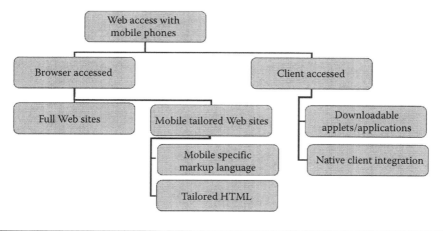

Figure 6.1   Different ways of accessing Web on mobile phones.

two alternatives; a site can be either identical to the one that the user accesses via a desktop computer or the content can be tailored for a mobile platform. Client-access means that applications connect to a service to fetch specific pieces of data from the Web: different approaches support different usage situations, and therefore one service can be accessed multiple ways.

6.2.1 Full Web on Mobile Phones

Full Web sites are sites developed with standard HTML for desktop computer use. The content on a mobile browser is (with some technical limitations) the same as the one the user sees when browsing the site on a desktop computer. Most mobile browsers do not support all audio and video formats; this means that a user may not be able to listen to background music or view video clips on the Web sites. In some cases, full Web site design is optimized for a specific browser, typically Internet Explorer. The layout of such sites may, therefore, look awkward on mobile (or other) browsers to a user who is familiar with the site on a specific browser on a desktop computer.

Full Web content on mobile devices is not really a new thing. It has been possible to access full Web content on mobiles for a long time; as long as it has been possible to access mobile-tailored content. For example, the Nokia Communicator provided a Web browser with HTML support as early as 1996. Especially with devices having relatively small screens, the visualization of the full Web content has been challenging and different solutions have been investigated. Kaasinen et al. (2000) demonstrated ways to render Web content to fit the screen of a mobile phone. Roto and Kaikkonen (2003) analyzed the problems users have when full pages are rendered to a narrow layout when viewed with mobile browsers. Currently, the narrow layout is no longer the only solution; as the size of the mobile phone screens has become bigger and screen resolution has increased, more devices are able to show the Web site layout in a more comparable manner to the layout seen on a desktop computer. Figure 6.2 shows how full Web pages of some social services, such as Facebook, Flickr, and Share on Ovi, look on a mobile device, Nokia N900.

Most Web sites are built with big computer screens in mind, but lately an increasing number of companies have started to take mobile browsers into consideration when building their full Web sites. The question now is "how do you best create Web sites that fit both desktop computers and mobile devices?" This is the question Sounders and Theurer (2008) from Yahoo! have investigated and defined guidelines to help developers to build full Web sites that also work well on mobiles.

6.2.2 Mobile-Tailored Browser Access

Tailoring Web content for mobile phones can be done in many ways, as Figure 6.1 shows. The mobile-tailored content accessed with mobile browser is the oldest way of tailoring content for mobiles. The companies that tailor their Web site allow users to access Internet content with a mobile browser and open Web sites that are

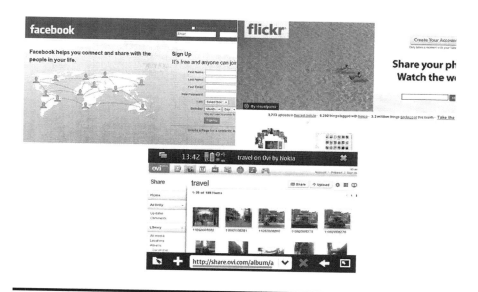

Figure 6.2   Screenshots of Facebook, Flickr, and Share on Ovi on Nokia N900.

Figure 6.3   Screenshots of Facebook, Flickr, and Share on Ovi mobile Web sites on Nokia N95.

designed for mobile phones. The tailoring is not only related to visualization of the site, but also to optimizing the data size and download times.

Figure 6.3 presents how the social networking sites, Facebook, Flickr, and Share on Ovi service may look on mobile browser, when the content is tailored for mobiles. It does not really matter to users how the mobile tailoring has been done; it is mostly invisible for an average user if the site has been developed by using a markup language designed for mobile devices (e.g., HDML, WML, cHTML, or

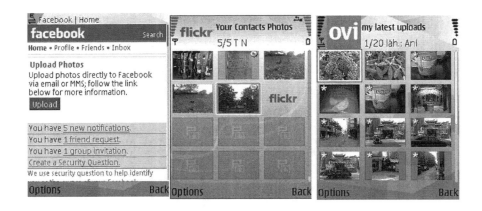

Figure 6.4   Screenshots of Facebook, Flickr, and Share on Ovi mobile applications on Nokia N95.

XHTML) or standard HTML. What is important is that the content and user interface is tailored to suit the mobile use.

6.2.3 Client-Based Access

The other way of tailoring Internet content for mobile use is to develop client applications that access the Internet. Figure 6.4 shows how the social networking sites, Facebook, Flickr, and Share on Ovi service mobile applications or widgets look. The use of this kind of applications has become more common during the past few years. The phone applications can support the upload or download of content to and from the Internet. The Web access can either be an integrated functionality in the phone's native applications (such as the calendar, photo gallery, music player, or phone home screen) or it can be a stand-alone application downloaded from the Web.

These downloadable applications can access specific data from a phone; the applications connect to a specific site for a specific information query or task. For example, they may be used for uploading photos to a photo blog, downloading a game to a mobile phone or to preview content in a music service. The client approach gives user possibilities that are not available in the browser accessed approach; for example, the photo application can automatically store the photos taken with mobile phone to an online service (Vartiainen et al. 2007).

6.2.4 Are There Differences between the Ways of Using Mobile Internet?

In 2007, I conducted a study on mobile Internet use (Kaikkonen 2008, 2009). The study was conducted in two parts: a global online survey with 390 respondents and contextual interviews for 23 users in Hong Kong, London, and New York. There is very little comparison information on mobile Internet use in different countries, but this study reveals some differences. All the survey respondents owned a device with a browser capable of accessing both full Web content and mobile-tailored content. The devices also supported mobile applications and widgets that were connected to Internet. Many of the respondents did take advantage of the browsing capability; almost 70% of the respondents browsed both full Web and mobile-tailored Web sites (very often, Wireless Application Protocol (WAP) portals). Rarely did a respondent report only browsing full Web sites: 14% of the respondents mentioned exclusively browsing sites that were available only as full Web sites. Browsing restricted to mobile-tailored sites was more common: 32% of respondents mentioned only browsing sites that are mobile-tailored. The use of downloadable applications, widgets, and native application-integrated solutions was not very common; 7% of the respondents mentioned use of applications that access the Web.

In my study, I found out that there were differences in the mobile phone Web access between users from different countries: majority of the respondents from Hong Kong browsed only mobile-tailored sites—typically, the WAP portal of their mobile operator. Overall, it was less common for Asian users to browse full Web sites compared to Europeans and North Americans. 23% of Asian respondents only browsed mobile-tailored sites. Similar browsing behavior was reported by 10% of North Americans and only 2% of European respondents. The interviews clarified the reasons behind the behavior: although users everywhere perceived that cost is an issue in terms of mobile Internet use, the interviewees in Hong Kong were especially conscious of the perceived difference in cost between accessing full Web and mobile-tailored Web sites from their mobile devices. Operators in Hong Kong had packaged their phone plans in such a way that the use of the operator mobile portal was part of the phone plan; users paid the same fee whether or not they used the portal but additional costs incurred when accessing other Web sites. As a result, only the users having wireless local area network (WLAN) in their phones browsed full Web sites. Due to the WLAN hotspots, the full Web use was more stationary than mobile site browsing, which happened also in true mobile situations (e.g., when walking or commuting).

The respondents of the online survey were asked to list up to five Web sites they had recently accessed with their mobile browser. Collectively, respondents identified 999 Web addresses: half of these sites were available only as full Web versions; 25% of the addresses led to sites that were clearly mobile-tailored, and the rest to sites available both in full Web and mobile-tailored formats, such as Google and Yahoo! Browsing on mobile-tailored sites was more common than on full Web sites. There was however more diversity with regards full Web sites; many users browsing mobile-tailored sites reported accessing the same operator portals. The users who reported accessing only mobile-tailored sites listed just one or two Web addresses, whereas the respondents who also browsed the full Web on mobile devices reported having viewed at least four sites.

The motivation of mobile Web browsing was mostly related to information search: 60% of the respondents mentioned accessing sites that supported clear information search, for example, the use of search engines, news sites, and news areas of operator portals. Communication (mostly Web-based e-mail) was mentioned by 20% of the respondents, and object handling, like adding text or photos to a blog, was mentioned by 30% of the respondents. All these three categories, information search, communication, and object handling, had both goal-oriented activities and "browsing for fun."

Recently, the display size of the mobile devices has increased, and users do not need to be as concerned about the cost as WLAN and flat fee subscriptions on mobiles have become more common. Even if academic research on subject is still missing, there is some evidence that users of such devices use more Web on mobiles than users of other types of devices. An Australian report (Cleary 2008) shows that iPhone users do not only browse more than users of other mobile phones, but they also browse richer Web sites. This indicates that they browse more full Web sites than the users of traditional mobile phones. The report also shows that iPhone users use more search engines than users of the other types of phones. The statistics are only capable of showing the data, but they do not reveal the reasons; what is the role of the WLAN, flat fee package, display size, interaction style on touch screen or the sales channel. Or can part of the reason rely on the demographic of the iPhone users.

6.2.4.1 Computer vs. Mobile Device

Most respondents in our survey accessed the Internet with both a desktop computer and a mobile phone. Some of the sites people browsed were the same on both PC and mobile. Some activities were the same no matter how users accessed a site, but there were also differences. When seeking information, users generally read news and searched information based on keywords in search engines; on mobile devices they read smaller amounts of text and browsed for a shorter period of time. Users also read e-mail on both mobile and desktop computers; on mobile devices, however, users read more e-mails than they wrote. Users also avoided reading very long e-mails on mobile phones if they were not essential; if they needed to write an e-mail on a mobile phone, their responses were typically shorter than on a desktop computer. Even though the responses were short on mobile devices, they were no less important than the longer ones written on a desktop computer. Many respondents reported that they followed blogs and discussion group conversations on mobile devices. Writing to, and active participation in, social sites was less common on mobile devices than on desktop computers. One could assume that this is mainly due to the small screen and numeric keyboard but although these do influence behavior, the interviews revealed more reasons: mobile Web sessions were shorter and more prone to interruptions than sessions on a desktop computer, so the latter was considered more appropriate for participating actively in social sites where one needs more time and peace. Even if some smart phones support multitasking, it was not common to leave the social Web site on the background while doing other chores, checking occasionally if there is something new and interesting, like people do on computers.

6.2.4.2 Full Web on Mobiles

When interviewing the smart phone users, we found out multiple reasons why they browse full Web sites on mobiles. Users may look for specific information that is only available as a full Web version (there is no mobile-tailored site or the mobile-tailored site does not, or users were not aware of a mobile-tailored solution). Information need on full Web browsing is often time critical, and the context in which information is needed does not allow the use of a desktop computer: either there is no desktop computer available, the social context does not allow the use of a desktop computer, or the user knows he or she will be changing location during the information search. When browsing full Web sites on a mobile device, users' information needs are often specific, and they know where to find the information. This applies also to the use of the social sites. Very often the depth of the information is more profound than when browsing on mobile-tailored sites. The motivation for use is less related to killing time than when browsing mobile-tailored sites. User experience when browsing on full Web sites with mobile is not always perfect: Vigo et al. (2008) note that browsing on full Web sites with mobiles can cause warnings and failures when pages are downloading. This kind of experiences can decrease the willingness to browse full Web sites later. The number of advanced devices is growing, but the service developer must keep in mind that there are lots of different kinds of devices in use with less advanced capability to show full Web sites.

6.2.4.3 Browsing Mobile-Tailored Sites

Even if information search is also common when browsing mobile-tailored sites, it was more often related to killing time or browsing interesting information than the full Web site browsing on mobiles. When browsing on mobile tailored site, it is likely that when a user sees something interesting, he or she delves deeper into the information. The mobile sites are smaller than full Web sites, so users were not concerned on wasting money if the new page did not contain that useful information. The main page of the mobile-tailored Web site was often regarded as the safe haven when browsing; users returned there to start the next exploration.

6.2.4.4 Internet Accessible Mobile Applications and Widgets

Internet use via mobile applications is becoming more popular and already in 2007, there were users using such applications. When using Internet via applications that can access the mobile services, a user has a limited set of tasks he or she can perform. Usually the user cannot browse outside the specific information source or service. The common use cases reported were related to time- and situation-critical activities, for example, uploading photos to photo-sharing sites or checking time-critical information. Often these situations were related to social activities, either with a group of friends physically present or with friends present online.

However, there were also time killing use cases, as some users had installed applets that fetched the comic of the day or other less goal-oriented material.

6.3 What Is Web 2.0?

Web 2.0 is often considered to have social approach to content generation and sharing, as well as participation in service development (O'Reilly 2005, 2007). Open communication and free distribution of the content are the key characteristics. In early days of Web 2.0, the social networking services were accessible only with computers, but an increasing number of them can be accessed with mobiles as well. They may have even mobile-tailored Web sites or Web accessible applets or applications developed to support the service use. Services like Flickr (www.flickr.com), Facebook (www.facebook.com), and even Wikipedia (www.wikipedia.com) are examples of Web 2.0 services.

6.3.1 Web 2.0 and Mobiles?

From the very beginning, mobile devices have been about social interaction. It is the ultimate reason for people to carry these devices with them, the reason why these devices are considered to be trusted and personal, just like our relationship with friends. First the voice call and later the text message changed the way we communicate and interact with friends. From this perspective the merge of mobile phones and the Web is very natural now, when the Web is increasingly about social interaction.

In early days of the mobile Internet, Odlyzko (2001) showed that content is not the real king. People have always been willing to pay more for communication than content. Even if the content does have value, the social interaction is even more important. This is why Odlyzko claims that the connectivity is the real king. Odlyzko predicted in 2001, that the future of the Internet will be about person to person communication. He did not predict the social networking in the way Internet provides today, but he was clearly on the right track. Odlyzko also pointed out that "growing storage and communication capabilities will be used often in unexpected ways." When given freedom of creativity, people start using the technical systems in unexpected ways.

The mobile devices of today allow users to create content in multiple ways; devices have cameras that can be shared with others, audio and video can be recorded, and there are means to edit pictures and draw your own pictures. This user-generated content, however, cannot be regarded as "content" in the way Odlyzko refers to the content. The user-generated content should be handled as a way of communication, rather than content as such. This is well expressed by comments from two interviewees in a mobile Web study I conducted in 2007 (Kaikkonen 2008, 2009):

> Picture's worth a thousand words, so you can take a picture and you don't need any comment on it at all.

> **Interviewee in a mobile Web study**

People communicate 70 percent with the bodies, and 30 percent with speech, so the pictures are something like body language, even if you send a flower, it's not just that, you need the language as well to make someone happy. We're sending electronic postcards, which is one of the things that make someone happy.

Another interviewee in a mobile Web study

As people are carrying mobile devices with them to multiple places, sharing moments with friends and loved ones can have a new dimension. You can share at that moment what you are doing and how you feel. In situations when you are in the middle of the activity, and not at home, work, or any other place where you can access the computer. The memory of exact emotion and experience fades fast. William James (1961) says that [religious] experience lasts 30 min or even less, and when the situation is over, it is difficult to bring back how it really felt. Having the possibility to share the meaningful experience in the moment with the help of mobile devices can change our communication patterns again, like mobile phones have done already twice in past, with voice calls and text messages.

6.4 Barriers to Internet on Mobile Devices

When designing for Internet on mobiles, the barriers to the use should be taken into consideration. The Forrester report (Lakshmipathy 2007) says that Internet on mobiles suffers from three major problems today: (1) Content is hard to find. The Forrester report talks here about the mobile-tailored content. (2) Usability is poor: typing with mobiles is difficult, mobile-tailored sites are not properly designed for mobile use, but rather modified slightly from computer Internet offering. There is also lack of consistency between the sites and within the site. The downloadable applications provide better usability, but the applications are hard to find. (3) Access to Internet on mobiles costs a lot and users do not understand how the cost is generated. This would mean that the services are not only difficult to use, but there is no reason to use them and using the services is like giving an open check to the network operator.

Browne (2007) has come to a similar conclusion and says that Internet on mobiles suffers from three major barriers in user friendliness, covering some of the problems pointed in the Forrester report: (1) mobile search is inaccurate, (2) carrier/operator portals are ambiguous, and (3) input is cumbersome. Browne gives five examples of how Japanese companies have made it easier for users to get mobile Internet content; most of these are related to finding the content. Japanese companies provide barcodes as shortcuts, applications are preinstalled to phones, Web addresses are simplified, and mobile phone input is taken in consideration in addresses. Tags are pushed to users and there are multiple paths to the sites. Links to mobile Web sites are sent to users by mobile e-mail, for example. According to Browne, these help mitigating all three barriers he mentions.

If the Internet on mobiles is looked from the developers' viewpoint, the scene has not improved much from the early days: the number of alternatives has grown, there are more devices than earlier, and the devices are even more diverse. There are devices from basic mobile phones to very powerful devices with full QWERTY keyboard and good quality display. These different devices and browsers are presenting user interface differently, and they support different application creation technologies. There are various ways to connect to the network, some phones have very slow connection, other phones have WLAN, and some others have fast networks. The network speed influences, for example, the kind of graphics that can be used in the service. A developer has to do lots of work when deciding how to develop a mobile service for as big a group of users as possible. In order to get enough content for Internet use in mobiles, it has to be easy enough for developers to create it by using standard tools that allow creation for multiple platforms.

6.5 About User-Centered Design Method Selection during Mobile Service Development Process

Since the early 1990s, interest in Internet on mobiles has been growing. Many researchers have conducted studies, from technical, business, and human factors perspectives. Information is available on technologies, business models' influence on consumer behavior, and on user interface design for small screens. There is also information on user-centric design process, in general, and specifically for services. When designing for mobiles, the same processes can be used, but at each step there are mobile-specific questions that need to be answered.

Many authors have shown that some user study methods work better in a specific phase of product or service development. Maguire and Sweeney (1989) have matched the methods and type of data captured with the method. Their evaluation approach table lacks contextual methods, which is understandable when taking into consideration the period of their work. Nielsen (1992), on the other hand, matched the task and the specific development phase when the activity is recommended to be done. Muller et al. (2000) have defined participatory design methods in relation to user activity and the position of the activity in development cycle.

The approach that is still missing is the one matching the specific questions that need to be answered during the service development, specific data on users, and methods used during the different phases. Ketola and Roto (2008) have shown that practitioners in different roles need different information on users. The reason why they need different information is that these professionals work in different phases of the product or service development or they are responsible for different issues. The practitioners need information on users to help in their decision making. It is clear that professionals in different positions make different kinds of decisions in their work. A product manager has his or her own decisions to make that differ from the decisions of a designer, an implementer, or a marketing manager. Different methods give answers to different questions; therefore, multiple methods are needed during the product development process. Table 6.1 shows examples of the information needs of humans and methods mapped to different phases of development process. The process used in the mapping is the one described by Garrett (2003).

Table 6.1   Decision Making Plane, User Information, and Method in Service Development Process

Decision Making Planes (According to Garrett)	Information on People	Sources of Information during Development	Example of Evaluation Methods
Surface plane: visual design	Psycho physiological information: function of hearing, visual system	Expert consultation, literature references	Usability test, expert evaluation, eye movement studies
Skeleton plane: interface/ navigation design	Cognitive information: information on memory, perception	Expert evaluations, literature	Usability test with verbal protocols
Structure plane: interaction/ information design	Cognitive information: perception	Expert evaluations, contextual studies, observations, task analysis	Usability tests, field pilots
Scope plane: functional specifications/ content requirement	Behavioral information, goals	Contextual studies, observations, diary studies	Contextual studies, observations, diary studies
Strategy plane: user needs/site objectives	Social behavior, emotions, consumption patterns, attitudes, behavior/ attitudes of large groups	Contextual studies, theme interviews, diary studies, group interviews, surveys trend analyses, customer segmentation	Contextual studies, theme interviews, diary studies, group interviews, surveys

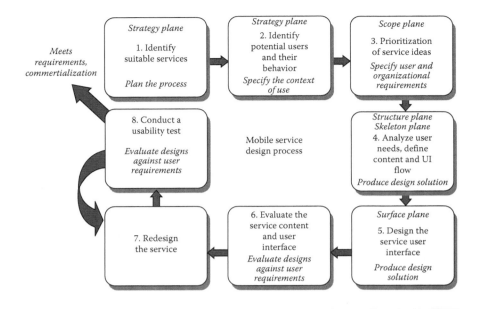

Figure 6.5   Consolidation of three user-centric design processes.

The service development processes advocated by Kaikkonen and Williams (2000) and Garrett (2003) look very different from each other at first look—and it might be difficult to match them with the framework provided by ISO 13407 (human-centred design processes for interactive systems). These three different approaches, however, only complement each other. The different planes of the Garrett model and phases of ISO 13407 can be matched to the mobile service design process by Kaikkonen and Williams. In Figure 6.5 the steps defined in original model of Kaikkonen and Williams are numbered. The planes of Garrett are in black italics and ISO 13407 phases are in grey italics. Garrett's model focuses on steps that include service-related decision making, whereas ISO 13407 takes the evaluation as part of the process. The mobile aspect of mobile service design process may not be evident in a high-level picture, but each step includes aspects that are relevant specifically in relation to mobile services: when identifying potential users and their behavior, we need to take into consideration also the mobile devices users own currently. When prioritizing the service ideas, the users' infrastructure and experiences with mobile devices and services play a big role in prioritization process. It is important to take these questions into consideration, as well as the core strategic question: what is the role of the mobile in the company's service portfolio. Very few companies have only mobile services, and the mobile service has to support the total offering.

6.6 Making Decisions during the Mobile Service Design Process

Each step in the design process described in Figure 6.5 has some mobile-specific questions that need to be answered. The existing information on users help in initial decision making, but evaluation of the decisions is always recommended, as technology developers should not rely only on their intuition. Heuristics, or predefined guidelines, help also in initial decision making; example of mobile service heuristics is the set defined by Väänänen-Vainio-Mattila and Wäljas (2009). According to them, a good mobile service allows a user to use and create a composite service; it allows cross-platform access, social interaction, and navigation, as well as dynamic service features, context-aware services, and contextually enriched content. Most of these heuristics are related to the first phases of the service creation process: service and target user identification and service idea prioritization. Or as Garrett (2003) defines it: the service strategy plane and scope plane.

6.6.1 Identifying Suitable Services

Most services have multiple ways of accessing the information: the company may have a full Web site that is designed to be used with a computer having reasonably large screen and fast Internet access. When designing mobile services, the decision to make is related to the company's service strategy: What is the role of the mobile service in the overall portfolio and is the service mostly about mobile, like Sports Tracker? Is the mobile just supporting the full Web use but the primary way of using the service is with computer, like Flickr or Facebook mobile versions? The company's service strategy should be leading the decision making, taking into consideration that people use mobile services differently depending on the way it is built. When developing the mobile services in markets where the role of the computer is minor, unlike Europe and North America, one cannot rely on availability of computers in a users' life.

6.6.2 Identifying Potential Users and Their Behavior

The key question in all development is to define who the target users are. You cannot have too much information about the demographic of the users, or too good insight on what they do in everyday life, how they do things currently, for example, how and when do they interact with other people, what kind of devices they have, especially mobile devices, and what is the overall infrastructure. It is very different to design a mobile service for people in a specific country than to design "a global" service. The daily chores are done differently in different countries; for example, commuting in North America may mean driving in a car solo, while in Netherlands it may mean riding a bicycle and in Singapore with public transportation. Users in African countries may not have computers or the wired network is less reliable than mobile network. The decision on whether the multiplatform access is required and in what form should be based on the target users and the environment they live in.

6.6.3 Prioritization of Service Ideas

The information on target users and their behavior should be used in service prioritization decision making. The best mobile services support the users' current behavioral patterns or answer the existing needs of the users. The mobile way of doing things should either remove the obstacles users have currently or combine relevant information to create new content. Good examples of mobile services are micro blogs like Twitter, Jaiku, and Sports Tracker. Twitter and Jaiku answer the social interaction needs humans have, while Sports Tracker supports current behavior of active people and combine the data gathered in a way that it creates new information. The questions in service prioritization should be as follows: Does the service idea save people's time or effort? Does it bring new information and value to everyday life? and Can other people interact and share, even if they did not have the latest gadgets?

6.6.4 Analyze User Needs, Define the Content and UI Flow

When the service ideas are prioritized, it is time to define the user scenarios related to services. This involves not only analyzing the content of the service, but also investigating what functions are necessary in different service platforms, what are the key scenarios in mobile usage situation, and what are the key scenarios when a user has other than a mobile device available. These are needed when defining the content in different platforms: Internet on computers, full Web on mobiles, and different mobile-tailored solutions. Elina Vartiainen describes in Chapter 2 what benefits the applications or widgets have compared to the browser accessible approach to the service. If there is both mobile-tailored Web site and application or widget, there has to be good reason for that and clear roles for both. Even if there are differences in different platforms—in interaction, possibly also in content—the usage flow needs to be the same in all platforms, and familiar elements should be used. When starting to design the service for mobiles, it would be good to define the core elements beforehand.

6.6.5 Design the Service User Interface

There is lots of information on mobile service user interface design or designing for small screens. Over the years, many researchers have addressed the mobile service user interface design, interaction design, and usability: Buchanan et al. (2001), Chittaro and Dal Cin (2002), Kim et al. (2002), Kaikkonen and Roto (2003), Hyvärinen et al. (2005), among many others. Elina Vartiainen describes in Chapter 2 the limitations that need to be taken into consideration, whether the approach is to design browser accessible service or an application.

Depending on the selected approach, the guidelines for this specific platform should be followed. It is always good to search for generic guidelines instead of a specific device/browser-related guidelines. Sounders and Theurer (2008) has defined guidelines to help developers to build full Web sites that also work well on mobile browsers even if the service has mobile-tailored version as well. These kinds of guidelines are good to follow.

6.6.6 *Evaluation of the Service Content and User Interface*

No matter how experienced the design team is, it is always good to evaluate the service with actual target users. It is well known that relying on the instincts of design team is not a good idea, as they may go wrong, as noted by Dan Russell in the article by Bill Slawski (2006). The basic procedure of usability tests applies also when testing on mobiles, but there is a particular question that needs to be taken into consideration in the mobile area: what can I test in the lab and when should I go to the field? Once miniature cameras became available, researchers have used them in different ways to study users and services in real-life environments. Several researchers have sought to answer the question, whether field testing is worth the effort or is it enough to test in the lab. Kjeldskov et al. (2004), Holtz Betiol and de Abreu Cybis (2005), Baillie and Schatz (2005), Kaikkonen et al. (2005), and Duh et al. (2006) are researchers who have published papers on comparative studies, where the answer to the question has been investigated. The outcome of the studies has been diverse: some researchers have found differences, some not. According to Kaikkonen et al. (2008) it is not worth the effort to conduct usability studies of mobile applications or devices in the field in order to find usability problems alone. Here, the usability problems are defined as in the ISO 9241-11 standard. However, when the goal is to gain knowledge about user behavior in a natural environment, for example, to understand where users might use the service, field tests can reveal information that cannot be gained in a laboratory. There are also applications and services that need to be tested in the field, for example, navigation systems cannot be tested properly in a laboratory environment.

6.7 Conclusions

Despite the need for further information on mobile service design and development, we already have material to work with. The main challenge when developing the services for Internet on mobiles is to use the existing user data in the decision-making process. During the decision-making process it is important to understand the granularity of the information: the different kinds of information on users are needed in different phases of the development process, and decisions in all phases should be in line with the strategic goal of the service. The right mobile-specific questions need to be answered in different decision-making situations. The decision making is especially challenging when designing the mobile service for multiple markets, as mobile devices have different roles in people's lives in different countries. It may be difficult for designers, developers, and management to make decisions when they have difficulties understanding that some end users use mobile devices and Internet services differently from what they see in their own environment. There is reason why there are differences between countries in the technical and business infrastructure and naturally users are evaluating what is good for them in their real context. Understanding users in different markets is challenging, as the development team has to learn to know when to rely on its instincts and when that is not a good idea.

References

Baillie, L. and Schatz, R. (2005). Exploring multimodality in the laboratory and the field. In *Proceedings of ICMI 2005*, Trento, Italy.

Browne, S. (2007). How Japanese companies guide their customers to mobile Internet experiences. *Forrester Best Practices*, April 23, 2007.

Buchanan, G., Farrant, S., Jones, M., Thimbleby, H., Marsden, G., and Pazzani, M. (2001). Improving mobile Internet usability. In *Proceedings of the WWW10*, Hong Kong, China, May 1–5, 2001.

Chittaro, L. and Dal Cin, P. (2002). Evaluating interface design choices on WAP phones: navigation and selection. *Personal and Ubiquitous Computing*, 6:237–244.

Cleary, J. (2008). Australian mobile Internet insight, impact of iPhone. Amethon Solutions Report, Downloadable in: http://www.amethon.com/SiteMedia/w3svc488/Uploads/Documents/9e1927e2–7e6f-4d1c-8f2a-46f06f55cfb7.pdf

Cui, Y. and Roto, V. (2008). How people use the Web on mobile devices. In *Proceedings of World Wide Web Conference*, Beijing, China, 2008, pp. 905–914.

Duh, H. B.-L., Tan G., and Chen, V. (2006). Usability evaluation for mobile device: A comparison of laboratory and field tests. In *Proceedings of MobileHCI 2006*, Espoo, Finland.

Garrett J. J. (2003). *The Elements of User Experience User-Centered Design for the Web*. New Riders, New York.

Hinman, R., Spasojevic, M., and Isomursu, P. (2008) They call it "surfing" for a reason: Identifying mobile Internet needs through PC deprivation. In *Proceedings of the 26th Annual SIGCHI Conference on Human Factors in Computing Systems, CHI'08 Extended Abstracts*, Florence, Italy, pp. 2195–220.

Holtz Betiol, A. and de Abreu Cybis, W. (2005). Usability testing of mobile devices: A comparison of three approaches. In *Proceedings of Interact 2005*, Rome, Italy.

Hyvärinen, T., Kaikkonen, A., and Hiltunen, M. (2005). Placing links in mobile banking application. In *Proceedings of the Seventh International Conference on Human Computer Interaction with Mobile Devices & Services (MobileHCI 2005)*, Salzburg, Austria, pp. 63–68.

ISO 9241-11 (1998). Ergonomic requirements for office work with visual display terminals (VDT)—Part 11: Guidance on usability, ISO, Geneva, Switzerland.

ISO 13407 (1999). Human-centred design processes for interactive systems, ISO, Geneva, Switzerland.

James, W. (1961). *The Varieties of Religious Experience*. Collier Books/Macmillan Publishing Company, New York.

Jaiku micro blog site, http://www.jaiku.com/ (accessed September 29, 2009).

Kaasinen, E., Aaltonen, M., Kolari, J., Melakoski, S., and Laakko, T. (2000). Two approaches to bringing Internet services to WAP devices. In *Proceedings of the Ninth International World Wide Web Conference on Computer Networks*, Amsterdam, the Netherlands, pp. 231–246.

Kaikkonen A. (2008). Full or tailored mobile Web—Where and how do people browse on their mobiles? In *Proceedings of the International Conference on Mobile Technology, Applications, and Systems* (*Mobility'08*), Yilan, Taiwan, Article No. 28

Kaikkonen A. (2009). Mobile Internet—Past, present and the future. *International Journal of Mobile Human Computer Interaction* (*IJMHCI*), 1(3):29–45.

Kaikkonen, A. and Roto, V. (2003). Navigating in a mobile XHTML application. In *Proceedings of Human Factors in Computing Systems Conference* (CHI'03), Fort Lauderdale, FL, pp. 329–336.

Kaikkonen, A. and Williams, D. (2000). Designing usable mobile services. In *SIGCHI Conference on Human Factors in Computing Systems* (*CHI2000*), The Hague, the Netherlands, Tutorial Notes (2000).

Kaikkonen, A., Kekäläinen, A., Cankar, M., Kallio T., and Kankainen A. (2005). Usability testing of mobile applications: A comparison between laboratory and field testing. *Journal of Usability Studies*, 1(1):4–16.8

Kaikkonen, A., Kekäläinen, A., Cankar, M., and Kankainen, A. (2008). Will laboratory test results be valid in mobile contexts? In Lumsden J. (ed.), *Handbook of Research on User Interface Design and Evaluation for Mobile Technology*, IGI Global, New York, pp. 897–909.

Ketola, P. and Roto, V. (2008). Exploring user experience measurement needs. In *Fifth COST294-MAUSE Open Workshop on Valid Useful User Experience Measurement* (*VUUM*), Reykjavik, Iceland, 2008.

Kim, H., Kim, J., Lee, Y., Chae, M., and Choi, Y. (2002). An empirical study of the use contexts and usability problems in mobile Internet. In *Proceedings of the 35th Hawaii International Conference on System Sciences 2002*, Big Island, HI, pp. 1–10.

Kjeldskov J., Skov, M. B., Als, B. S., and Høegh, R. T. (2004). Is it worth the hassle? Exploring the added value of evaluating the usability of context-aware mobile systems in the field. In *Proceedings of the 6th International Mobile HCI 2004 Conference*, Glasgow, Scotland, pp. 61–73.

Lakshmipathy, V. (2007). What's wrong with the mobile Web? *Forrester Best Practices*, February 12, 2007.

Maguire, M. and Sweeney, M. (1989). System monitoring: Garbage generator or basis for comprehensive evaluation system. In A. Sutcliffe and L. Macaulay (eds.), *People and Computers V* (*Proceedings of the HCI'89 Conference*, Nottingham, U.K., September 5–8, 1989), Cambridge University Press, Cambridge, U.K., pp. 375–394.

Muller, M., Lafenière, D., and Dayton, T. (2000). Card games for participatory analysis and design: Variations on a theme. In *SIGCHI Conference on Human Factors in Computing Systems* (*CHI 2000*), The Hague, the Netherlands, Tutorial Notes.

Nielsen, J. (March 1992). The usability engineering life cycle. *Computer*, 25(3):12–22.

O'Reilly, T. (2005). What is Web 2.0, O'Reilly Media, Inc., http://oreilly.com/web2/archive/what-is-web-20.html

O'Reilly, T. (2007). What is Web 2.0—Design patterns and business models for the next generation of software, *Communications & Strategies*, 1:17–37.

Odlyzko, A. (February 2001). Content is not king. *First Monday*, 6(2), URL: http://firstmonday.org/issues/issue6_2/odlyzko/index.html (accessed April 13, 2009).

Roto, V. and Kaikkonen, A. (2003). Perception of narrow Web pages on a mobile phone. In *Proceedings of International Symposium on Human Factors in Telecommunications 2003*, Berlin, Germany, pp. 205–212.

Slawski, B. (2006). Why do people Google Google? Understanding user data to measure searcher intent, Search Engine Land, http://searchengineland.com/why-do-people-google-google-understanding-user-data-to-measure-searcher-intent-10091 (accessed September 23, 2009).

Sounders, S. and Theurer, T. (2008). High performance Web sites 14 rules for faster-loading pages, Yahoo guidelines, http://www.techpresentations.org/High_Performance_Web_Pages (accessed April 13, 2009).

Taylor, C. A., Anicello, O., Somohano, S., Samuels, N., Whitaker, L., and Ramey, J. A. (2008). A framework for understanding mobile internet motivations and behaviors. In *Extended Abstracts of Human Factors in Computing Systems Conference CHI 2008*, Florence, Italy, April 2008, pp. 2679–2684.

Väänänen-Vainio-Mattila, K. and Wäljas, M. (2009). Development of evaluation heuristics for Web service user experience. In *Extended Abstracts of CHI 2009*, Boston, MA, April 4–9, 2009, pp. 3679–3684.

Vartiainen, E., Roto, V., and Popescu, A. (2007). Auto-update: A concept for automatic downloading of Web content to a mobile device. In *Proceedings of the Fourth International Conference on Mobile Technology, Applications, and Systems and the First International Symposium on Computer Human Interaction in Mobile Technology*, Singapore, 2007, pp. 683–689.

Vigo, M., Aizpurua, A., Arrue, M., and Abascal, J. (2008). Evaluating Web accessibility for specific mobile devices. In *W4A2008—Technical, Co-Located with the 17th International World Wide Web* Conference, Beijing, China, April 21–22, 2008, pp. 65–72.

Chapter 7

UrbanWeb: A Platform for Mobile, Context-Aware Web Services

Frank Allan Hansen and Kaj Grønbæk

Contents

7.1 Introduction

On the traditional Internet, social computing or so-called Web 2.0 applications have become widespread in recent years. The term "Web 2.0" was coined by Tim O'Reilly in 2005 [34] and is often associated with applications such as wikis, blogs, photo- and video-sharing sites, social-networking sites, Web-based communities, Web services, mash-ups, and folksonomies. All of these Web 2.0 applications share some common aspects, including collaboration, interoperability, user-created information, and information sharing on the World Wide Web. The same trend is now spreading to the mobile Internet, for example, through moblogs [1,11], various location-based mobile hypermedia services [12], and pervasive games [16,37]. At the same time, advances in bandwidth and new mobile devices are promising access to information on the Web from anywhere, at anytime, and on any device, thus allowing users to access and produce information even when mobile. While this promise sounds tempting, it has become evident that mobile devices do not support the same ease of interaction as we have come to know and expect from desktop computers. This is mostly due to the smaller screens sizes and limited input capabilities, resulting in less than optimal user experiences when accessing Web services through the traditional interfaces. Furthermore, a user is never anywhere, anytime, or using any device, that is, users are always at a specific place, at a certain time, and using a particular device, and our activities are anchored within a physical and digital *context*, which changes as we move about or engage in different activities. Based on this observation, we focus in this chapter on how to develop *context-aware* Web applications that take the user's context into account in order to create applications that are more optimized toward mobile devices and that make it easier to present information that are relevant to the user in his or her current situation. Cues about the user's context can be acquired from a number of places. We provide several examples on how to utilize location information (e.g., from GPS sensors) and physical TAGs (e.g., 2D-barcodes or RFID-codes) to support Web 2.0 applications that are anchored into a user's context and provide information tailored to the physical environment (Figure 7.1).

Another aspect that has become important with the dawn of Web 2.0 is the "Web as a platform." While some regard the mobile Web as being restricted to browser-based access [19] (and indeed much work has been done in this area, e.g., in the form of several document formats for mobile browsers such as inline HyperText Markup Language [iHTML] for information Mode [iMode], wireless markup language [WML] for wireless application protocol [WAP], compact HTML [cHTML], and extensible HTML [XHTML] Mobile Profile), it has become very central to Web 2.0 that services should provide interfaces for both the human-readable Web and for the programmable Web. The human-readable Web typically provides an HTML representation, which can be rendered directly in Web browsers, whereas the programmable Web is based on formats such as extensible markup

Figure 7.1 A 2D-barcode as physical link marker to information.

language (XML), JavaScript Object Notation (JSON), custom formatted plain text or binary documents. Some of this data may also be used directly to render HTML pages and some may be consumed by other services and "mashed up" to create new kinds of presentations or application. The similarity between these services, however, is that they all communicate over the hypertext transfer protocol (HTTP)—if a service does not use HTTP it is not a Web service—other than that, the Web as platform is very open and extensible.

The idea of the programmable Web or the machine-readable Web is not new to Web 2.0, but has been one of the foundations in the Semantic Web visions [2] as well, where software agents consume information from various interlinked sources and compute new connections and information and present this to the Web users in meaningful ways. In the context of the mobile Web, the World Wide Web Consortium (W3C) has also emphasized the idea of "One Web" [18]. One Web means that the Web should not be regarded as a number of separate systems, for example, one for the desktop, one for mobile devices, and one for other classes of devices, but should be seen as a single unified platform where the same information and services are available to users irrespective of the device they are using. In the recommendations for One Web, W3C stress, however, that this does not mean that exactly the same information should be available in the same representation across all devices. Both the user's context, the capabilities of the mobile device, bandwidth issues, and the nature of the service or information are factors that may affect the representation ([18], Section 3.1). As a simple example, consider the variation between Web user-agents on current mobile devices, such as mobile phones. Some phones come with rather limited HTML or WAP browsers, while others come with full-featured implementations that support rendering of entire Web pages designed for the desktop and have full JavaScript support (e.g., many of the browsers based on the open source WebKit browser engine [39], like Apple's Mobile Safari, Goggle's Android browser, and Nokia's S60 browser). Other phones come preloaded with "Web Widget" frameworks [20,21], which often support close integration with the underlying mobile platform for special purpose Web applications. And finally, other platforms support development of native (or nonbrowser-based) applications in language such as C, Java, or Adobe Flash, which have built-in HTTP support and thus can easily utilize the Web infrastructure. These applications may require totally different representation of the data raging from pure HTML to XML or JSON, but nevertheless they are all valid entry points to the mobile Web and should therefore be supported by mobile Web services.

In this chapter, we describe a framework, UrbanWeb, for mobile context-aware Web services. UrbanWeb has been designed to support urban computing applications. Urban Computing [29] is an emerging research field that focuses on architecture, social interaction, and design of computer systems for use in urban areas where users may be living, working, or visiting and thus has very different needs for information. Thus, urban computing shares many qualities with Web 2.0 applications and we try to combine these in the UrbanWeb framework, for example, by applying Web technologies such as mash-ups (service aggregation), collaborative tagging [17], social filtering [8], etc.), in order to create new social services for urban environments such as blog and Wiki-inspired Web applications, where information is *consumed, produced, and shared* by users and—unlike their traditional Web-only counterparts—are anchored in the physical contexts they are concerned with.

UrbanWeb consists of a conceptual model for context and context-awareness, based on an extension of the notion of tagging in Web 2.0, and an implementation of the model in a framework that makes it easy for developers to create new service and application. The framework has been designed to provide a number of lightweight Web services for different mobile devices and in order to provide a rich variety of ways to anchor digital information in the physical environment, the framework design has been inspired by previous work on open hypermedia [22], geo-spatial hypermedia [9], and context-aware hypermedia [10,12]. Thus, the framework model and structuring mechanisms have been very much informed by the work in the hypermedia community. In the following sections, we describe the framework design and a number of services that have been built with the framework.

The remainder of the chapter is structured as follows: Section 7.2 introduces the notion of context and context-awareness and develops the conceptual model for context that is used in the UrbanWeb framework. Section 7.3 describes the implementation of the model and the UrbanWeb framework. Section 7.4 discusses the use of the framework and presents a number of different context-aware Web applications that has been developed with UrbanWeb and finally Section 7.5 concludes the chapter.

7.2 Defining Context for Mobile Services

The notion of computers responding according to their users' implicitly stated context is an intriguing and challenging one. In this section, we describe our conceptual model for context-awareness for mobile services. However, before presenting the model we discuss a number of previous efforts to define and model context.

7.2.1 Context-Aware Computing

Context-aware computing refers to software systems that can adapt their behavior, interface, and structures according to a user's context. The idea of utilizing information about the user's context and specifically location information in computing systems was advocated by Mark Weiser in his seminal 1991 paper on ubiquitous computing [40] and some of the early research on location-aware and context-aware computing was performed on the Olivetti Active Badge systems in 1992 [39]. However, the term "context-aware computing" was not defined until 1994 by Schilit et al. [26,27].

Information about a user's context can be made available to the system either explicitly by the user (e.g., by entering a login name) or implicitly through associated sensory systems. Adding context-aware capabilities to systems is desirable for several reasons: first of all, interaction with applications can be greatly improved if an application adapts to a user's situation and only provides information relevant to that situation. For example, filtering information about a city based on a user's location can make it much easier for users to gain an overview, especially given the limitations in terms of screen size and interaction with Web browser user interfaces on small, mobile devices [12]. Second, knowledge about the physical environment surrounding a user can be used to create new powerful applications. The Web, for instance, has traditionally been concerned with linking digital resources stored on computers, but the ability to identify physical resources through tagging or sensor input enables Web systems to also support linking to locations and physical objects, as we discuss in this chapter.

Generally, context-aware computing functionalities or *context-awareness* can be categorized as belonging to one or more classes of functionality. Schilit et al. [27] present a taxonomy of context-aware applications that are split into two orthogonal dimensions. The first dimension focuses on whether the task at hand is getting information or carrying out a command, and the second dimension concentrates on whether the action is effected manually or automatically. The taxonomy includes the functions: *proximity selection, automatic contextual reconfiguration, contextual information,* and *context-triggered actions.* Proximity selection is described as a user-interface technique where objects nearby are emphasized or otherwise made easier to choose. Automatic contextual reconfiguration is the process of adding or removing components (e.g., servers) or changing connections between components based on context information. Contextual information and commands describes the classes of applications, which use context information to decide what information to present to the user and what commands to execute (or how to execute them). Finally, context-triggered actions are based on the idea of expert systems with simple IF-THEN rules used to specify how context-aware systems should adapt. Pascoe [23] proposes another taxonomy of context-aware capabilities. This taxonomy contains four core capabilities: *contextual sensing, contextual adaption, contextual resource discovery*, and *contextual augmentation*. Contextual sensing is the most basic capability allowing the system to detect environmental states through a sensory system and present them to the user (e.g., as Schilit et al.'s proximity selection), contextual adaption utilizes the context information to adapt their behavior to the current situation, contextual resource discovery describes the systems capabilities to use context information to find other systems in the same context and exploit these resources, and contextual augmentation is the process of augmenting a (physical) context with (digital) information by associating context information to the data. Dey [7] combines and generalizes Schilit's and Pascoe's definitions even further and proposes three general types of context-aware application support: *presentation of information and services, automatic execution of services*, and *tagging of context to information to support later retrieval*. Dey's notion of presenting information and services not only captures Schilit et al.'s categories of proximity selection and contextual information and commands, but also Pascoe's contextual sensing (and presentation) capability. Automatic execution of services corresponds to Schilit et al.'s context-triggered actions and Pascoe's contextual adaption. Dey's last category, tagging of context to information to support later retrieval, is also a part of Pascoe's contextual augmentation. Finally, Pascoe's contextual resource discovery is a requirement for adding or removing components as described by Schilit et al.

As we find Dey's categories to be the most general, we have focused on supporting these mechanisms in the UrbanWeb framework, especially on how to tag context information to other information and on mechanisms that allow us to present relevant information and services to users, based on the their current context. Before describing how we represent context information and our context-aware functionality we discuss previous definitions of context and different context models in order to motivate the approach taken in the UrbanWeb framework.

7.2.2 Previous Definitions of Context

When first introducing the notion of context-aware computing, Schilit et al. listed three important aspects of context: where you are, whom you are with, and what resources are nearby [26,27]. Schilit et al. defined context as a location and its dynamic collection of nearby people, hosts, and devices. A number of similar definitions exist that all try to define context by enumerating examples of context elements [3,6,24]. However, as discussed by Dey [7], these types of definitions may be too specific and hard to apply when trying to define general support for context-awareness: for example, how is it decided whether a type of information can be regarded as context if it is not listed in the definition, and how does a system handle new types of contexts if its design is based on a fixed set of types? Later, definitions of context became more general. Schmidt et al. [28] use the following definition: "[Context is defined as] knowledge about the user's and IT device's state, including surroundings, situation, and to a less extent location." Chen and Kotz [4] define context as "the set of environmental states and settings that either determines an application's behavior or in which an application event occurs and is interesting to the user." Similarly, Dey [7] provides the following definition: "Context is any information that can be used to characterize the situation of an entity. An entity is a person, place, or object that is considered relevant to the interaction between a user and an application, including the user and application themselves." When used as guidelines for implementations, these latter definitions stress that systems should be able to handle many heterogeneous context-types while still support developers in specifying and implementing applications that utilize specific types of context that are relevant for their specific application.

7.2.3 Previous Models of Context

The different notions of context have lead to a number of varied approaches to modeling and handling context in computer systems. Context defined as a fixed set of attributes can be represented by a hard-coded, optimized context model, whereas a more general definition will lead to more generic and flexible models. Typically, data structures include key/value pairs, tagged encodings, object-oriented models, and logic-based models [4].

In Web and hypermedia systems the prevalent way to model context has typically been as either composites serving as partitioning mechanisms on the global network or as key/value pairs—associated with links, nodes and other hypermedia objects—that describe parameters of the context the hypermedia objects belong to. Context modeled as composites is often a purely structural partitioning concept constraining browsing and linking to some kind of context given by the user explicitly or implicitly. Key/value pairs associated with objects are typically used to describe in which context objects are visible. Neptune's Contexts [5], Intermedia's Webs [41], and the Webvise Context composites [25] belong to the first category while Schilit's and Theimer's located objects [26] and environment variables in the context-aware Web browser Mobisaic [35] belong to the latter.

7.2.4 Model for Context and Context-Awareness

The conceptual model for context and context-awareness, which are used in the UrbanWeb framework, is derived from the previous definitions of context discussed above, especially the *operational definitions* that emphasize that context is the set of both digital and physical variables that are relevant for the interaction between the user and the system (e.g., the definitions by Chen and Kotz [4] and Dey [7]). These definitions implies that it is important to model context as a generic construction that can capture context-data in its many different forms (i.e., on the framework level), but at the same time it should be relatively easy for developers to specify specific types of data that are considered as relevant context information for a given application or service (i.e., on the application level).

We have achieved this by modeling context as an extension to the notion of Web 2.0 tagging conceptually; resources in UrbanWeb can be described as a set of properties. Each property consists of a name, a type, and a value. Some of these properties may be used to describe the internal structure of an object, that is, to model the object, while other properties describe the resources' external properties, that is, their context. In this way, resources are modeled by a completely uniform interface of properties. As a consequence, a single class of objects models all resources. These objects have a unique ID, some meta-data common to all resources, and a set of properties that can be tagged onto the objects as needed. If a resource, for example, has a physical location that can be described by a GPS-coordinate, a direction that can be described by compass data, or an ID that can be retrieved from a radio-frequency identification (RFID)-TAG or 2D-barcode, these data can be added as properties to the object in order to describe the context of the resource as illustrated in Figure 7.2.

The important aspect of this approach is that none of these properties are hard-coded into the model, but can be added when the developer needs to specify the context. And this can happen even on runtime. This approach supports the need for a general context mechanism that can represent many different types of data and that also makes it easy for developers to specify and handle specific types of context in their applications.

While this may sound as a very simple way to model data, it is actually similar to the approach taken by many modern programming languages. For example,

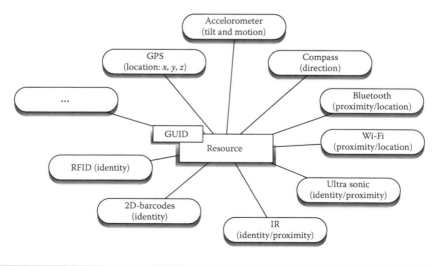

Figure 7.2   Context modeled as an extension to Web 2.0 tagging.

the basic construction in many object-oriented programming languages is the class, which specifies classes of objects as a set of properties (variables) with names, types, and values, and a set of methods that operate on the properties (e.g., the Java programming language [31] or Microsoft's C# language [30]). Our tagging approach is similar, except that we do not include methods, but only data in our model, but this still provides us with a general and extensible way to model data. However, as discussed above, the important difference between traditional class-based approaches and the tagging-based approach is that context is not modeled by predefined classes, but as extra properties on instances of a single class of objects and in this way it resembles a more prototype-based programming approach (as seen in, e.g., JavaScript/EcmaScript [32] or the Self language [33] developed by Xerox Parc and Sun Microsystems).

While modeling resources and their context is very important, we also need to add context-aware functionality to the services and applications in order to utilize the context data. The general class of context-aware functionality that supports association of context-information with data is very well handled by our context model as discussed above. The other class of functionality that focuses on contextual augmentation, that is, on presenting relevant service and information to the user is the main functionality that is provided by the UrbanWeb framework and is handled by *context matching*. Context matching is a function that given a context instance can find resources that have a similar context. Similarity is in this case defined as the same set of properties with the same values or values within a certain range of the target value. This is illustrated in Figure 7.3.

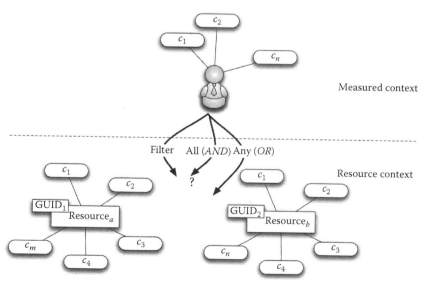

Figure 7.3   Context matching.

The figure illustrates a case with two unique resources, $Resource_a$ and $Resource_b$, both of which, have a set of properties that describes their context: $(C_1, C_2, C_3, C_4, C_m)$ and $(C_1, C_2, C_3, C_4, C_n)$. The user's application has measured its current context to include the properties (C_1, C_2, C_n). To find out which resources are relevant in the current context, the user's measured context is matched against the resources' context. This is done either by requiring that *all* context properties should match (corresponding to an AND query and a very precise match) or by requiring that *any* of the properties should match (corresponding to an OR query). In the example in Figure 7.3, the *"any"* query would return both resources, $Resource_a$ and $Resource_b$, since both match context properties C_1 and C_2. On the other hand, an *"all"* query would only match $Resource_b$ as this is the only resource that has a context described by the all the properties (C_1, C_2, C_n). $Resource_b$ is returned even though it has more context properties than described in the user's measured context (e.g., C_3 and C_4). However, this just means that the resource's context is described in more detail than the user's context or that the resource is modeled by a number of further properties that is not part of its context, and thus, this does not influence the context match.

Even though it is useful to match for equality between the measured context properties and the resources' context properties, it is also sometimes useful to be able to formulate a context match that includes a range (or fuzziness) around the target values. We support this by allowing a filter to be supplied with each property. The filter is a simple operator that is applied to the context match, for example, the "$<$," "$\leq$," "$>$," "$\geq$" operators can be used to find properties with values smaller or greater to the target value. A property can be included multiple times in the context match in order to search for properties within a range. For instance, if we have a measured property x with a value of 55.8700, we could search for other properties within a range of 100 by supplying a property x with value 55.8650 and operator "$<$," and another property x with value 55.8750 and operator "$>$."

The tagging based model and the context matching mechanism is the core of the UrbanWeb framework and give developers an easy yet powerful platform to develop mobile, context-aware Web services. We return to the implementation details of the framework in Section 7.3.

7.2.5 Context for Mobile, Web 2.0 Services

As discussed above, we have applied a very general definition of context to our design, in order to create a general and extensible foundation for the UrbanWeb framework. However, when developing Web applications and services for current mobile devices, we need to consider what specific mechanisms exist for capturing context information. One approach to add context-aware capabilities to mobile devices is by adding new sensor hardware or modifying the device in other ways, as done by, for example, Schmidt et al. [28]. While this may be a reasonable approach in a research setting or when adding new devices to the marked, it is much more difficult to apply to the general case where users just have their mobile device with

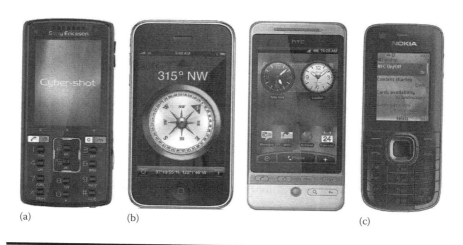

(a) (b) (c)

Figure 7.4 **Off-the-shelf mobile devices with sensor capabilities. (a) Camera phone. (b) Sensor enabled smart phones. (c) RFID enabled mobile phone.**

its built-in capabilities. So let us instead look at the sensor capabilities of current off-the-shelf mobile devices.

Figure 7.4 illustrates three classes of devices. Class (a) is a "standard" mobile phone. As a minimum these phones typically have manual input via a keypad, a built-in camera and microphone, Bluetooth connectivity, and a high-speed mobile network connection (e.g., 3G or EDGE). While many of these "sensors" are geared toward communication features they can also be used to provide context information to the applications. For instance, both 3G cell information and Bluetooth- or Wi-Fi access points can be used to provide (course grained) location information, the camera can be used to scan 2D-barcodes, which in turn can be used to identify objects or provide location information, and finally the user can manually provide information by typing on the keypad [10]. So most mobile phones are actually able to provide some context information to the system. Class (b) of mobile devices in Figure 7.4 implements more sensors, which are designed toward context-aware applications. These sensors typically include a GPS sensor, a digital compass, and a 3D accelerometer. These additional sensors make it much easier to, for example, get location information without relying on other infrastructures such as the 3G cell network, and these devices are obvious candidates for advanced mobile applications. The final class of mobile devices (c) is mobile phones that come with some special purpose sensor. In the example in the figure, we have a Nokia 6212 NFC phone with RFID capabilities, which allow the phone to scan RFID and thus support easy object identification and location measurements. While these special purpose phones might not be widely available, they point in the direction of the next generation mobile phones and the sensor capabilities they will offer.

Even though these potential sources of context information are available on the mobile phones, they may not be accessible to Web applications. Traditional Web browsers provide a sandboxed-environment for the Web pages and applications, which does not allow the applications to access any part of the platform running the browser. In these browsers, the only way to provide context information is probably through manual entry by the user. However, new Web standards open the possibility for better platform integration. As an example, Apple's Mobile Safari browser (OS3, Figure 7.4b) includes support for W3Cs Geolocation application programming interface (API) [36], which provide a location API directly in JavaScript. Other vendors provide "Web Widget" framework that makes it possible to create specialized, single purpose Web application for mobile phones. Even though most of the frameworks are based around a standard browser engine, the framework supports custom extensions that allow widgets to access the underlying mobile platform and utilize, for example, the GPS or the phone's camera. Both Nokia Web Runtime Widgets [20] and Opera's Web Application Platform [21] implement deep integration into the underlying platform and thus enable the developer to utilize the context information in the applications. Finally, most of the mobile platforms support development of some kind of "native" application by offering C, C++, Objective C, or Java software development kits (SDKs) for developers. While these applications are not running in a browser environment, they may be easier to integrate with the underlying platform and they may still utilize the programmable Web interfaces and thereby the same service and infrastructure as pure HTML-based Web applications.

7.3 The UrbanWeb Framework

In this section, we discuss the implementation of the UrbanWeb framework and the concepts presented in Section 7.2.

In its current implementation, the UrbanWeb architecture, as seen in Figure 7.5, is implemented in PHP 5 and uses MySQL as the storage backend.

The framework consists of a basic data model (which we will discuss below) that handles persistence through a custom Data-Object interface to the underlying database. All services use this basic data model, as it also implements the context tagging mechanism. Services in Urban Web are implemented as components, which exhibit a Web interface to the functionality. This approach has several advantages: first, it is quite easy to reuse existing functionality since this is encapsulated and implemented on the component level, and second, a component may have several different services that use its functionality, but provide different service interfaces or data formats to the clients. By implementing multiple interfaces we are able to support a "One Web" approach to the UrbanWeb service functionally. This means that it has been quite easy to support clients ranging from pure browser-based clients that render HTML to Web Widget-based application and "native" applications developed with for example, Java Micro Edition, as

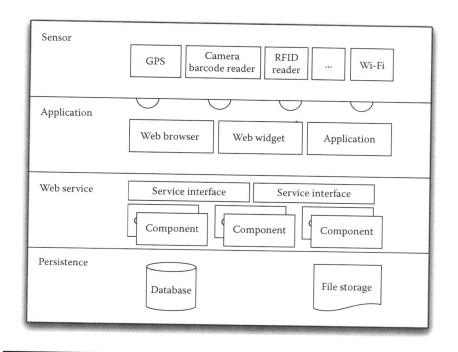

Figure 7.5   The UrbanWeb architecture.

we describe in Section 7.4. Context information is handled at the application-level in UrbanWeb. While the framework has the capability to model and represent context, the actual measurement of sensor data that is used as cues for describing the context is handled by the applications themselves. This may include acquiring location information from a GPS, using a Bluetooth location infrastructure or scanning an RFID chip or 2D-barcode TAG.

7.3.1 Model Implementation

The UrbanWeb data model is one of the core components in the framework and is used across all services. The model is depicted in a simplified version in Figure 7.6.

The model is based around a single PHP class, the `UrbanWeb_BaseObject`. The `UrbanWeb_BaseObject` provides the modeling primitives for resources and context information: a Universally Unique Identifier (UUID) identifies a resource within the system, a number of meta data attributes, including creator information and time stamp information for creation and last access, and a content property that makes it easy to create text-typed resources such as document resources or annotation resources. Other types of resources can also be modeled, which reference other materials such as pictures or videos that are stored on the file system or on the Web. Each `UrbanWeb_BaseObject` includes a list of

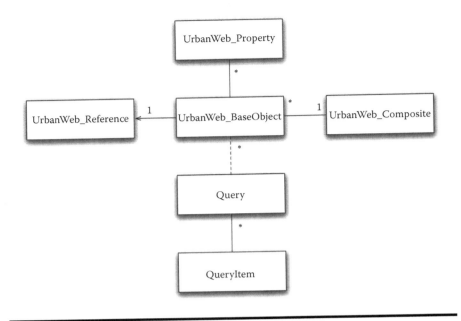

Figure 7.6   The UrbanWeb data model.

UrbanWeb_Property objects, which are properties with a name, a user-select-able type, and a value. The UrbanWeb_Property objects are the foundation for the context tagging system and for modeling UrbanWeb objects. The properties are also directly supported by the persistence system, so all properties that are added to an UrbanWeb_BaseObject are automatically saved to the database with the object. Internally, the UrbanWeb_BaseObject represents its properties as a list of UrbanWeb_Property objects. However, to make it easy for developers to use the model, the interface to the properties are handles through PHPs reflection mechanisms and by intercepting getter and setter calls to PHP class properties. This effectively means that developers can access the UrbanWeb_Property objects in much the same way they would access normal PHP class properties. For example, if a developer creates an instance of an UrbanWeb_BaseObject and tries to set a field:

```
UW_BaseObject->propertyname = somevalue;
```

The UrbanWeb_BaseObject instance will first check if propertyname is a PHP class property. If it is, the property will be set, otherwise the call will be handled as if propertyname is the name of an UrbanWeb_Property object and this object will get its value property set to somevalue. All this is handled in the background by the model, and the developer can thus use the tagging system in a similar manner as the PHP class system would normally be used.

If the developer wants better tool support, for example, auto completion on property names, the easiest way to archive this is by implementing subclasses of the `UrbanWeb_BaseObject` class with custom getter and setter methods for each property. By doing this, PHP code editors will be able to provide auto completion on the method names, while the developer still retains the advantages of the freeform tagging approach and persistence features.

The UrbanWeb model also includes a couple of other useful constructions. The `UrbanWeb_Reference` class implements simple references between two `UrbanWeb_BaseObject` instances and the `UrbanWeb_Composite` class implements a one-to-many relationship between two or more `UrbanWeb_BaseObject` instances. Both of these relationships are also supported by the persistence system. The `UrbanWeb_Composite` class is implemented as a set data structure, but is easy to extend to represent other composite types such as lists or trees simply by subclassing the `UrbanWeb_Composite` class.

Context Matching is implemented in two different ways in the model. The first way is provided by the Data-Object implementation in the model, which support a `find()`-method on instances of the `UrbanWeb_BaseObject` class. Developers may create an `UrbanWeb_BaseObject` instance and populate it with properties with names, types, and values. By calling the `find()`-method the model will inspect the object and automatically generate an SQL-request that return all other objects with same set of properties and the same values. This is quite useful for finding objects by ID, objects of a certain type, or objects that have been tagged with some identifying context information, such as a 2D-barcode identifier or a RFID id. If the developer needs to create a fuzzy match that searches for objects within a range of their values, the model provides a query mechanism (see Figure 7.6). A `Query` contains a number of `QueryItems` and each `QueryItem` has a name, a type, a value (corresponding to the structure of an `UrbanWeb_Property` object), and an operator that is applied as a filter on the properties. When a `Query` has been populated with `QueryItems` it can be used to automatically generate an SQL-request that searches for `UrbanWeb_BaseObject` in the database with corresponding property names and property values. It should be noted here that even though the query mechanism is separated from the Data-Object model, the `Query` objects also provide an interface to the underlying database and automatically creates the SQL-query, so developers do not have to worry about the database structures, but can perform context matching through the models various methods.

To summarize, the main advantages of the UrbanWeb framework for developers are

- Context modeling—a dynamic tagging model that supports freeform tagging and dynamic addition of properties on classes of objects, even on runtime.
- Context-aware functionality—integrated context matching implemented as part of the model.

- Free persistence—the Data-Object implementation provides a convenient interface to the underlying database.
- Component-based service framework—the UrbanWeb framework architecture makes it easy for developers to create new components and service and to deploy new instances of the framework.

Having now described the UrbanWeb framework and its implementation, we will turn to services that have been developed with the framework.

7.4 Context-Aware Mobile Web Services

A number of different services have been built with the UrbanWeb framework. These range from purely browser-based applications to native applications for mobile phones that utilize Web services built on the infrastructure provided by UrbanWeb. In this section, we discuss some of these services and how they utilize contextual information when presenting information. We discuss how to provide location-based information for urban computing applications, location-based moblogs that support social computing, location-based poll services, and Mobile Urban Dramas which are special purpose applications that allow users to experience audio dramas, which use context-awareness to anchor a story in the streets of a city.

7.4.1 Location-Based Information in the City

The first applications discussed here were designed as mobile, urban Web application for use during the annual Aarhus city Festival in Denmark. The applications have been developed and used from 2007 to 2009. The Aarhus Festival* is among the largest cultural events in Scandinavia and showcases local, national, and international artists. The festival is a thematic event and the theme of the year determines the artistic program. The 2007 theme was "in motion." As part of this theme a 5 km long route called the "Red Route" was established through the center of the city to allow festival participants to experience the city, its architecture, and its activities from new angles. A number of 2D-barcodes were placed along the "Red Route," and the UrbanWeb applications relied on these visual tags as link anchors to location based services and existing Web resources.

2D-barcodes are 2D-patterns that encode a piece of text, typically an address or a URL. The 2D-barcodes can be used with camera phones that include software to decode the barcode from a scanned image. 2D-barcodes such as Semacodes† or QR-codes‡ have the advantage that they clearly visualize the link endpoint in the physical world, and when decoded, provide the link address directly.

---

* http://www.aarhusfestuge.com/
† http://semacode.com/
‡ http://www.denso-wave.com/qrcode/index-e.html

While typical scanning programs do not handle the URLs themselves, but launch the mobile phone's built-in browser and redirect it to the decoded URL, the UrbanWeb applications utilize the URLs to integrate Web resource with other services, such as the commenting facilities in the blogging service or poll services as discussed below (see Figure 7.7a). In terms of the context model, the URL may bear meaning in itself (i.e., it is a Web address), but it is also used as property describing part of the context for services and information. In this way, the URL becomes an identifier for the 2D-barcode's physical location and in turn the location of the information and services. Figure 7.7a depicts how Web pages are assigned a unique URL to make them available from Web browsers. Each URL is encoded as a 2D visual tag and placed as a link marker at a physical location. However, the URL of the Web page is also used as identifier in other services, for example, to identify an associated mobile blog or poll. This approach provides a loose coupling between Web resources and the new social services, since the link between them is handled by the application and not by the existing Web site. It is therefore quite easy to both anchor existing Web pages in the physical world and augment them with commenting facilities or other services without having to modify them or move them to a different system.

However, not all mobile devices have a camera or the capabilities to decode 2D-barcodes and for these devices we introduce a "TinyTag" identifier. The TinyTag is a simple numeric code, which is associated with each 2D-barcode (and correspondingly with each URL). These codes can be sent to a TinyTag mapping service (provided as part of the UrbanWeb infrastructure), which in turn will return the corresponding URL. While most mobile clients can use the TinyTags, this technique requires an extra network lookup compared to the 2D-barcodes, which are decoded directly on the mobile phones. Nevertheless, TinyTags offer a convenient input method for mobile phones that is quite similar to typing an SMS message and it relieves the user from having to manually input long or cryptic URLs. Furthermore, as the numeric code is generated and printed as part the barcode, we can still rely on the 2D-pattern as an easily recognizable physical anchor for the link, but the interaction is more manual and similar to systems such as Yellow [42], which also rely on manual specification of the anchor.

7.4.2 Location-Based Mobile Blogging

During the Aarhus Festival, a number of 2D-barcodes TAGs were placed along the "Red Route" as depicted in Figure 7.8. Each TAG was placed at a specific location in the city and the festival secretariat linked location-specific information to each of the TAGs. The information ranged from dynamic updated Web pages describing concerts and events to more static pages, which described the specific locations. Artist-created other Web pages and tags were placed nearby their exhibitions to inform about the displayed art. All this information was handled by and published in the Aarhus Festival's existing CMS.

While the TAGs provide *access* to location-specific information, we also wanted to support social computing aspects and allow users to *create* and *share* information in

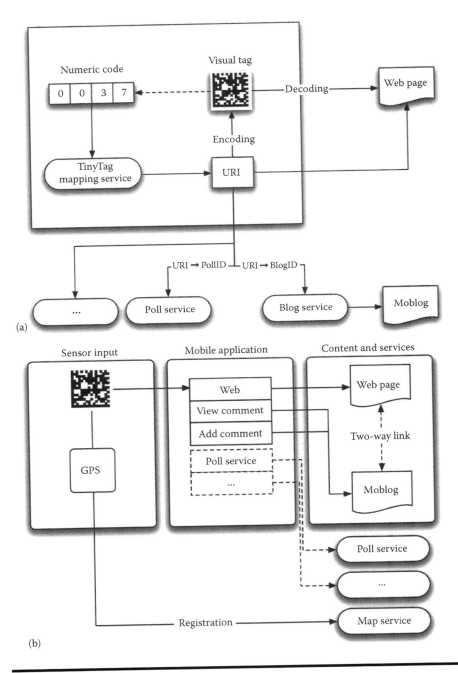

Figure 7.7 **Service aggregation in UrbanWeb. (a) Aggregating services using 2D-barcode TAGs. (b) Identifying services with multiple sensors.**

Figure 7.8 **The mobile phone blogging user interface. (a) Scanning a TAG to identify the context. (b) choosing the action. (c) Creating a multimedia blog post. (d) Viewing the blog that is anchored on the TAG.**

situ as well. Therefore, a mobile application, the *TagBlogger*, was developed with the UrbanWeb framework. To be compatible with the largest number of different mobile phones from different manufacturers, the mobile application was developed on the Java JME platform and themed to match the festival theme as illustrated in Figures 7.8 and 7.9. The application allowed users to scan TAGs (Figure 7.8a), browse the linked information, see comments by other users, and leave comments on the TAGs themselves (Figure 7.8b). Comments could include text, images taken with the phone, or existing images or video clips stored in the phone's memory (Figure 7.8c and d). Thus, this application utilized 2D-barcodes not only to access location-sensitive information, but also as the anchor point for new user-generated content in the city.

2D-barcodes and TinyTags were used as the basic context parameter, since these technologies were the ones most likely to be supported on the majority of mobile phones. However, experiments with a combination of GPS and 2D-barcodes were also performed. The basic idea here was that if a user had a GPS-enabled mobile phone, we would acquire the user's position, whenever a TAG was scanned. The GPS data would then be added as a context property onto the resources that already had the 2D-barcode as a property of their context. In this way, resources identified by a 2D-barcode could also be geo-tagged. These geo-tagged Web resources and moblogs could then be displayed on a map in a Web browser and inspected when the user returned home. The interesting aspect of this technique is that context information is added to the resources in a collaborative manner. Some users may not even have

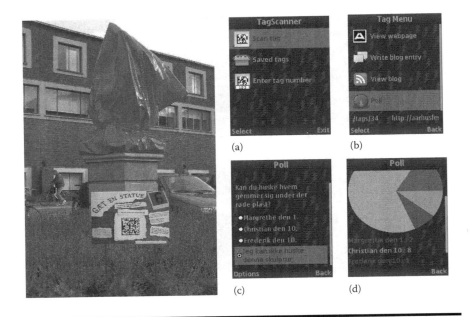

Figure 7.9   The mobile phone poll service user interface (Danish UI). (a) Scanning a TAG or entering a TinyTag number to identify the context. (b) Selecting a service associated with the particular context (poll). (c) Taking the poll. (d) Viewing the poll results.

a camera in their mobile phones, but can still access the services by manually entering a TinyTag identifier, while others may have a GPS and can thereby help geo-tag the resources. The geo-tagged information will, however, be available to all users, so even the first user will be able to see his or her own comments on the map if somebody has added GPS information to the particular moblog.

The blogging service was implemented as a "static" service in the infrastructure, in the sense that each 2D-barcode TAG had, per definition, always an associated blog, whether this blog actually existed or not. If a user tried to read a blog or create a new blog post on a blog that did not exist, the blog would dynamically be created and presented to the user. Thus, by using the UrbanWeb application, any URL encoded in a 2D-barcode could be augmented with the commenting facilities of the moblog service and the usage of the system could be revealed on the map through collaborative geo-tagging.

7.4.3 Location-Based Polls

Another urban Web service developed for user under the Aarhus festival was a location-based poll service. Like the moblog service, the poll service allowed active user participation. Instead of letting users create new content, the poll service was used to let users vote and participate in predefined polls and quizzes created by the festival secretariat.

Most of the 2D-barcodes along the "Red Route" had associated polls, but the secretariat also created a number of special quiz signs, inviting users to participate. Each of these signs was placed near statues, which had been wrapped in red opaque plastic. In the quiz, the user is asked to recall the material, structure or theme of the hidden statue (see Figure 7.9).

The poll service is an example of a "dynamic" service. Each 2D-barcode does not automatically have a poll associated with it, but polls can be tagged with the URL encoded in the barcode to create the link between the barcode and the poll. When the user scans a 2D-barcode, the application queries the poll service for any available polls. If a poll has been linked to the barcode, it is presented in the user interface, but if no poll is found, the poll entry in the service menu is not visible (see Figure 7.9b).

Dynamic services are more general than static services, since arbitrary dynamic services can be linked to a physical context. Then, whenever the mobile application registers a change in the user's context (e.g., when the user scans a tag) it can query the UrbanWeb infrastructure for available services that are compatible with the mobile application and present the available services to the user.

7.4.4 Mobile Urban Dramas

The final example of applications built on the UrbanWeb framework that we discuss here are *Mobile Urban Dramas* [13]. Mobile Urban Dramas are interactive audio plays that let the user be the main character in a drama where the real urban environment becomes the scenography. In the play, users are equipped with mobile phones, headsets, and maps, and experience a drama in the streets of a city, where they trigger different scenes of the play (audio, video, and text) through location-based technology such as GPS, 2D-barcode TAGs, or TinyTags. Furthermore, they may receive what appears to be SMS messages and phone calls, and encounter real-life actors as part of the play that takes place along a prespecified, possibly branching path through a cityscape or landscape as illustrated in Figure 7.10. From 2007 to 2009, four different plays with dramatic, educational, and historical content have been developed in cooperation with a local theater "Katapult" from Århus, Denmark [15]. These plays are: "CORRIDOR" (2007), "GAMA—ON THE TRAIL OF UNKNOWN LAND" (2008), "HASLE INTERACTIVE" (2009), and "HIKUINS VENDETTA" (2009) [13,14].

Unlike the *TagBlogger*, the Mobile Urban Dramas did not open up for user contributions and social computing, and was thus a somewhat different kind of application and more restricted than the *TagBlogger*. Mobile Urban Dramas utilized URLs encoded in 2D-barcodes in a special manner: as the play included several hundred megabytes of audio files, these files were not downloaded from the Web over the mobile Internet, but preloaded on either a borrowed set of phones or on the user's own phone that had to be configured with data during installation. Each encoded URL was mapped to a specific audio file in the application and when

(a)   (b)

(c)

Figure 7.10 Experiencing a mobile urban drama. The user photographs TAGs as he walks through the city, which enables him to listen to the audio drama. (a) Listening to a Mobile Urban Drama in the city. (b) TAGs trigger scenes in the play. (c) The TAGs are easily visible and provides link anchors for the particular location.

a TAG was scanned it would trigger the corresponding audio file. However, the URLs also pointed to theater Katapult's Web page and if scanned by a third-party application, they would redirect the browser to an information page describing the audio theater and explaining how to buy tickets. Thus, the Mobile Urban Drama application utilizes the URLs to identify the local resources, but as the URLs still encode Web addresses, they also serve as normal URLs for people not using the theater application. The Mobile Urban Drama application is a much more specialized application than the festival services discussed above, but nevertheless, it clearly illustrates how 2D-barcodes can serve as a lightweight infrastructure for accessing information in various ways.

7.5 Conclusion

With the advance of new mobile devices with full Web support and cheap high-speed mobile Internet connections, the development of advanced Web service is spreading to the mobile Internet. Even though many existing Web services may scale to the mobile Web and provide access to the services from anywhere at anytime, other services may benefit from being augmented with context-aware functionality, which can help filter information based on the user's context and present relevant information and services to the user.

This chapter has discussed the design and implementation of the UrbanWeb framework for mobile, context-aware Web services. UrbanWeb is built on a conceptual model for context, which defines context as an extension to the notion of Web 2.0 tagging. This model supports freeform tagging of resources with context information, in order to describe both the physical and digital context of the resources. At the same time, the model supports context-awareness in the form of *context matching* based on various filtering mechanisms on the tagged context information. UrbanWeb provides a lightweight implementation of the conceptual model that makes it easy for developers to utilize context information in their service and to develop new services for a variety of different mobile Web clients.

A number of applications have been developed with the framework. The chapter discusses how to present location-specific information in urban environments, how to create social moblog services that allow users to browse, create, and share information that is produced in situ, and how to aggregate multiple Web services by utilizing context information. Finally, we discuss a special purpose mobile application that utilizes the context-aware capabilities in the framework to implement a mobile urban drama. We also discuss how various sensor technologies can be used to provide cues about the user's and application's contexts, such as GPS, 2D-barcodes, or manual entry, and how these cues can be utilized and combined to create novel Web services for both mobile and desktop use.

Acknowledgments

This work was supported by ISIS Katrinebjerg, Center for Interactive Spaces and the Alexandra Institute. The UrbanWeb project is funded by the Danish Agency for Science, Technology and Innovation, project 274-07-0218. We also wish to thank our center colleagues and our partners who have been involved in the realization of the UrbanWeb applications.

References

1. Bamford, W., Coulton, P., and Edwards, R. 2007. Space-time travel blogging using a mobile phone. In *Proceedings of the International Conference on Advances in Computer Entertainment Technology* (Salzburg, Austria, June 13–15, 2007), ACE '07, ACM, New York, vol. 203, pp. 1–8.
2. Berners-Lee, T., Hendler, J., and Lassila, O. May 2001. The semantic web. *Scientific American*, 35–43.
3. Brown, P., Bovey, J., and Chen, X. October 1997. Context-aware applications: From the laboratory to the marketplace. *IEEE Personal Communications*, 4(5):58–64 (doi:10.1109/98.626984).

4. Chen, G. and Kotz, D. 2000. A survey of context-aware mobile computing research. Technical Report TR2000-381, Department of Computer Science, Dartmouth College, November 2000.

5. Delisle, N. M. and Schwartz, M. D. 1987. Contexts—A partitioning concept for hypertext. *ACM Transactions on Information Systems*, 5(2):168–186.

6. Dey, A. K. 1998. Context-aware computing: The CyberDesk project. In *Proceedings of the AAAI 1998 Spring Symposium on Intelligent Environments* (Stanford, CA), AAAI Tech. Report SS-98-02, pp. 51–54.

7. Dey, A. K. 2001. Understanding and using context. *Personal and Ubiquitous Computing*, 5(1):4–7 (doi:10.1007/s00779017 0019).

8. Espinoza, F. et al. 2001. Geonotes: Social and navigational aspects of location-based information systems. In *Proceedings of the 3rd International Conference on Ubiquitous Computing*, Atlanta, GA, Springer-Verlag, London, U.K., pp. 2–17.

9. Grønbæk, K., Vestergaard, P. P., and Ørbæk, P. 2002. Towards geo-spatial hypermedia: Concepts and prototype implementation. In *Proceedings of the 13th ACM Hypertext Conference* (College Park, MD, June 11–15, 2002), ACM Press, New York, pp. 117–126.

10. Hansen, F. A. 2006. Ubiquitous annotation systems: Technologies and challenges. In *Proceedings of the 17th Hypertext Conference* (Odense, Denmark, August 22–25, 2006), ACM Press, New York, pp. 121–132.

11. Hansen, F. A., Christensen, B. G., and Bouvin, N. O. 2005. RSS as a distribution medium for geo-spatial hypermedia. In *Proceedings of the 16th ACM Hypertext Conference* (Salzburg, Austria, September 06–09, 2005), ACM, New York, pp. 254–256.

12. Hansen, F. A. et al. 2004. Integrating the web and the world: Contextual trails on the move. In *Proceedings of the 15th ACM Hypertext Conference* (Santa Cruz, CA), pp. 98–107.

13. Hansen, F. A., Kortbek, K. J., and Grønbæk, K. 2008. Mobile urban drama—Setting the stage with location based technologies. In *Proceedings of 1st Joint International Conference on Interactive Digital Storytelling* (Erfurt, Germany, November 26–29, 2008), Springer-Verlag, Berlin, Germany, pp. 20–31.

14. Hikuins Vendatta. Mobile Urban Drama. Website 2009: http://www.visitaarhus.com/international/en-gb/menu/turist/hvad-sker-der/hikuins-blod/hikuin.htm

15. Katapult Theatre. http://www.katapult.dk/ (only in Danish)

16. Magerkurth, C. et al. July 2005. Pervasive games: Bringing computer entertainment back to the real world. *Computers in Entertainment*, 3(3):4–4.

17. Marlow, C., Naaman, M., Boyd, D., and Davis, M. 2006. HT06, tagging paper, taxonomy, Flickr, academic article, to read. In *Proceedings of the 17th Hypertext Conference* (Odense, Denmark, August 22–25, 2006), ACM Press, New York, pp. 31–40.

18. Mobile Web Best Practices 1.0. Basic Guidelines. W3C Recommendation 2008. Website: http://www.w3.org/TR/mobile-bp/#OneWeb

19. Mobile Web. Wikipedia article. 2009. http://en.wikipedia.org/wiki/Mobile_Web

20. Nokia Web Runtime widgets. Website 2009: http://www.forum.nokia.com/Technology_Topics/Web_Technologies/Web_Runtime/

21. Opera Widgets. The Opera Web applications platform. Website 2009: http://www.opera.com/business/solutions/widgets/technology/

22. Østerbye, K. and Wiil, U. K. 1996. The flag taxonomy of open hypermedia systems. In *Proceedings of the 7th ACM Hypertext Conference* (Bethesda, MD, March 16–20, 1996), ACM, New York, pp. 129–139.

23. Pascoe, J. 1998. Adding generic contextual capabilities to wearable computers. In *Proceedings of the 2nd IEEE International Symposium on Wearable Computers* (Pittsburgh, PA, October 19–20, 1998), ISWC, IEEE Computer Society, Washington, DC, p. 92.

24. Ryan, N. S., Pascoe, J., and Morse, D. R. 1998. Enhanced reality fieldwork: The context-aware archaeological assistant. In *Computer Applications in Archaeology*, Oxford, U.K., October 1998.

25. Sandvad, E., Grønbæk, K., Sloth, L., and Knudsen, J. L. 2001. A metro map metaphor for guided tours on the Web: The Webvise guided tour system. In *Proceedings of the 10th International World Wide Web Conference* (Hong Kong), W3C, ACM Press, New York, pp. 326–333.

26. Schilit, B. N. and Theimer, M. 1994. Disseminating active map information to mobile hosts. *IEEE Network*, 8:22–32 (doi:10.1109/65.313011).

27. Schilit, B., Adams, N., and Want, R. 1994. Context-aware computing applications. In *Proceedings of the Workshop on Mobile Computing Systems and Applications* (*WMCSA'94*) (Santa Cruz, CA), pp. 85–90.

28. Schmidt, A. et al. 1999. Advanced interaction in context. *LNCS*, 1707.

29. Shklovski, I. and Chang, M. F. September 2006. Guest editors' introduction: Urban computing—Navigating space and context. *Computer*, 39(9): 36–37.

30. The C# programming language. Website 2009: http://msdn.microsoft.com/en-us/vcsharp/aa336809.aspx

31. The Java programming language. Website 2009: http://java.sun.com/

32. The JavaScript programming language. ECMA-262 and ISO/IEC 16262. Website 2009: http://www.ecma-international.org/publications/standards/Ecma-262.htm

33. The Self programming language. Website 2009: http://selflanguage.org/

34. Tim O'Reilly. 2005. What is Web 2.0. design patterns and business models for the next generation of software. O'Reilly Network. Website: http://www.oreillynet.com/pub/a/oreilly/tim/news/2005/09/30/what-is-web-20.html

35. Voelker, G. M. and Bershad, B. N. 1994. Mobisaic: An information system for a mobile wireless computing environment and engineering. In *Proceedings of the Workshop on Mobile Computing Systems and Applications* (Santa Cruz, CA), IEEE Computer Society, Santa Cruz, CA, pp. 85–90.

36. W3C Geolocation API Specification. Website 2009: http://www.w3.org/TR/geolocation-API/

37. Walther, B. K. July 2005. Atomic actions—Molecular experience: Theory of pervasive gaming. *Computers in Entertainment*, 3(3):4–4.

38. Want, R. et al. 1992. The active badge location system. Technical Report 92.1, Olivetti Research Ltd., Cambridge, U.K.

39. WebKit Open Source Project. Website 2009: http://webkit.org/

40. Weiser, M. 1991. The computer for the 21st century. *Scientific American*, 265(3):66–75.

41. Yankelovich, N., Haan, B. J., Meyrowitz, N. K., and Drucker, S. M. 1988. Intermedia: The concept and the construction of a seamless information environment. *IEEE Computer*, 21:81–96.

42. Yellow Arrow. 2005. Product page: http://yellowarrow.net

Chapter 8

MyMobileWeb 4: A Framework for Adaptive Mobile Web 2.0 Applications and Portals

José Manuel Cantera Fonseca, Cristian Rodríguez de la Cruz, Ignacio Marín Prendes, and Miguel Jiménez Gañán

Contents

8.1 Introduction

Making the Mobile Web 2.0 a reality is not only a matter of using highly capable devices and enhanced Web browsers. End users are expecting more from the new generation of the Web on the move. First of all, they want to use applications and portals adapted and optimized for every device. Another emerging requirement is the availability of rich user interfaces (UIs) (AJAX [1], graphics, maps) supporting higher functionalities while optimizing traffic costs. Furthermore, context awareness (e.g., enabling location-based services) is demanded to obtain the most relevant information or content depending on the situation or environment the user is in.

In addition, application developers are seeking powerful, standards-based (and possibly open source) technologies that make possible the creation of advanced, high-quality mobile Web applications together with the reduction of both production costs and time to market.

This chapter describes the latest version of MyMobileWeb (version 4), which provides the technology for the creation and spreading of mobile Web applications that can adapt seamlessly to multiple devices without extra effort from the developer. MyMobileWeb is an open source, standards-based software framework that simplifies the rapid development of mobile Web applications and portals. MyMobileWeb encompasses a set of technologies, which enable the *automatic adaptation* of applications to the target delivery context [2] (browser, device, network, location, etc.), thus offering a harmonized user experience.

Nowadays, there are approaches for creating universal content [3,4], which can be rendered by any device. However, they do not exploit the capabilities of the target delivery context at a maximum, resulting in a poor user experience.

The MyMobileWeb project advocates for the development of mobile-specific versions of applications and contents rather than pruning or adapting a version intended to be used from desktop PCs [5,6]. To this aim, MyMobileWeb offers the ability to incorporate contents from standard content management systems (CMS) or to connect to JavaEE information systems, thus making it easier to develop a Web-based, multi-channel access layer to legacy or existing systems.

The remainder of this chapter is structured as follows: first, we provide a general overview of MyMobileWeb, its main technologies, and its architecture. Then, we describe how applications are designed using MyMobileWeb by developing an example. We end the chapter with the main conclusions.

8.2 MyMobileWeb Key Technologies

The key technologies offered by MyMobileWeb are

- The IDEAL2 [6A] language for the declarative description of device-independent UIs and adaptation policies. The section below describes this language in more detail.
- The SCXML-based language for describing application flows modeled as state machines. State Chart XML [7] (SCXML) is a W3C standard. Application Flows describe the behavior to be performed in reaction to the interaction of the user with the system. The set of actions to be performed by an application flow does not only depend on the events raised by the UI but also on the application state. For example, the next page to be loaded when a link is activated could depend not only on the activated link per se, but also on the user profile.
- The "Device Description Framework" concerned with obtaining information about the characteristics of devices and Web browsers by interfacing with different device description repositories (DDRs). MyMobileWeb is able to connect to any DDR supporting the W3C's *DDR Simple API* recommendation [8]. Particularly, connectors with the most popular DDRs such as WURFL [9], Device Atlas [10] or UAProf [11] are provided off the shelf. In addition, our framework supports multiple vocabularies of properties, including the W3C's DDR Core Vocabulary [11A] and the WURFL's vocabulary.
- The "Adaptation and Rendering Engine," in charge of selecting and generating the final markup, script, style sheets, and other resources (images, audio, video) to be delivered to the mobile device. A flexible an extensible architecture allows both to add and to extend the rendering components.

■ The "Client-side Framework" (a.k.a. "Mobile AJAX Framework"), which enables rich interactions in different Javascript-enabled browsers. Such a framework provides (cross-browser) convenience methods for asynchronous HTTP requests, DOM [12] manipulation and advanced UI components (calendar, range, tabs, etc.).

8.2.1 IDEAL2

IDEAL2 [6A] (Interface Description Authoring Language, version 2) is the result of 4 years of research and experience in languages for the description of device-independent UIs. This language, based on different W3C standards (XForms 1.1 [13], DISelect 1 [14]), is modular and extensible. The main elements of IDEAL2 are containers and UI components. To help Web authors, the syntax of IDEAL2 is similar to XHTML's. Nonetheless, as an authoring language, it incorporates abstractions and extensions that make it a more powerful and higher level language than XHTML, as the latter is only intended for browser consumption. In addition, IDEAL2 describes UIs in an abstract manner, that is, without commitment on how such a UI will be finally rendered. For example, IDEAL2 defines a UI component named "select1," which can be rendered as a drop-down list, a set of radio buttons, or as a nice navigation list with hyperlinks. The decision on how an abstract UI element will be finally rendered will depend on the device and Web browser identified at runtime. The rendering engine can make this decision automatically on behalf of the developer. Nonetheless, developers can force a specific rendering by means of adaptation policies.

IDEAL2 is only intended to provide the description of a UI from a structural and behavioral point of view. All the aspects that have to do with look-and-feel and layout are specified by means of CSS2 [15] and extensions. Using this approach IDEAL2 helps authors to separate the different presentation concerns, making it easier to adaptat to multiple devices without code duplication. For instance, an author can vary the layout of an application by assigning different style sheets depending on the device.

By using IDEAL2, developers can concentrate on the application functionality without worrying about markup languages or scripting capabilities. MyMobileWeb's rendering engine will take care of such details, performing graceful degradations for those less capable devices. For example, if tables are not properly supported, a list will be presented instead.

8.3 MyMobileWeb Core Functionalities

Figure 8.1 depicts a functional view of MyMobileWeb. The main feature provided by MyMobileWeb is "content and application adaptation." Adaptation is a process of selection and generation of the application's UI or contents in order to (a) accommodate to the restrictions imposed by the target delivery context and

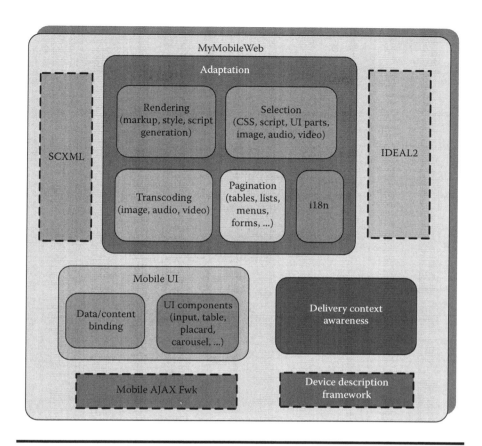

Figure 8.1   MyMobileWeb functional view and main enabler technologies.

(b) to ensure that the resulting user experience is sufficiently harmonized with such a delivery context. Adaptation encompasses a number of sub-functionalities:

■ *Automatic generation* of the most suitable markup, style sheets, and Javascript to realize the different UI elements (specified using IDEAL2). MyMobileWeb adapts and renders to a wide spectrum of Web technologies, from legacy (WML [16]) to the most modern (WebKit-based) enhanced browsers present on iPhone or Android devices. Particularly, our framework is capable of generating mobileOK [4] content, following the best practices and recommendations dictated by W3C [3] and endorsed by mTLD, the company which runs the .mobi domain, dedicated to mobile-optimized Web content.

■ *Selection of CSS style sheets* to customize the look and feel or layout for different devices. For instance, the iPhone native look and feel are achieved, thanks to this technique.

- *Selection of UI parts* (including Javascript code) depending on the characteristics of the delivery context. For example, a table can show more or less columns depending on the available resolution width. These adaptation policies can be conveyed by authors by means of the *selection* attribute, as mandated by the W3C DISelect 1 specification.

- *Pagination of long content* such as tables with many rows, menus with many options, or forms with many fields.

- *Selection or transcoding of different multimedia content* (images, audio, video). While selection consists of choosing the best from a set of variants, transcoding implies a transformation process from a source to a target resource. MyMobileWeb incorporates a simple image transcoding component and can interoperate with any OMA-STI [17] compliant transcoding service, such as the open source Alembik [18] system.

- *Internationalization and literal management,* which consists of selecting literals depending on the delivery context (e.g., to accommodate a short space by choosing an acronym) or the user preferences (preferred language).

MyMobileWeb provides runtime support for a wide variety of UI components (specified at design time using the IDEAL2 language) targeted to the mobile environment: input (including automatic completion), date, time, selection, menus, chained menu, table, list, range, placard (for combining text and images), and carrousel (for showing sequentially a list of items). Depending on the capabilities of the browser used, the realization of UI components is done by server-side markup code generation, client-side DOM manipulation, or a combination of them. Finally, an extension library provides the capability to render statistical graphics for multiple devices in different formats: vector (SVG) or raster images.

Data and content binding are also important functionalities provided by MyMobileWeb. They are intended to the (declarative) association of UI components with the data or contents (and possibly their constraints) they are going to display or manipulate. For example, a table component can be bound to a Java collection. As a result, such a table will have as many rows as items the collection has. Likewise, a selection component could be two-way bound: to a collection, which will contain the set of available options and to a string variable, which will store the user's selected item. Data binding and their constraints are declared by using a subset of the W3C's XForms 1.1 syntax within IDEAL2.

Content binding follows a similar approach. An image, text, or similar content stored in a content repository can be bound to an element of a page (paragraph, list, table, etc.). The MyMobileWeb runtime will be in charge of connecting to the content repository, extracting the right resource, and finally rendering it as part of the UI. MyMobileWeb can interoperate with any JSR-170 [19] compliant content repository, such as Alfresco [20].

The awareness of the delivery context is another key functionality provided by MyMobileWeb. It marks a step toward increasingly context-aware mobile Web applications. The detection of the device and Web browser used takes place at server side, thanks to a "DDR Service" implemented using the *device description frame-work* technology previously described. A DDR service can provide information about the "a priori known," that is, static, characteristics of the delivery context. For instance, a DDR can provide the display width of a device when it is in its default orientation. However a DDR cannot know whether the device has been rotated or not. To overcome this issue, MyMobileWeb implements an AJAX-based mechanism devoted to notify the server when a (dynamic) property changes at client (device) side during a session. The same notification system can be used in other situations. For example, to notify changes in the user's location or the battery level. These ideas are still in a preliminary and experimental development phase. Nonetheless, we are aimed at evolving the component to have a complete context management infrastructure for mobile Web applications. Such infrastructure will allow to obtain contextual information at client or server side using the same API. We are now in the first implementation phase, in which we have implemented a generic API at the client side based on the W3C's delivery context client interfaces (DCCI) [21]. Finally, the "delivery context model" used by our technology is compliant with the W3C's "delivery context ontology" [2] specification.

As an experimental functionality, IDEAL2 incorporates an extension, which allows to add semantic annotations to UI descriptions. One possible application of this feature is the automatic completion of forms. For instance, the owner information stored in the user's device can be used to complete an annotated form, which is asking for personal data. This functionality can be helpful, especially in those devices with more limited input mechanisms.

On the other hand, all functionalities aimed at exploiting information present in the user's device have security and privacy issues still to be solved. We envisage that such issues will be overcome by future developments made by industry initiatives, OMTP-Bondi [22], or standards bodies (W3C has recently launched a new working group called DAP [23], device APIs and policies).

8.4 High-Level Architecture

Figure 8.2 describes graphically the high-level architecture of applications developed on top of MyMobileWeb. The main elements are a mobile Web browser and a Web server, which embeds a Java Servlet 2.3 [24] container. The server subsystem is responsible for UI rendering and adaptation and controlling the application flow. The client (browser) can have no intelligence (as it occurs in micro-browsers, which are neither AJAX nor scripting capable) or can incorporate Javascript components to support rich interactions in enhanced browsers.

The server typically calls backend business services (or databases), which provide the business logic or data necessary to implement the functionality of the application (e.g., a service that allows to perform a bank transaction). Other

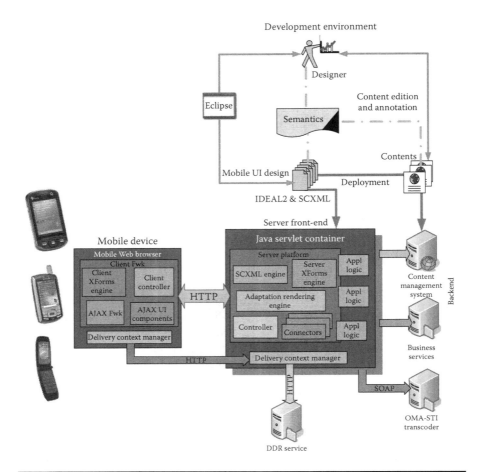

Figure 8.2   MyMobileWeb high-level architecture.

applications might be based on a content management system (compliant with the JSR-170 specification), which will provide the contents to be showed and adapted by MyMobileWeb.

On the top of the figure is depicted the development dimension: authors create applications by using the specification languages (IDEAL2 and SCXML) plus Java code that implements the application logic. An Eclipse plug-in facilitates the development and test activities. Once an application has been developed, it should be deployed on a Java Servlet container such as Tomcat 5 [25] or superior. The deployment process is two phased. During the first phase, the IDEAL2 specifications are translated by means of XSLT [26] to a series of JSP pages. Such JSP pages will be in charge of rendering the UI in collaboration with the adaptation and rendering engine. This pre-generation step is intended to improve performance at runtime.

Then, the pre-generated JSP's plus all the MyMobileWeb libraries and application Java code have to be deployed to the servlet container.

8.4.1 Client-Side Architecture

At the client side (browser), we can distinguish two main subsystems: the **client framework** and the (client-side) **delivery context manager**.

The **client framework** enables rich interactions between the user and the application, exploiting the Web browser capabilities at a maximum and leveraging the traditional Web page model. It includes functionalities typically present in AJAX frameworks but tailored to the restrictions imposed by the mobile environment. A number of different components are part of the client framework: the *XForms engine* in charge of performing validation and data binding without requiring authors to write Javascript code, the *AJAX Framework*, which provides: (a) convenience, cross-browser functions for controlling dynamically the presentation (DOM manipulation) and issuing asynchronous HTTP connections to the server side and (b) the client runtime support for UI components. All of them are coordinated by the *client controller*, which coordinates all the actions executed in response to user events.

The (client-side) **delivery context manager** is in charge of monitoring the dynamic properties of the delivery context, which are known at the device side. As such, it is capable of notifying changes (for instance, screen orientation changes) to the server via HTTP (AJAX).

8.4.2 Server-Side Architecture

At the server side two subsystems are present: the **server platform** and the (server-side) **delivery context manager**.

The server platform subsystem can be decomposed into different modules: the (server) *XForms engine*, the *SCXML engine*, the *adaptation and rendering engine*, the *controller* and several *connectors*. Once a new HTTP request is received, the *controller* starts to perform the actions to be taken. First of all, if the delivery context is not known yet, the delivery context manager subsystem is called to start the delivery context identification. After this operation, which usually occurs only when a new user session is created, the *XForms engine* is called to perform data validation and binding. Then, the *SCXML engine* is invoked. Depending on the event coming from the client side, the next actions to be taken will be decided: to get more data/contents by invoking the application logic (OAs), to go to the next presentation, to send new data/contents to the client framework (e.g., as a response to an AJAX request). Once the application data model has been modified the *XForms client engine* is called one more time. This time, it will bind the data that is stored in the server data model with the markup or data that will be sent to the client side. Then, the controller will call the *adaptation and rendering engine*, which will perform all the contextual adaptations (guided by the adaptation policies expressed by authors) as per the delivery context characteristics, provided by the **delivery context management** subsystem. During the adaptation and rendering process, it might be needed to call content repositories or transcoding systems. The *CMS connector* module is responsible for connecting MyMobileWeb with any JSR-170-compliant content repository. This module enables the usage of content-binding technology by MyMobileWeb authors. The *STI transcoding connector* offers connectivity to any OMA-STI compliant transcoding server or library, such as the open source Alembik transcoder. Once the adaptation has concluded, the render process takes place. This last step will produce the final markup and scripting code that will be sent to the browser.

8.4.2.1 Delivery Context Management Subsystem

This subsystem is responsible for informing the adaptation processes about the characteristics of the target delivery context. It is a façade, which hides all the complexity related to gathering contextual information. It is prepared to interact with different context providers, particularly the DDR service. Such a service offers a simple HTTP (REST) API for giving access to the information stored in different DDRs (WURFL, UAProf, Device Atlas). In fact, it is a wrapper of the device description framework previously described. To this aim, the DDR service offers interfaces, which give access to two functionalities: (1) identification of the device and Web browser used by examining the content of the HTTP headers sent by the device and (2) provision of all the information about the invariant characteristics of the device and Web browser used. Nonetheless, the delivery context management subsystem can interoperate with any DDR that implements the W3C's DDR simple API recommendation. Finally, the delivery context management subsystem receives periodic (through HTTP) notifications of property changes produced at the device side. For example, when an orientation change occurs, a notification is received and the value of the "currentOrientation" property is updated. Later, this information will be made available to the adaptation processes, which will act accordingly. For example, in landscape mode, a table might display additional columns.

8.5 Developing Applications with MyMobileWeb: An Example

This section is intended to give an overview of the development of applications using MyMobileWeb version 4.0. We will show some of the most important functionalities provided by this framework and how developers get access to such functionalities. We will demonstrate this by developing a (minimalist) multi-device mobile Web portal dedicated to the Spanish National Soccer League (*La Liga*).

8.5.1 Steps for Developing a Mobile Portal

As usual, the first step to start developing an application is to gather all the requirements that need to be met. Then the next activity is to design the information architecture of the portal, which will allow to determine the number of pages (presentations in MyMobileWeb jargon) needed and the navigation flow between them. Afterward, the structural and presentational aspects (style) of pages will be designed. The subsequent step is to create the business logic (application operations in MyMobileWeb's terminology) that will give functionality and content to your application. Once the application has been developed, it has to be tested on multiple devices (or emulation environments), making the necessary adjustments and corrections. Finally, the application has to be deployed in a production environment, making it available to users.

8.5.2 Requirements Analysis

The requirements for our mobile soccer portal are simple and concise:

R1 Users should be able to know the latest news about *La Liga*.
R2 Users should be able to know the latest news about their favorite clubs.
R3 Users should be able to know the rounds, games, and standings.
R4 Users should be able to navigate through a gallery of images of their favorite clubs.
R5 The portal must provide statistics about the competition, such as the number of leagues won by the different clubs throughout history.
R6 The portal contents must be presented both in English and Spanish (internationalization requirement).

Apart from these functional requirements, our application has to provide a harmonized user experience for multiple delivery contexts, especially for three "hero devices": Apple's iPhone, Windows Mobile 2005–6-enabled smartphones equipped with an IE mobile browser and Nokia Series 40 6212 feature phones. These three devices have been chosen because they have different form factors and input mechanisms: multi-touch screen and advanced gestures, stylus, or a classical four-way scroller.

8.5.3 Information Architecture

The proposed information architecture is depicted in Figure 8.3. The application will have an initial menu, which will give access to the different functionalities provided by the portal: the latest news of *La Liga*, the information about rounds, the information dedicated to specific teams, and the global statistics. The latest news about *La Liga* or about any specific team will be presented as a master detail list. We will also dedicate a specific presentation for club image galleries. At this abstraction

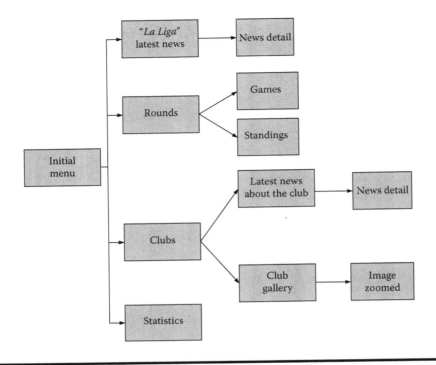

Figure 8.3   Information Architecture for the mobile soccer portal.

level, we have divided the presentation of the information about rounds in two different views: one for the games and another for the standings. The statistics will be presented in a single view, which will contain a nice bar diagram.

8.5.4 Setting Up the Development Environment

Now that we have a clear view of what we want to achieve and what is the information architecture of the application, the development activities can start. First of all, it is needed to configure the development environment. For doing so, the following products must be installed:

- Java Development Kit v1.5 or superior
- Eclipse IDE for Java EE Developers

Then, the MyMobileWeb SDK v4 and the MyMobileWeb Eclipse plug-in compatible with such a version have to be installed and configured. Both of them can be downloaded from the MyMobileWeb's Web site [27]. The application can also be developed without the Eclipse plug-in, but we strongly recommend using it as it simplifies and accelerates the development process. The following step is to create a new MyMobileWeb project using the Eclipse plug-in. This new project is already set up and configured to start the development and testing cycles of the new application. Due to space limitation reasons, we are not giving here all the configuration details that can be found in the corresponding user manuals [28].

8.5.5 Designing the Application Flow

The first step that needs to be completed in the development process is the design of the flow of the application in accordance with the information architecture previously presented. An application flow in MyMobileWeb is divided into the so called presentation operations (OPs). An OP represents a use case and it is composed by the set of presentations and the flow between them that realizes such a use case. Thus, we can identify the following OPs for our portal:

- **InitialMenu**: This is the first OP of the application and will be called once the application is invoked by the user. This OP will contain a presentation with a menu that will allow users to choose between the different functionalities offered by our portal.
- **News**: It corresponds to the use case responsible for presenting a list of news and the detail of each news presented, so at least two presentations will be needed.
- **Rounds**: This OP is in charge of presenting the games and standings for the latest rounds. As the standings and games are strongly related, we have decided to include both in the same presentation but in different sections.
- **Team**: This use case is concerned with the selection and presentation of information about a team. A presentation with a menu will allow users to select the team they want to know more about.
- **Gallery**: The Gallery OP is intended to present an image gallery allowing the user to select and display. Although this OP could be merged with the "Team" OP we have separated them for modularization and reusing purposes, as in future versions of the portal might be needed in another context.

8.5.5.1 Using SCXML as a Specification Language

The next step is to specify our application flow using W3C's SCXML [7]. Figure 8.4 graphically depicts the SCXML state machine corresponding to our application. In the following paragraphs, we describe the SCXML formalisms and how they are mapped to the elements that intervene in an application flow: OPs, presentations, events resulting from user interaction, and the application logic:

- *States*: Each OP is modeled as a state that has as many substates as presentations such an OP is going to have. Following this rule, it can be seen that, for instance, the "News" OP is modeled as a state identified as "News" that

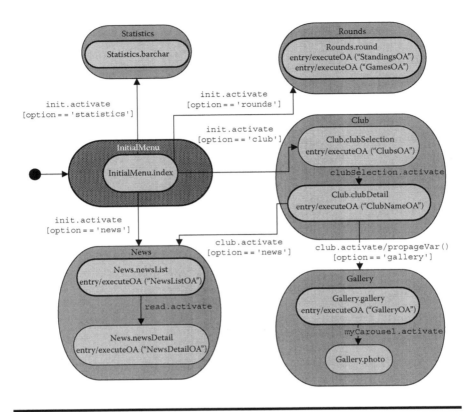

Figure 8.4 Application flow modeled as a finite state machine.

has two substates: "newsList" and "newsDetail." Thus, there are two kinds of states: OP states and presentation states. When the state machine is in a presentation state, it means that the user is actually viewing and interacting with such a presentation. The framework guarantees that when a presentation state is reached the application will navigate to the corresponding page.

■ *Transitions*: The flow between OPs and presentations within an OP is modeled as transitions triggered by events raised by the interaction of the user with the application and (optionally) controlled by logical conditions that operate over the data model. For instance, you can see that the *InitialMenu* state (corresponding to the *InitialMenu* OP) has transitions to the other OPs it gives access. Such transitions are triggered when a menu option is selected and are conditioned by the value of the menu option actually selected.

■ *Actions*: Actions are executed when a new state is reached or when a transition is triggered. Actions can be used basically to express what application logic, application operations (OAs) in my MyMobileWeb terminology, should be executed. For example, when the state "newsDetail" is reached, an OA called

"NewsOA" will be executed. Such an OA will obtain from an RSS service a list with the latest news, which will be displayed by the corresponding presentation. Actually, OAs are Java classes provided by the application programmer.

Apart from these transitions defined by the application programmer, MyMobileWeb adds automatically other transitions that deal with error conditions that may arise. For space limitation reasons, we are not presenting here the XML excerpts that declare the state machine, which implements our application flow. Interested readers can find it in the public SVN repository [29] that contains the source code of the example application.

8.5.6 Designing the Presentation Layer

The presentation layer in MyMobileWeb is developed using the IDEAL2 language. We will start looking into IDEAL using the presentation, which will give access to the different functionalities provided by our portal (index-xml). The XML corresponding to such presentation can be seen in Figure 8.5. There are two main blocks: the resources block in which dependencies are declared (script code, CSS sheets, etc.) and the "ui" section devoted to the structure of the UI. IDEAL2 is concerned only with the specification of the presentation structure. The look and feel and layout are controlled by means of CSS. This approach has the advantage of customizing the UI for different devices without duplicating the XML code.

A presentation in IDEAL is structured in sections. There are two (optional) special sections, the *header* and the *footer*. If declared, these two sections will always appear at the top or the bottom of the final page regardless of pagination. As in our portal, the header and the footer are always the same; they include common content to avoid duplication. A *section* typically contains one or more div blocks, which act as second-level containers. Different UI elements can appear inside a *div*. In our first example, we can see a menu that contains as many options as functionalities are provided by our portal. It is important to observe that the identifier assigned to the menu options is the same as the identifier used by the SCXML state machine.

8.5.6.1 Using Images

We would like to decorate our menu with nice images. Images in a mobile environment can be problematic for two reasons: not all devices or Web browsers support all image formats and the optimal image size can vary depending on the screen size available. To deal with these issues, MyMobileWeb provides two different mechanisms: *selection* and *transcoding*. Selection is a functionality that consists of choosing the best image from a set of alternatives (media set). Transcoding is the transformation of a source image provided by the developer in a target image, which satisfies all the restrictions in terms of formats and optimal size. The usage of transcoding or selection depends on the availability of different sizes and formats or in the nature of

```
<?xml version="1.0" encoding="UTF-8" ?>
<!DOCTYPE ideal2>
<ideal id="index" title="La liga">
 <resources>
  <link id="icon" rel="shortcut icon" expr="!mymw:belongsTo
     ('iPhone')" type="image/x-icon" href="${myFavIcon}" />
  <link id="iconIPhone" rel="apple-touch-icon"
     expr="mymw:belongsTo('iPhone')" href="${myFavIcon}" />
  <link rel="stylesheet" id="soccerStyle" href="soccer.css" />
 </resources>
 <ui>
  <body>
   <header id="header">
    <include content="Common/generic/common/header" />
   </header>
   <section id="main">
    <div>
     <menu id="init" ref="option" class="soccer">
      <a id="news" class="decorate" longtitle="Soccer News"
        resourceid="news">News</a>
      <a id="club" class="decorate" longtitle="My clubSelection
        club" resourceid="team">My club</a>
      <a id="rounds" class="decorate" longtitle="Standings and
        games" resourceid="table">Rounds</a>
      <a id="statistics" class="decorate" longtitle="Winners
        statistics" resourceid="statistics">Statistics</a>
     </menu>
    </div>
   </section>
   <footer id="footer">
    <separator class="line" />
    <include content="Common/generic/common/powered" />
   </footer>
  </body>
 </ui>
</ideal>
```

Figure 8.5   IDEAL2 presentation specifying the application initial menu.

the images themselves, as the transformation process may be degrading. Nonetheless, selection and transcoding are not exclusive, that is, images can be first selected and then transcoded. In our example, it is specified that an identifier (resourceid) of the media set has to be used. The images that are part of each media set have to be left in a configurable directory in the application space.

8.5.6.2 Defining the Look and Feel

To specify the look and feel and layout of presentations, IDEAL2 is used in conjunction with CSS. CSS sheets and styles are bound to presentations using the same

```
menu.soccer{
    layout:vertical;
    align:left;
    colored:true;
    background-color1:#ffd8ad;
    background-color2:#f7f6ef;
    white-space:wrap;
    width:100%;
}
a.decorate {
    img-display:both;
}
separator.line {
    display-as:hr;
}
```

Figure 8.6   CSS styles for the initial menu.

syntax as in HTML. Apart from the standard CSS 2 properties, MyMobileWeb defines extended properties. Figure 8.6 shows the style we have assigned to our menu. It will be displayed as a vertical menu, that is, one option per line, and will have alternative background colors. It is noteworthy that MyMobileWeb provides default CSS styles (optimized for different device families) for all the UI elements, which has been proved to be extremely helpful for developers.

Other elements of our page are also affected by this style sheet. For example, hyperlinks (a elements) are going to be displayed with their accompanying images. This example also demonstrates another important functionality of MyMobileWeb, which is how abstract UI elements can be rendered in different ways using the display-as attribute. In this particular case, the separator element is going to be realized as a horizontal rule. Other options are a line break or an image. This abstract UI rendering decision can also be made automatically by the adaptation engine (taking into account the characteristics of the device and the Web browser).

Figure 8.7 shows how our presentation is displayed in different devices and browsers. You can see how MyMobileWeb adapts the interface and look and feel to be the most suitable for each device. For example, the iPhone menu is displayed in big fonts to allow to make a touch selection.

8.5.7 Making a Presentation Dynamic

The previous example showed a presentation, the structure of which is totally defined at design time. Now we are going to demonstrate (Figure 8.8) how to create a dynamic presentation, which will depend on data or content calculated at runtime. In this case, we also have a menu for club selection. As the number of clubs in the national league can vary depending on the season, we would like to

Figure 8.7   Initial Menu screenshots (Windows Mobile, iPhone, Nokia 6212).

```
<menu id="myMenu" ref="club" class="clubs center">
  <a id="header" repeat-nodeset="clubList" src="${clubList.current.
     image}" href="${clubList.current.href}">${clubList.current.
     name}</a>
</menu>
```

Figure 8.8   IDEAL2 excerpt with a dynamic menu.

obtain the name of the clubs from a remote service. Thus, we specify a menu as well, but instead of putting the options directly in the presentation, we insert a template option with a repetition structure (`repeat-nodeset` attribute). This repetition structure will be controlled by a collection provided by the application logic described below. In this example, it is also shown how the attribute values can be made dynamic (using the JSP 2.0 "${x}" syntax).

8.5.8 Creating the Application Logic

Application logic in MyMobileWeb is specified by means of the so-called application operations. An OA is a Java class that implements a well-known interface, which defines an operation to be provided by developers: *execute*. Such an operation has an input parameter, which represents the current application data model (`Context` class). The `Context` class acts as an object container. Each contained object is associated with an identifying key. In the example below, first we obtain the list of clubs by calling a remote service, being each club a Java Bean structure with an attribute per field we need at the presentational level (Figure 8.9). The whole club list is represented by a Java List container. Once we have the container filled, we store it in the data model (*setElement* method) under the key "clubList." It can be observed that this key is exactly the same as the specified in the `repeat-nodeset` attribute

```
public class Club {
  private String name;
  private String href;
  private String image;
  // Constructor, getters and setters
}
public class ClubsOA extends BasicApplicationOperation {
  public void execute(Context the_context) throws OAException {
    List<Club> data = new ArrayList<Club>();
    // Calling the service...
    List<ClubInfo> clubs = DataHolder.getInstance().getInfo();
    for (ClubInfo clubInfo : clubs) {
      Club club = new Club();
      club.setName(clubInfo.getClubName());
      club.setHref(clubInfo.getURL());
      club.setImage(clubInfo.getPicture());
      data.add(club);
    }
    the_context.setElement("clubList", data);
  }
}
```

Figure 8.9   **Application logic to obtain the list of clubs.**

in the IDEAL2 authoring unit. Also, the names of the Java Bean attributes have to be equal to the names used in the presentation (image, href, name).

8.5.9 Customizing the Presentation Layer Depending on the Delivery Context

One of the most important functionalities provided by MyMobileWeb is the awareness about the device and Web browser used, that is, the delivery context, thanks to the information provided by the DDR service. This functionality can be exploited by authors to customize their application for a broad range of devices. In the example above, you can see how we provide a specific icon for the iPhone. Related to this is the style overriding feature intended to override properties of a style for specific delivery contexts. Using such a feature we can for instance, change the layout and appearance of our menu: vertical for iPhone devices, grid for the rest. Developers can specify by configuration their own device families or reuse those provided by the platform (Figure 8.10).

8.5.10 Using Tables

We are going to show how to use tables in the MyMobileWeb framework by focusing on the "Rounds" functionality. The idea is to provide information about each

Figure 8.10   Screenshots with the team selection menu.

round by displaying the games and the standings. To demonstrate these features, we have designed the presentation below. Such presentation has two sections: one for the games and other for the standings. Each section has a dynamic table controlled by a data model collection specified by means of the `repeat-nodeset` attribute. As previously noted, such an attribute indicates that there will be as many table rows as indicated by a collection that has to be present in the application data model. Our tables are dynamic because they will display different information depending on the round considered. Both tables belongs to the `paginate` class. This means that if the number of rows exceeds the maximum number of rows that device can manage reasonably, they should be automatically paginated by MyMobileWeb. Some of the columns of the standings table are subject to a condition over the delivery context to avoid having too many columns in small devices. The corresponding OAs that obtain the games and standings from a remote service are not shown due to space limitations, but they follow the same structure as the one previously described (Figures 8.11 and 8.12).

Figure 8.11 shows the final result obtained. It can be observed that the two sections are presented in two different tabs. The "tab layout" is one of the predefined layouts that MyMobileWeb provides. This functionality can be easily achieved by means of a CSS style.

8.5.11 Placard: Advanced Layout Combining Text and Images

The next example shows how a list of news can be presented to the user. Each news item will contain a title, text content, an accompanying image and optionally other

```
<section id="main">
  <section id="games" title="Games">
    <div>
       <table id="gamesTable" class="soccer paginate">
         <th class="header">
            <td>Date</td>
            <td>Games</td>
         </th>
         <tr repeat-nodeset="games">
            <td><output ref="date" /></td>
            <td><output ref="game" /></td>
         </tr>
       </table>
    </div>
  </section>
  <section id="standings" title="Standings">
    <div>
       <table id="standingsTable" class="soccer paginate">
         <th class="header">
           <td>Pos</td>
           <td>Name</td>
           <td>W</td>
           <td expr="dcn:deviceWidth() gt 300">L</td>
           <td expr="dcn:deviceWidth() gt 300">T</td>
           <td>Pts</td>
         </th>
         <tr repeat-nodeset="standings">
           <td><output ref="pos" /></td>
           <td class="bold"><output ref="name" /></td>
           <td><output ref="won" /></td>
           <td><output ref="lost" /></td>
           <td><output ref="tied" /></td>
           <td class="bold"><output ref="pts" /></td>
         </tr>
       </table>
    </div>
  </section>
</section>
```

Figure 8.11   Declaring multiple sections and dynamic tables in IDEAL2.

metadata such as author, ratings, etc. To display this item set properly, we have decided to use a placard component. A placard element enables the combination of text, images, and other elements in different positions. As usual, for those less capable devices MyMobileWeb will gracefully degrade the placard. MyMobileWeb has different placard layouts that correspond to the most common use cases. The role attribute is used to convey the purpose of each element within the placard, conditioning the place in which it will be finally rendered. Our placard is affected by a repeat-nodeset attribute, as the news are dynamically obtained from

Figure 8.12   Screenshots with standings and games.

an RSS service. Due to space limitation reasons we are not describing the application operation, which obtains the list of news. Interested readers can find the source code in the SVN repository. In addition, this example demonstrates how the range component can be used to display ratings, which are very common these days in Web portals.

With regards to the CSS styles, it is worth having a look at the style properties assigned to the image within the placard. The transcode property indicates that the image associated to each news item will be automatically transformed to fit in the area reserved to it. It is noteworthy that in this particular case, there is no better option than transcodification because such an image comes in an arbitrary size from an Internet RSS feed. Lastly, to avoid image deformation, we have indicated that the aspect ratio should be conserved.

Figure 8.13 shows how the placard has been finally rendered on different devices. When the "read more" hyperlink is activated the entire news item will be showed. Such functionality has been implemented using a simple IDEAL2 authoring unit with a (transcoded) image and a paragraph with dynamic contents (Figure 8.14).

8.5.12 Carousel: Displaying a Dynamic Catalogue of Objects

MyMobileWeb provides an abstract component called carousel intended to display a sequential catalogue of objects to be selected by the user. This feature can be used in our portal to show the clubs' image galleries. Figure 8.15 is an IDEAL2 excerpt declaring a carousel of images about clubs. As our images are provided by an external service (Flickr) we have opted to use a dynamic template affected by the repeat-nodeset attribute.

```
<placard id="placard" class="${newsCard}" repeat-nodeset="news">
      <p role="mymw:subtext" class="news bold">
       <span>${subject}</span>
       <a id="read" value="${value}"><read more...</a>
      </p>
      <p expr="dcn:deviceHeight() gt 300" role="mymw:subtext"
        class="news">
       <span class="sub">Date:<</span>
       <span class="sub italic">${date}</span>
       <span expr="!empty(author)" class="sub">  and author:
         </span>
       <span expr="!empty(author)" class="sub italic">${author}</
        span>
      </p>
      <p expr="dcn:deviceHeight() gt 300" role="mymw:subtext"
        class="news">
       <range id="rating" expr="dcn:deviceHeight() gt 300"
        class="rating" start="1" end="5" step="1" ref="rating" />
      </p>
      <image expr="dcn:deviceWidth() gt 300" role="mymw:icon-left"
        class="news" alt="img" id="img" src="${image}" />
</placard>
CSS

placard.newsCard {
      layout: card;
      width: 100%;
      border-radius: 6px;
      margin: 1;
}

placard.newsOddCard {
      background-color:#ffd8ad;
}

placard.newsEvenCard {
      background-color:#ec894b;
}

image.news {
      transcode:true;
      weight-width:25;
      aspect-ratio:true;
      vertical-align: top;
}
```

Figure 8.13   IDEAL2 excerpt declaring a placard component to display news.

Figure 8.14   A list of news realized by means of the placard component.

```
<section id="main">
  <div class="center common title.common" id="container" title="by
    Flickr">
    <carousel id="myCarousel" ref="myImage">
      <image id="header" repeat-nodeset="gallery" src="${gallery.
        current.src}" class="thumbnail" alt="${gallery.current.
        title}" value="${gallery.current.value}" />
    </carousel>
  </div>
</section>
```

Figure 8.15   IDEAL2 excerpt declaring a carousel element.

Once we have declared the carousel, we have to specify the application logic, which will call the Flickr service and will feed the presentation with the necessary data. As it can be seen in Figure 8.16, the Flickr service is invoked and then we fill a Java collection with all the data of the images to be displayed.

Figure 8.17 depicts how a carousel will be finally realized in different devices. It can be observed that the rendering is slightly different. Furthermore, the number of images per page and the optimal image size has been automatically calculated. In addition, big images have been transcoded into thumbnails. The carousel component exploits, when available, the AJAX capabilities of devices, thus the page is not reloaded during pagination. For iPhone devices the carousel supports multi-touch gesture events, which facilitate user interactions. The carousel functionality is an example of the powerfulness of the MyMobileWeb

```
public class GalleryOA extends BasicApplicationOperation {

 public void execute(Context the_context) throws OAException {

  String club = (String) the_context.getElement("club");

  Club club = DataHolder.getInstance().getClubInfo(club);

  Flickr f = new Flickr("f9cafaaa85466f34d41bfbbbf5c671d9");
  SearchParameters keyword_search = new SearchParameters();
  String[] search_parameters = { club.getName(), "futbol" };
  keyword_search.setTags(search_parameters);
  keyword_search.setTagMode("all");

  List<Map<String, String>> aux = new ArrayList<Map<String,
    String>>();
  try {
    PhotoList photolist = f.getPhotosInterface().search(keyword_
      search,30, 1);
    int cont = 0;
    for (Iterator<Photo> iterator = photolist.iterator(); iterator.
      hasNext();) {
      Photo photo = iterator.next();
      Map<String, String> map = new HashMap<String, String>();
      map.put("src", photo.getThumbnailUrl());
      map.put("title", photo.getTitle());
      map.put("value", photo.getMediumUrl());
      aux.add(map);
    }

  } catch (Exception e) {
    log.error(e);
  }

  the_context.setElement("gallery", aux);
  }
}
```

Figure 8.16   OA that invokes Flickr to obtain the list of images for the carousel.

framework and demonstrates how developers can create compelling mobile Web
applications without so much effort.

8.5.13 Developing Statistical Graphics for Mobile Devices

MyMobileWeb includes an extension library, which provides the capability
to render statistical graphics for multiple devices in different formats: vector
(SVG) or raster images. The library supports different kind of graphics: pie,
bar, scatter, etc.

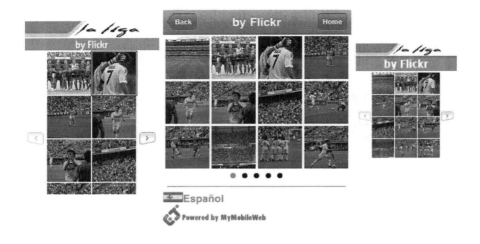

Figure 8.17   Carousel rendering.

```
<div id="container" class="center nowrap">
   <barchart id="barchart" alt="Leagues" src="resource/charts/
     teams/barchart.xml" ref="selectedBar" class="myChart">
     <label>List of winners</label>
     <caption axis="x">Club</caption>
     <caption axis="y">Number</caption>
   </barchart>
</div>
```

Figure 8.18   IDEAL2 excerpt containing a statistical graph component.

The example below demonstrates how easily we can support the statistics functionality required for our portal. A component called `barchart` must be included. Then, we use the `src` attribute to convey the data source (XML file) to be used. In this particular case, we want to show a bar chart that depicts the number of championships (national and European) won by the different clubs throughout history. We also need to specify the captions in each axis and the corresponding legends (Figures 8.18 and 8.19).

Figure 8.20 depicts the result obtained in our three hero devices. For instance, iPhone 3.0 owners can also experience an animated graphic, as the WebKit version included in such a device supports SVG animation capabilities.

8.5.14 Internationalization

The internationalization requirement of our application can be easily met. In fact MyMobileWeb provides a nonintrusive mechanism for creating multi-lingual

```
<chart id="data" type="BarChart">
 <data>
   <serie name="Leagues">
     <item y="31"/>
     <item y="19"/>
     <item y="9"/>
     <item y="6"/>
     <item y="2"/>
     <item y="1"/>
   </serie>
   <serie name="Champions">
     <item y="9"/>
     <item y="3"/>
     <item y="0"/>
     <item y="0"/>
     <item y="0"/>
     <item y="0"/>
   </serie>
 </data>
 <axes>
   <axisx name="Club">
     <value>Real Madrid</value>
     <value>Barcelona</value>
     <value>Atletico</value>
     <value>Valencia</value>
     <value>Deportivo</value>
     <value>Sevilla</value>
   </axisx>
 </axes>
</chart>
```

Figure 8.19   IDEAL2 XML data source for the statistical chart.

applications. Such a mechanism works as follows. Developers design the IDEAL2 authoring units (presentations) using their mother tongue. Later, all the literals used within presentations can be externalized (using a tool called "Literal Extractor") to XML files and sent to the translation department. At deployment time, the XML files with the translations in the different languages supported must be provided. At runtime, presentations will be rendered in accordance with the user's preferred language.

8.5.15 Deploying the Application

First of all it is it is important to remember that before deployment, all the IDEAL2 presentations must be pre-processed and converted to JSP pages. Then, the deployment process of a MyMobileWeb application is similar to any other Java servlet application. Nonetheless, additional steps might be required such as setting

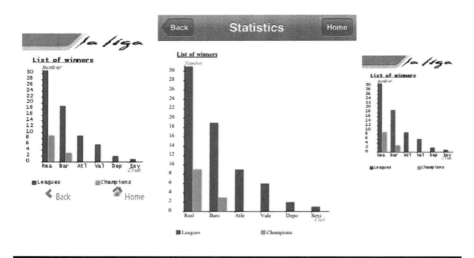

Figure 8.20   Statistical chart rendering.

up a DDR Service or an OMA-STI transcoder such as Alembik [18]. The detailed steps are not covered in the present work and can be found in online manuals and tutorials [28].

8.6 MyMobileWeb: The Community

MyMobileWeb was launched at the end of 2004 as an internal project at Telefónica I+D, a R&D&i company owned by Telefónica S.A., the Spanish leading telecommunications operator. The project was aimed at creating the technology to provide multi-channel access to mobile field service applications. At the end of 2005, Telefónica I+D decided to release the code in Morfeo [30]. Morfeo is an open source software community, whose members are enterprises, public administrations, universities, technological centers, and small and medium enterprises. They support the open source as an effective strategy for research, development, and innovation processes aimed to develop software technologies.

After the code release, a consortium was formed to ensure the continuous development of the product aligning it with new requirements and market demands. In 2006, the project was labeled as Eureka-Celtic, a European brand, which has guaranteed public funding to help the investment made during these years.

At the moment, MyMobileWeb is a product used successfully in several portals and information systems in different sectors, such as tourism, banking, or public administrations.

Finally, MyMobileWeb has made outstanding contributions to different W3C standards, namely, the DDR simple API recommendation and the delivery context ontology.

8.7 Conclusions and Outlook

MyMobileWeb makes the Mobile Web 2.0 a reality by providing the technology that enables the creation of compelling applications that offer a harmonized user experience in disparate delivery contexts. In addition, MyMobileWeb is an integral framework, which simplifies and accelerates the development of mobile solutions. Finally, MyMobileWeb is based on Web standards, which have an important competitive advantage in comparison with proprietary (closed) solutions. As a corollary, it can be said that by using MyMobileWeb, the time to market and the development costs are considerably reduced while at the same time improving the user experience.

MyMobileWeb is a product in continuous development, driven by an open source community focused on innovating in the area of mobile Web applications. Furthermore, for the next years, we plan to improve the technology to satisfy new market demands, as we believe that the Web browser will be the platform for application front-ends in the future mobile Internet. In fact, we are planning to develop a new set of components to enable the creation of richer, context-aware mobile Web applications offering an increasingly better user experience. To this aim, our medium-term roadmap is targeted to create (1) a framework to support offline Web applications (exploiting the upcoming HTML5 [31] offline capabilities) and (2) the runtime infrastructure for supporting context-aware applications that are respectful with user's privacy and security. To achieve these ambitious goals, we will adopt the upcoming open standards coming from the OMTP-Bondi and W3C DAP initiatives.

Acknowledgments

The authors wish to express their gratitude to the members of the MyMobileWeb Consortium for their support and contributions to the project. At the time of writing, they were Yaco Sistemas, Answare Technologies, Tisco GmbH, Germinus, Universidad de Valladolid, Universidad de Castilla La Mancha, Symbia IT, Internet Web Serveis, Fundación CTIC, Universidad Politécnica de Madrid, and Telefónica Investigación y Desarrollo.

References

1. J. J. Garrett, Asynchronous javaScript + XML (AJAX), February 18, 2005. Available at http://www.adaptivepath.com/ideas/essays/archives/000385.php. Retrieved on October 5, 2009.
2. R. Lewis and J. M. C. Fonseca (eds.), Delivery context ontology, W3C Working Draft (work in progress), June 16, 2009. http://www.w3.org/TR/2009/WD-dcontology-20090616/. Latest version available at http://www.w3.org/TR/dcontology/

3. J. Rabin and C. McCathieNevile (eds.), Mobile web best practices 1.0, W3C Recommendation, July 29, 2008. http://www.w3.org/TR/2008/REC-mobile-bp-20080729/. Latest version available at http://www.w3.org/TR/mobile-bp/

4. S. Owen and J. Rabin (eds.), W3C mobileOK basic tests 1.0, W3C Recommendation, December 8, 2008. http://www.w3.org/TR/2008/REC-mobileOK-basic10-tests-20081208/. Latest version available at http://www.w3.org/TR/mobileOK-basic10-tests/

5. Instant Mobilizer. Available at http://instantmobilizer.com. Retrieved on October 5, 2009.

6. Merkur. Available at http://merkur.fundacionctic.org. Retrieved on October 5, 2009.

6A. J. M. Cantera, C. Rodriguez, and J. L. Díaz, IDEAL2. Core language. Available at https://files.morfeo-project.org/mymobileweb/public/specs/ideal/. Retrieved on October 5, 2009.

7. K. Reifenrath et al. (eds.), State chart XML (SCXML): State machine notation for control abstraction, W3C Working Draft (work in progress), May 13, 2010. http://www.w3.org/TR/2010/WD-scxml-20100513/. Latest version available at http://www.w3.org/TR/scxml/

8. J. Rabin, J. M. C. Fonseca, R. Hanrahan, and I. Marín (eds.), Device description repository simple API, W3C Recommendation, December 5, 2008. http://www.w3.org/TR/2008/REC-DDR-Simple-API-20081205/. Latest version available at http://www.w3.org/TR/DDR-Simple-API/

9. WURFL (Wireless Universal Resource File). Available at http://wurfl.sourceforge.net. Retrieved on October 5, 2009.

10. Device Atlas. Available at http://deviceatlas.com/. Retrieved on October 5, 2009.

11. Open Mobile Alliance, UAProf specification, Open Mobile Alliance, October 2001. Available at http://www.openmobilealliance.org/tech/affiliates/wap/wap-248-uaprof-20011020-a.pdf

11A. W3C Working Group, Device description repository core vocabulary, W3C Working Group Note, April 4, 2008. Available at http://www.w3.org/TR/2008/NOTE-ddr-core-vocabulary-20080414/

12. World Wide Web Consortium, Document object model informative web site, World Wide Web Consortium. Available at http://www.w3.org/DOM/. Retrieved on October 5, 2009.

13. J. Boyer (ed.), XForms 1.1, W3C Recommendation, October 20, 2009. http://www.w3.org/TR/2009/REC-xforms–20091020/

14. M. Froumentin, R. Lewis, and R. Merrick (eds.), Content selection for device independence (DISelect) 1.0, W3C Candidate Recommendation (work in progress), July 25, 2007. http://www.w3.org/TR/2007/CR-cselection-20070725/. Latest version available at http://www.w3.org/TR/cselection/

15. B. Bos, T. Çelic, I. Hickson, and H. W. Lie, Cascading style sheets level 2 revision 1 (CSS 2.1), W3C Candidate Recommendation (work in progress), September 8, 2009. http://www.w3.org/TR/2009/CR-CSS2–20090908/. Latest version available at http://www.w3.org/TR/CSS2/

16. Open Mobile Alliance, Wireless markup language version 2 specification, Open Mobile Alliance. Available at http://www1.wapforum.org/tech/terms.asp?doc=WAP-238-WML-20010911-a.pdf

17. Open Mobile Alliance, OMA standard transcoding interface V1.0 specifications, Open Mobile Alliance. Release date: May 15, 2007. Available at http://www.openmobilealliance.org/technical/release_program/sti_v10.aspx. Retrieved on October 5, 2009.

18. Alembik. Available at http://alembik.sourceforge.net/. Retrieved on October 5, 2009.
19. D. Nueschler et al., JSR 170: Content repository for java technology API. Available at http://jcp.org/aboutJava/communityprocess/final/jsr170/index.html. Retrieved on October 5, 2009.
20. Alfresco. Available at http://www.alfresco.com. Retrieved on October 5, 2009.
21. R. A. Hosn, M. Womer, M. Froumentin, R. Lewis, S. Sathish, and K. Waters (eds.), Delivery context: Client interfaces (DCCI) 1.0., W3C Candidate Recommendation (work in progress), December 21, 2007. http://www.w3.org/TR/2007/CR-DPF-20071221/. Latest version available at http://www.w3.org/TR/DPF/
22. OMTP Bondi. Available at http://bondi.omtp.org. Retrieved on October 5, 2009.
23. W3C Device APIs and Policies Working Group. Available at http://www.w3.org/2009/dap/. Retrieved on October 5, 2009.
24. D. Coward et al., JSR 53: Java servlet 2.3 and Javaserver pages 1.2 specifications. Available at http://www.jcporg/aboutJava/communityprocess/final/jsr053/. Retrieved on October 5, 2009.
25. Apache Tomcat. Available at http://tomcat.apache.org. Retrieved on October 5, 2009.
26. J. Clark (ed.), XSL transformations (XSLT) version 1.0, W3C Recommendation, November 16, 1999. http://www.w3.org/TR/1999/REC-xslt-19991116. Latest version available at http://www.w3.org/TR/xslt
27. MyMobileWeb's web site. http://mymobileweb.morfeo-project.org
28. MyMobileWeb's online manuals. http://forge.morfeo-project.org/wiki_en/index.php/MyMobileWeb_Platform#Documentation
29. MyMobileWeb's public SVN repository. https://svn.morfeo-project.org/mymobileweb/
30. Morfeo open source community. http://www.morfeo-project.org
31. A. van Kesteren and I. Hickson (eds.), Offline web applications, W3C Working Group Note, May 30, 2008. http://www.w3.org/TR/2008/NOTE-offline-webapps-20080530/. Latest version available at http://www.w3.org/TR/offline-webapps/

Chapter 9

Mobile Search and Browsing

Berna Erol, Jiebo Luo, and Vidya Setlur

Contents

9.1 Introduction

Mobile phones are increasingly being used for functionalities other than voice calls and texting, such as e-mailing, collecting and managing multimedia, playing games, browsing the Internet, and performing many other business and personal tasks. Although the computation and memory capabilities, battery life, and connection speeds of mobile phones are increasing at a rapid pace, they still have some limitations that are likely to stay at least in the foreseeable future: small display size and limited peripherals for text input. Until foldable displays, mobile projectors, and

195

rollable keyboards become mainstream, these limitations will continue to make data entry and visualization of information on mobile phones difficult.

Nevertheless, mobile phones have some features that make searching for information easier compared to other devices. These features include

- *Integrated sensors*: Mobile phones have sensors such as GPS, digital compass, and accelerators that give applications/services more metadata to make intelligent decisions on user's intentions with search. For example, there are many mobile applications that rank search results based on location information.
- *Always connected*: The main functionality of mobile phones is to provide connectivity. Therefore, typically a mobile phone has connection to the Internet more often than the other devices, such as laptops.
- *Instant on*: When a user needs instant information, such as looking up a restaurant phone number or a location, it would take a user only a few seconds to turn on a mobile phone, where a computer would take several minutes to be turned on.
- *Personal*: The mobile phone is personal. Therefore it is likely that mobile phones are customized to user preferences, information about user contacts, and their preferred social networks that include links to friends. This information can be utilized in mobile search for better and more targeted results.
- *Always accessible*: Most people carry their mobile phone with them all the time. When a user needs information, particularly outside of their home and work, it is much likely that they have access to a mobile phone than they have access to a computer.
- *Integrated audiovisual capture*: Most smartphones have audio and visual capture capability. In the past, many content-based retrieval systems were developed, but they were not widely used. Submitting an image or an audio query to a personal computer requires many steps. Mobile phones make it simpler to submit audiovisual queries to content-based retrieval systems.

In recent years, many exciting mobile search applications have been introduced to take advantage of smart phone functionalities listed above [1–29]. These applications address a real pain point for most mobile users, allowing them to search with minimal or no text entry, using multimodal information such as location, and visual and audio capture.

Searching and browsing capability could be an integral part of many mobile applications. In this chapter, we review some of the state-of-the-art mobile search, browsing, and visualization techniques that could be useful for any researcher or developer who is developing applications and services for mobile phones.

9.2 Using Sensors and Context for Search

Continued advances in sensor technology are changing the way humans interact physically or virtually with the physical world around them. Some of these sensors first found uses in military applications and recently they have been increasingly adapted to consumer products such as mobile phones. In addition to visual and audio sensors, these sensors include

- Location sensors: GPS, A-GPS, magnetometer (digital compass)
- Motion sensors: accelerometer
- Ambience sensors: temperature, air pressure, humidity, heat/IR, light sensor, proximity sensor
- Affective state sensors: pulse, body temperature, blood pressure
- Display sensors: multitouch sensors

Algorithms and software were also developed that take advantage of this sensory information. Such software determines the context of usage in order to predict the user's intentions and retrieves relevant information. Mobile context can be categorized in many ways. Here we categorize it into three categories as shown in Figure 9.1: spatial context, temporal context, and social context.

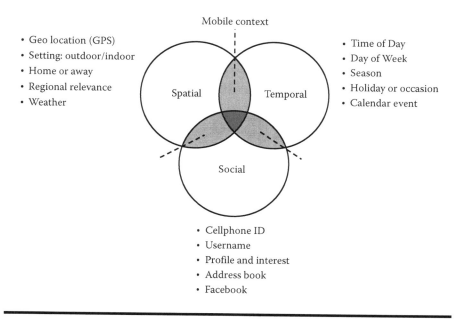

Figure 9.1 **Mobile context categorization: spatial context, temporal context, and social context.**

Mobile applications can utilize context information to intelligently search and retrieve information relevant to the spatial, temporal, and social situation of the user, for example they allow a user to do the following:

■ Get driving directions (spatial context)
■ Listen to the stories of historic landmarks by merely pointing a mobile phone at them (spatial context)
■ Locate friends who are nearby (spatial and social context)
■ Locate the nearest open pharmacy (spatial and temporal context)
■ Locate an open restaurant nearby that serve user's favorite cuisine and see comments made by user's friends on the food, service, and atmosphere (spatial, temporal, and social context)

Location-based search can assist users as they navigate in unfamiliar locations and enhance their experience as they visit popular points of interest. Many research prototypes and commercial systems were developed that use location information from GPS, A-GPS, and digital compass. For example, Loopt [5] is an application that allows users to share their location, status, and photos with everyone or just with their friends. As users move around, Loopt uses GPS to automatically update their location and status. Another application Start Walk [6] uses GPS and digital compass to provide astronomy information to users, such as names of the stars that they point their mobile phone to.

Nokia's Mobile augmented reality application MARA [1] enables sensor-based, video see-through mobile augmented reality (Figure 9.2). As users point their mobile phone camera to their surroundings, they can receive information that is overlaid on

Figure 9.2   MARA system explores video see-through mobile augmented reality. (From http://research.nokia.com/research/projects/mara/index.html. With permission.)

the video input. The application uses an accelerometer (for orientation), a compass (for direction), and GPS (for location) to add information to the images viewed on the screen. For example, it enables pointing your mobile phone to a historical landmark such as Empire State Building and getting all the related information.

For developers who plan to integrate augmented reality in their applications, Refs. [2,3] give a good overview of what it takes to make augmented reality practical on mobile phones. They also give an overview of software architecture of these systems and talk about development and debugging strategies.

Some systems, such as Zonetag system by Yahoo [7], use location information to label/tag photos with metadata. For example, tagging the picture of Golden Gate Bridge with "Golden Gate Bridge" and "San Francisco." They use location information to suggest tags for user's photos based on geotagging and existing community tags for the same location, making the tedious image tagging task much easier. Such tags are then used for categorizing and retrieving images in personal photo collections and in the Internet image/photo sharing sites.

Even though the latitude and longitude information from GPS is accurate enough to provide driving directions, it may not be enough to help identifying the exact building a person is standing in front of. Also GPS signal may not be available at indoor locations. Some of the image tagging systems use location information (context) and visual information (content) together for identifying tags for photos. For example, if it is identified that the person is taking a photo in San Francisco, content of the photo can be searched (e.g., using features such as color and shape) in a database that contains only the landmarks only from San Francisco.

Tagging of media content can facilitate later search based on textual terms, and location context is very useful for recognizing the semantic content in geotagged media. Such a premise is demonstrated [8], where the researchers obtain satellite images corresponding to the picture location data and investigate their novel use for recognizing the picture-taking environment. Moreover, this inference is combined with classical vision-based event detection methods that employ both color- and structure-based visual vocabularies for characterizing ground and satellite images, respectively. Modeling and prediction involve some of the most interesting semantic event-activity classes encountered in consumer pictures, including those that occur in residential areas, commercial areas, beaches, sports venues, and parks. The fusion of the complementary views achieves significant performance improvement over the baseline ground view.

Another important form of context information available on a mobile device is social context. On a cell phone, additional social context would be available, for example, from personal contact list or connections social networking accounts where the social connections among individuals and their friends are stored. These friendship links and other social network context can boost the accuracy of automatic face recognition in new photographs [9,10]. Certainly, time and location information (graduation, birthday, anniversary, Christmas at grandma's house, etc.) can also be used to help recognize people and their social relationship. This line of research is expected to push event recognition and media annotation to the next level.

9.2.1 Visual Search for Mobile

Most mobile phones have very limited keyboards and text entry is difficult. However, most of them support a very rich way for inputting search queries—via visual and audio capture. In this section, we will present examples of systems that initiate search via visual input, which is also known as visual search. Search initiated by audio capture will be covered in the next section.

Visual capture of mobile phones can be used for linking the physical world with the digital world in two ways: (1) using machine identifiable codes such as barcodes and QR codes and (2) through visual content-based search.

Today many mobile devices support QR code reading. There are many magazines and catalogs that link the paper articles to online information through QR codes. ZXing "Zebra Crossing" [11] is an open source QR code project where developers can use to embed QR code reading functionality into their own applications. Also ARToolkit is popular open source mobile software that allows linking of physical objects to digital information via tracking of machine readable codes [12].

Machine-readable codes offer a practical solution for linking physical media to online information. However, they may not be desirable in some cases, because of visual esthetics or difficulty of pre-associating QR codes with physical object or a media. In such cases, images/video of the physical object or media can be used to retrieve information via visual similarity search.

Visual search is synonymous with a field called content-based image retrieval (CBIR). It covers any technology that in principle helps to organize digital picture archives by their visual content [13]. Perhaps the biggest problem with all the current CBIR approaches is the reliance on visual similarity for judging semantic similarity, which may be problematic due to the so-called "semantic gap" between low-level content and higher-level concepts. This intrinsic difficulty in solving the core problem poses a great challenge for CBIR to become useful for real-world applications. In a broad sense, visual search must deal with potentially large-scale ontology and dataset, physical variations (scaling, distortion, lighting, pose, viewpoint, etc.), background clutter, and subjectivity in user intent. Using mobile phones, visual search can be made easier by using auxiliary information beyond pixels, facilitated by the mobile sensors and mobile context as we discussed in the previous section. A series of recent papers show that camera metadata and location sensors can be used in conjunction with visual content analysis to boost media understanding and media search [7,8,14,15].

Recently many practical applications and services using visual search became available. Some example applications include taking pictures of the following:

■ Books to perform price comparison
■ Products in catalogs to access product usage information
■ Real-life objects, such as shoes and handbags, to find similar products
■ Newspaper articles to receive electronic articles in the e-mail

SnapTell [16] is a popular commercial application of visual search on mobile phones where a user can submit a picture of a product (e.g., the cover of a book, CD, or DVD) and retrieve useful information such as price, seller rating, and product reviews. SnapTell's search technology is patent pending proprietary technology called "accumulated signed gradient" (ASG) and support accurate retrieval from a database of millions of images.

HotPaper is a technology developed at Ricoh Innovations [17,18] that enables users to *electronically interact* with paper documents by pointing their camera phones at a document. This is illustrated in Figure 9.3a. Electronic interaction is in the form of reading and/or writing images, audio and video clips, URLs, notes, etc. to the paper. HotPaper does not require presence of any machine-readable codes.

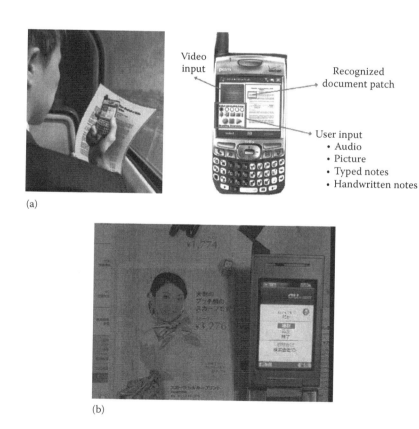

(a)

(b)

Figure 9.3   Examples of accessing online information from paper documents: (a) Ricoh HotPaper allow associating audio and video clips to anywhere on a paper document and (b) NetRicoh Catalog plays back a video clip that shows how to use a product. (Courtesy of Ricoh Innovations Inc, Menlo Park, CA.)

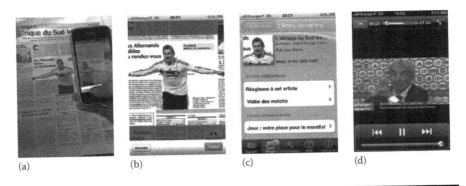

(a) (b) (c) (d)

Figure 9.4 DooG application allows a user to (a) capture an image from a newspaper, (b) submit this as a search query, (c) retrieve relevant information in the form of links, and (d) play/view relevant information. (Courtesy of E. Krzyzosiak, LeParisien & DooG, France, http://www.doog.mobi/)

It utilizes an algorithm called brick wall coding [17], which uses a small document patch image to retrieve electronic versions of documents without performing optical character recognition.

Other similar apps include Vodafone Otello and NetRicoh catalog. The latter goes one step further and can provide rich information on the product, for example, instructions in the form of a video or on how to wear a scarf. NetRicoh application is shown in Figure 9.3b.

DooG [19] is also another application that connects paper to online information. DooG application supports taking pictures of newspaper articles and retrieving relevant information. Screenshots of the DooG application are given in Figure 9.4.

LikeThis [20,21] is mobile application that allows taking pictures of products, such as shoes, handbags, dresses, and finds a list of visually similar products. This is illustrated in Figure 9.5. Users can then refine their search by brand name, shape, and color of the products.

9.2.2 Audio Search for Mobile

Audio is one of the most natural ways for humans to inquire about information. When people are mobile, such as walking and driving, audio input can be used for initiating queries and retrieving information. Many mobile phones support accessing contact information, placing phone calls via simple audio commands, and directory search. Recently, use of audio queries has been employed for more sophisticated mobile applications, ranging from singing a music tune for finding music albums featuring a particular song to making generic online search queries [22–29].

Shazam [22,23] is a popular mobile music search application. It supports recording few seconds of music clips from a sound source such as radio and identifies the matching songs. Shazam extracts a unique audio fingerprint and searches

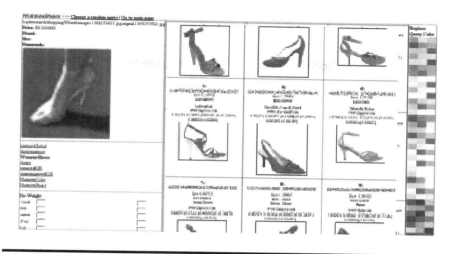

Figure 9.5 LikeThis mobile application and Like.com retrieves visually similar products given a query image of the product. (Courtesy of B. Gokturk, Like.com)

a very large database of previously computed audio fingerprints of songs. Shazam reported in 2008 that their database contains 8 million songs. Another system for mobile music search is [24], where Lu et al. from Microsoft Research proposed a system for "mobile ring tone search" via humming a melody.

Mobile applications that support Internet search via speech recognition have also been widely available. Examples include Google voice search [25] and Yahoo OneSearch [26], where voice is transcribed at a server and search results are displayed on the phone. Live Search Mobile [28] is an application from Microsoft that uses speech recognition to locate information related to local business such as phone numbers, directions, reviews etc. They report to achieve 60%–65% recognition accuracy. The accuracy of speech recognition and the voice queries depends on many factors, such as the size of possible search terms and the query language. A good introduction to voice search technologies can be found in Ref. [29].

9.3 Information Visualization and Browsing

The small display size of mobile devices makes it challenging to visualize and browse information, including browsing of mobile search results. Therefore, any mobile application and service provider needs to be aware of the state-of-the-art display and visualization techniques for the effective communication of such information. In this section we give an overview of the three directions for visualizing media on mobile phones:

1. Media retargeting for adapting photographs and video to smaller display sizes
2. Techniques for browsing and searching the mobile Web
3. Data visualizations on mobile devices, such as maps

9.3.1 *Media Retargeting*

For many display devices, a specialized art form has evolved in order to best author media for that device. Prime examples of such specialized art forms are television and movie production. However, the need for independently authored imagery for every display type becomes prohibitive as the number of display types grows. In this section we will present some of the tools that allow to author imagery once, and then automatically *retarget* that imagery for a variety of different display devices.

In most imagery, the story is communicated to the viewer via the presence of a few *key objects*. The remaining objects in the imagery provide a context for the story to occur, and are referred to as *contextual objects*. In Figure 9.6, the key objects are the boys and the soccer ball, while the contextual objects are the green

(a)

(b)   (c)

Figure 9.6 **Preserving functional realism rather than photo-realism by image retargeting: (a) The source image containing three areas of higher importance, the two boys, and the ball. (b) The source image retargeted to fit a PDA display. (c) The source image retargeted to fit a cell phone display. (Courtesy of Nokia, Palo Alto, CA.)**

grass and background crowd. The contextual objects are less important. Imagery that was effective for a particular task, message, display, or context may be ineffective when some of those aspects alter. Adapting imagery that was effective in one scenario to be effective in another scenario is called *retargeting*. Retargeting tends to preserve the story of the imagery by preserving *both* key and contextual objects in the same scene.

Several researchers have explored automating image retargeting through automatic cropping processes [14,30–32]. Some of these techniques [30] use automatic cropping based on a visual attention model, which uses saliency maps [31] and face detection to determine interesting areas in an image.

Recently, a nonphotorealistic technique for retargeting images was proposed in [33], which was the extended to video in [34]. Wolf et al. presented a system in [35] to retarget video that uses nonuniform global warping. They concentrate on defining an effective saliency map for videos that comprises of spatial edges, face detection, and motion detection. Seam carving for image retargeting was presented in [36] and later extended to video [37]. Media retargeting continues to be an active and important area of graphics research pertaining to mobile displays.

9.3.2 Techniques for Browsing and Searching the Mobile Web

The fastest growing community of Internet users is made up of people who use various mobile devices to access the Internet, such as mobile phones. As the amount of information available on these small-screen devices continues to grow, it is essential to prevent users from wading through irrelevant content to find a single piece of relevant information.

There have been attempts to automatically adapt Web content for mobile browsing that include creating thumbnails of Web pages that are readable in small displays [38–44]. Summary thumbnails help users identify viewed material and distinguish between visually similar areas [40]. Multimedia thumbnails algorithm converts documents and Web pages into audiovisual stream for utilizing the time and audio channel of mobile devices and compensate for the small display size [42]. Fishnet is a Web browser that displays Web pages in their entirety by using a fisheye view, that is, showing a focus region at readable scale and spatially compressing page content above and below the region [43]. More recently, the work of "Visual Snippets" explores how different representations are used in a variety of Web related contexts and presents a representation that supports help users with revisitation of Web pages [44]. Figure 9.7 shows an example of a dynamically generated icon that is viewable when a mobile user hovers on a hyperlink of a Web page.

9.3.3 Information Visualization on Mobile Devices

Visualizing of informational images, such as maps, on mobile devices involves a process of reducing information volume, while at the same time preserving the

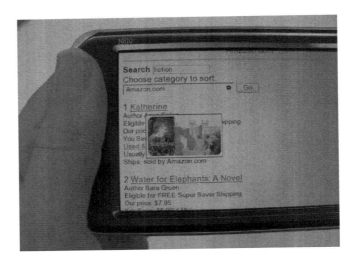

Figure 9.7 Viewing an automatically generated preview icon to aid directed search on a Nokia N810 mobile device.

significant information to be communicated. One of the most technically and intellectually challenging aspects of visualizing such information is a proper evaluation of the interrelationships between the different objects for a given application, the objectives or goals of the process, and the specific spatial and attribute transformations required to effect the changes. Map retargeting is a process of optimizing original map content for a specific purpose or display need. For example, retargeting map data to serve the navigational needs of mobile users or retargeting map data for a tourist wanting to explore a new city are different.

There has been some research in the area of combining physical and digital media for mobile map navigation. Reilly et al. [45] present a technique for relating paper tourist maps and electronic information resources using radio frequency identification (RFID) technology, called marked-up maps. Kray et al. describe a method of presenting route instructions on a mobile device depending on various situational factors such as limited resources and varying quality of positional information [46].

9.4 Conclusions

Mobile platform is emerging as the best platform for media search. The key reason is the availability of rich and valuable context information—right here, right now, and about the right people. Today, mobile phone's multimodal input interface and ubiquitous Internet connectivity allow people access information with unprecedented efficiency.

What does the future hold for mobile search? We imagine that more sensors will be integrated with mobile devices, such as indoor location sensors, RFIDs, making them even more powerful in sensing the world around us. Increasing connection speeds will allow mobile phones to communicate more effectively with cloud servers that have massive computation powers. This will remove any consideration for limited computation capacity of mobile devices, enabling even more powerful applications and experiences.

Mobile phones are becoming a second set of eyes, ears, sensing even before we do with accuracy better than we have. They are aware of our environment (in both the physical and virtual worlds) and providing exact information to us exactly when we need it. One cannot help but think: Future has arrived and our relationship with information and our environment will be forever changed.

References

1. M. Kahari and D.J. Murphy, MARA-sensor based augmented reality system for mobile imaging. In *Proceedings of ISMAR*, Santa Barbara, CA, 2006.
2. D. Wagner and D. Schmalstieg, Making augmented reality practical on mobile phones, Part 1, *IEEE Computer Graphics and Applications*, 29(3), 12–15, 2009.
3. D. Wagner and D. Schmalstieg, Making augmented reality practical on mobile phones, Part 2, *IEEE Computer Graphics and Applications*, 29(4), 6–9, 2009.
4. M. Etoh, Cellular phones as information hubs. In *ACM SIGIR Workshop on Mobile Information Retrieval*, Singapore, 2008.
5. Loopt, http://www.loopt.com
6. Star Walk, Vito Technology Inc., http://vitotechnology.com/star-walk.html
7. S. Ahern, M. Davis, D. Eckles, S. King, M. Naaman, R. Nair, M. Spasojevic, and J. Yang, Zonetag: Designing context-aware mobile media capture to increase participation. In *Workshop on Pervasive Image Capture and Sharing*, Irvine, CA, 2006.
8. J. Luo, J. Yu, D. Joshi, and W. Hao, Event recognition with a third eye. In *ACM Multimedia Conference*, Vancouver, Canada, 2008.
9. A. Gallagher and T. Chen, Using context to recognize people in consumer images, *IPSJ Transactions on Computer Vision and Applications*, 1, 115–126, 2009.
10. Z. Stone, T. Zickler, and T. Darrell, Autotagging facebook: Social network context improves photo annotation. In *First IEEE Workshop on Internet Vision*, Los Alamitos, CA, 2008.
11. ZXing ("Zebra Crossing"), Google Open Source Project, http://code.google.com/p/zxing/
12. ARToolkit, http://www.hitl.washington.edu/artoolkit/
13. R. Datta, D. Joshi, J. Li, and J.Z. Wang, Image retrieval: Ideas, influences, and trends of the new age, *ACM Computing Surveys*, 40(2), 1–60, 2008.
14. H. Liu, X. Xie, W. Ma, and H.J. Zhang, Automatic browsing of large pictures on mobile devices. In *Proceedings of ACM International Conference on Multimedia*, Berkeley, CA, pp. 148–155, 2003.
15. J. Luo, M. Boutell, and C. Brown, Exploiting context for semantic scene content understanding, *IEEE Signal Processing Magazine*, 23(2), 101–114, 2006.

16. SnapTell, http://www.Snaptell.com
17. B. Erol, E. Antúnez, and J.J. Hull, HOTPAPER: Multimedia interaction with paper using mobile phones. In *ACM Multimedia Conference*, Vancouver, Canada, 2008.
18. J.J. Hull, B. Erol, J. Graham, Q. Ke, H. Kishi, J. Moraleda, and D.G. Van Olst, Paper-based augmented reality. In *International Conference on Artificial Reality and Telexistence*, Esbjerg, Denmark, pp. 205–209, 2007.
19. DooG, http://www.doog.mobi/
20. Like, http://www.like.com/
21. X. Lin, B. Gokturk, B. Sumengen, and D. Vu, Visual search engine for product images, *Proceedings of the SPIE, Image Retrieval Applications*, 2008.
22. A. Wang, The Shazam music recognition service, *Communications of the ACM*, 49(8), 44–48, 2006.
23. Shazam Music Search, http://www.shazam.com/
24. L. Lu and F. Seide, Mobile ringtone search through query by humming. In *Proceedings of the ICASSP*, Las Vegas, NV, 2008.
25. Google Mobile App with Voice Search, http://www.google.com/mobile/apple/app.html
26. Yahoo OneSearch with Voice, http://mobile.yahoo.com/onesearch/voice
27. X. Xie, L. Lu, M. Jia, H. Li, F. Seide, and W.Y. Ma, Mobile search with multimodal queries, *Proceedings of the IEEE*, 96(4), 589–601, 2008.
28. A. Acero, N. Bernstein, R. Chambers, Y.C. Ju, X. Li, J. Odell, P. Nguyen, O. Scholz, and G. Zweig, Live search for mobile: Web services by voice on the cellphone. In *Proceedings of ICASSP*, Las Vegas, NV, 2008.
29. Y.Y. Wang, D. Yu, Y.C. Ju, and A. Acero, An introduction to voice search, *IEEE Signal Processing Magazine (Special Issue on Spoken Language Technology)*, 25(3), 28–38, 2008.
30. B. Suh, H. Ling, B.B. Bederson, and D.W. Jacobs, Automatic thumbnail cropping and its effectiveness. In *Proceedings of the 16th Annual ACM Symposium on User Interface, Software and Technology*, Vancouver, Canada, pp. 95–104, 2003.
31. L. Itti, C. Koch, and E. Niebur, A model of saliency-based visual attention for rapid scene analysis, *IEEE Transactions on Pattern Analysis and Machine Intelligence*, 20(11), 1254–1259, 1998.
32. L. Chen, X. Xie, X. Fan, W.Y. Ma, H.J. Zhang, and H. Zhou, A visual attention model for adapting images on small displays, Technical Report MSR-TR-2002–125, Microsoft Research, Redmond, WA, 2002.
33. V. Setlur, M. Neinhaus, T. Lechner, and B. Gooch, Retargeting images for preserving information saliency, *IEEE Computer Graphics and Applications*, 27(5), 80–88, 2007.
34. C. Tap, J. Jia, and H. Sun, Active window oriented dynamic video retargeting. In *Proceedings of the Workshop on Dynamical Vision*, 2007.
35. L. Wolf, M. Guttmann, and D. Cohen-Or, Non-homogeneous content-driven video-retargeting. In *Proceedings of IEEE International Conference on Computer Vision*, Rio de Janeiro, Brazil, pp. 1–6, 2007.
36. S. Avidan and A. Shamir, Seam carving for content-aware image resizing, *ACM Transactions on Graphics*, 26(3), 10, 2007.
37. M. Rubinstein, A. Shamir, and S. Avidan, Improved seam carving for video retargeting, *ACM Transactions on Graphics*, 27, 1–9, 2008.
38. S. Bjork, L.E. Holmquist, J. Redstrom, I. Bretan, R. Danielsson, J. Karlgren, and K. Franzen, West: A web browser for small terminals. In *Proceedings of ACM Symposium on User Interface Software and Technology*, Asheville, NC, pp. 187–196, 1999.

39. J.O. Wobbrock, J. Forlizzi, S.E. Hudson, and B.A. Myers, Webthumb: Interaction techniques for small-screen browsers. In *Proceedings of ACM Symposium on User Interface Software and Technology*, Paris, France, pp. 205–208, 2002.

40. H. Lam, and P. Baudisch, Summary thumbnails: Readable overviews for small screen web browsers. In *Proceedings of SIGCHI Conference on Human Factors in Computing Systems*, Portland, OR, pp. 681–690, 2005.

41. A. Woodruff, A. Faulring, R. Rosenholtz, J. Morrsion, and P. Pirolli, Using thumbnails to search the web. In *Proceedings of the SIGCHI Conference on Human Factors in Computing Systems*, Seattle, WA, pp. 198–205, 2001.

42. B. Erol, K. Berkner, and S. Joshi, Multimedia clip generation from documents for browsing on mobile devices, *IEEE Transactions on Multimedia*, 10(5), 711–723, 2008.

43. P. Baudisch, B. Lee, and L. Hanna, Fishnet, a fisheye web browser with search term popouts: A comparative evaluation with overview and linear view. In *Proceedings of the Working Conference on Advanced Visual Interfaces*, Gallipoli, Italy, pp. 133–140, 2004.

44. J. Teevan, E. Cutrell, D. Fisher, S.M. Drucker, G. Ramos, P. Andre, and C. Hu, Visual snippets: Summarizing web pages for search and revisitation. In *Proceedings of CHI*, Boston, MA, 2009.

45. D. Reilly, M. Rodgers, R. Argue, M. Nunes, and K. Inkpen, Marked-up maps: Combining paper maps and electronic information resources, *Personal Ubiquitous Computing*, 10, 215–226, 2006.

46. C. Kray, C. Elting, K. Laakso, and V. Coors, Presenting route instructions on mobile devices. In *Proceedings of the Eighth International Conference on Intelligent User Interfaces*, Miami, FL, pp. 117–124, 2003.

Chapter 10

Ontologies for Context Definition on Mobile Devices: Improving the Mobile Web Search

Jesús Vegas and Daniel Aréchiga

Contents

10.1 Motivation

Web pages are virtually the most useful and known Internet service, essentially because it brings the users any kind of information in an enormous volume. In the latest 1980s, when the Internet was in development, nobody was able to visualize the impact it will have a few years later. It is in the early 1990s that the Web revolution brought an exponential increase in servers, services, pages, and, specially, users.

Easy to use is a key factor responsible for the success of the World Wide Web. In addition, whatever be the interest of the user, it is very probable that there are many Web sites that offer related information. The real issue now, however, is how to find this information without getting lost in the attempt. The usual way to find the needed information within this sea of possibilities is through the use of Web search services like Google (Google, 2009), Yahoo (Yahoo, 2009), and Altavista (Altavista, 2009).

Web search engines are designed to provide an environment specially designed to find Web pages. These pages have been previously analyzed, indexed, and prioritized to finally show Web sites that fulfill certain elements to be considered like user's query results, that is, sites that possibly user wished to access.

Later, the user needs to manually analyze and discard the Web pages within the provided results to get the desired information, a tedious or even frustrating task when ambiguous or unrelated information is provided. When this occurs, the user must restructure the keywords used, adding or modifying these words, which in some cases ends with abandoning the information search.

On the other hand, modern mobile devices have evolved in their screen size, processing power, and connectivity possibilities, transforming them into powerful Internet access devices, consequently bringing the Web search capabilities to our palms.

There were nearly 4 billion mobile phone subscribers around the world in early 2009 (Americas, 2009). This implies a big potential market for Internet services for mobile users. Recent studies (Church et al., 2007) reveal that Web search will become a meaningful tool for mobile users in the next years.

Mobile Web search represents a different paradigm compared to a desktop search. In a mobile environment, the user needs a time and cost-efficient Web search as well as obtain significant results, taking into account the context that surrounds him or her, all in a limited interface device with a higher bandwidth cost.

A Web search engine could provide better fitted results to the user if the information that surrounds the circumstances of the query is taken into account. These elements are the context of the query and are formed by device and network characteristics, user preferences, location, and any other related element that could help to improve the search query.

To enhance and enrich the Web search queries, we have proposed the creation of an ontology that conceptualizes the context of the search query; this ontology will model concepts that provide information about the characteristics related to the user and device and its situations, providing an opportunity to improve the search results.

This work is based on a previous project, the MYMOSE Project (MORFEO— My Mobile Search, 2008), developed by collaboration between GRINBD research group of the University of Valladolid and Telefónica I+D.

10.2 Fundamental Concepts about Ontologies and Its Creation

The basic element in all the semantic applications is the use of ontologies. The most commonly agreed definition of ontology is as follows: "An ontology is an explicit and formal specification of a conceptualization of a domain of interest" (Gruber, 1993). This definition comprises two central elements:

- The conceptualization of concepts is formal and allows making some reasoning by computers.
- An ontology is designed for some particular domain of interest or knowledge.

Staab and Studer (2004) also define ontology as a 4-tuple (C, R, I, A) where "C" is a set of concepts (also known as classes), "R" a set of relations (properties), "I" a set of instances, and "A" a set of axioms (role restrictions).

The classes are the description of concepts from a domain of knowledge. Properties are characteristics and attributes from these concepts. The instances are specific and particular "real life" elements that populate the ontology. The role restrictions are the restrictions on the properties and classes, some rules that allow identifying when a class or property could or could not be linked, instantiated or used.

10.2.1 Creation of an Ontology

Because ontology is a conceptualization of knowledge, there is no "one and only" way to create it. If two different people create two different ontologies related to a certain domain, these ontologies will be different in few or more aspects, but they will not be equal, because the structure and elements of the ontology are very subjective.

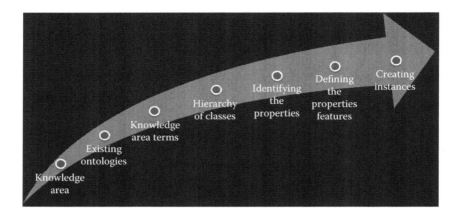

Figure 10.1   Ontology creation basic steps.

Even when creating ontologies, the fundamental reason to develop may change from different points of view or requirements. For example, if we create a "coffee" ontology, this could be different from the producer's and distributor's points of view and from the consumer's view as well. Maybe the distributors will be more interested in the market and prices or the wholesale inventory, that is, coffee management. From the consumer point of view, the information of interest could be about the area of origin, kind of coffee, taste and smell, and any other particular coffee characteristics. If we will develop an ontology for coffee inventory and production systems, it will be different from a coffee gourmet store ontology.

As we can see, there are some elements that should be taken into account during ontology creation. There are many coincidences on the ontology creation method according to some different researches (Grüninger and Fox, 1995) (Uschold and King, Towards a Methodology for Building Ontologies, 1995) that we can summarize as seven steps proposed by Noy and McGuinness (2000) that an ontology developer should follow (Figure 10.1). Note that the steps should be followed in order; the ontology creation is an iterative and cyclic process, especially in the last four steps.

10.2.1.1 First Step: Which Is the Knowledge Area We Want to Conceptualize?

As a first and most important step, we need to answer some fundamental questions that enable us to define the scope of the project. This step allows us to identify the real objective of the ontology, which we will create and define. The focus and subsequent steps follow:

- What domain will cover the ontology we are creating? This means the knowledge that we will want to conceptualize and make available for a system or knowledge sharing. As an example, is it a coffee, travel planner, or car parts ontology?
- What will be the use of ontology? In the coffee example, if the ontology aims at producers or consumers, then we have to define the focus on which the ontology will be used.
- For what types of questions the information in ontology should provide answers? It is an important question, because this only allows us to determinate how the relations will be structured to represent the related domain to support the use it will have.
- Who will maintain and use the ontology? This is not a secondary issue, because ontologies tend to grow and are in a constant development. It is important to establish who will follow this task and who can use it.

The answers provided to these questions will be changing as the development progresses, but this allows to define the scope of the ontology being developed.

10.2.1.2 Second Step: Are There Some Other Related Existing Ontologies?

As a second step, when creating an ontology, it is important to include the analysis and reuse of existing related ontologies. It is important to identify how some others work, because existing ontologies could provide a solid and proven base. This task becomes important and could be a requirement when our intended ontology will interact with other applications.

When searching for existing ontologies, there are some online libraries that provide many reusable ontologies. Also, some commercial libraries include ontologies that are publicly available.

It is possible to find an ontology representing a specific knowledge domain that is close to our needs; depending on the case, we can use it partially or use the structure adapting it to our requirements.

10.2.1.3 Third Step: Brainstorm Terms Related to the Area of Knowledge

The third step is related to the enumeration of domain terms. Creating a list of terms allows us to take the dimension and delimitating of the knowledge, which will be represented in the ontology. It does not matter at this point if a term will be a class or property and how will it interact with others. These issues will be determined in the following steps.

This list of terms will act as a feedstock to the next steps, which are most important. Some of the initial members may be deleted from the list or new members added based on the analysis of class, property, and axiom definition tasks.

10.2.1.4 Fourth Step: Create the Hierarchy of Classes

From the concepts of the list obtained in the previous step, we will determinate which will be classes and their hierarchy. Uschold and Grüninger (*Ontologies: Principles, Methods and Applications*, 1996) propose three approaches to solve this task:

- Top-down. Initiate with identifying the most general concepts to define it as top-level classes, and then define the next level of classes from the rest, and so on.
- Bottom-up. Start to discover the concepts that are more specific, which will be the lowest-level classes, and then group these classes under more general concepts to finally obtain the highest-level classes.
- Combination. This approach is necessary to identify the top-level and lowest-level classes and to later find the middle-level classes on the hierarchy structure.

There is no one preferred method for any case; the way to solve this task will be selected according to the ontology creation problem and needs, and even the developer skills and preferences.

The list generated in step three is used to discover not only the terms that identify objects, which are independent, but also terms that describe these objects. It is on this step that the relation between classes is defined, such relations as "is-a" or "kind of."

There are some extensive methods to evaluate the term characteristics that allow us to identify if an object is a class or a property, and how to ensure that the hierarchy structure is consistent and well done, but these are not described in this chapter.

10.2.1.5 Fifth Step: Identifying the Properties for the Previous Defined Classes

Once the classes are defined in the previous step, it is necessary to determinate which of the remaining terms will be properties in our ontology. The properties must be attached to classes because they represent characteristics of a related concept. For example, on the coffee ontology, the term "coffee" could be a class and the characteristics as "acidity," "taste," and "aroma" could be properties of the coffee.

It is important to know that all the inherited classes will derive their properties from their parent class. We should consider assigning the property to the most general class or the higher level class possible.

In this step, we found and added some terms to the ontology that were not considered in the previous steps. We need to remember that ontology creation is an iterative and cyclic process. Many methods to identify which terms will transform to properties and which class will be associated are available and these methods can be found in the literature related to ontology creation.

10.2.1.6 Sixth Step: Defining the Features of Properties

This step describes the characteristics of the previously defined properties with reference to the values they can take, such as the type, cardinality, allowed values, and other elements.

Continuing with the coffee example, the property "acidity" will be a *string* type, with a single cardinality because coffee can have one and only "acidity" characteristic, and finally the allowed values could be "strong," "moderate," and "light."

These definitions allow the developer to delineate limits and restrictions when creating instances in the ontology. It is necessary to take some time and analyze what restrictions should be imposed on the ontology and how these restrictions should be implemented.

10.2.1.7 Seventh Step: Creating Instances for the Ontology

This is the final step in the ontology creation method, but cannot be considered the last one because the ontology creation is an iterative process. In this step, we will create instances from the classes defined in the ontology.

An ontology is a structure and conceptualization of concepts of a knowledge domain. By means of the instantiation, we are creating a knowledge base that includes "real life" elements.

Some of these instances could be created before their use in applications in ontology development, but many of them will be created by the applications that require such ontology. The instances represent the content in the knowledge structure defined by the ontology.

Finally, there are some naming conventions for the classes and properties proposed by different experts, but even when these conventions are not a standard that everybody should follow. The main recommendation is to determinate which convention will be followed when developing an ontology and then using it consistently in all the ontologies. It is a good idea to analyze the requirements of the applications that will use the ontology or the reused ontology naming convention.

Most of the naming conventions are similar in some aspects and they are described as follows:

- Use the case of the letters consistently. The most useful and clear is to use lower case for all the words and upper case for the starting letter of any word (i.e., *FlavorQuality* or *TypeOfRoast*).
- Use complete words, without use of abbreviations (i.e., use *CoffeeShop* instead of *CS*).

- Avoid using symbols like accents or commas. It is even preferable not to use spaces because some systems cannot interpret it adequately, and the premise of ontologies is to share knowledge.
- Use singular for all the names and use plural only when required (i.e., use *TypeOfCoffee* instead of *TypesOfCoffee*).

10.3 Conceptualizing the Context on Mobile Devices

Understanding the context as the interrelated condition in which something exists or occurs (*Merriam-Webster's* online, "context," 2009) or the surroundings, circumstances, environment, background, or settings, which determine, specify, or clarify the meaning of an event (Wikitionary, a wiki-based Open Content dictionary, "context," 2009), we can create a semantic structure that supports the conceptualization of the context in mobile devices to be used to enhance the mobile Web search.

This semantic structure is defined as a context model. The context model (Arias, Cantera and Vegas, 2008) is a formal representation of all the attributes required to identify and manipulate the context and has been layered into three levels of properties or attributes (Figure 10.2).

- Directly fetched properties. These are properties that can be automatically gathered from context information sources. For example, the location coordinates obtained from a GPS device, or the date and time.
- Derived properties. These are "implicit" properties that can be inferred or calculated directly from fetched properties. They constitute an abstraction layer on top of the directly fetched properties layer. For example, the name of the country and region, taking into account the location coordinates provided by a GPS, or the season and whether it is day or night from the date and time.

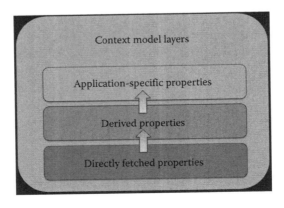

Figure 10.2 Context model properties layers.

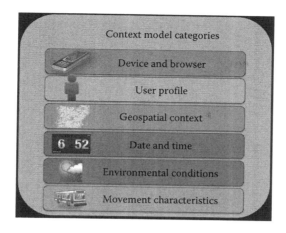

Figure 10.3 Context model categories.

- Application-specific properties. Applications might define or redefine additional properties on top of the basic layers by using their own rules, as long as they ensure formal consistency.

The context model is completed with categories of information required to conceptualize the context on mobile devices (Figure 10.3). From these categories must emerge the definition of classes, properties, and their relationships.

The defined categories classify the concepts relative to a certain domain of data according to the characteristics of the following:

- Device and its capacities: a qualification regarding device's display capacities, processing power, and connection characteristics, among others.
- User and his or her preferences: the historic preferences on queries and results.
- Location: the physical situation and its characteristics.
- Date and time, and any other temporal items, which could be valuable as context elements.
- Environmental information, like weather and daylight hours.
- User's movement and its characteristics.

10.4 Creating an Ontology to Conceptualize the Mobile Context

The previously described ontology creation process is a useful tool that can be applied to propose the construction of the context conceptualization on mobile devices. This conceptualization will allow us to provide a structured knowledge management related to the context with the objective of improving the Web search from mobile devices.

At this time, we are working on the final details in the context ontology creation, a complex and long work in progress. In this chapter, we will only explain the first three steps in ontology creation and the rest of the process will not be described.

10.4.1 Determining the Domain

As a first step, we need to determinate the domain in which the definitions will be used. This is clearly identified as the description of the *context* to support Web search *on mobile devices*. From this premise, we continue with the ontology creation process. Although this is a simple step, a clear and well-defined domain will help us obtain better results.

10.4.2 Identifying Existing Ontologies

The second step when starting an ontology-based project is the search for any standardized and tested ontology structure. We find some other ontology or groups of ontologies related to the context definition in the mobile environment. The focus of these ontologies are different and they are based primarily on their creation motivation, and even when there are some similarities with our needs, the differences are lot more relevant.

We analyzed the existing mobile ontologies, their motivation, and a brief analysis of structure to determinate if they could be functional for our purpose. Although the previous context ontologies were used to enhance the Web search (Gauch, Chaffee, and Pretschner, 2003) (Sieg, Mobasher, and Burke, Ontological User Profiles for Representing Context in Web Search, 2007) (Pretschner and Gauch, 1999), these efforts were focused on the desktop environments and do not include the conceptualizations required for the mobile search characteristics.

Also there are previous proposals that use ontologies for mobile applications using the context as an improvement (Weiβenberg, Voisard, and Gartmann, 2004), but with a significant difficulty to adapt to the Web search.

Basically, we have three ontologies related to the context definition on mobile devices. These are the Spice Project (SPICE, the Mobile Ontology, 2008), the W3C Delivery Context Ontology (World Wide Web Consortium, 2006), and the MYMOSE Project ontologies (Cantera Fonseca, García Bernardo, and Vegas Hernández, 2008).

10.4.2.1 Spice Project's Mobile Ontology

The Mobile Ontology (SPICE, the Mobile Ontology, 2008) (Villalonga et al., 2007) was proposed after the Spice Project on the European Community's Sixth Framework Programme (FP6).

The SPICE project was created as a framework for mobile services. The ontology use was required to obtain a shared and agreed view. Ontologies are a semantic base for shared interfaces and exchanged data. Their premise was to develop a standardized mobile ontology for every need required for mobile devices. The ontology is formally specified in the Web Ontology Language.

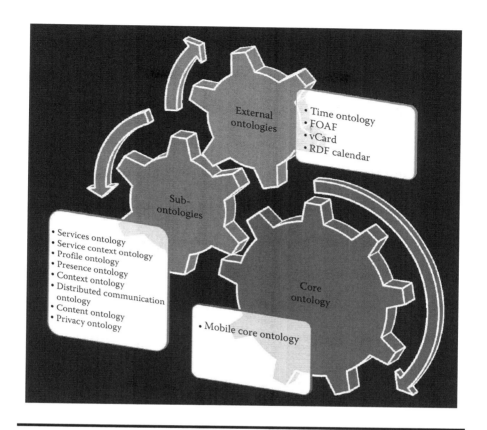

Figure 10.4 **SPICE ontology overview.**

The architecture of the SPICE ontology is structured as three levels of ontologies (Figure 10.4): one main or core level, which includes the base ontology; another level of subontologies, which includes all the function-specific ontology definitions, that inherits from the core ontology; and finally external level ontologies that group all the ontologies not created by this project. These classification levels and ontologies are structured as follows:

Mobile Ontology Core
- Mobile ontology core

Subontologies
- Services
- Service context
- Profile
- Presence

- Context
- Distributed communication sphere
- Content
- Privacy

External ontologies imported into the mobile ontology
- Time ontology
- FOAF
- vCard
- RDF calendar

In the scope of SPICE, the Mobile Ontology is designed to serve as data exchange format between the components of the platform as well as a reasoning tool.

These sets of ontologies are focused on mobile environments, but the fundamental concept was to establish a stretch collaboration between the user, device, and services. There are some class definitions with similar interests to our needs, but the delivery context and user context have different priorities and focus on the ones we are looking for a context definition. Following are some examples of the ontologies that describe these differences.

- The SPICE's Context Ontology contains more than 200 class definitions, and most of them are basically take from an external source, the Amigo Project (AMIGO, Ambient intelligence for the networked home environment, 2008), another European project related to the home networking with definitions focused on the home context, which is not compatible with the mobile environment as we visualize it.
- The context definitions contain basic location classes and properties, based primarily on the geographical position, including physical situation inside buildings, obtained from sensors like radiofrequency identification (RFID) tags, limiting the possibility of obtaining the location of the device to buildings that support this infrastructure.
- The context definitions are concentrated on the device characteristics (Figure 10.5) based on the environmental information about the device. For example, there are class definitions like "Speaker" derived from another class called "OutputDevice," or another, which indicates the "ProcessingUnit" characteristics.

This approach is a complete and a well-structured effort, but does not comply with the philosophy and needs required for a context-supported search because it does not take into account other contextual elements like type of place (i.e., *airport* or *at the beach*), season or weather concepts (i.e., *rainy day* or *winter*), or even time factors (i.e., *early morning*).

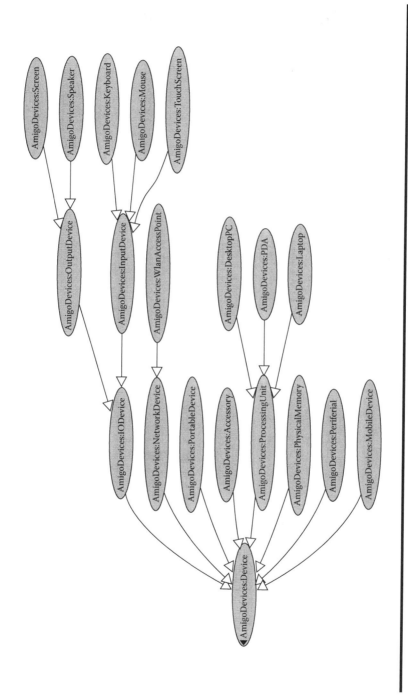

Figure 10.5   SPICE context ontology example.

10.4.2.2 W3C's Delivery Context Ontology

Another important element in the mobile Web search is the delivery context. This is the combination of characteristics of the device where the search results will be delivered, including the device distinctiveness, which could be relevant for the best content adapting and displaying experience.

This is a well-understood issue by the W3C and is reflected in their Delivery Context Ontology definition (W3C's Delivery Context Ontology, 2008). This ontology contains fine structured classes and properties, which include the characteristics of the device, the software used to access the service and the network providing the connection among others.

Every concept is very exhaustively analyzed and modeled in the ontology, which includes eight base classes; four of them include an extensive set of subclasses, properties, and relation definitions (Figure 10.6).

For example, the "Delivery Context Location Entity" class definition (Figure 10.7) not only contains many subclasses that describe obvious information like geographical coordinates (GeoCordinates class) but also contains classes to model concept data about how the location was acquired and the provider.

10.4.2.3 MYMOSE Project Ontologies

These set of ontologies are perfectly focused on the conceptualization of context on mobile devices to enhance the Web search and are divided into two parts:

■ One related to the domain concepts that the system could use for a query enhancement, containing associations between terms and context in which this terms are allowed. Also contains properties related to the terms, e.g., the term "Restaurant" is linked to "Japanese," "Italian," and "Mexican" as types of restaurants. This is conceptualized by the Domain Ontology.

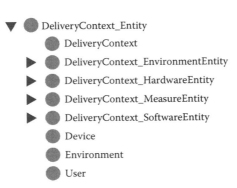

Figure 10.6 **W3C's delivery context ontology base classes.**

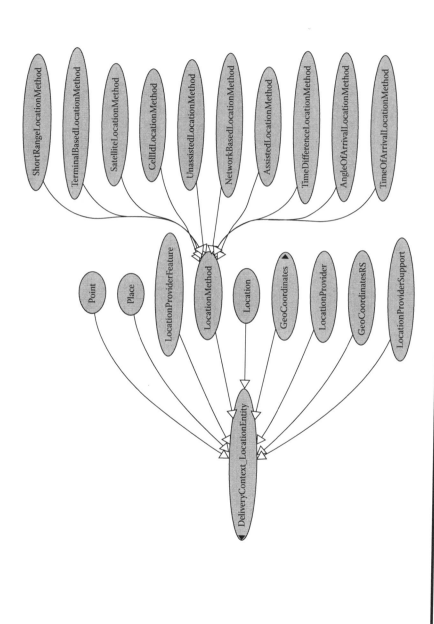

Figure 10.7 W3C's ontology delivery context location entity class example.

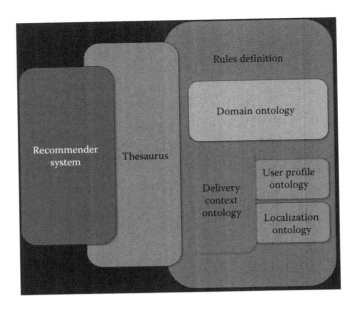

Figure 10.8   MYMOSE context management structure.

■ Another part associated with the context definition. Here the context is conceptualized by means of three ontologies, the Delivery Context, User Profile, and Localization ontologies.

All the MYMOSE ontologies interact with the user interface by means of a set of rules and a thesaurus. This ontology contains the terms used by a word recommender interface (Figure 10.8). Also this ontology considers a classification of the delivery context or device, based on the Mobile Web Initiative (W3C's Mobile Web Initiative Working Group, 2008) specifications.

These ontologies are well structured and designed with two fundamental strengths: the mobile context definition and a well-built device physical and technical characteristic classification. Even when the mobile context definition is focused on the mobile Web search enhancement, there are some elements that are not fully developed and exploited and can be improved or optimized such as the user profiling and interests and concepts related to the movement, taking into account that we are working in the Web search enhancement in mobile environments.

The first two discovered ontologies are well structured and complete, but do not include elements required for our needs and could not be used as context definitions to enhance the Web search. The last one is designed for the task, but has some missing elements. This analysis allows us to identify some ideas and concepts that could be taken into account in the next step.

10.4.3 *Identifying the Knowledge Domain Terms*

In the third step, we discovered the terms related to the domain of study that should be considered. We raised a question about the context issues that could be considered in the context definition:

■ What are the main elements to consider as context in a mobile environment?

This question led us to identify six categories of concepts that could be relevant to consider as a context when enhancing a query search. These categories are as follows:

1. The device and network characteristics
2. The user preferences
3. The physical situation
4. The temporal conditions and circumstances
5. The environmental conditions and circumstances
6. The user movement characteristics

These will be the main groups of concepts that determinate the terms that need to be conceptualized for the ontologies required for context definition. Based on these top level concepts, we discovered a second level of terms.

In the following, we show and describe the second level of terms discovered for each of the top level groups of concepts. These terms will be equivalent to classes in the proposed ontologies. Each of the next tables describes each group of top level concepts. Lower levels could exist, but in this research, we are trying to identify the first two levels as a base structure for an ontology development.

In the first group of terms, the device and network characteristics (Table 10.1) are described. This group identifies the terms related to the device processing and displaying capabilities and the network characteristics where the user is connected from.

Initially we do not consider the concepts related to the more specific characteristics about the device like battery charge or if Bluetooth enabled.

The second group contains the user preferences concepts (Table 10.2). These concepts represent the user's preferences relative to the previous searches or visited Web pages. This allows us to identify the preferred type of page, results, and page format among other concepts.

The physical situation (Table 10.3) is represented by another group of terms. These concepts characterize the location of the user and device and its characteristics.

The following group of concepts are the related to the temporal conditions and circumstances (Table 10.4). In this group, the characteristics of the date, time, and its relativeness are conceptualized.

Table 10.1   Device and Networks Characteristics Concepts

Concept	Term	Description
Kind of browser (determinate the browser type and characteristics)	BrowserType	Describe the browser type (Opera, Safari, etc.)
	BrowserVersion	Browser's version
	BrowserJava	Supports java?
	BrowserFlash	Supports flash?
	BrowserWAP	WAP enabled?
	BrowserWEB	WEB enabled?
	BrowserCookie	Cookie's enabled?
	BrowserPDF	PDF support?
Kind of processor (determinate the device's processor characteristics)	CPU	Describes the processor characteristics
	CPUSpeed	Represents the processor's speed
Screen (determinates the screen characteristics and capabilities: size, resolution, displaying colors, etc.)	ScreenSize	Screen size in pixels
	ScreenType	Screen type
	ScreenResolution	Screen resolution
	ScreenImages	Images support?
	ScreenColor	Color support characteristics
Network connection (describes the characteristics of the connection)	ConnectionType	Describes if the device is connected upon Wi-Fi, GPRS, 3G, etc.
	ConnectionSpeed	Actual connection speed
	ConnectionCost	Actual type of connection cost
	SignalStrength	Measuring of signal actual strength

Table 10.1 (continued)   Device and Networks Characteristics Concepts

Concept	Term	Description
Device capabilities description (characteristics of the page that can be displayed on the device)	DeviceID	Identification of the device characteristics (WURFL)
	DeviceLevel	Device classification according their technical characteristics
Localization method (GPS, aGPS, Wi-Fi, etc.)	LocalizationMethod	Identify the method to obtain the localization
Sensors (describes if another kind of sensors exists)	Sensor	Instantiates the existing additional device sensors (biometric, accelerometer, pressure, etc.)

Table 10.2   User Preferences Concepts

Concept	Term	Description
User identification	UserID	Identification of the user
Preferred Web sites (stores the Web sites that user usually navigate: cinema pages, sports pages, etc.)	PageURL	Visited Web site page URI
	PageType	Characteristic of the page visited
	PageNoOfVisits	Stores the number of visits to determinate URL
	PageTypePreferred	Store the preferences about recreational activities obtained from previous searches
Preferred results (obtained from previous search queries and results)	PageFormatPreferred	Store the preferred type of page format (HTML, SHTML, No-Flash, etc.)
	PageDistancePreferred	Store the preferences about distance obtained from previous search results

Table 10.3   Physical Situation Concepts

Concept	Term	Description
Geographic location	GeographicLocation	Location in degrees decimal format
	UTMCoordinates	Location in UTM format
Political geographical location (describes the political geographic actual location)	Country	Actual country location
	AdmRegion	Actual administrative region (state, province, canton, etc.)
	Town	Actual town location
	Street	Actual street location
	Number	Actual number of street location
	PostalCode	Actual location postal code
Kind of place	TypeOfPlace	Airport, downtown, beach, rural, etc.
User's location Characteristics	TypeOfLocation	Indoor, outdoor, etc.
	TypeOfSituation	Business, pleasure, etc.

Table 10.4   Temporal Conditions and Circumstances Concepts

Concept	Term	Description
Date and time	Time	Actual time in location
	Date	Actual date in location
	Season	Actual season of the year
	TimeZone	Actual time zone
Relative time	RelativeTime	Early morning, at night, etc.
Relative date	RelativeDate	Weekday, early month day, middle of the year, etc.

Another group was created to represent the environmental conditions and circumstances (Table 10.5), where information related to the environment such as climatic or daylight elements are conceptualized.

Finally, the group of user movement characteristic concepts (Table 10.6) are proposed. This helps us conceptualize the terms relative to the movement of the user/device and its characteristics.

Table 10.5 Environmental Conditions and Circumstances Concepts

Concept	Term	Description
Weather and forecast	WeatherActual	Description of actual weather
	WeatherForecast	Description of weather forecast
Day characteristics	SunsetTime	Time of sunset for today
	SunriseTime	Time of sunrise for today

Table 10.6 User Movements Characteristics Concepts

Concept	Term	Description
Kind of movement (slow, fast, etc.)	MoveSpeed	Actual movement average speed
Relative distance	MoveType	Actual type of movement (inter-city, inside city, etc.)
Type of transport	TransportType	Actual type of transportation (feet, bus, train, etc.)

Finally, these proposed terms were considered as a base collection of terms, which enable us to conceptualize the Web search context from a mobile device.

Currently, subsequent processes of ontology creation are in their final stages, in a series of iterative tasks between the last layers of development. From this process, we will obtain the first release of the complete ontology or set of ontologies, which will be published for free use.

10.5 Using the Context to Improve the Web Search from Mobile Devices

Even when only a little fraction from the total of people with mobile devices are connected to the Internet, the mobile Internet market is growing faster (Church et al., 2007). Statistics show how the increasing number of mobile Internet–connected devices will double in a couple years (Americas, 2009), making the mobile Internet one of the fastest-growing sectors.

The initial idea of improving the Web search in mobile devices using ontologies comes from the MYMOSE Project (Cantera Fonseca, García Bernardo, and Vegas Hernández, 2008) (Aréchiga and Vegas, 2009). This project was developed with the main objective of developing tools to provide an enhanced Web search, which brings mobile users a new experience, giving the user the most related, context-aware, and better device-adapted results.

Although this proposal already contains ontologies, we are trying to improve and enhance these semantic tools to strengthen the support provided by creating a new and more complete semantic structure.

The current progress in the Web search project offers an interface, which provides easy-to-use features like thesaurus-based word recommender supported by context definitions and a distance and device characteristic clustered results. The ontologies are a fundamental support, creating a semantic base to conceptualize the required elements.

The search system offers an interface specially designed to fit any kind of mobile browser, with a strong item in a word recommendation system, which allows the user to reduce the number of keystrokes needed to introduce the search terms.

The word recommendations are made by the use of a thesaurus, which provides its recommendations based on rules and ontologies. The thesaurus contains words that can be recommended to the user while the ontologies (supported by rules) determinate if some word should be recommended or not according to the actual context characteristics.

A fundamental element to obtain the desired behavior in this proposal are the implemented rules. Rules are defined to integrate the recommender system, the thesaurus, and the ontologies. These rules allow us to specify whether a certain concept is suitable or not in the current context, after determining if a relation is valid (if concept is suitable in context). Then the recommender system will show it as a recommendation and after a term is selected will show the related properties from the domain ontology. This relation can be described as follows:

$$\text{concept}(?c) \wedge \text{context}(?t) \wedge \{\text{conditions}(?t)\}$$
$$\rightarrow \text{isSuitableInContext}(?c; ?t).$$

where ?c is a *concept* that will be evaluated against the current *context* and *conditions* specify which conditions should the context satisfy for that concept to be suitable. The conditions term is a logical equation representing the conditions that the delivery context must satisfy for that specific concept to be suitable in the context.

We are working on these context definitions with a special focus for our efforts in the leisure and travel concepts. Some previous studies related to the mobile environment identifies that the user's preferences of information are associated with leisure and entertainment (Kamvar and Baluja, Deciphering Trends In Mobile Search, 2007) (Baeza-Yates, Dupret, and Velasco, 2007).

10.5.1 Scenario of Using a Web Search Supported by Ontologies

Now, we can easily imagine a scenario for the mobile search situation: A user goes to an unknown city for professional reasons or on vacation. As a stranger to the city, he or she will need to know about the services he or she will require such as accommodation, food, transportation, and entertainment activities.

This user takes his or her Internet-connected cell phone and opens the search page on the browser to elaborate a query searching for the information he or she needs. The user expects to receive information suited to his or her needs, taking into account the actual context and its preferences.

In the search box, the user starts to type some letters. For example, when the "r" and "e" letters are typed, the word recommender shows some words starting with "re" such as *restaurant* and *residence*. These word recommendations are made from terms in a thesaurus, but the words are selected according to the actual context and preferences, conceptualized in the ontologies.

If the desired search word appears, the user will select it from the list; if it does not appear, then he or she will continue with a regular search query. If a word from the list is selected, then the interface will show some other selectable options with related concepts that could complement the search query.

Following with the previous example, when *restaurant* is selected, then the screen will show some other properties related to the restaurants such as the kind of restaurant (Japanese, Italian, Mexican, etc.) or restaurant category (five stars, fast food, etc.). All these related properties are obtained from the ontologies also, based on the stored related terms and the user preferences. These terms, if selected, will be used to enhance the query that will be delivered to the search engine (Figure 10.9).

The ontology use then must provide environment, device, and preference considerations that will modify the search behavior, suggesting words to the user and even related properties when the recommended word is selected.

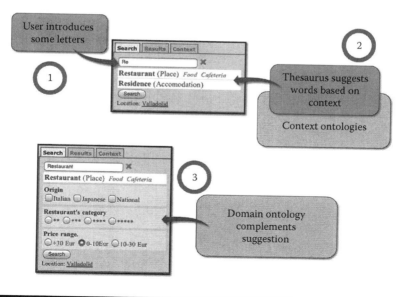

Figure 10.9   Enhancing the Web search with ontologies.

It is expected that the system must be capable of not recommending the "snowboard" term in summer or at a side of the beach, "movies" at midnight, or "Japanese" as a type of restaurant if it is not the user's preference.

10.6 Conclusions

The mobile context ontology creation remains as a work in progress; there is a lot of work to be done. Previous efforts provide hope for obtaining acceptable results in a short time.

Creating ontologies requires a strong methodological knowledge to obtain a well-structured, focused, and clear semantic structure. This knowledge must be applied with a focus on the creation of a conceptual framework to define the context in mobile devices to support Web search enhancement.

Related existing ontologies have been analyzed. On this basis, a set of concepts have been proposed as a foundation for the mobile context ontology creation. Actual smartphones offer some applications or widgets that allow finding some kind of the information we intend to bring to the user, but the ontology use for query search enhancement should be a more efficient way to spread the use of Web search because it is an independent and device-adapted platform.

References

Altavista. 2009. http://www.altavista.com (accessed February 16, 2009).

Americas, 3G. 2009. http://www.3gamericas.org/ (accessed September 16, 2009).

AMIGO, Ambient intelligence for the networked home environment. 2008. http://www.hitech-projects.com/euprojects/amigo/index.htm (accessed February 12, 2009).

Aréchiga, D. and J. Vegas. Ontology supported personalized search for mobile devices. In *Proceedings of the 3rd International Workshop on Ontology, Conceptualization and Epistemology for Information Systems, Software Engineering and Service Science*. Amsterdam, the Netherlands: CEUR-WS.org, 2009, pp. 1–12.

Arias, M., J. M. Cantera, and J. Vegas. Context-based personalization for mobile web search. In *Proceedings of the 2nd International Workshop on Personalized Access, Profile Management, and Context Awareness Databases, PersDB 2008*. Auckland, New Zealand (in conjunction with the 34th VLDB Conference), 2008, pp. 33–39.

Baeza-Yates, R., G. Dupret, and J. Velasco. A study of mobile search queries in Japan. In *Query Log Analysis: Social and Technological Challenges. A Workshop at the 16th International World Wide Web Conference*, Banff, AB, 2007.

Cantera Fonseca, J. M., G. García Bernardo, and J. Vegas Hernández. An automatic page classification method to improve user experience on the MobileWeb. In *Proceedings of the Second International Conference on Mobile Ubiquitous Computing Systems, Services and Technologies, UBICOMM 2008*. Valencia, Spain: IEEE Press, 2008, pp. 171–177.

Church, K., B. Smyth, P. Cotter, and K. Bradley. Mobile information access: A study of emerging search behavior on the mobile Internet. *ACM Transactions on the Web* (ACM Transactions on the Web), 1(1), 2007.

Gauch, S., J. Chaffee, and A. Pretschner. Ontology-based personalized search and browsing. *Web Intelligence and Agent Systems* 1(3–4): 219–234, 2003.

Google. 2009. http://www.google.com (accessed September 16, 2009).

Gruber, T. R. A translation approach to portable ontology specifications. *Knowledge Acquisition* 5(2): 199–220, 1993.

Grüninger, M. and M. S. Fox. Methodology for the design and evaluation of ontologies. *Workshop on Basic Ontological Issues in Knowledge Sharing (IJCAI 1995)*. Montréal, QC, 1995.

Kamvar, M. and S. Baluja. Deciphering trends in mobile search. *Computer* (IEEE Computer Society), 58–62, 2007.

Merriam-Webster OnLine "*context*." 2009. http://www.merriam-webster.com/dictionary/context (accessed June 15, 2009).

MORFEO—My Mobile Search. 2008. http://mymobilesearch.morfeo-project.org/mymo-bilesearch/lng/es (accessed September 06, 2008).

Noy, N. F. and D. L. McGuinness. Ontology development 101: A guide to creating your first ontology. *Knowledge Systems Laboratory, Stanford University*, 2000. http://www-ksl.stanford.edu/people/dlm/papers/ontology-tutorial-noy-mcguinness-abstract.html (accessed March 03, 2009).

Pretschner, A. and S. Gauch. Ontology based personalized search. In *Proceedings of the 11th IEEE International Conference on Tools with Artificial Intelligence*. Chicago, IL: IEEE Computer Society, 1999, p. 391.

Sieg, A., B. Mobasher, and R. Burke. Ontological user profiles for representing context in web search. In *Proceedings of the 2007 IEEE/WIC/ACM International Conferences on Web Intelligence and Intelligent Agent Technology*. Silicon Valley, CA: IEEE Computer Society, 2007, pp. 91–94.

SPICE, the Mobile Ontology. 2008. http://ontology.ist-spice.org/index.html (accessed March 02, 2009).

Staab, S. and R. Studer. *Handbook on Ontologies*. Berlin, Germany: Springer, 2004.

Uschold, M., and M. Grüninger. Ontologies: Principles, methods and applications. *Knowledge Engineering Review* 11(2): 93–155, 1996.

Uschold, M. and M. King. Towards a methodology for building ontologies. *Workshop on Basic Ontological Issues in Knowledge Sharing, held in conjunction with IJCAI 1995*. Montréal, Canada, 1995.

Villalonga, C. et al. Mobile Ontology: Towards a standardized semantic model for the mobile domain. *Service-Oriented Computing—ICSOC 2007 Workshops. ICSOC 2007*. Vienna, Austria: Springer-Verlag, 2007, pp. 248–257.

W3C's Mobile Web Initiative Working Group. 2008. http://www.w3.org/Mobile (accessed March 23, 2009).

W3C's Delivery Context Ontology. 2008. http://www.w3.org/TR/dcontology/ (accessed June 10, 2009).

Weißenberg, N., A. Voisard, and R. Gartmann. Using ontologies in personalized mobile applications. In *Proceedings of the 12th Annual ACM International Workshop on Geographic Information Systems (ACM-GIS 2004)*. Washington, DC: ACM, 2004, pp. 2–11.

Wikitionary. A Wiki-based open content dictionary "context." 2009. http://en.wiktionary.org/wiki/context (accessed June 15, 2009).

World Wide Web Consortium (W3C). Delivery Context Overview for Device Independence. 2006. http://www.w3.org/TR/di-dco/ (accessed January 26, 2009).

Yahoo. 2009. http://www.yahoo.com (accessed September 09, 2008).

Chapter 11

Deploying Intelligent Mobile Applications: Server or Device?

Ben Falchuk

Contents

11.1 Introduction

In the last several years, the mobile applications landscape has been transformed in dramatic fashion. From dazzling new enabling hardware and human–computer interaction technologies, to attractive new devices, to powerful programming interfaces, and much improved ubiquitous access to the mobile Internet. These factors—together with a growing marketplace for mobile applications, the possibility of significant return on investment from application development, and an ever-growing need for always on communication—have helped to make the development and distribution of mobile applications a compelling exercise. While there has been both a market and various mechanisms that enable mobile application ecosystems for many years, a perfect mix of desirable new devices, development tools, user-generated content, and social media have created something of a gold rush [1].

This chapter describes the issues whose interplay influences whether mobile application developers choose to design their app for deployment upon a mobile device* or for deployment upon a server accessible to the mobile handset (or perhaps a combination of the two). In many ways, the decision can change almost every subsequent equation, from how the application will generate revenue, to how it will be maintained, to how it should be designed from a human–computer interaction point of view. Indeed, many now feel that applications will dictate the winning platforms [2]. While in some ways the end-user is not likely to care where service logic resides, in other ways the decision can affect the end-user in a profound way. After reading this chapter, the reader should have a greater appreciation of a wide variety of deployment-affecting issues, "gotchas," and philosophies. Our hope is that readers of this chapter will come across both familiar and surprising issues and as a result will be better equipped to get to the bottom of the deployment question—should they be faced with it themselves—or to understand why a given mobile application is deployed as it is.

The chapter first takes a brief look at how mobile application deployment has changed from the recent past to the present. It then describes a class of

---

* In this chapter, we largely assume that the mobile device is of the "smartphone" class, i.e., a mobile device capable of significantly more than just telephony, SMS, and contacts.

mobile applications called *proactive notification applications*, which we use as a basis for discussion. It then analyzes a wide range of specific issues related to deployment; for each issue, both background and one or more concise points are made about whether the issue skews deployment choice to one alternative over another. The issues are drawn mainly from the technical, financial, and other loosely related and intangible realms. Finally, we summarize and conclude.

11.2 Brief History and Summary of Today's Landscape

In the very early days of mobile services, the NTT DoCoMo's i-mode* service made access to the Internet possible from the NTT line of mobile phones. The information and services that were accessible, however, were initially controlled only by NTT in a so-called "walled garden." i-mode services were written specially for the NTT i-mode handsets using a mix of Web (Compact HTML) and proprietary implementations. The i-mode information services were initially mini Web sites (e.g., local news, weather), but as i-mode expanded, these sites could be created by third parties through the operator and would be deployed onto customer mobile handsets as convenient icons (shortcuts). The Wireless Application Protocol (WAP)[†] played a similar role in opening up service development but with a somewhat more global feel than the NTT-centric i-mode given that hundreds of companies contributed to its definition. WAP-enabled services (mobile sites) were built using the Wireless Markup Language (WML), not cHTML and were loaded to the mobile handset as needed. During the early 2000s it was common to debate the comparative merits of WAP versus i-mode.

Later, two newer technologies changed the game for mobile service programmers: Java Platform Micro Edition (originally named Java 2 Micro Edition, or J2ME for short)[‡] and BREW. These technologies were not intended to improve access to the mobile Internet but rather to standardize the development of mobile binary applications that run directly on the device. J2ME was developed by Sun Microsystems to extend the reach of Java to classes of mobile handsets supporting particular capability profiles. The Binary Runtime Environment for Wireless (BREW) was created by Qualcomm as a development platform for mobile services for Qualcomm devices. Much like J2ME, the BREW suite included software emulators to improve the application development process, an application loader, and software development kits. The Verizon Wireless implementation of BREW called

---

* http://www.nttdocomo.com/services/imode/index.html

† Open Mobile Alliance (OMA) Wireless Application Protocol page, http://www.openmobile-alliance.org/ tech/affiliates/wap/wapindex.html

‡ Sun Java ME homepage: http://java.sun.com/javame/index.jsp

Get It Now* was a marketing success and allowed mobile customers instant access to mobile applications written by third parties but hosted in the walled garden and downloaded on-demand by end-users.

The principal distribution method for both BREW and J2ME MIDlets then was referred to as over-the-air (OTA), in which a mobile subscriber uses the mobile handset to choose, initiate download, and confirm local installation of—and get charged for—the binary program. Typically the running application is then either stand-alone (needing no resources outside of the handset) or partially network-dependent (relying on access to a server for information to run; e.g., a stock-quote application). With variations between operators, most application development and deployment cycles (including J2ME and BREW) have rigorous testing and certification processes. Testing, done largely by the developer, attempts to rid the application of unwanted bugs. Certification—in the case of Qualcomm, for example—is performed by the operator or a third party, and "evaluates your application against the established levels of stability, quality, and compliance with standard platform requirements that Qualcomm's carrier partners demand."† In addition, developer registration is sometimes a required step on a per-manufacturer basis, and sometimes incurs one-time or recurring costs. For practical and economic reasons, many of these and other gates still exist today [3].

Currently, the landscape is busier and even more exhilarating, subject to the rapid pace of change of both technological innovations and business drivers. The once seemingly exclusive right to mobile service innovation previously enjoyed by mobile operators and manufacturers is now broken, ceding to marketplaces—both large and formal or grassroots and ad hoc—in which applications are posted, rated, exchanged, and purchased. Analogous to the Web 2.0 effect in which power shifted from centralized corporate content-creators to individual contributors, power has shifted toward the mobile developers who can create *and* market the most interesting applications and create buzz in the media (traditional and Web 2.0). Indeed, in some ways, that mobile Web 2.0 social media have quickly become woven into mobile users' lives is hardly a surprise given our fundamental need to be social and communicate.‡ A significant proportion of the creative energy of mobile development is currently directed toward mobile social applications to facilitate linking, posting, updating, reading, and communicating with and amongst social resources. The tools and methods supporting the creation and hosting of mobile Web services and content has also seen fundamental changes. Open-source software packages have made supporting

---

* Verizon Wireless Get it Now, http://wirelesssupport.verizon.com/features/data_services/get_it_now.html?t=3
† http://brew.qualcomm.com/brew/en/developer/newsletter/2001/september_2001.html
‡ Most social theorists (such as Locke) agree that "man is a social animal" with the predisposition to create societies and communicate.

both desktop *and* mobile Web applications far easier and more affordable than ever before. The so-called LAMP stack—Linux-Apache-MySQL-Perl/PHP/Python—for example, is a suite of free and open source software that is both scalable and sophisticated enough to support the server-side operations of even the most highly trafficked mobile Web sites.* Software Development Kits (SDK) and platform emulators for mobile development that integrate with *Eclipse* and *NetBeans*, for example, are both commonplace and highly effective (for a relatively seasoned developer a "hello world!" mobile application can usually be brought to life on an emulator in a matter of minutes, regardless of the platform). Mobile applications are now delivered to the mobile device in any number of ways, including

- Over-the-air (e.g., BREW and J2ME support this)—Applications are found: on-deck, off-deck, or off-portal, and subsequently transferred over the radio network to the device
- Synced from PC and installed (e.g., iTunes and iPhone)
- Mobile Web only (e.g., browser-based)
- Custom (e.g., streamed from Internet)
- Preloaded on handset
- Bought on storage media, loaded onto handset

Today's mobile application developers are empowered with tools and software that they could only dream about a decade ago. More and more, they are limited only by their own imagination.

11.3 The Setting: Proactive Notification Applications as a Driving Example

Context-aware systems (CAS) [4] are those systems that take into account information about their environment—the physical, network, and informational layers—by receiving or reading data from an information source such as a sensor, location server, or a GPS unit, and fundamentally adapting, configuring, or customizing their actions accordingly. CAS systems that are accessible invisibly, for long durations, and in many locations, are the crux of so-called ubiquitous computing [5] (both CAS and ubiquitous computing have rich research heritages). Context, the machine-readable information surrounding a given event or time, can include: location, proximity to places or other digital resources, physical activity, attentional direction, preferences, meteorological information, and so on. Contextual data can also come from applications and include: calendar appointments, to-do's or tasks,

---

* See "Shining a light on LAMP" by P. Yared, available at http://www.developer.com/

time, music currently playing, etc. To users, effective CAS systems sometimes feel intelligent because they seem to take "the right action at the right time." They achieve this effect in part by first fusing contexts together and then computing over their representation to determine the best action.

The perfect storm of mobile applications, Web 2.0, and programmatic APIs has recently yielded a particular type of CAS that we call proactive notification applications (PNA). PNAs are widely popular because they help users handle the ever increasing complexity of day-to-day life through timely information delivered at the *right* moment and place [6]. They can be found useful in both business and personal settings. Simple PNAs arrived on the Web in the mid-2000s—Remember the Milk, ReQall, and Wakerupper are examples at the time of writing.* In addition, PNA-like applications appear prominently on lists of the top iPhone, MIDP, and Android community applications. The best of these tools generally feature

- An ability for users to inject future scheduled reminders manually and quickly from the Web or from mobile Web, often using a short-hand notation.
- Varying ability to deliver reminders across multiple transports, such as e-mail and SMS.
- Varying ability to read and exploit user location on a moment to moment basis (e.g., location-sensitive suggestions).

In the large, these systems generally did not do the following:

- Take actions that were not pre-scripted ahead of time by the user (e.g., User: "System—send me a reminder 10 min before this appointment").
- Make any sort of sophisticated inference by fusing the context from various information sources.

A grander vision for truly proactive PNAs more closely resembles that of intelligent personal assistants in which the software components make sophisticated inference over a large set of metadata. PNAs will run on an ongoing basis and provide advice, information, and notifications that are relevant to the user's current context without overwhelming or annoying the user. Indeed, one main use case of PNAs will be to serve as an intelligent information filter, dropping information that is not useful or relevant to the user at the moment while notifying and helping the user with information that is relevant in context. The underpinnings will include rules, machine learning, natural language parsing, and

---

* Representative of the high churn rate in this realm, other systems absorbed by larger rivals, that is, by 2008, Twitter had bought out I Want Sandy gaining rights over its elegant user syntax and algorithms.

artificial intelligence (AI). In our research lab, we have designed and built such systems [7–10]. Our approach is specifically intended for mobile service providers and intended to scale to support large numbers of customers (e.g., millions). Our research and development have allowed us to crystallize several emblematic use cases of PNAs:

- Notify a user when she is near a location, which might satisfy a to-do item. For example, in our system, a notification is generated for the user when she is near a Chinese restaurant and one of her objectives was to "pickup dinner" (the system applies preferences). Alternately, we have experimented with medical objectives such as popping up reminders when a user is near a park where she can complete today's nurse-created "exercise program."
- Perform some basic natural language understanding and one level of task resolution by correlating complex to-do's such as "plan John's party" to a series of more basic inferred to-do's (e.g., "stop at grocery store") and subsequently remind users of these tasks (see above).
- Propagate selected system-generated information about user context into the user's Twitter (and other) social network as a tweet.
- Infer location and reason for travel (i.e., an imminent appointment at a location within driving distance) and remind the user when she is "running late" for such an appointment (by computing estimated time of arrival at the appointment location).
- Integrate with vehicular telematics* systems and provide proactive reminders when nearby to automotive service locations (e.g., gas stations, garages) and the vehicle requires that particular service.
- Allow users to set preferences for notification and to mark tasks as completed or satisfied.

Figure 11.1 illustrates aspects of one of our systems in which a notification appears overlaid upon a map on the user's mobile device. The notification is system-generated through inference over location and both imminent and nearby to-do entries (e.g., in the figure, a steakhouse is suggested in order to satisfy a user-entered task with the text "pickup dinner").

In summary, while PNA is just one example of a compelling mobile application, it is an excellent one for our subsequent discussion because (a) it is topical and on the horizon, (b) it has a nontrivial logic and use case, and (c) it is a viable candidate for either standalone *or* server-based deployment.

---

* See Telematics, in Wikipedia. Retrieved September 25, 2009 from http://en.wikipedia.org/wiki/Telematics#Vehicle_telematics

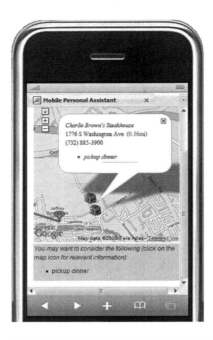

Figure 11.1   Telcordia PNA pilot application.

11.4 The Factors

This section now outlines an interesting set of issues which, in part, determine how a mobile PNA (or other mobile application) might be deployed.* The main problem at hand, then, is whether to (a) host the application in the network or on a distributed (home) platform, or (b) deliver a principally device-resident application. We will use the following terms in describing the location of deployed mobile applications. "In Network" applications refer to those whose primary service logic and data reside off of the device in one or more networked servers. These applications do not install additional binary components onto the device, but may rely on a Web browser as platform.† They may also have a limited ability to "launch" primitive

---

* We will make reference to real products, phones, mobile operating systems, news, and so on in order to make our points. However, we hope to keep the discussion more broadly informative and will resist the urge to describe too many specific applications, be they for iPhone, Blackberry, Android-based phones, or what have you. We also do not claim to present exhaustive example sets or proofs for any particular issue. Rather, the reader should use the text as a starting point for further investigation.

† IE Mobile, Mobile Safari, Skyfire, Opera Mini, BOLT, and Opera Mobile are some of the most popular and powerful mobile Web browsers (see "Have Web Browser Will Travel" by J. Lendino in *PC Magazine*, September, 2009).

on-device applications (such as SMS sending or Contacts) vis a vis user interactions. "On Device" applications refer to those whose primary data and computing occurs on the handset platform. These applications install one or more binary components onto the device and use the device's storage to keep long-term state. They may also have—to some degree—an ability to "launch" primitive on-device applications (such as SMS sending or Contacts) and to read from local sensor hardware. "Hybrid" applications are a blend of both of the previous two types.*

11.4.1 Financial Issues

11.4.1.1 Issue: Reaching and Serving a User Base

Is the potential user base of the mobile application large? Will it grow rapidly?

"On-Device" Deployment: Mobile phone penetration is high but there is no single dominant mobile operating system (OS) or platform. Delivering a binary application across multiple platforms and through multiple channels becomes costlier, in principle, than choosing a single platform and channel. On the other hand, a multichannel approach can help reach more of the user base compared to a single mode approach (e.g., movie studios deploy promotional movie–related applications across many channels). In the case of connected collaborative applications (e.g., PNAs that make use of social network input) the on-device application will need to integrate with cloud services for social network updates. In turn, this network usage may incur costs to the user or complications to the developer who may sometimes create a business agreement with wireless service providers to provide this access as preferred traffic. For a growing user base and an unchanging application, delivering the application through a central store front or Web site makes sense. When the application updates often, maintaining versions for customers and making those available are more difficult.

"In-Network" Deployment: On the server side, meeting the needs of a mobile application that might grow virally (and in fact depends on a large user base for content) causes less alarm than meeting them on a device. As the user base grows, server-side features and content can be added fairly easily and content—when driven by end-users—comes for free. Server-side components can be scaled "up and out" and can perform aggregation and filtering to reduce information clutter from the device point of view. A server can also better accommodate application updates and a changing set of backend partners (as opposed to updating software across every user device). It is noted that suddenly changing the feature set of a Web application can result in end-user confusion and anger.

---

* We use the term "hoster" to signify the business entity actually managing the availability of the service on a data network, and "deployer" to signify the actor controlling/making the deployment choice.

11.4.1.2 Issue: Digital Rights, Copying, and Disruption

Can the mobile application be easily copied, stolen, modified, or disrupted by malicious users?

"On-Device" Deployment: Developing for on device deployment has benefits, such as exploiting platform specific native features (e.g., touch-screens). Applications can also be secured, by, for example, obfuscating their code and locking them with digital rights management (DRM) (e.g., iTunes). Yet, much to developers' dismay, device-resident applications can still be both hacked and copied illegally. For example, Crackulous is a tool to strip DRM from iPhone applications while i-hacked. com offers techniques to break iTunes.* De-compiling mobile Java applications— which reveals the source code from which the application was built—is also technically possible. Thus, the application codebase becomes at risk when the binary is in any way accessible to users on their devices. A breach in the codebase could result in an illegal "bastardized" version of the application implemented by somebody else that siphons revenues away from the inventing developer.

"In-Network" Deployment: Deploying a PNA application in network puts vital service logic onto network accessible servers. This creates a risk: the servers could be hacked or brought down by, for example, distributed denial of service (DDoS) attacks. Also, servers can be scaled "up and out," but this comes at additional cost to the deployer or hoster and is a non-trivial factor.† In addition, every service outage angers customers and negates customer goodwill. Attacks can target big and small applications; for example, in June 2008 Internet Movie Database (an Amazon company) was temporarily brought down by a DDoS attack.‡ Even the distinct look and feel of a network-resident PNA is subject to stealing but the application source is probably safer on the backend server than the mobile device (i.e., protected by more firewalls). Indeed, copyright laws are confusing; some Web aspects—such as the execution (HTML and look) of an application—can be copyrighted, but look and feel is much harder to protect.

11.4.1.3 Issue: Hosting

"On-Device" Deployment: For PNA applications installed directly on mobile platforms there is sometimes no need to host additional server-based logic—in other words, the PNA logic is co-located on the device with any and all the data it needs (e.g., it might need only to read local calendar appointment and contacts list). This saves a great deal of running costs to the deploying entity. However, the

---

* See "Crackulous released, promises to bust iPhone app protection scheme", by C. Ziegler on Engadget.com, February 2009, available at http://www.engadget.com
† See "Scaling Twitter: Making Twitter 10000 Percent Faster" by T. Hoff at http://highscalability. com
‡ http://news.cnet.com/8301-10789_3-9962676-57.html

"delivery" mechanism becomes important as well. Is the PNA delivered to end-users from a mobile Web site run by the developer or from a third-party hosted platform (e.g., iTunes)? If we let A = *cost of hosting the app on one's own site*, and X = *income from hosted advertisements on your own site*, and let B = *the slice of each sales revenue that the hoster takes* (e.g., 0.3 for 30%), and n = *expected number of unit sales*, then a simple comparison of $(A - X)$ to nB may help decide on a course of action.

"In-Network" Deployment: Hosting an application requires infrastructure and costs are incurred depending on whether the developer fronts this infrastructure or a third-party hosting service does. A rare exception may be a PNA that can be configured to use the infrastructure already owned by the user (e.g., his home wireless network, a Web server running on his home desktop, and his cellular data plan). A PNA might employ a low-cost hosting service that inserts advertisements into the service in exchange for supporting development and hosting costs (many mobile applications take this approach). An in-network PNA that needs to read and write "cloud" data or to accommodate a relatively large number of user will require a systematic way to accommodate incoming and outgoing traffic (one way or another, increased traffic past key thresholds will trigger costs). Costs incurred by increasing PNA-originating traffic will be passed on to the deploying entity and the end-user.* The teaser model, such as "free to sign-up, $10 to have full features" is a common way to hook users on features before beginning to bill them, and revenues can be used to offset hosting costs.

11.4.1.4 Issue: Maintenance

Who maintains the PNA mobile application and how?

"On-Device" Deployment: A PNA deployed on a mobile device can be fairly stable to OS platform changes and is likely to run effectively through OS version changes. Sometimes, however, a major mobile OS upgrade might require the re-engineering of the application software or, at the least, costly bug-fixes.[†] With on-device applications there is also a remote risk that the end user will accidentally corrupt or delete the application or its dependent device-resident information. This will result in costly service calls to the PNA administrator or re-seller. That the typical mobile phone life is a scant 1.5 years, binary application sellers can, in some cases, re-sell their wares to the same user multiple times (i.e., each time he upgrades to a new device or platform), but over several years.

---

* 2009 is the year that Twitter is creating paid accounts as a part of their business model, after several years of free unfettered mobile access: http://www.dbtechno.com/industry/2009/08/21/paid-accounts-coming-to-twitter-service/

† One needs only follow the development forums for Android, iPhone, and most all of the mobile OS's to discover developers whose application has "broken" from an OS change.

"In-Network" Deployment: Supporting a highly trafficked in-network application can imply costly infrastructure and maintenance. Expensive and time-consuming events such as system upgrades, fixes, and software patching will be necessary to protect servers and middleware against obsolescence, operating system bugs, security vulnerabilities, and so on. In-network applications may also store application-data, which in turn implies the design and maintenance of databases or file systems that can scale with the application user-base. Open source databases such as MySQL are free to own, but require human effort to create a design and to maintain operations. Server-based PNAs that mashup data from yet other Internet sites through connectors (e.g., one module for Facebook data, one for Flickr, and so on) may be faced with API changes and connector re-design.*

11.4.1.5 Issue: Marketing and Channels

Channels can be thought of as the means for mobile applications and mobile marketing to customers using the most customer-appropriate media and transmission modes. Businesses are now greatly concerned with creating unified multichannel marketing plans to foster business. The actors in a channel include the developer, publisher, mobile operators, aggregator, portal, content partners (e.g., advertisers), end-user, and so on. A short channel often refers to one with three or fewer actors. In short channels, each actor may have a good deal of control and business agreements (e.g., between developer and distributor) may be formal or informal. As the value chain is shorter, it may be that fewer cumulative costs are passed down to the consumer, while each actor has potentially fewer IT and integration costs. Four or more actors is considered a long channel. Multichannel approaches can be advantageous for getting a product to people across multiple media and formats, but the costs of these strategies (e.g., a specialized team for each channel type) may get passed on to end-users.

As a related example, the Apple iPhone can be thought of as having three "channel modes." A first is the App Store, aimed squarely at the iPhone end consumer base. A second is referred to as ad hoc distribution, in which a limited set of users get a direct line to the product (e.g., testers). A third is referred to as the enterprise channel in which an enterprise specifies and authorizes particular iPhone applications for its needs and then writes the in-house applications that will run only on selected phones. The in-house applications are distributed and installed using iTunes.

"On-Device" Deployment: For a developer choosing to deploy a given PNA or other application on-device, there is nothing to prevent using either a very long or short chain to bring that deployment to reality and to market the application. Two diverse examples are (1) using iTunes to deploy and distribute an independently

---

* Sites like ProgrammableWeb.com assist in tracking the capabilities and promise of Website API's.

developed PNA (only a few actors in this chain) and (2) developing a preloaded game for particular hardware and particular mobile platform and distributing it via Walmart (several actors need to coordinate to make this happen).

"In-Network" Deployment: In-network PNA applications can also employ short or long channels, and is a function of the business arrangements. For example, OhDontForget.com is an online PNA using the Web as a platform. The actors in a longer multichannel approach for Web-based PNAs might include mobile Web actors, standard Web content hosters, content providers, advertisers, analytics providers, billing and accounting entities, and aggregators who all work together to create a multichannel—yet in-network- mobile application.*

11.4.1.6 Issue: Revenue Models

There are several models by which mobile PNA applications create revenue streams for the various actors in the channel. The PNA might be free [11], given as a loss leader, as purely a marketing scheme or gimmick, or might be laden with advertising (making it free to the end-user, but not to the advertiser). It might cost a one-time fee to install the PNA, a monthly (intermittent) fee, or a per-usage fee, or some combination. In addition, where the application ultimately resides is a factor: on-deck, off-deck, or off-portal. For on-deck PNAs, the operator hosts and bills for the service, as well as taking a piece of the sale price (which can range from 6% (Japan) to 50% (United States), to 90% (Europe). Here, deck placement is key and favors big guns (with big pocketbooks to influence placement), not individual developers. Off-deck PNAs are found by end-users searching on the operator's portal. Typically the operator provides billing but their cut is smaller than for on-deck. For off-portal applications, independence from the operator is gained, but only about 4% of users, for example, have managed to buy a game in this manner [12].

"On-Device" Deployment: Applications running directly on mobile devices really can implement any of the above types of models. Ad-supported mobile applications can keep the ecosystem going.† Pay-per-use, subscription, and billing are not trivial technical initiatives.

"In-Network" Deployment: In-network applications could use any of the above-mentioned models, implemented in part with Web technologies. Numerous examples of in-network services and innovative or fine-grained charging models exist. It is worth noting that while the PNA may be free to install, the end-user pays monthly or per-kilobyte network access fees to a communications service provider (CSP).

---

* See the YouTube Fast Forward channel for further details at http://www.youtube.com/user/FastForward
† See "1.6 Million Ad-Supported Games Downloaded in 1st Month" at wirelessandmobilenews.com, 2008.

CSPs themselves are in an excellent position to bundle free applications with their communications plans*—a "teaser" model sometimes hooks new subscribers on features or applications and then later implements fees for a more "full-featured" version. In addition, the market for both Web and mobile ads is relatively strong (compared to traditional media) and this is a model that can sustain both in-network and on-device PNAs.

11.4.1.7 Issue: Power Costs

Application design and deployment create power-related concerns for several of the actors in the system, including end-user and hoster.

"On-Device" Deployment: A device's battery ultimately costs money to charge and replace. Also, certain uses of the mobile device's modules by applications can cause greater rates of battery dissipation and even though batteries are rechargeable, their lifetimes are shortened in different ways by different device usage patterns. In PNAs, long-running background processes[†] are highly desirable (necessary for pervasive, asynchronous informative messaging with the user) but are a major concern because of their potential to drain batteries (continued processing implies continued battery usage). The use of graphics and the device screen affect power as well. Still, for the developer, power utilization is not really an issue that is carefully considered nowadays, though a very power-naïve application will surely get a bad "reputation," just as consumers often consider battery life when purchasing new devices.

"In-Network" Deployment: A PNA deployed in-network does not directly use the battery of a device, of course, but if the application is well architected, can have advantages over on-device ones. Because in this case, PNA logic runs on a server, the mobile device need not be always computing but only be ready at any time (or particular times) to receive information (e.g., via SMS) from the server. Device readiness (e.g., the ability to receive an SMS) is not as costly to battery life than application running and maximized (for one thing, the screen need not be used). A networked server, however, itself costs money to power and these costs are invisibly factored into hosting costs and subsequently passed on to the end-user.

11.4.1.8 Issue: Use Case—Realities of iPhone Development

At the time of writing, certified applications delivered via iTunes for iPhone give up 30% of each unit sale to Apple. As a developer of iPhone applications, an Apple nondisclosure agreement limits what one can discuss about the app in public.

---

* Messaging services such as Yahoo!®, Windows Live®, AIM®, navigation programs, and social network applications are commonly bundled.
† As of 2009, Apple promises iPhone "push notification," a method that, while not background processing per se, does allow server applications to send messages anytime to individual iPhones, see http://reviews.cnet.com/8301-19512_7-10263277-233.html

In addition, 10%–20% of revenue might have to be ceded by the developer for credit card processing fees. Thus for each sale of a $2 iPhone application, Apple gets $0.62, credit card processor (e.g., Visa) gets $0.25, and the developer gets $1.12. App prices seem to be dropping at dramatics rates: "…today, the average top-100 app sells for $2.55. A month ago, it was $2.78. Two months ago, it was $3.15. (The 50 most popular applications have dropped in price even faster: The average top-50 app now sells for $2.39, down 34% from $3.63 two months ago.). Most noticeable difference: $10 premium apps have vanished from the bestseller charts" [13]. The uncertainty and risks rise. What if someone copies your application? What are the financial issues? Can you afford to pursue it in court? The opacity of certification and timing are an issue, sometimes resulting in long, silent waits for applications to appear in the store. Apple can also reject applications at their will and sometimes without reason.* On the other hand, success stories from independent developers who have made relatively good revenue from their iPhone offerings are now relatively plentiful.†

11.4.2 Technical Issues

11.4.2.1 Issue: Who Do You Serve?

The PNA might serve only a single user and no requirements related to contexts beyond what it has access to on the device. It might serve a single user and exploit multiple diverse contexts requiring connectivity and fusion. It might serve a large number of users and compute over multiple diverse contexts, including cultural divides.

"On-Device" Deployment: The easiest path through the design, test, deploy cycle is sometimes to build a well-isolated single-user application. These applications have few to no interactions and dependencies on other data, making them easier to test (and sometimes market). However, apart from games, these kinds of applications are now less popular and the market currently demands "community" or "social" or "communicative" applications that interwork with social media sites and require data connections to various cloud sources. Given resource constraints of individual mobile devices, an on-device application serve a computing "hub" for a large number of users from a single point (e.g., the peer to peer model), nor network efficiently to remote resources in any very large-scale fashion (i.e., the device cannot feasibly be used as the gateway or server for many users). Mobile peer-to-peer technologies and research will continue to drive this.‡ Understanding internationalization may also require additional resources if the user base comprises many nationalities or cultures.

---

* *See "Apple's Capricious Rules for iPhone Apps"* http://bits.blogs.nytimes.com/2008/09/16/apples-capricious-app-policy/

† See "Coder's Half-Million-Dollar Baby Proves iPhone Gold Rush Is Still On," http://www.wired.com/gadgetlab/2009/02/shoot-is-iphone/

‡ See, for example, IEEE workshops on mobile peer-to-peer computing; for example, http://www.samrg.org/mp2p/2009/

"In-Network" Deployment: There is little point in building a network-centric server if either the user base is very small or if there is no need to interwork with Internet data sources (assuming the logic could be implemented just as easily for a device). If the mobile application is to incorporate Internet data, require any sort of global view, or fuse many user contents together, then in-network is desirable. Communication between networked servers has few "gotcha's" besides server downtime. Communication from server to end device can be problematic on financial and technical dimensions (to reach a device from server one usually sends an SMS since the device does not have a fixed network address). Finally, internationalization issues would seem to be comparable on devices versus servers.

11.4.2.2 Issue: Asynchronicity

PNA applications will continue to be exemplified by their need to deliver intelligence at "just the right moment." Often this moment will be unexpected but important—low tire pressure notifications, stock price notifications, and so on. Mobile devices, however, are not always turned on, and even when powered on they may not be reachable (e.g., no signal) or they may be busy (e.g., user in a voice or data session already). SMS is one of the most used asynchronous messaging mode, providing store-and-forward relatively low latency message transmission.

"On-Device" Deployment: The usefulness of asynchronicity is limited by device readiness to receive input. To get around this, PNAs can be implemented as "pull" applications, which update only when they start up and subsequently when the user asks them to. This is obviously a compromise on user convenience. Asynchronicity, we see, is an issue with a strong tie to *background processing*. Windows Mobile "push" e-mail was apparently implemented by the device creating a long-running HTTP connection to the mail server; if there is mail for the user, a server response comes back and the client pulls the mail. If not, the request eventually times out.*

"In-Network" Deployment: A PNA application running "in network" on a cloud server is always available and runs independently of the mobile devices it may serve. It can therefore act asynchronously all it likes but the matter remains of communicating resulting information to the user (see previous discussion). The server could use a push mechanism providing it knew the network address, e-mail address, or MSISDN (i.e., SMS) address of the device. The latency of different push mechanisms differs and application designers need to carefully consider the ramifications.

---

* See http://www.techatplay.com/?p=482 for more comparison.

11.4.2.3 Issue: The Service Logic

In proactive notification systems, a certain sophistication of service logic is usually required in order to manage context, manage, and process triggers or rules, parse messages, make inference, communicate with other data services, and so on. At an overall level, the issues important to developers and deployers include questions like: (a) Are coding libraries readily available to implement the logic I have in mind? (b) Do the operating systems I am targeting support the nonfunctional requirements (security, reliability, quality, etc.) that I require?

"On-Device" Deployment: In reality, computing libraries to support complex software are "hit and miss" on device platforms. On the other hand, emerging examples include the following:

- "Raccoon Project"—a Nokia port of Apache to Symbian OS; still requires a gateway for IP translation/registration but promises an instance of Apache on each mobile phone*
- Workflow engines for mobile OS—a Nokia Research Center (Finland) initiative
- Simple Object Access Protocol (SOAP) clients (e.g., kSOAP[†])
- Nokia—PAMP—Personal Apache MySQL and PHP (open source Web platform)[‡]

Not long ago, memory and heap size were limiting factors. With high-end mobile devices now containing 128 MB RAM and more, this is less of an issue. For PNA developers with eyes on rich graphics—such as the use of 3D rendering—there are emerging mobile technologies. For example, "3D maps… [are] no longer hindered by lack of 3D programming interfaces" [14], OpenGL ES is supported on a range of devices, and JSR-184, Direct3D Mobile, and VRML (Pocket Cortona[§]) are available for Windows Mobile. It is noted, however, that continued use of the device screen or the device's graphics processing engine uses up the battery much faster than baseline activity. Also, nonstatic or inaccessible network address of mobile devices poses a challenge if the device is to play the server role to networked clients (a gateway may be required). Finally, on-device deployment is the preferred choice when a requirement is to sense information from the user locale.

---

* Mobile Web Server Symbian (O), in Wikipedia, Retrieved September 25, 2009 from http://en.wikipedia.org/wiki/Mobile_Web_Server_(Symbian_OS)
† http://ksoap2.sourceforge.net
‡ http://wiki.opensource.nokia.com/projects/PAMP
§ http://www.parallelgraphics.com/products/cortonace

"In-Network" Deployment: Deploying PNA logic on a server allows a wide swath of software modules to support the PNA with few to no limitations in terms of computing libraries. In addition, some examples of libraries found primarily for workstations but not for mobile OS's include: semantic Web libraries, AI, mobile agents, mathematical, statistical, formal logic, database, reliable storage mechanisms (though this is changing with SQLite). A question one poses is, *Where are the resources needed in computation?* When the answer is "largely on the device," it may make little sense to compute from server over the radio access network (i.e., additional costs and latency). When the answer is "largely in the network," the in-network model is very advantageous because there are (a) far fewer scalability and control issues, (b) far fewer availability issues, (c) far fewer complexity issues (server software can be arbitrarily complex, almost, compared to on device). For example, the AT&T Watson speech recognition service is, "so complex that it is more practical to run the software on centralized servers than to install, manage, and maintain it on countless mobile devices" [15]. The software uses acoustic model, feature extraction, pattern classification, language, and word models; most of these are better off running on more proficient CPUs. So while the I/O happens on the device, the collected speech data is transferred to a server where most of the analysis occurs (results are sent back to the device).

11.4.2.4 Issue: Background Processing

Mobile applications tend to need background processes in order to know about things in the outside world, things happening locally, and to communicate and receive communications on an on-going basis. In order to preserve battery life and present rogue or forgotten processes from doing harm, OS vendors have not always included support for background processing. In an ideal PNA, notifications would be generated in the background and should be either (a) ambient, in which notifications come calmly and in ways that do not distract from other tasks; however devices are not always on or "ready" to receive notifications and they must be on or near the user to be effective in this case, or (b) active, in which the user takes an action to see and review new notifications. The following is a high-level summary of some OS and technology level support for background processing at the time of writing.

- ■ Android platform*—Allows for "services," which can run in the background and perform tasks. The system keeps the service running as long as either it is started *or* there are one or more connections to it. Once neither of these situations hold, the service is effectively terminated. Background services may be killed by the system if under heavy memory pressure. If this happens, the system will later try to restart the service.

---

* http://www.android.com

- iPhone—"Only one iPhone application can run at a time, and third-party applications never run in the background. This means that when users switch to another application, answer the phone, or check their e-mail, the application they were using quits."* The Apple Notification API (ANA) promises to allow the sending of (a) counters/badges, (b) audio alerts, (c) popups to iPhone via a Server in the cloud. ANA is not quite the same as full-fledged background processing, does not permit the same rich use cases, but does enable pleasantly effective notification scenarios.

- Sun Mobile Information device Profile 2.0 (MIDP)†—With MIDP, the visible interface of a MIDlet can be set to invisible mode, thus pushing the application off of the screen and into the background. The MIDP Push Registry (in which a MIDlet can be started in the mobile platform upon receipt of some external event), is a weaker form of background processing in which a MIDlet can be activated by a message sent from a remote entity. The support of various Java JSR's and background processing differs greatly from MIDP phone model to model.

- Symbian‡ (see Linux mobile devices as well)—Symbian uses process/threads model where each user process has its own private address space and can launch threads. "Long running tasks usually run in the background. They use a low priority thread and do not directly respond to user interaction. They generally use all the CPU time available to them and run for a period of a fraction of a second to a few seconds. Occasionally long running tasks, critical to the currently selected UI application, run at a higher priority for a few seconds."§ Again, in reality, Symbian support for these paradigms differs between models.

As many mobile developers know, it is absolutely necessary to test mobile applications such as PNAs directly upon the real targeted devices.

"On-Device" Deployment: At the time of writing, devices tend to have hit and miss support for background processing. Most mobile developers are unhappy in one way or another with either the mechanisms or the spotty support for "true" background processing (not a weaker form). However, technology is now appearing for most major mobile OSs to deliver something close to true—if not indeed true—background processing.

---

* "The iPhone OS Platform: Rich with Possibilities" http://developer.apple.com/iphone/library/documentation/userexperience/conceptual/mobilehig/developingsoftware/DevelopingSoftware.html
† http://java.sun.com/products/midp/
‡ http://www.symbian.org/
§ "S60 5th Edition C++ Developer's Library v2.1," see http://www.forum.nokia.com/

"In-Network" Deployment: Relatively few to no limitations on background process-ing on behalf of a user or group (due to scalable servers), and by definition, a server and client are decoupled so the server can process arbitrarily to—and in parallel with—a client. Apple Mobile Me* subscription service, for example, is a network-based background process that syncs user data and messages with the iPhone asyn-chronously. Notification API is a network-based server through which third-party services reach individual iPhones.

11.4.2.5 Issue: Radio Network (Latency and Game-Like PNAs)

Many smartphones access the Mobile Web which is, in many cases, a compromised version of the desktop Web (i.e., reformatted and simplified, for example, WAP). Mobile Web has been designed so that it is delivered efficiently over the radio net-work. Nonetheless, contention at the radio access level is an ongoing traffic engi-neering challenge for mobile network operators. Increased blocking, dropping, and poor service can mar user experience and cause customer churn (e.g., recognizing their initial poor user-experience may have been network-based Apple has taken action†).

It turns out, in fact, that in many "fun" applications, especially those involv-ing multiple users or sources, "latency is the enemy of fun" [16]. Gaming and/or fluid applications tend to have to be "hybrid" applications with components on both device and the network. Information stored on the device may allow reduction of both: (a) number of calls to server and (b) amount of information in calls but when transmitting from client to server utilizing the network protocol efficiently is important (e.g., forcing a PDU into a single packet can improve overall performance). In gaming applications, the network entity is required for both technical and social reasons: (1) technical—the server is the relay and keeps the master state, and (2) social—the server is the "arbiter of truth" (detects and discourages cheating). Serious server-side scaling (esp. of mobile multi-player games) is challenging and can be done on the basis of: (a) geography of the game, where each server may host an individual isolated part of the game (this requires a sort of prediction of what parts of the game will be most popular, or (b) shards of the game where a copy of a part of the game residing on a server; players see gamers in the same shard but not from others (which might limit interactions and usability). As an example, in building Project Darkstar [16], an enormous open source virtual world project, developers chose to use a server and server-side database (but not SQL), and decided that data caching would be a key technique to ensure overall efficiency.

---

* See http://www.apple.com/mobileme/
† "AT&T Plans Network Upgrade to Delivery Faster 3G Mobile Broadband Speeds," see http://www.iphonehacks.com/2009/05/att-plans-network-upgrade-to-delivery-faster-3g-mobile-broadband-speeds.html

"On-Device" Deployment: PNAs running on the device may (a) play the role of server to event generators in the environment; in this case, network performance may be jeopardized, depending on the number of such devices in a given region and the event-rate from generators (e.g., imagine a team of users all getting streamed events from the Internet, all in the same cellular sector), or (b) play a strictly local role, creating and consuming only local data; in this case, the radio network is not in jeopardy of being overused.

"In-Network" Deployment: PNA applications running in the network may depend only on a few classes of events that emanate directly from devices or may rely heavily on information from the Internet (such as social network metadata, networked calendars, etc.). A naïve in-network PNA might send notifications to the mobile device in such a fashion as to overwhelm the radio network. A smarter approach might use aggregation and filtering to send only critical and/or intermittent information, or might infer and take into account the current context or attention of the end-user. Certain classes of PNAs that make heavy use of data streaming could be a jeopardy to the radio network load [17].

11.4.2.6 Issue: Interfacing with the Environment (Including Location Sensing)

For PNAs, the state of the outside world and the relative position of the device to it, is often of importance; for example, to enable the sensing of (a) proximity of people (e.g., many tens of "find my buddies applications"*), (b) current location, bearing, and speed, (c) trends and "tribes,"† (d) for interfacing with data embedded in the environment (e.g., networked devices, scan-able codes, kiosks,…) and with other mobile users (e.g., P2P-enabled devices such as Microsoft Zune).

"On-Device" Deployment: On-device sensors are co-located with the user and the device is obviously in the best position to understand its local environment and to process locally sensed information. However, for this to be feasible we ask: Is the device open and extensible? Can new sensors be attached? Can software be written on them?‡ Limitations of the mobile device may constrain which kind of sensors and processing are enabled, but nowadays almost all modern mobile OSs have APIs allowing software development against onboard sensors (e.g., particularly Bluetooth, but also camera, pedometer) A variety of interesting mobile applications integrating on-board scanning, gyroscopes, and microphones have recently been made available on almost all platforms.

---

* See, Google Latitude, IRL Connect (the "first visual social networking site"), Loopt, and Buddy Beacon.
† See Sense Networks suite of products, http://www.sensenetworks.com/
‡ For Java phones, for example, consider the Mobile Sensor API (JSR 256) and others.

"In-Network" (*Sensing*) *Deployment*: The only way in-network services can sense local activities is through direct or indirect fusion of local information. For example, in partnership with a mobile service provider, in-network probes on base stations might listen-in on the interactions between mobile devices and the station (and infer location and other activity). Alternatively, specialized sensors may "sense" the device's presence (e.g., through Bluetooth) and start service flows in the network while interworking with the server. In general, a seeding process must occur first in which devices have uniquely identifiable codes (e.g., MSISDN) to the hardware that is sensing them. Computing and processing of sensed data may surely take place in the network after it is gathered or sensed locally, but this requires that the device first senses the data and then transmits it (see AT&T's *Watson* voice-recognition service). As long as position—or some other attribute—can be sensed and transmitted, the in-network approach could infer other environmental metrics near the user by considering what other sensors and people have reported, or through learning or statistical approaches.

11.4.2.7 Issue: Interfacing with Data

Interfacing with various information sources is a key use case of PNA's (e.g., for reading, correlating, mashing-up information). To read cloud information from the device requires: (a) a data connection (e.g., HTTP) as needed, and (b) support for "always" on—for example can the application pull important data on a long-running basis? What if the user's in a voice session? Thus the location of the information that drives the application has a great impact on the technical approach. There are several equally legitimate scenarios: (1) information is self-contained in an application or on the device or PC (e.g., for financial or other reasons the developer may have decided this), (2) information is in the network (e.g., Google Calendar, e-mail, contacts, photos, blogs), or (3) information flows between the network and the device (e.g., Active Sync, Google Calendar Sync).

"On-Device" Deployment: If all the data that the PNA needs is device-resident, then the best option is to access it from a device-resident application. If some of the data is in the network then on-device deployment might yet still be best, but other factors come to bear. If all of the PNA data is in the network, then the "thin client" approach may be the best, allowing the in-network components to interface with the networked data and (a) report back to the device or (b) build a Web page accessible to the Web browser on-board the device.

"In-Network" Deployment: There are great advantages to interfacing to data via the network: (a) core bandwidth is very likely higher than radio access network bandwidth, (b) reliability is high (chances of disconnections low, and, compared to the device, no chance of "out of service area"), (c) you can "see the whole Internet," through TCP/IP, HTTP, and other transport (e.g., this gives rise to mashup possibilities), and (d) all the computing advantages of using infrastructure hardware

are also true. There are some disadvantages too. If some of the information you need is device-resident, one of these approaches is necessary: (a) a software component on the device to push data to the network and (b) a server on the device that responds to requests and uses the device APIs in order to acquire the data it needs to transmit (it is now becoming possible to run HTTP servers directly on mobile platforms*).

11.4.3 Intangible Issues

Here we note that end users mostly do not know or care where their services run. Most would not care if their voice session was carried on the Internet (VoIP) or on a carrier's infrastructure. When it comes to devices though, many users expect "personality" and, as such, device customization is an emerging business model (see custom ring-tones, wallpapers, custom phone design,† built-to-order‡). This section draws some high level (and somewhat light-hearted) comparisons between those who tend to favor network services versus on-device applications and is vaguely akin to the Net-heads versus Bell-heads argument [18] in that we draw on some fairly coarse stereotypes.

"On-Device" Deployment Arguments: Passionate prodevice folks are fairly often "maverick" Net users, early adopters, programmers, and do-it-yourselfers. They have a general mistrust or dislike of carriers (though device manufacturers (e.g., LG, Apple) are more envied) and they *tend to* rebel against centralized control and servers. They relish applications that (a) they do not have to pay for or are free of digital rights restrictions, (b) run independently of "big business" (e.g., see the "anti-Microsoft" angle, or free community hot-spots as in Whisher.com and others). This tends to make these users lean toward using client-side applications and development as well as jail-broken and open handsets and networks. To cater to these folks mobile PNA developers should deploy onto the device, make the PNA essentially free, highly customizable, and feature-laden.

"In-Network" Deployment Arguments: On the other hand, the Network is both the communications fabric and the data cloud and a large number of end users are "Googlers," whose main interactions are with Web data and who trust centralized control of these services (e.g., Flickr, Facebook, Google, eBay, etc., all of whom store a great deal of information for and about the user). This is slightly different than trusting the operators, who seem to be at the lowest ring. For these users, access to user profiles, mashups, social network data, and their own personal and business data stored on centralized servers, is important. Device-resident data is not

---

* See Nokia Mobile Web Server (beta), at http://mymobilesite.net/
† See http://www.modumobile.com/#/catalog/
‡ Nearly 60% of phone buyers cite size, shape and color as key features when choosing a phone (see: http://www.pcmag.com/article2/0,2817,2350615,00.asp for more details).

necessarily key; the device need only support a Web browser. For users who trust the cloud, the risk of a lost device (and all local data) compels them to prefer the network-centric operations and storage. In applications that involve multiple entities, the central server is an "arbiter of truth," which net-heads appreciate. To cater to these folks, developers should build a mobile PNA application that sits on a Web site, mashes-up many Internet data sources, and does not fail, crash or go offline for long periods.

Of course, many end-users exhibit characteristics of both of these higher order types and there are no solid rules on how to satisfy this majority all the time.

11.4.4 *Conclusions*

The dramatic rise in mobile cell phone penetration and the cultural phenomena of mobile applications—especially to enhance social communications and groups—is one of the most dramatic changes in human history to how we communicate. Intelligent proactive notification applications (PNA) (part of a larger personal productivity category) will remain an important class of mobile services and stand to benefit greatly from innovations in processing, storage, and information models. Reminders are a subset of so-called location-based services, which have, and (many think) will continue to generate billions in revenue for operators and providers. Indeed, PNA and mobile advertisements are "sibling," almost equivalent, technologies. The mobile advertising market itself is said to be worth billions of dollars. While at the time of writing the iTunes model for distribution has emerged as most effective, mobile PNA—and indeed all—application developers need an understanding of the trade-offs between developing their idea as a standalone mobile application, as a mobile Web site accessible to the mobile device, or as some hybrid of the two. The decision has a potentially large impact on what happens next with the mobile application: how it succeeds, how it is maintained and updated, and how much revenue it may yield for the developer and the rest of the actors in the value chain. On the other hand, perhaps it does not matter much, as most mobile applications are simply not used for very long ("after 20 days, less than 5% of those who downloaded an application are actively using it" [19]). Table 11.1 summarizes some of the issues that were discussed in the main body of this chapter. We have found that there are compelling reasons for both in-network and on-device deployment modes. In reality, a hybrid approach that uses applicable techniques from both sides may be desirable, depending on the unique characteristics of the particular PNA being developed.

While this chapter has mainly explored some of the issues underpinning deployment decisions for a compelling class of mobile applications, we referred to as proactive notification applications, the discussion has broad applicability to other classes as well.

Table 11.1 Summary

| Issue | Suitability, Applicability ("✓" Implies Suitability, "×" Implies Not Suitable, and "–" Implies "No Advantage, or Hybrid Approach") | |
	In-Network	*On-Device*
User base (size)	✓ (serving large user base can be done)	× (platform choices and connectivity)
Digital rights, copying, disruption	✓ (keep all binary and logic far from user)	× (application logic could be hacked, copied illegally)
Hosting	× (ongoing hosting costs for the service deployer)	✓ (no hosting, end-user assumes cost of network)
Maintenance	× (hosted applications will have ongoing IT and upgrade costs)	× (maintaining and updating across multiple platforms hard)
Marketing and channels	– similar across both	– similar across both
Revenue models	✓ (ad-driven is attractive)	– (getting rich achieved only by a small percentage)
Power	– in-network apps still need to use device Web browser	– apps need to use screen and sensors which drains power
iPhone	× (simply creating a site not as powerful as having a custom app)	✓ (works well; many apps use network access as well)
Who is the user base	– communications, internationalization	– communications, internationalization
Asynchronous messaging	✓ (server operates all the time, can push to device as needed)	× (device not always one or ready)
Service logic	✓ (few to no constraints, except reading local data is hard)	× (can't run 24 × 7 anyways)

(continued)

Table 11.1 (continued) Summary

Issue	Suitability, Applicability ("✓" Implies Suitability, "×" Implies Not Suitable, and "–" Implies "No Advantage, or Hybrid Approach")	
	In-Network	*On-Device*
Background processing	✓ (few to no constraints, except reading local data is hard)	– (most phones will soon be able to perform background jobs)
Radio access network	– (may or may not transmit large amount of data to devices)	– (may or may not require data network)
Interfacing with the environment	× (possible but difficult to infer location-sensitive sensing)	✓ (in the right place with the ability to sense)
Interfacing with data	✓ (pulling from the "cloud" easy)	– (some constraints and penalty for pulling data)

Acknowledgments

The author acknowledges past helpful collaborations with colleagues Shoshana Loeb and Thimios Panagos as related to mobile reminder applications.

The trademark and trade names DoCoMo, i-mode, J2ME, Java, Qualcomm, Brew, Get it Now, Verizon Wireless, Apache, Linux, MySQL, python, Eclipse, NetBeans, iTunes, iPhone, Remember the Milk, reQall, Wakerupper, Android, Twitter, Telcordia, Blackberry, Mobile Safari, Internet Explorer, BOLT, Opera, Skyfire, Internet Movie database, Amazon, Facebook, Flickr, Wal-mart, OhDontForget, Apple, Yahoo!, Windows Live, AIM, Windows Mobile, Symbian, Nokia, OpenGL, Direct3D, SQLite, AT&T, Watson, MobileMe, Microsoft, Zune, Sense Networks, Sense Networks, Google, Google Calendar, Active Sync, eBay, Bluetooth, and YouTube belong to their respective owners.

References

1. S. Loeb, B. Falchuk, and T. Panagos, *The Fabric of Mobile Services: Software Paradigms & Business Demands*, Wiley-Blackwell, New York, 2009.
2. S. Levy, Betting on the store: In the battle of the smartphones, applications will determine the winner, *Wired Magazine*, 17(09), September 2009.
3. O. Leitner, So, you want to deploy a J2ME app in the US?, Next Generation Mobile Content blog, at http://www.nextgenmoco.com/2008/01/so-you-want-to-deploy-j2me-app-in-us.html

4. M. Baldauf, S. Dustdar, and F. Rosenberg, A survey on context-aware systems, Technical Report TUV-1841-2004-24, Technical University of Vienna, Vienna, Austria, 2007.
5. M. Weiser, R. Gold, and J. S. Brown, The origins of ubiquitous computing research at PARC in the late 1980s, *IBM Systems Journal*, 38(4), 693–696, 1999.
6. S. Loeb, T. Panagos, and B. Falchuk, Actionable user intentions for real-time mobile assistant applications, *Workshop on Personalization in Mobile and Pervasive Computing (in Conjunction with UMAP 2009)*, Trento, Italy, 2009.
7. B. Falchuk, A. Misra, and S. Loeb, Server-assisted context-dependent pervasive wellness monitoring, *Proceedings of the ICST International Workshop on Wireless Pervasive Healthcare (WiPH) (in Conjunction with Pervasive Health'09)*, pp. 1–4, London, U.K., 2009.
8. B. Falchuk, K. Sinkar, A. Dutta, and S. Loeb, Mobile contextual mashup service for IMS, *Proceedings of the Second IEEE International Conference on Internet Multimedia Services Architecture and Applications*, pp. 1–6 Bangalore, India, December 2008.
9. B. Falchuk and S. Loeb, Towards guardian angels and improved mobile user experience, *Proceedings of the IEEE GLOBECOM 2008—Communications Software and Services Symposium*, pp. 1–5, New Orleans, LA, 2008.
10. B. Falchuk, T. Panagos, and S. Loeb, A deep-context personal navigation system, *Proceedings of ITS America 15th World Congress on Intelligent Transportation Systems*, New York, 2008.
11. C. Anderson, *Free: The Future of a Radical Price*, Hyperion, New York, 2009.
12. M. Callow, P. Beardow, and D. Brittain, Big games, small screens, *Power Management*, 5(7), 40–50, December 2007.
13. D. Frommer, iPhone App Prices Tanking (AAPL), Business Insider.com, February 2009.
14. A. Nurminen, Mobile 3D city maps, *IEEE Computer Graphics*, 28(4), 20–31, 2008.
15. S. Cass, Want an easy way to control your gadgets? Talk to them, *Discover Magazine*, December 2008.
16. J. Waldo, Scaling in games and virtual worlds, *CACM*, 51(8), 38–44, August 2008.
17. J. Motti, Wireless devices may be future P2P battleground, Datamation.com, August 2008.
18. S. Steinberg, Netheads vs Bellheads, *Wired Magazine*, 4(10), 144–147, 206–213, 1996.
19. T. Krazit, Most iPhone applications gathering dust, CNET news, http://news.cnet.com/most-iphone-applications-gathering-dust/, February 20, 2009.

Chapter 12

Hosting Web Services on Mobile Devices

Muhammad Asif and Shikharesh Majumdar

Contents

A Web service (WS) is a software component that can be accessed and executed over the Internet using standard protocols and well-defined interfaces (Web Services Architecture, 2004). The protocols used are the Simple Object Access Protocol (SOAP Specifications, 2007) and the Hyper Text Transfer Protocol (HTTP). WS interfaces are described in a well-defined format using the Web Service Description Language (WSDL) (Web Service Description Language, 2001). Typically, service providers publish their Web services to a registry such as the Universal Description, Discovery, Integration (UDDI) registry (UDDI Specifications). WS clients can search for Web services by querying the UDDI registry.

Web services have been traditionally used in a wired network of desktop computers. Hosting of services on mobile devices is a relatively new concept. There are a number of WS toolkits or WS execution environments (WSEE) available for hosting Web services on desktop machines that are connected to wired networks. A WSEE can be defined as a platform that facilitates the execution of Web services. Mobile devices such as netbooks, mobile phones, and Global Positioning System (GPS) devices have resource constraints that make it difficult to host Web services using existing solutions for desktop computers. In this chapter, the details of the design and the implementation of two WSEE are presented. These two WSEE are devised for resource-constrained devices.

12.1 Overview

Because of the interoperability Web services provide in a distributed heterogeneous environment, Web services technology is becoming popular in a number of domains, including business-to-business, electronic commerce, and in automating information exchange between business processes. Because of the large variety in the available mobile devices, interoperability provided by a Web service can lead to an effective integration of these devices with the client. There are many interesting applications where hosting of Web services on handheld mobile devices is useful. A few examples of such applications are described next:

- A shipping company can provide a facility to its customers for tracking the shipped items in real time. The tracking is possible if the vehicle carrying the shipping items is equipped with a mobile device, a GPS receiver, and a tracking WS hosted on the mobile device. The universal resource identifier of the hosted WS can be shared with the customer. The customer can directly access the hosted WS to track the shipped items in real time and estimate their delivery time.

- A supply chain management system is a network of different companies or departments for producing, handling, and/or distributing a specific product to the consumer. A person running a small business and using a small tablet PC or a netbook in the field can be a part of a supply chain system. The services offered by such a person in the field can be available through Web services hosted on his or her device. The reason for hosting Web services on his or her device is that for such applications, the data to be used in a WS needs to be the most recent. Since he or she is always updating the data while working in the field, it makes sense to host Web services on his or her device.

- Tracking skilled persons, such as doctors, in an emergency situation is another example. For such application, a WS is assumed to be hosted on his or her smart phone equipped with a tracking sensor (e.g., a GPS receiver). The WS can be invoked at a nursing station in the hospital, and the exact location of the doctor can be determined.

- A smartly built GPS device in a vehicle can host a traffic information service for peer GPS-based devices to access. Any peer GPS-based device in another vehicle can access such services (on multiple devices) to collect information about traffic congestions, roadblocks, or accidents, and can suggest an alternate route to the driver.

- In pervasive computing, interaction among different devices can be managed conveniently if the devices are offering services using WS technologies. The use of WS technology for interaction of devices is useful to integrate devices from different manufacturers even when they are devised using different platforms and different programming languages.

A number of other applications such as a wallet service, a parcel tracking system, sharing the time or the dictionary are discussed in the literature (Berger et al., 2003; Srirama et al., 2006). There are a number of challenges in hosting such Web services on mobile devices. These challenges are discussed next.

12.1.1 Challenges

The key challenges include resource constraints, diversity in hardware and operating system used for mobile devices, nonavailability of WS execution environments, and weak wireless signals.

12.1.1.1 Limited Resources

The WS application may require complex algorithms to achieve its goals; complex cryptographic algorithms are executed if the WS application requires an end-to-end security at the message level. In many cases, significant CPU processing is required to provide the desired functionality. For example, image format conversion requires a considerable amount of CPU and memory resources. Moreover, accessing or providing a WS always incurs the overhead of SOAP/XML processing. The required resources for WS applications include CPU power, memory, battery, and bandwidth. Unfortunately, handheld devices have very limited resources, and this represents a major challenge for hosting WS applications on handheld mobile devices.

To minimize the resource demand on mobile devices, two novel WS execution environment (WSEE) architectures are proposed in this chapter. For computationally intensive Web services, the partitioning of WS applications and executing parts of such applications on a remote computing node may be necessary. The details of WS partitioning are beyond the scope of this chapter. A comprehensive analysis and proposed partitioning algorithms are available in Asif et al. (2008, 2009).

12.1.1.2 Diversity

Another challenge in the domain of handheld devices is the diversity in hardware architectures that include Advanced RISC Machine (ARM) and Microprocessor without Interlocked Pipeline Stages (MIPS). There is also diversity in the available operating systems for mobile devices. These include Windows Mobile, iPhone OS, Symbian, and Palm OS.

Micro edition of Java (JavaME, 2006) is selected for implementation because Java is a platform-independent language. The other reasons for selecting Java ME include

- Java is a popular programming language.
- Most of the mobile devices come with built-in support for Java.
- Java ME has very good support for multithreading and communication with other nodes.

Binary Runtime Environment for Wireless (BREW) (Brew Framework, 2008) is another option that is available for running C/C++ as well as Java code on mobile devices. BREW is not very popular, however, because it is not supported by a large variety of mobile devices and uses a proprietary framework for implementation of BREW.

12.1.1.3 Nonavailability of WSEE

The popular WSEEs available for desktop computers are the Apache Axis2 (Apache Axis2 Project) and Glassfish Project (Glassfish Project) for a Java runtime environment, and AlchemySOAP (Alchemy SOAP 1.0.0) for a C/C++ runtime environment. However, these WSEE are not suitable for mobile devices because of the reasons presented next. First, these WSEE are too large and resource-demanding for handheld devices. Second, these WSEE are developed for such runtime environments which are currently unavailable for mobile devices. For example, Apache Axis2 requires Java runtime environment 1.4 or later, which is only available for a desktop machine. Third, limited CPU power, communication bandwidth, battery and available memory on handheld devices are often inadequate for using such heavyweight WSEEs. Mobile devices need a lightweight environment for hosting Web services, so that devices can have enough free resources to perform the other core functions, such as making or receiving phone calls.

It can be argued that the solution developed in the native code of the device will be the most efficient. However, writing optimized programs in low-level native codes for a large variety of handheld devices is an expensive solution because of the diversity of available handheld devices. Using a WS is attractive, because WS technology provides interoperability in a heterogeneous environment in which the WS client and the WS provider may be implemented using different languages and may run on diverse platforms.

The following two sections present details of two WSEEs for the hosting of Web services on mobile devices. One of these WSEEs uses lightweight components and can be used to host Web services that need support for a few WS standards. The second WSEE works as a distributed application, a part of which is deployed on a mobile device and another part on an intermediate node.

12.2 Lightweight WSEE

A WSEE essentially requires a set of software components that facilitate hosting of Web services. The required software components include a transport component for data exchange with WS clients, a SOAP engine component for processing XML/SOAP messages, and an invocation engine that can execute requested WS applications on behalf of WS clients.

As already mentioned, the Java ME environment is selected based on its ability to be operable on diverse platforms for mobile devices such as Windows Mobile OS, Symbian OS, and Palm OS. Java ME has two configurations: the Connected

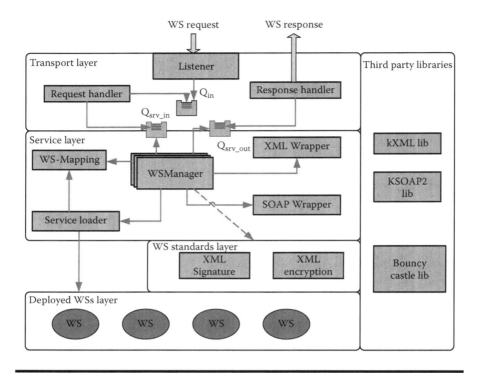

Figure 12.1   A lightweight WSEE.

Device Configuration (CDC) (JSR 218) and the Connected Limited Device Configuration (CLDC) (JSR 139). The prototype of the lightweight WSEE that is implemented for mobile devices is capable of running under both the CDC and the CLDC configurations.

The architecture of the proposed lightweight WSEE is based on four layers: a transport layer, a service layer, a WS standards layer, and a WS applications layer. A high-level architectural overview of the proposed lightweight WSEE is shown in Figure 12.1. For resource-constrained devices, lightweight versions of each of these components are devised. The details of these layers and their key components are discussed next.

12.2.1 Transport Layer

For the proposed lightweight WSEE, the transport layer is capable of exchanging SOAP messages using HTTP or TCP protocols. Note that there is no specific transport protocol associated with the exchange of SOAP messages. SOAP messages can be exchanged between nodes using any transport protocol. This allows

WS applications to select any appropriate transport mechanism according to the availability of resources and the quality of service requirements.

12.2.1.1 Initial Setup

As the lightweight WSEE application starts, the transport listener (Listener in Figure 12.1), which is the main program, performs the following steps for initial setup of the WSEE.

- Load WSEE environment variables from a java property file ("wsee_config. properties"). These environment variables include
  - Number of minimum (N_{min}) and number of maximum (N_{max}) threads of WSManager (explained in the next subsection).
  - Transport mode: HTTP or TCP.
  - Port number at which the transport listener is listening.
  - Timeout after which the transport listener disconnects from a nonresponsive client.
  - Root directory name for the deployed Web service applications.
- Create a thread (request handler) for handling WS requests.
- Create a thread (response handler) for handling outgoing WS response messages.
- Create WSManager threads according to the value provided through environment variable N_{min}.
- Initialize the Web Service Mapping (WS-Mapping) component. During the initialization stage, details of all the deployed Web services are loaded in memory. (The details of WS-Mapping component are described in the next section.)
- Create a server socket and start listening at port specified in the properties file.

On receiving an incoming request from a WS client, the listener creates an object of Java Socket class. This newly created object is put in the Q_{in} queue, and the request handler is notified. The listener starts listening again at the specified port.

12.2.1.2 Request Handler

The request handler thread is either in a wait state or in a run state. The request handler is in the wait state when there is no Socket object in Q_{in}. On arrival of a new Socket object in Q_{in}, the request handler is notified by the listener. On receiving the notification, the request handler goes into the run state. The request handler remains in the run state when it is in the process of handling the new request. At the completion of processing the request, the request handler checks Q_{in}. If Q_{in} is empty, the request handler goes into the wait state and waits for a new request. If Q_{in} is not empty, the request handler selects the next request (Socket object) waiting in the queue, and starts its processing. The key objective of the request handler is to separate the SOAP message from the request. The steps of separating the SOAP message from an HTTP request are as follows:

- Fetch all data bytes of HTTP request into a memory buffer.
- Parse the buffer data to get values of the following key HTTP headers.
  - Method: A WS request can be sent using either the GET method or the POST method. For the current implementation of the lightweight WSEE, the WS client is required to send the SOAP message using the POST method. Note that in case of the POST method, the SOAP message data is passed in the body of HTTP request.
  - SOAP-ACTION: This is an application-specific header attribute. It is used to identify the requested Web service on the server.
  - Content-length: This represents the length of the body of an HTTP request. It is used to extract the SOAP message from the body of the HTTP request.
  - Content-type: This represents the type of the data in the body of an HTTP request. Commonly used values are "text/plain" (for plain text), "text/html" (for HTML data) and "image/gif" (for image data). For SOAP messages, the content-type is set to "text/xml."
- Based on the content-length, the SOAP message is extracted from the body of the HTTP request.

Once the SOAP message is extracted, an object of SoapRequest class is created. The SoapRequest is an application data object that holds information about the SOAP

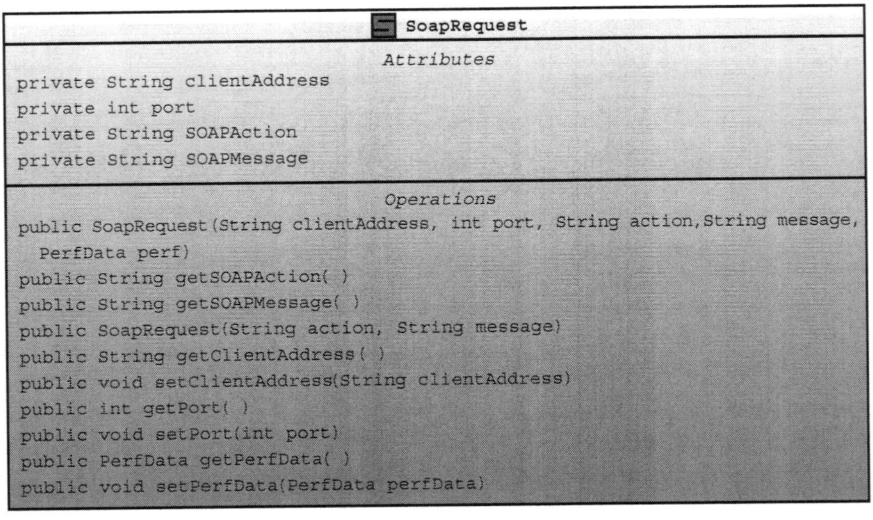

```
SoapRequest

Attributes
private String clientAddress
private int port
private String SOAPAction
private String SOAPMessage

Operations
public SoapRequest(String clientAddress, int port, String action,String message,
  PerfData perf)
public String getSOAPAction( )
public String getSOAPMessage( )
public SoapRequest(String action, String message)
public String getClientAddress( )
public void setClientAddress(String clientAddress)
public int getPort( )
public void setPort(int port)
public PerfData getPerfData( )
public void setPerfData(PerfData perfData)
```

Figure 12.2   Class diagram of SoapRequest.

request message. The class diagram of SoapRequest is shown in Figure 12.2. Attributes of SoapRequest class include the WS client's address, the WS client's port, SOAP action (URI of requested WS), and the SOAP message. Operations shown in the class diagram of SoapRequest include constructor, accessors, and modifiers. The SoapRequest object is put into Q_{srv_in} queue. After this, the request handler fetches a new request from Q_{in} (if Q_{in} is not empty) or goes into the wait state (if Q_{in} is empty).

12.2.1.3 Response Handler

Once the WS request is processed by the service layer, the response SOAP message (if any) is placed in a separate queue, Q_{srv_out}. The response handler works in a similar manner as the request handler. The response handler thread is in a wait state when there is no response message in Q_{srv_out}. On arrival of a new response message in Q_{srv_out}, the response handler is notified by the service layer. On receiving this notification, it fetches the response message from Q_{srv_out} and starts its processing. The response message contains the response SOAP message and information about the WS client. A WSEE application class, WS response, is used to represent the response message. The class diagram of WS response is shown in Figure 12.3. Note that the operations shown in the class diagram are the constructors, accessors, and modifiers.

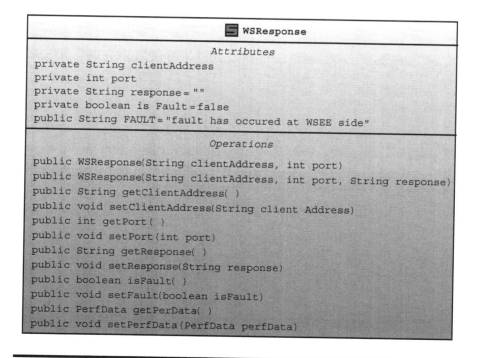

Figure 12.3   Class diagram of WSResponse.

The WS client address and the port number at which the client is expecting to receive the response are stored in WS response object by WSManager. The attribute "response" contains the response SOAP message. There are two attributes (isFault and FAULT) for indication of any fault that occurs while invoking the requested WS. Currently, these two attributes are not being used. After fetching WS response object from Q_{srv_out}, the response handler opens a socket connection with the WS client and sends the response SOAP message. Once the response message is sent, the response handler fetches a new response SOAP message from Q_{srv_out} (if it is not empty) or enters into the wait state.

12.2.2 Service Layer

This is a core layer of the lightweight WSEE. The primary responsibilities of this layer are the parsing of the incoming SOAP messages, executing the WS application, and then wrapping the results of the WS application into a response SOAP message.

Currently, the service layer uses Remote Procedure Call (RPC) style as a WSDL binding. The WSDL binding describes how a WS is bound to a messaging protocol. The document style binding is another popular style of the binding, and can be added in future versions. The reason for using RPC style of binding is that it is easy to implement, and the WSDL generated using the RPC style binding is straightforward and easy to understand. Note that RPC style and document style bindings are not programming models. These bindings only help to translate a WSDL to a SOAP message. The subcomponents of the service manager are described next.

WSManager (shown in Figure 12.1) is the main controller of the service layer. Multiple threads of WSManager are available in a thread pool. These threads are used to process WS requests waiting in Q_{srv_in}. The maximum number of threads that can be created for WSManager is provided as an input parameter in "wsee_config.properties" file. The WSManager uses different components such as XML Wrapper, SOAP Wrapper, a service loader, WS-Mapping, and components for conforming to different WS standards. Interaction of WSManager with other components is captured in a UML sequence diagram, which is presented in Figure 12.4. The different operations that are shown in the sequence diagram (Figure 12.4) are explained next.

```
getReqMsg( ):
```

This is the first operation shown in the sequence diagram (see Figure 12.4). In this operation, WSManager gets the WS request object from Q_{srv_in} if the queue is not empty. In case Q_{srv_in} is empty, WSManager waits until a new WS request arrives in Q_{srv_in}.

```
createSOAPEnv( ):
```

This operation is used by WSManager to create an object of SoapSerializationEnvelope class (see KSOAP documentation (KSOAP 2 Project, 2003) for more details). WSManager interacts with SOAP Wrapper for this operation. SOAP Wrapper

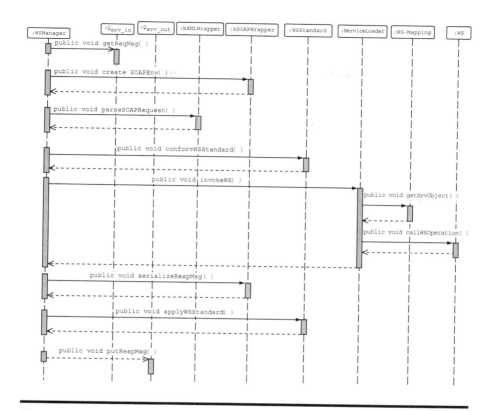

Figure 12.4 Interaction of different components of the service layer.

uses an open source KSOAP2 library to create a SoapSerializationEnvelope object. SoapSerializationEnvelope is a comprehensive object that can separately hold the SOAP request header, the SOAP request body, the SOAP response header, and the SOAP response body.

```
parseSOAPRequest( ):
```

As mentioned previously, the SoapSerializationEnvelope object keeps different parts of SOAP request message separate. As shown in Figure 12.4, WSManager uses XML Wrapper, which in turn uses an open source kXML library to parse the SOAP request message. A sample code that is used for parsing of a SOAP request message is shown in Figure 12.5.

```
ConformWSStandard( ):
```

This operation is used by WSManager (see Figure 12.4) to interact with the WS standards layer. For the current implementation of the lightweight WSEE, this operation performs the tasks that are required for the verification of XML Signature (WS-Security) of WS clients and signing of response SOAP messages.

```
import java.io.Reader;
import java.io.StringReader;
import org.ksoap2.serialization.SoapSerializationEnvelope;
import org.kxml2.io.KXmlParser;
import org.xmlpull.v1.XmlPullParser;
import org.xmlpull.v1.XmlPullParserException;

public sate void parseSOAPReq(SoapSerilizationEnvelope env, String soapReq){

        XmlPullParser parser=null;

        try {

                parser = new KXmlParser();

        }

        catch (Exception e) {
                e.printStackTrace();

        }

        try {

                Reader reader = new StringReader(soapReq);
                parser.setInput(reader);
                parser.setFeature (XmlPullParser.FEATURE _ PROCESS _
                  NAMESPACES, true);
                env.parse(parser);

        }
        catch (XmlPullParserException ex) {
                ex.printStackTrace();
        }
        catch (Exception ex) {
                ex.printStackTrace();

        }
}
```

Figure 12.5   Sample Java program for parsing SOAP message using kXML library.

`invokeWS ( ) :`

After performing the tasks that are required for the conformation with the given WS standard(s), if any, WSManager uses this operation to interact with the service loader for invoking a particular method of the requested WS class.

`getSrvObject ( ) :`

This operation is used by the service loader to interact with WS-Mapping to locate the Java class of the requested WS.

`callWSOperation( )`:

Once the Java class of the requested WS is identified, the service loader first uses this operation to create an instance of the class, and then invokes the requested method of the Java class.

`serializeRespMsg( )`:

In this operation, WSManager serializes the response of the requested WS application into a SOAP message. For this step, WSManager uses XML Wrapper and SOAP Wrapper for serialization.

`ApplyWSStandard( )`:

As shown in Figure 12.4, this operation is called by WSManager to perform the actions required to conform to a given WS standard. For example, the response SOAP message may be required to be encrypted (using XML encryption). Signing the response SOAP message is another example of performing an action to conform a WS standard before sending it to the WS client. In the current implementation, this operation is used to sign the response SOAP message using XML Signature specifications.

`putRespMsg( )`:

WSManager uses this operation to store the final response SOAP message in Q_{srv_out} and to notify the response handler (transport layer).

In the next subsections, different components of the service layer are described.

12.2.2.1 XML Wrapper

For XML parsing and serialization, a lightweight XML parser, kXML (KXML 2 Project), is used. The kXML parser is based on a pull parsing technique and is an implementation of the XML PULL parser API (XML Pull Parsing). XML Pull Parsing is a process of parsing XML as a stream, rather than building a Data Object Model (DOM) tree or push parsing in which events are pushed out to a client code. The pull XML parsers are fast and more memory-efficient in comparison to the parsers that are based on DOM. To access this third-party library, a wrapper component (XML Wrapper) is introduced in the service layer. The XML Wrapper provides an interface to WSManager for accessing XML parsing methods.

12.2.2.2 SOAP Wrapper

For SOAP processing, a kSOAP (KSOAP 2 Project) library is used. kSOAP is an open source library for WS clients to access Web services. To use it on the server side for processing of SOAP message, the open source library is extended in this research. The extension is introduced for the invocation of Web services using the remote procedure call (RPC) style binding. For accessing core methods of the kSOAP library and its extension (devised for the lightweight WSEE), a wrapper (SOAP Wrapper) is introduced.

12.2.2.3 Service Loader

This component is used for invocation of the requested Web services. WSManager supplies the SOAP action parameter and the parsed SOAP request message to the service loader. The service loader interacts with WS-Mapping to get the Java class name (with path) based on the value of the SOAP action. The sample code used in the service loader is presented in Figure 12.6.

In the first block (B1) of the sample code shown in Figure 12.6, the service loader gets the parameters from the SOAP request message. In this step, two arrays are created: one array is for the types of the WS parameters, and the other array is for values of the WS parameters. These two arrays are required by Java Reflection API for invocation of the requested method on an instance of the WS application class. In the next step (see block B2 in Figure 12.6), the Java class name fetched from WS-Mapping is used to create the instance (WS Class object in Figure 12.6) of the class using Java Reflection APIs. In the last step (see block B3 in Figure 12.6), the service loader creates an instance of Java Method class using WS Class object and the two arrays. The Java Method class instance is used to invoke the requested method of the WS. The comments shown in Figure 12.6 explain the different blocks of the sample code. Note that this sample code is only for a simple invocation of a WS method using RPC style binding.

12.2.2.4 WS-Mapping

WS-Mapping is a static component that manages an in-memory list of deployed Web services. The lightweight WSEE assumes that all deployed Web services are available in directory whose path is provided in "wsee_config.properties" file. Under the Web services root directory, individual Web services are assumed to be available in subfolders. A WS is required to provide Java implementation of the WS application and a "ws.xml" file in a folder under the root directory of Web services. The ws.xml file is required to follow the following XML grammar:

```
<webservice>
    <ws-uri> … </ws-uri>
    <ws-class> … </ws-class>
    <ws-standards> … </ws-standards>
</webservice>
```

In this XML format, the `<webservice>` tag is the root node of "ws.xml" document. `<ws-uri>` represents the SOAP action that is used as a key for identification of the Web service. Value inside the `<ws-uri>` node is required to

```
/**
 * parsedReq is a SoapObject (KSOAP Lib class) of WS request and it is
 * passed by WSManager. wsClassName is fetched from WS-Mapping based
 * on the value of SOAP Action parameter.
 **/

    String methodName = parsedReq.getName();
    Class types[] = new Class[parsedReq.getPropertyCount()];
    Object[] args = new Object[parsedReq.getPropertyCount()];
    PropertyInfo arg = new PropertyInfo();                      } B1
    Hashtable properties = new Hashtable();
    for (int i = 0; i < types.length; i++) {
       parsedReq.getPropertyInfo(i, properties, arg);
       types[i] = (Class) arg.type;
       args[i] = parsedReq.getProperty(i);

    }

/**
 * A new instance of the class is created using Java reflection.
 * Note 1: This is only a sample code. For example Web services,
 * advanced features of Java reflection can be used
 * Note 2: catch blocks are not shown. Exceptions to be catched are
 * ClassNotFoundException, IllegalAccessException,
 * Instantiation Exception
 **/

    Object result = null;
    Class wsClass = null;
    Object service = null;
    try{                                                        } B2
       wsClass = Class.forName(wsClassName.getClassName());
       service = wsClass.newInstance();

    }
    // catch block ....

/**
 * Here first an object of the requested method is created using
 * Java reflection and then it is invoked. Again, only try block
 * is shown. Exceptions to be catched are IllegalArgumentException,
 * NoSuchMethodException, IllegalAccessException,
 * InvocationTargetException, Security Exception
 */

    try{
       Method method = service.getClass().getMethod(methodName,
         types);                                                } B3
       result = method.invoke(service, args);

    }
    // catch block ....
```

Figure 12.6 Sample Java program for invocation of a method on an instance of a class using Java Reflection APIs.

be unique. `<ws-class>` node refers to the name of the Java class of the WS application. `<ws-standards>` node can contain multiple child nodes. Each child node corresponds to the actions that need to be performed to conform to a specific WS standard. The support for the custom data types as WS parameters is planned for a future version. The current version of the lightweight WSEE supports primitive data types and collections as WS parameters. At the time of initialization, WS-Mapping reads "ws.xml" files for all deployed Web services, and loads the relevant data of the deployed WSs in memory.

After invoking the requested WS application, WSManager serializes the results into a response SOAP message. If the requested WS requires conformation with additional WS standards such as the one that requires the verification of a XML Signature, then WSManager interacts with the component that provides the verification of the XML Signature.

12.2.3 WS Standards Layer

To support additional WS standards, another layer is introduced in the system design of the lightweight of WSEE. This layer is assumed to contain the components that perform actions that are required for conformation with different WS standards. For testing purposes and as a proof of concept, support for verification of the XML Signature of incoming SOAP messages and signing of the outgoing SOAP messages are provided in the current version of the lightweight WSEE. For verification and signing of XML Signatures, a lightweight version of a third-party cryptography library (Legion of the Bouncy Castle) is used. A brief overview of the steps involved in verification of the XML Signature and signing of SOAP messages are discussed next.

12.2.3.1 XML Signature Verification

The first step of verification of the XML Signature is the extraction of the digest, the signature, and the public key elements form the SOAP header. Note that the digest, the signature, and the public key elements are placed in the SOAP header by WS clients. The digest of a message is a unique number that is created by using a hashing algorithm for representing the message. If the message is changed, the digest value will also be changed. The signature is an encrypted form of a message that is obtained by using the private key of a user and the message itself. In the second step, the digest of the message is computed and compared with the digest value already extracted from the SOAP header. In the third step, the signature extracted in the first step is verified by using the public key information. Note that this key information is also extracted in the first step. If the computed digest does not match the digest value extracted in the first step or if the signature is not verified by using the public key, the XML Signature verification is said to have failed, and a SOAP fault is sent back to the WS client. A SOAP fault is an optional part of a SOAP envelope (in addition to the header and the body) used for reporting errors.

12.2.3.2 XML Signature Signer

Signing a SOAP message is also a three-step process. In the first step, the digest for the contents to be signed (the body or any element of a SOAP message) is computed. In the second step, the signature of the contents to be signed is calculated using the private key of a user. In the last step, the digest, the signature, and the public key information are inserted in the SOAP header. The XML Signature specifications (XML Signature: Syntax and Processing, 2008) describe structure of elements that can be used to insert the digest, the signature, and the public key information in the SOAP header.

12.2.4 Performance Analysis

A simple WS, "SearchProducts," is used for performance analysis. This sample WS has a "Search" operation that provides a list of products (from its local database) based on price information provided by the WS client. The WS extracts the input parameter "price" from the incoming SOAP message, after which it searches a local database to get a list of products that has a price less than the input price parameter, and then returns the list as a response. The database query involves operating on two tables to get the desired result set.

The WS client program is written using standard Java, and is run on a machine that is equipped with a 1.86 GHz single core Intel processor and a memory of 1.5 GB. The lightweight WSEE and the sample WS are deployed on a Dell Axim x51v PDA. The intercommunication between the client machine and the server PDA is based on the wireless local area network (LAN) that uses IEEE 802.11 standard. The measurements were made on a dedicated network where the experiments ran without any interference from other applications.

The two performance metrics of interest are end-to-end mean response time (R) and mean processing time (P). *Response Time* is the difference between the time when a WS response is received by the client and the time when the SOAP message is sent by the client to the server. *Processing time* is the time taken by the WSEE to process a request. It is the difference between the time when a WS response message is sent by the transport manager of the WSEE back to the WS client and the time when the WS request is received by the transport manager of the WSEE.

The prototype of the lightweight WSEE is analyzed for different cases. The prototype uses such values of the "price" parameter that result in SOAP messages of different sizes. The prototype is also analyzed from the concurrency point of view. A number of clients send SOAP requests to the WS concurrently. Key characteristics of the lightweight WSEE that are observed based on the implementation and experimental analysis are summarized as follows:

■ The proposed WSEE has a very small memory footprint that is less than 100 KB.
■ The system can process concurrent SOAP requests from multiple clients, and the mean response time seems to be in an acceptable range for a dozen concurrent clients or less. As the number of concurrent clients increases, the mean

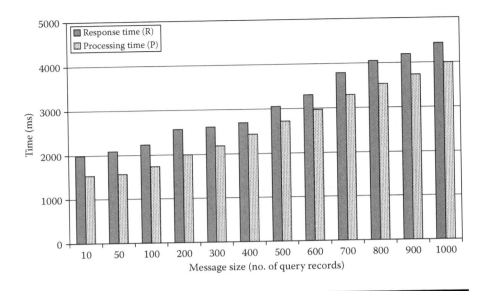

Figure 12.7   Effect of size of response SOAP message on the response time (R) and the processing time (P).

response time (and the mean processing time) for each request also increases. This is because of the increase in the number of active SOAP manager threads that compete for the processor. The WS deployed on a mobile device is not expected to be accessed by a very large number of clients concurrently, but the support for handling a small number of multiple clients concurrently is an important feature for hosting Web services in any environment.

■ The lightweight WSEE is tested for a number of scenarios. The system is observed to be scalable with the response message size: it has been observed that the mean response time R increases in a sub-linear fashion with the message size (see Figures 12.6 and 12.7).

■ As the size of the message increases, the average response time and the average processing time also increases. For this experiment, the processing time includes the times for invoking the Web service, for preparing the SOAP response message, and for signing of the SOAP response message.

12.3 Partitioned WSEE

In the lightweight WSEE, only the verification and the signing of the XML Signature are supported as a proof of concept. There are many other WS standards proposed by the WS community for security, reliability, and transactions. A WSEE that supports a large number of WS standards is difficult to deploy on mobile devices with limited resource capabilities. For such resource-demanding requirements, the proposed technique is to partition the execution environment and deploy the two partitions on two different nodes (an intermediate node and the mobile device node). The proposed technique of WSEE partitioning is based on the splitting of the SOAP engine (WSManager in Figure 12.1) functionality. The functionality of WSManager is divided into two partitions only: one for the intermediate node and the other for the mobile device. The intermediate node is the one that receives the WS requests from WS clients on behalf of the mobile device.

12.3.1 Motivation of the Proposed WSEE Partitioning Technique

WSManager performs a series of tasks for invoking a WS. To explain the motivation and concept of a distributed WSManager, the sequence of tasks related to the invocation of a WS is described. The typical tasks that are performed by a WSManager thread are decryption of incoming SOAP message (T1), verification of the identity of the WS client (T2), verification of the integrity of the message (T3), invocation of the WS (T4), signing the response message with the service provider's certificate (T5), and encryption of the response message (T6) before sending it back to the WS client. The sequence of tasks as they are performed by a WSManager thread is captured in Figure 12.8a. Figure 12.8b shows the execution of tasks T1–T6 by a distributed WSManager.

As mentioned previously, the functionality of WSManager is divided into two partitions: one of the partitions is handled by an *intermediate WSManager* and the other by a *mobile WSManager*. Note that both the nodes (intermediate as well mobile device node) will have the same implementation of WSManager. But WSManager on the intermediate node does not need to interact with the service loader and WS-Mapping components. The objective of having the same implementation of WSManager on the two nodes is to make the partitioning of WSManager tasks configurable. The two WSManager implementations will differ only in their executions. One WSManager will perform one set of tasks, and the other will execute another set of tasks. The intermediate WSManager can be assigned tasks that demand more resources. The mobile WSManager is responsible for processing the tasks that require local resources on the mobile device or use confidential information available on it. For example, if encrypting or signing a response message requires a security certificate of the device's owner, then it is not a good practice to delegate such tasks to an intermediate node. With a distribution of tasks shown in Figure 12.8b, the intermediate WSManager decrypts the incoming message (T1), and verifies the identity of the WS client (T2) and the integrity of the message (T3). The mobile WSManager executes the task for invoking the requested WS (T4) and the task for signing the response message (T5). In the end, the intermediate node performs the task for encrypting the response message (T6).

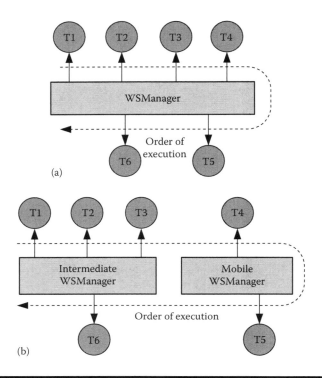

Figure 12.8 Tasks of a WSManager (a) performed by a single WSManager and (b) performed by a distributed WSManager.

12.3.2 Configurable Partitioning Scheme for Execution of WSManager Tasks

To make it possible to allocate different sets of tasks for execution to the intermediate WSManager and the mobile WSManager for a given WS, a configurable partitioning scheme is introduced. A *partitioning scheme* can be defined as an agreement between the two distributed components of WSManager. This partitioning scheme helps the two components to identify which tasks need to be executed by which component. The partitioning scheme is required to be submitted at the time of deployment of a new WS. The submitted partitioning scheme is available to both components of WSManager (the intermediate WSManager and the mobile WSManager) at the time of execution. A sample partitioning scheme for execution of WSManager tasks is presented in Figure 12.9. Note that this arrangement has the flexibility of using different partitioning schemes for different Web services hosted on the mobile device.

This scheme uses an XML schema that defines the XML elements with attributes for each WSManager task. The scheme shown in Figure 12.9 is designed in such a way that the most of the tasks related to security are processed by the intermediate

```
<?xml version= "1.0" encoding="ISO-8859-1"?>

<!- A sample partitioning scheme to split task execution between an
    intermediate node and a mobile node

-->

<webservice uri= 'http://www.carleton.ca/sce/mobileweb/sample 1'>
  <!-- tasks assigned to static SOAP engine -->
  <intermediate>
        <DECRYPTION required="false"/>
        <IDENTITY required="true"/>
        <INTEGRITY required="true"/>
  </intermediate>

  <!-- tasks assigned to mobile SOAP engine -->
  <mobile>
        <WS _ OPERATION class-nake=""/>
        <SIGNATURE required="true"/>
  </mobile>
  <intermediate>
        <ENCRYPTION required="false"/>
  </intermediate>
</webservice>
```

Figure 12.9   A sample partitioning scheme.

WSManager running on an intermediate node. In this scheme, the intermediate WSManager is assigned decrypting of the incoming SOAP message, verifying of identity of the WS client (IDENTITY), validating the integrity of the message (INTEGRITY), and encrypting the response message (ENCRYPTION). The mobile WSManager is assigned the task of a WS invocation (WEBSERVICE) and the task of signing the response message. The sequence of the tasks performed by the two components of WSManager is in the same order as specified in the partitioning scheme. In Figure 12.9, there are two "intermediate" and one "mobile" XML blocks of elements. An "intermediate" block includes a list of tasks that are assigned to the intermediate WSManager. A "mobile" block represents a set of tasks for the mobile WSManager. The "intermediate" block that comes after the "mobile" block contains a list of tasks that have to be performed on the intermediate node after the execution of tasks by the mobile WSManager.

The partitioning scheme is made configurable because the partitioning of tasks depends on both the nature of a WS and the resource capabilities of a mobile device and the intermediate node. For example, if the device is not capable of performing tasks related to security due to limited resources, an intermediate node can execute those tasks. In another case, if the owner of a more powerful mobile device likes performing these security-related tasks on the device to avoid any security risks or because of business requirements, a different configuration file enabling the required partitioning is to be used. The two components of WSManager use the same implementations of the WSEE prototype, but they differ only in terms of the tasks they perform at run time.

12.3.3 System Overview and Design

The lightweight WSEE (discussed earlier) is enhanced to devise the partitioned WSEE. The execution of different WSManager tasks is achieved through a chain of handlers. Each handler is represented by an XML element (under "intermediate" or "mobile" blocks of elements in Figure 12.9) in the partitioning scheme. Each handler contains a set of operations for performing one particular WSManager task. After performing the assigned task, the handler passes the control of execution to the next handler in the chain. If the XML element attribute "required" is set to "false" for a handler in the partitioning scheme, then the control of execution is passed to the next handler without executing any code for that handler.

After processing part of the incoming SOAP message, the intermediate WSManager delivers the rest of the message to the mobile WSManager. Note that the parts of a SOAP message that have been processed and not required by mobile WSManager tasks can be eliminated from the message before forwarding it to the mobile WSManager. This will add processing overheads, but it can save bandwidth. The current implementation of partitioned WSEE forwards a complete SOAP message to the mobile WSManager.

The internal details of the partitioned WSEE based on two components of WSManager are shown in Figure 12.10. On receiving a new WS request, the transport layer of the WSEE deployed on an intermediate node passes the incoming SOAP message to the intermediate WSManager. The intermediate WSManager processes the message, and performs tasks that are assigned to it in the partitioning scheme. The typical tasks assigned to the intermediate WSManager are the actions that are required for conformation to different WS standards. It is important to mention that there are WS standards such as WS-Security that requires execution of resource-demanding algorithms. Even if a mobile device is capable of executing such resource-demanding algorithms, the repeated invocations of these algorithms will not only hinder the device from performing its core functions (such as voice calls), but it will also consume its battery power significantly. Note that the service layer deployed on the intermediate node does not include WS-Mapping and the service loader components. These components are required for the invocation of requested Web services. The service layer of the WSEE deployed on the intermediate node is mainly responsible for performing actions that are required for conformation to resource-demanding WS standards.

In the next step, the intermediate WSManager forwards the SOAP message to the part of the WSEE deployed on the mobile device node. On receiving the SOAP message from the intermediate node, the transport layer of the WSEE deployed on the mobile node passes it to its service layer. The mobile WSManager executes the tasks that are assigned to it by the partitioning scheme (see partitioning scheme

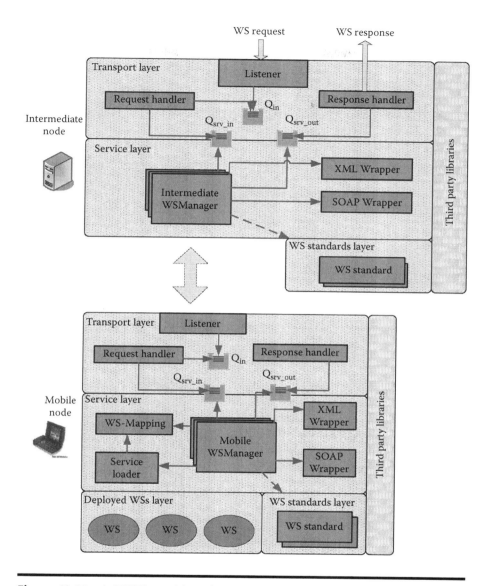

Figure 12.10 A WSEE partitioned across an intermediate node and a mobile device node.

example in Figure 12.9). Note that the service layer of the WSEE deployed on the mobile node includes WS-Mapping and the service loader. After performing the actions required for conformation to WS standards (if required), the mobile WSManager uses the service loader for invocation of the requested WS. The steps of executing the WS were explained in the discussion of the lightweight WSEE.

The results of executing the requested WS are serialized into a SOAP message (response SOAP message) by the mobile WSManager. The response SOAP message is returned to the intermediate WSManager. On receiving the response SOAP message, the intermediate WSManager can execute rest of the task(s) (if any) assigned to it. In the last step, the WSEE hosted on the intermediate node sends the final response message to the original WS client.

Note that the WS clients only communicate with the intermediate node. The communication between the intermediate WSManager (intermediate node) and the mobile WSManager (mobile node) is transparent to the WS clients. Thus, a use of a wireless environment-friendly transport mechanism for exchanging SOAP messages between the two components of WSManager is expected to improve system performance.

12.3.4 Performance Analysis

A prototype of the partitioned WS execution environment is implemented. The prototype is tested for a number of scenarios. Two sample Web services are implemented for the experimentation: Image WS and Tracking WS. The two services use the same partitioning scheme (shown in Figure 12.9). The details of these sample Web services are presented next.

12.3.4.1 Image WS

This Web service is designed to fetch image data from a mobile device, based on a set of input parameters: "location" and "keyword." At the time of saving an image on the device captured through a built-in camera, different characteristics such as location information, one to three keywords, and a textual description of the image (such as what kind of event or scene this image is capturing) are stored with the image. This WS returns data of only one image that is found to best match the input parameters. If an appropriate match is not found, a null response is returned.

12.3.4.2 Tracking WS

This Web service running on a GPS-based mobile device is designed to fetch the location information of the mobile device (hosting the WS). The Web service is useful to track skilled persons (such as doctors) or vehicles without requiring a human response. This WS collects the GPS coordinates (latitude and longitude) from the GPS receiver. The GPS coordinates are used to search the GeoNames database file (GeoNames) that is available locally on the device. GeoNames is used to provide the nearest address and the details of the location for which GPS coordinates are collected from the GPS receiver. GeoNames is a geographical database that covers all countries and contains more than eight million places. GeoNames database used in this research has a collection of nearly one thousand sample locations in Canada, and each record in the database has attributes such as location name, GPS coordinates, population, elevation, time zone, etc. In the experiments presented in the chapter, the step of obtaining GPS coordinates is replaced with a random selection of a database record. In this synthetic application, a database record is selected randomly for every new request, and every record has the same probability of being selected. This WS requires no input parameters and responds back with details of the location. This Web service is designed to be a lightweight application and is less data-intensive when compared to Image WS.

12.3.4.3 Experimental Setup

The WS client program used to access the sample Web services is written using standard Java. The WS clients are run on a machine equipped with 1.5 GB RAM and a Pentium M processor with a speed of 1.86 GHz. A desktop computer equipped with a Pentium IV processor is used as an intermediate node. The CPU speed is 3.0 GHz, and 1 GB of RAM is available on the node. Both machines run the Windows XP professional operating systems. The lightweight and the partitioned WSEEs are deployed on a Dell Axim x51v PDA with sample Web services. The PDA as the mobile device has an Intel XScale ARM processor (PXA270) that can be run at multiple speeds (208, 520, and 624 MHz). The PDA used is equipped with a RAM of 64 MB and is running Windows Mobile 5.0 as an operating system. The Java ME environment (J9) available on the PDA is a Java Virtual Machine (JVM) provided by IBM for Java ME CDC devices. The installed J9 is based on the specification of CDC 1.1. The client machines or the intermediate node communicate with the server PDA using a wireless LAN. The client machine, the intermediate node, and the mobile device are equipped with wireless adaptors that use the IEEE 802.11 standard. Note that the experiments are performed in a lab, and the wireless environment used is wireless LAN that is based on the IEEE 802.11 standard. The type of the wireless environment is not expected to affect the relative performance of the two WSEEs that are discussed in this chapter.

The performance of the system is analyzed by measuring the end-to-end response time. Experiments that are performed to investigate the performance of the partitioned WSEE are described as follows.

- In the first experiment, the performance of WSEE using a distributed WSManager is compared with that of the non-partitioned lightweight WSEE.
- In the second experiment, the effect of the transport mechanism on performance is investigated.
- In the third experiment, the partitioned WSEE is analyzed from a scalability point of view.

Note that only one of the two Web services is used at a time in all experiments. A closed system model is used. Each client operates cyclically and sends one request at a time. As soon as the response is received, the client repeats the cycle. The system is stressed by increasing the number of concurrent WS clients. For the results presented in this paper, each client generates 1000 requests. For an experiment with 10 WS clients, for example, the response time is calculated by taking the average of response times of 10,000 requests. Such numbers of requests were found to provide adequate measurement accuracy required for analyzing the relative performances of the two WSEEs (partitioned and non-partitioned). The results of these experiments are presented in next subsections.

12.3.4.4 Comparison of Lightweight WSEE and Partitioned WSEE

In this experiment, the Web services are run on the Dell Axim x51v PDA. In the first step, these Web services are accessed through the lightweight WSEE and the response time is measured. In the next step, the same Web services are executed using the partitioned WSEE. For the partitioned WSEE, HTTP is used to exchange SOAP messages between the two components of WSManager. A comparison of the mean response times for the two versions of the WSEE is presented in Figure 12.11.

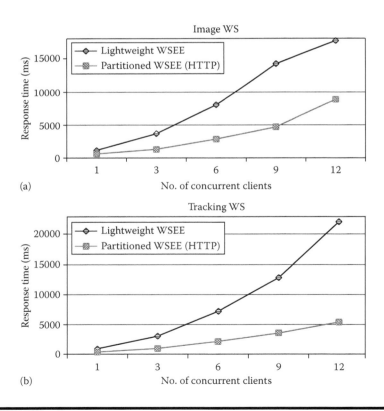

Figure 12.11 Comparison of the lightweight WSEE and the partitioned WSEE with (a) Image WS and (b) Tracking WS.

As expected, for the both WSEE versions, contention for resources and queuing delay increases with an increase in the number of concurrent clients, and the mean response time increases (see Figure 12.11). The results presented in Figure 12.11 show that the mean response time for the partitioned WSEE is significantly lower than the mean response time for the non-partitioned lightweight WSEE. Second, the mean response time for Tracking WS is lower than that for Image WS for a given number of concurrent clients and a given WSEE version. This is expected, because Tracking WS is not as data-intensive as Image WS.

12.3.4.5 Effect of the Transport Mechanism

The exchange of SOAP messages between the intermediate WSManager and the mobile WSManager can be private, providing the opportunity of using a transport mechanism with a lower overhead in comparison to using HTTP. In this experiment, the effect of using HTTP and TCP Sockets as a transport mechanism on system performance is investigated. The performance comparison of the two transport mechanisms for invocation of Tracking WS is presented in Figure 12.12. A similar trend is observed when Image WS is used.

The partitioned WSEE with TCP Sockets as a transport mechanism performs better than a partitioned WSEE using HTTP as a transport mechanism. This experiment demonstrates that a heavyweight protocol such as HTTP may not be suitable for use on systems with resource constraints and having a short response time requirement. Modified versions of TCP for wireless environment that are discussed in the literature (Avancha et al., 2002) can also be used as a transport mechanism between an intermediate and a mobile node for further improvements of response time.

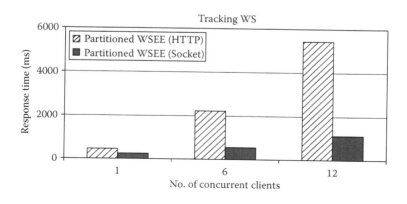

Figure 12.12   The effect of transport mechanism on performance of Tracking WS.

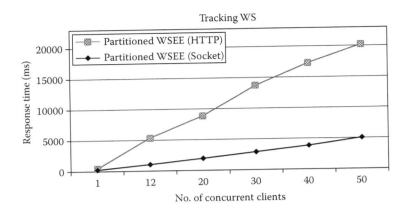

Figure 12.13 Effect of the number of concurrent clients accessing Tracking WS using the partitioned WSEE.

12.3.4.6 Scalability

The experiments performed for accessing sample Web services using the lightweight WSEE have shown that the lightweight WSEE is capable of handling a dozen concurrent clients with a reasonable response time (Asif et al., 2007). It is not advisable to use too many concurrent connections on mobile devices because of their limited resources, and sometimes only a limited number of connections are allowed by the device's operating system. The deployment of a part of the WSEE on an intermediate node is also useful to improve the scalability of the system. In this experiment, the effect of the number of concurrent WS clients accessing Tracking WS using the partitioned WSEE is investigated. The partitioned WSEE is used with both HTTP and TCP Sockets. Comparison of the performances of the two techniques is presented in Figure 12.13.

The number of WS clients is varied from 1 to 50. The partitioned WSEE with TCP sockets is observed to be more scalable with the number of concurrent clients in comparison to the partitioned WSEE with HTTP (see Figure 12.13). As the number of clients is increased, the response time increases sub-linearly for both the transport mechanisms.

12.4 Summary

Hosting Web services on resource-constrained mobile devices is a new concept. Very little work seems to have been done in this area. In this chapter, two WS execution environments for hosting Web services on mobile devices are discussed.

The design of the lightweight WSEE is such that it can be implemented in any object-oriented language. The current implementation of the lightweight WSEE can be deployed on any device that has support of a Java ME runtime environment.

The lightweight WSEE is observed to be scalable. It is also capable of processing concurrent SOAP requests from multiple clients, and the performance seems to be in acceptable range for a dozen concurrent clients or less when no partitioning is performed.

In the partitioned WSEE based on a distributed WSManager, the two components of WSManager (the intermediate WSManager and the mobile WSManager) execute different sets of configurable tasks. The sets of tasks executed by the two components are complementary to each other. A partitioned WSEE has led to a significant improvement in mean response time. The partitioning approach has also improved the system's capability of handling a significantly higher number of concurrent clients in comparison to the non-partitioned lightweight WSEE.

Representational state transfer (REST) (Fielding, 2000) is an emerging architectural style for WS-based systems. Although limited in the level of security and flexibility in comparison to SOAP, its simplicity may be attractive for mobile devices. Investigating the deployment of RESTful Web services on the WSEE forms an important direction for future research.

References

Alchemy SOAP 1.0.0. C++ open source SOAP-based web services framework. http://www.orch8.net/dev/asoap.html

Apache Axis 2 Project. User's guide. http://ws.apache.org/axis2/1_4/userguide.html

Asif, M., S. Majumdar, and R. Dragnea. Hosting web services on resource constrained devices. Paper presented in the *Proceedings of 2007 IEEE International Conference on Web Services (ICWS-2007)*, Salt Lake City, UT, July 9–13, 2007, pp. 583–590.

Asif, M. and S. Majumdar. Performance analysis of mobile web service partitioning frameworks. Paper presented in the *Proceedings of the Sixteenth International Conference on Advanced Computing and Communication*, Chennai, India, December 14–17, 2008, pp. 190–197.

Asif, M. and S. Majumdar. A graph-based algorithm for partitioning of mobile web services. Paper presented in the *Proceedings of 17th Annual Meeting of the IEEE/ACM International Symposium on Modelling, Analysis and Simulation of Computer and Telecommunication Systems (MASCOTS 2009)*, London, U.K., September 21–23, 2009.

Avancha, S., V. Korolev, A. Joshi, and Y. Yesha. On experiments with a transport protocol for pervasive computing environments. *Journal of Computer Networks* 40(4): 515–535, 2002. http://ebiquity.umbc.edu/paper/html/id/125/On-Experiments-with-a-Transport-Protocol-for-Pervasive-Computing-Environments

Burger, S., S. McFaddin, C. Narayanaswami, and M. Raghunath. Web services on mobile devices—Implementation and experience. Paper presented in the *Proceedings of the 5th IEEE Workshop on Mobile Computing Systems & Applications (WMCSA'03)*, Monterey, CA, October 9–10, 2003, p. 100.

Fielding, R. Architectural styles and the design of network-based software architectures. PhD thesis, University of California, Irvine, CA 2000.

GeoNames. Geographical database system. www.geonames.org (accessed October 15, 2009).

Glassfish Project. A reference implementation of JAX-WS. https://jax-ws.dev.java.net/ (accessed October 15, 2009).

JavaME. Java micro edition by Sun Microsystems 2005. http://java.sun.com/javame/ (accessed October 15, 2009).

JSR 218. Connected device configuration 1.1, 2005. http://jcp.org/aboutJava/community-process/mrel/jsr218/index.html

JSR 139. Connected limited device configuration 1.1, 2005. http://jcp.org/aboutJava/communityprocess/final/jsr139/

KSOAP 2 project. http://ksoap.objectweb.org/ (accessed October 15, 2009).

KXML 2 project. http://kxml.sourceforge.net/kxml2/ (accessed October 15, 2009).

Legion of The Bouncy Castle. www.bouncycastle.org (accessed October 15, 2009).

Qualcomm. BREW framework. http://brew.qualcomm.com/brew/en/

Srirama, S. N., M. Jarke, and W. Prinz. Mobile web service provisioning. Paper presented in the *Proceedings of the Advanced International Conference on Telecommunications and International Conference on Internet and Web Applications and Services (AICT/ICIW 2006)*, Guadeloupe, French Caribbean, February 19–25, 2006, p. 120.

SOAP Specifications. 2007. http://www.w3.org/TR/soap/ (accessed October 15, 2009).

UDDI Specifications. Universal description, discovery, and integration (UDDI) version 3.0. http://www.uddi.org (accessed October 15, 2009).

Web Services Architecture. 2004. http://www.w3.org/TR/ws-arch/ (accessed October 15, 2009).

Web Service Description Language. 2001. http://www.w3.org/tr/wsdl (accessed October 15, 2009).

XML Signature: Syntax and Processing. 2008. http://www.w3.org/TR/xmldsig-core/ (accessed October 15, 2009).

XML PULL Parser API. http://www.xmlpull.org/ (accessed October 15, 2009).

Chapter 13

Mobile Web Service Provisioning in Cellular Enterprise

Satish Narayana Srirama and Matthias Jarke

Contents

13.1 Introduction

It is well accepted by now that the Internet can be seen as a large-scale distributed information system with numerous information providers and users. We are also entering a new generation of an open and dynamic Web, with peer production, sharing, collaboration, distributed content, and decentralized authority in the foreground. This new Web generation is often termed "Web 2.0" (O'Reilly, 2005). From the information systems engineering viewpoint, the Internet and Web 2.0 have led the evolution from static content to Web services. Web services are distributed software components that can be accessed over the Internet, using well-established Web mechanisms and XML-based open standards and transport protocols such as Simple Object Access Protocol (SOAP) (Box et al., 2000) and HyperText Transfer Protocol (HTTP) (IETF, 1999). Public interfaces of Web services are defined and described using W3C (World Wide Web Consortium)-based standard and Web Service Description Language (WSDL) (Christensen et al., 2001). Web services have a wide range of applications, and are primarily used for enterprise integration. The biggest advantage of Web services lies in their simplicity in expression, communication, and servicing. The componentized architecture of Web services also makes them reusable, thus reducing development time and costs (Gottschalk et al., 2002).

Concurrently, the capabilities of high-end mobile phones and personal digital assistants (PDAs) have increased significantly, in terms of both processing power and memory capabilities. Smart phones are becoming pervasive, and are being used in a wide range of applications such as location-based services, mobile banking services, ubiquitous computing, social networking, and the like. The higher data transmission rates achieved in wireless domains with Third Generation (3G) (3GPP, 2006) and Fourth Generation (4G) (Thomas, 1999) technologies and the fast creeping of all-IP broadband-based mobile networks has also boosted this growth in the cellular market. The situation drives a large scope and demand for software applications for such high-end smart phones.

To meet the demand of the cellular domain and to reap the benefits of the fast developing Web services domain and standards, the scope of mobile terminals as both Web services clients and providers is being observed. *Mobile Web services (MWS)* enable communication via open XML Web service interfaces and standardized protocols also on the radio link, where today still proprietary and application- and terminal-specific interfaces are required. There are several organizations to support mobile Web services: Open Mobile Alliance (OMA) (OMA, 2004), Liberty Alliance (LA) (LibertyAlliance, 2006) on the specifications front; some practical data service applications such as OTA (over-the-air provisioning), application handover etc. on the commercial front; and IBM and SUN toolkits (IBM, 2006; Sun Microsystems, 2006) on the development front. Thus, though this is the early stage, we can safely assume that mobile Web services are the road ahead. Mobile Web services lead to manifold opportunities for mobile operators, wireless equipment vendors, third-party application developers, and end users.

Mobile terminals accessing Web services cater for anytime and anywhere access to services. Some interesting mobile Web service applications are the provisioning of services such as e-mail, information search, language translation, and company news for employees who travel regularly. There are also many public Web services accessible from smart phones, such as the weather forecast and stock quotes. *Mobile Web service clients* are also significant in the geospatial and location-based services (Benatallah and Maamar, 2003). While mobile Web service clients are common, the research on providing Web services from smart phones is still sparse. Since 2003, the scope of *mobile Web service provisioning* was studied by two projects at RWTH Aachen University (Srirama et al., 2006a; Gehlen, 2007), where *Mobile Hosts* were developed, capable of providing basic Web services from smart phones.

Mobile Hosts enable seamless integration of user-specific services to the general enterprise. Moreover, services provided by the Mobile Host can be integrated with larger enterprise services, bringing added value to enterprise services. For example, services can be provided to the mobile user, based on his up-to-date user profile and context. Profile details such as device capabilities, network capabilities, and location details can be obtained from the mobile at runtime, and can be used in providing the most relevant services, such as maps specific to devices and location information. Besides, Mobile Hosts can collaborate among themselves in scenarios like *Collaborative Journalism* and *MobileHost CoLearn System*, and bring value to the enterprise and the community in general (Srirama, 2008a).

Once the Mobile Host was developed, extensive performance analysis was conducted to prove its technical feasibility. While service delivery and management from the Mobile Host were thus shown to be technically feasible, the ability to provide proper quality of service (QoS)—especially in terms of security and reasonable scalability—for the Mobile Host is observed to be very critical. Similarly, the huge number of Web services possible, with each Mobile Host providing some services in the wireless network, makes the discovery of these services quite complex. Proper QoS and discovery mechanisms are required for successful adoption of mobile Web services into commercial environments. Moreover, the QoS and discovery analysis of mobile Web services has raised the necessity for intermediary nodes helping in the integration of Mobile Hosts with the enterprise. Based on these requirements, a *Mobile Web Services Mediation Framework* (*MWSMF*) is designed and established as an intermediary between the Web service clients and the Mobile Hosts within the Mobile Enterprise, using the Enterprise Service Bus (ESB) technology (Chappell, 2004; Schulte, 2007). This chapter summarizes the establishment of the *Mobile Enterprise* along with a scenario, where the concept was used.

The sections are organized as follows: Sections 13.2 and 13.3 introduce the concept, implementation, and performance details of mobile Web services and the Mobile Host, respectively, while Section 13.4 briefly introduces different means of addressing and accessing the Mobile Host. Section 13.5 discusses the applications of the Mobile Host, while Section 13.6 addresses one such application developed in the m-learning domain, along with a detailed usability analysis. Section 13.7 summarizes the research challenges associated with establishing the Mobile Enterprise. Section 13.8 addresses the QoS realization issues of mobile Web services in terms of their security and scalability aspects. Section 13.9 summarizes the details of publishing and discovery of mobile Web services in peer-to-peer (P2P) networks. Section 13.10 discusses the deployment, realization, and evaluation details of our mobile Web services mediation framework. Section 13.11 discusses a case study where the research has been applied, and the chapter concludes with a short summary of the results in Section 13.12.

13.2 Mobile Web Services

Service Oriented Architecture (SOA) is a component model that delivers application functionality as services to end-user applications and other services, bringing the benefits of loose coupling and encapsulation to the enterprise application integration. Services encapsulate reusable business function and are defined by explicit, implementation-independent interfaces. SOA is not a new notion, and many technologies like Common Object Request Broker Architecture (CORBA) (OMG, 2004) and Distributed Component Object Model (DCOM) (MicrosoftCorporation, 1996) at least partly represent this idea. Web services are recent in these developments and by far the best means of achieving SOA. Using Web services for SOA provides certain advantages over other technologies. Specifically, Web services are based on a set of still-evolving, though well-defined W3C standards that allow much more than just defining interfaces. Web services have a wide range of applications, and range from simple stock quotes to pervasive applications using context-awareness, such as weather forecasts, map services, etc. The biggest advantage of Web service technology lies in its simplicity in expression, communication, and servicing.

Concurrent to the SOA developments, the capabilities of today's high-end mobile phones and PDAs have increased significantly. The developments in the cellular world are twofold: first, there is a significant improvement in device capabilities, such as better memory and processing power; second, with the latest developments in mobile communication technologies with 3G and 4G technologies, higher data transmission rates in the order of few Mbs were achieved. Moreover, smart phones are becoming pervasive and are being used in a wide range of applications such as location-based services, mobile banking services, ubiquitous computing, community and social networking, etc. The main driving force for the rapid acceptance of such small mobile devices is the capability to get services and run applications at any time and at any place, especially while on the move. The experience from Japanese market shows that the most important factor in this development is that the terminals are permanently carried around, and thus people can use so-called "niche-time" to use the devices for various things (Ichikawa, 2002).

These developments have triggered a large scope and demand for software applications for smart phones in high-end wireless networks. Many software markets have evolved, such as NTT DoCoMo, capturing this demand of this large mobile user base. Many nomadic services were provided to mobile phone users. For example, DoCoMo provides phone, video phone, *i-mode* (Internet), and mail (i-mode mail, Short Mail, and Short Message Service or SMS) services. i-mode is NTT DoCoMo's proprietary mobile Internet platform. With i-mode, mobile phone users can get easy access to thousands of Internet sites, as well as specialized services such as e-mail, online shopping, mobile banking, ticket reservations, and restaurant reviews (NTTDoCoMo, 2009). Similarly, a free mapping, search and navigation application for mobile phones is being provided by LocatioNet Systems. The company's free service called *Amaze* looks like a hybrid between the popular *TomTom GPS* (Global Positioning System) system and *Google Maps*. Apart from these services, many location-based services have been developed to improve the general tourism experience (ETC, 2007).

These nomadic services bring benefits to all the participants of the mobile Web. The *mobile user* benefits from these mobile services, and the mobile phone becomes the network computer and wallet Personal Computer (PC) for him. *Enterprises* can benefit as they can support technologies and services that allow for anywhere and anytime connectivity of office information sources. *Mobile operator networks* can increase their revenues with "open" models. For example, NTT DoCoMo with its i-mode portal has proved this success, where the operator provides a framework and environment in which third-party content developers can deploy their services. The *content providers* can in turn get incentives from these open models.

From the analysis of most of these nomadic mobile services, each operator provided some set of services, applicable to a specific group, over specific platforms. But most of these approaches were proprietary and followed specific protocols. For example, if we consider a company trying to advertise itself, it can use the mobile push services that are run over the Global System for Mobile communications (GSM) network. Then the advertisement has to be shaped in such a way that it fits the terminals and platforms by mobile operators and vendors. This makes the services un-interoperable, and the integration of services becomes highly impossible. In order to overcome the interoperability issues and to reap the benefits of the fast developing Web services domain and standards, mobile Web services domain came to the picture. In the mobile Web services domain, the resource-constrained mobile devices are used as both Web service clients and providers, still preserving the basic Web services architecture in the wireless environments.

Mobile terminals accessing Web services are common these days (Balani, 2003; Benatallah and Maamar, 2003), and mobile Web service clients cater for anytime and anywhere access to services. Some interesting mobile Web service applications are the provisioning of services for employees who travel regularly, such as e-mail, information search, language translation, company news, and the like. There are also many public Web services accessible from smart phones like the weather forecast, and stock quotes, among others. They are also significant in geospatial and location-based service applications. Moreover, there is significant support for mobile Web service clients from several organizations such as OMA, LA on the specifications front, and IBM and SUN toolkits (IBM, 2006; Sun Microsystems, 2006) on the development front. Thus, though these are early days, we can safely assume that mobile Web services are the road ahead.

13.3 Mobile Web Service Provisioning

While mobile Web service clients are common these days, and while many software tools already exist in the market easing their development and adoption, the research on providing Web services from smart phones is still sparse. A mobile device in the role of a service provider enables entirely new scenarios and end-user services. Moreover, the paradigm shift of smart phones from the role of service consumer to the service provider is a step toward practical realization of various computing paradigms such as pervasive computing, ubiquitous computing, ambient computing, and context-aware computing. For example, the applications hosted on a mobile device provide information about the associated user (e.g., location, agenda) as well as the surrounding environment (e.g., signal strength, bandwidth). Mobile devices also support multiple integrated devices (e.g., camera) and auxiliary devices (e.g., GPS receivers, printers). For the hosted services, they provide a gateway to make their functionality available to the outside world (e.g., providing

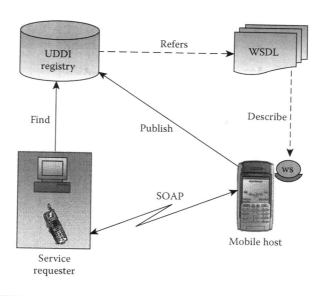

Figure 13.1 Basic mobile Web services framework with the Mobile Host.

paramedical assistance). In the absence of such provisioning functionality, the mobile user has to regularly update the contents to a standard server with each update of the device's state.

Since 2003, the scope of *mobile Web service provisioning* was studied by two projects at RWTH Aachen University (Srirama et al., 2006a; Gehlen, 2007), where *Mobile Hosts* were developed that were capable of providing basic Web services from smart phones. Figure 13.1 shows the basic mobile Web services framework with Web services being provided from the Mobile Host. The Mobile Host is a light-weight Web service provider built for resource-constrained devices like cellular phones. It has been developed as a Web service handler built on top of a normal Web server. Mobile Web service messages can be exchanged using SOAP over different transportation protocols such as HTTP, Block Extensible Exchange Protocol (BEEP), User Datagram Protocol (UDP), and Wireless Application Protocol (WAP). In the Mobile Host's implementation, the SOAP-based Web service requests sent by HTTP tunneling are diverted and handled by the Web service handler component. The Mobile Host was developed in PersonalJava on a SonyEricsson P800 smart phone. The footprint of the fully functional prototype is only 130 kB.

Open source kSOAP2 (kSOAP2, 2008) was used for creating and handling the SOAP messages. kSOAP2 is thin enough to be used for resource-constrained devices, and provides a SOAP parser with special type mapping and marshaling mechanisms. Considering the low-resource constraints of smart phones, no deployment environment can be easily provided. Hence, all services have to be deployed at the installation of the Mobile Host. Alternatively, the Mobile Host can be configured to look for services at other locations apart from the main JAR location, where the services could then be deployed at runtime. There is also support for OTA and dynamic deployment of new services to the Mobile Host. Along with these basic features, a lightweight Graphic User Interface (GUI) was provided to activate and deactivate the deployed services as and when necessary, so as to control the load on the Mobile Host. The GUI also has support for providing memory usage details of the smart phone and basic server operations like start, stop, and exit, thus helping in evaluating the performance analysis of the Mobile Host.

The key challenges addressed in the development of the Mobile Host are threefold: to keep the Mobile Host fully compatible with the usual Web service interfaces such that clients will not notice the difference; to design the Mobile Host with a very small footprint that is acceptable in the smart phone world; and to limit the performance overhead of the Web service functionality such that neither the services themselves nor the normal functioning of the smart phone for the user is seriously impeded. Even though the Web service provider is implemented on the smart phone, the standard WSDL can be used to describe the services, and the standard Universal Description, Discovery, and Integration (UDDI) registry (Bellwood, 2002) can be used for publishing and un-publishing the services. Figure 13.1 basically illustrates this idea of advertising mobile Web services to a UDDI registry. An alternative for the UDDI-based discovery is also studied, where the study tried to realize the Mobile Host in a P2P network (Milojicic et al., 2003), thereby leveraging the advertising and searching of WSDL documents to the P2P network. The approach is addressed in detail in Section 13.9, while discussing the discovery issues of mobile Web services.

Mobile Host realization is also possible with other Java variants like Java 2 Micro Edition (J2ME). We have also developed a J2ME-based Mobile Host apart from the PersonalJava-based one, with almost similar performance measurements. Nevertheless, Web services are not the only studied means of providing services from devices like smart phones and PDAs. The provisioning can also be based on any distributed communication technology, such as Java Remote Method Invocation (RMI) or Jini, if the device supports the respective platform. Van Halteren et al have addressed nomadic mobile service provisioning, based on Jini technology (van Halteren and Pawar, 2006). The approach proposes the Mobile Service Platform (MSP) as a supporting infrastructure, which extends the SOA paradigm to the mobile device. The MSP design is based on the *Jini Surrogate Architecture Specification* which enables a device which cannot directly participate in a Jini Network to join a Jini network with the aid of a third party (Sun Microsystems, 2001). Using this architecture, a service provided from the device is composed of two components: (1) a service running on the mobile device, referred to as a *device service* (DS) and (2) a *surrogate service* (SS), which is the representation of the device service in the fixed network. The surrogate functions as a proxy for the device service and is responsible for providing the service to the clients. The MSP supports the communication between the device service and the surrogate service.

Thus, by using a mobile service platform, a service hosted on a mobile device can participate as a Jini service in the Jini network (Pawar et al., 2007).

However, splitting a service into a device service and surrogate introduces a state-synchronization problem. The surrogate must be aware of the change in the state of a device service. The most serious limitation of this approach is that it is based on a proprietary protocol. The technology (Jini) is also fixed. The client should be therefore be aware of Jini technology. Moreover, the services are to be developed both for the surrogate and the device, and changes are not propagated. The approach thus tightly fixes the service provided by the mobile device to platform (Java), protocol (HTTPInterconnect), technology (Jini), and to the surrogate host, thereby seriously affecting the interoperability of the provided services. The main benefit with our developed Mobile Host is the achieved integration and interoperability for the mobile devices. It allows applications written in different languages and deployed on different platforms to communicate with Mobile Hosts over the cellular network; of course, the benefits it acquired from the Web services domain in general (Srirama, 2008a).

13.3.1 Performance Evaluation of the Mobile Host

Once the Mobile Host was developed, it was extensively tested for performance issues such as memory load, server-processing load, etc. The evaluation of the system was conducted using services like the mobile photo album service, which helps in fetching and browsing through pictures taken by the smart phone, the location information service that provides the exact location information of the mobile terminal as GPS data, and other auxiliary services such as 'Echo' and 'ls' services, among others. The test setup comprised a Mobile Host developed and deployed on the SonyEricsson P800 smart phone and a standalone Apache Axis Web service client (Apache Software Foundation, 2009a). The smart phone had an internal memory of 12 Mb and a 16 MB memory stick duo card. The ARM9 processor of the device clocked at 156 MHz. The Axis client invoked different services (within the context of this discussion, it is assumed that the client knows the exact location—Uniform Resource Identifier (URI)—of the service and the service description) deployed on the Mobile Host, and the performance of the Mobile Host was observed, by taking timestamps and memory foot prints while the Mobile Host was processing the Web service request. The tests were conducted both in High-Speed Circuit Switched Data (HSCSD) and General Packet Radio Service (GPRS) environments. Further details of addressing Mobile Hosts in different environments are discussed in the following section.

The detailed performance evaluation of this Mobile Host clearly showed that service delivery as well as service administration can be done with reasonable ergonomic quality by normal mobile phone users. As the most important result, it turns out that the total Web service processing time at the Mobile Host is only a small fraction of the total request-response invocation cycle time (<10%) in a GPRS network, the rest all being transmission delay. We have also observed that the increase in transmission rates can increase the processing capability of the Mobile Host. This makes the performance of the Mobile Host directly proportional to achievable higher data transmission rates. Thus, the high data transmission rates achieved, in the order of few Mbps, through advanced mobile communication technologies in 2.5G, 3G, and 4G, help in realizing the Mobile Host in commercial environments.

In terms of performance of the Mobile Host, the key question was whether a reasonable number of clients could be supported with an overhead that would not prevent the main mobile user from using his or her smart phone in the normal fashion (either to supply the services or just for the usual local phone functions). This study was also required, since it would define the limit for the number of concurrent participants in the collaborative application environments. Concurrent requests were generated for the services deployed on the Mobile Host, simulating multiple clients. The results of this regression analysis for checking the scalability of the Mobile Host are very encouraging; the Mobile Host was successful in handling up to eight concurrent accesses for reasonable service like location data provisioning service with response size of approximately 2 kb. However, it was observed that the increased concurrent access of the Mobile Host affected the Mobile Host's ability to access internal and external resources (Srirama, 2008a).

13.4 Mobile Terminal Access

Once a Web service is developed and deployed with the Mobile Host, the mobile terminal that is registered and connected within the mobile operator network requires some means of identification and addressing, which allows the Web service to be accessible from the Internet as well. Generally, computers and devices in a TCP/IP network are identified using an IP address. The IP address that is required for the data transfer to and from smart phones (as for any other IP communication client such as Web servers, Intranet workstations, etc.) is assigned during the communication configuration phase. Typically, the IP address assigned to mobile devices using a GPRS connection is only temporarily available and is known only within the mobile operator's network, which makes it difficult to use the IP address in client applications. Very few operators provide smart phones with public IP in a GPRS network that can be directly used to reach a smart phone from the Internet. The operational setup for accessing the mobile terminal in a GPRS network is shown in Figure 13.2, with the interaction numbered 1. The mobile TCP/IP connection between the Web service client and the Mobile Host is deployed on top of a GPRS link into the mobile operator network. From there, the traffic is routed through the Internet to/from the Web service client. The problem of addressing each mobile node with IP is being addressed by Mobile IP version 6 (Mobile IPv6) (Johnson et al., 2004). The key benefit of Mobile IPv6 is that even though the

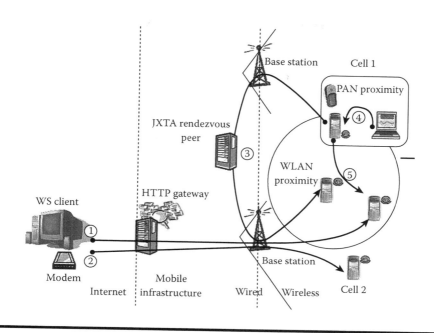

Figure 13.2   Mobile Web service provisioning and interactions.

mobile node changes locations and addresses, the existing connections through which the mobile node is communicating are still maintained.

The mobile Web service provisioning project also has identified other means of addressing the Mobile Host in HSCSD and P2P environments. In the HSCSD addressing scenario, a HSCSD connection is established between the smart phone and the prototyping network, which is connected to the Internet. HSCSD is an enhancement of Circuit Switched Data (CSD) data services of current GSM networks. HSCSD allows the access of non-voice services with a data rate about three times higher than that of CSD. Higher rates are achieved by using multiple channels for the data transmission. With this technology, subscribers can send and receive data from their portable computers or mobile devices at a speed of up to 28.8 kbps. The HSCSD connection uses a Public Land Mobile Network (PLMN) and the Public Switch Telephone Network/Integrated Services Digital Network (PSTN/ISDN) for making the data call to the server. The connection is setup by using Point-to-Point Protocol (PPP) over a circuit-switched data call to a modem that is connected to one of the servers in the network. On top of this PPP link, a TCP/IP end-to-end connection between the mobile terminal and the dial-in server is established. Hence, as long as the data call persists, the mobile terminal can be addressed using the IP address assigned to it by the dial-in server. Thus the Web service deployed on the mobile terminal can be accessed from any client within the network environment (Srirama et al., 2006b). The interaction is shown in Figure 13.2.

As mentioned previously, the need for public IP for each of the participating Mobile Hosts was observed to be the major hindrance for commercial success of the Mobile Host. Therefore, alternative architectures were studied for addressing mobile Web services. In a JXTA network, each peer is uniquely identified by a static peer ID, which allows the peer to be addressed independent of its physical address, such as a DHCP-based IP address. This peer ID will stay with the device forever, even though the device supports multiple network interfaces like Ethernet, Wi-Fi, or Bluetooth to connect to the P2P network. Hence, the scope of the Mobile Host in the P2P networks was studied so as to address the Mobile Host with peer ID. A virtual P2P network can be established by connecting the Mobile Hosts to JXTA super peers, as explained in Section 13.9.1. Now, by using the peer ID, the Mobile Host does not have to worry about changing IPs or operator networks, and it is always visible to the Web service client. Mapping the peer ID to the IP is taken care of directly by the JXTA network, thus eliminating the need for public IP. The JXTA-based P2P network also helps in better discovery of the huge number of Web services possible with Mobile Hosts, by acting as a dynamic cache of advertisements of the mobile Web services. The approach is explained in detail in Section 13.9. This Mobile P2P interaction in Figure 13.2 is numbered 3.

Provisioning of mobile Web services in a totally decentralized manner is even more challenging. This kind of interaction between peers is also referred as pure P2P. *Pure P2P* is a setup like the classic Gnutella file sharing network. The interactions numbered 4 and 5 in Figure 13.2 represent this pure P2P network idea. In our case of mobile Web services, this means that discovery, invocation, and integration of Web services occur between mobile devices directly, without any centralized entities such as base stations. We had not studied how to provide mobile Web services according to this kind of technical usage scenario, but this approach promises to have the best cost-effectiveness as long as interactions between clients and providers of mobile Web services do not involve proprietary mobile networks. Bluetooth could be a perfect technical solution for establishing such a pure P2P network. This kind of interactions tends to enable personal computing using various devices in Personal Area Network (PAN), partially or fully based on mobile Web services (Srirama, 2008a).

13.5 Applications of the Mobile Host

The paradigm shift of smart phones from the role of service consumer to service provider is a step toward practical realization of various computing paradigms such as pervasive computing, ubiquitous computing, ambient computing, and context-aware computing. For example, the applications hosted on a mobile device provide information about the associated user (e.g., location, agenda) as well as the surrounding environment (e.g., signal strength, bandwidth). Mobile devices also support multiple integrated devices (e.g., camera) and auxiliary devices (e.g., GPS receivers, printers). For the hosted services, they provide a gateway to make available their functionality to the outside world (e.g., providing paramedical assistance).

Primarily, with the Mobile Host, the smart phone can act as a multi-user device without additional manual effort on part of the mobile carrier. Several applications were developed and demonstrated with the Mobile Host. For example, in a remote patient tele-monitoring scenario, the Mobile Host can collect remote patient's vital signs such as blood pressure, heart rate, temperature, and related information from different sensors and provide them to doctors in real time. In the absence of such a Mobile Host, the details are to be regularly updated to a server, from where the doctor can access the details. The latter scenario causes problems with stale details and increased network loads (Srirama and Jarke, 2009b). The scenario is illustrated in Figure 13.3. A second example is that in case of a distress call the mobile terminal can provide a geographical description of its location (as pictures), along with location details. Another interesting application scenario involves the smooth coordination between journalists and their respective organizations while covering events like the Olympics (Srirama et al., 2006b).

The Mobile Host in a cellular domain is of significant use in any scenario which requires polling that exchanges significant amount of data with a standard server, for example, a mobile checking for Really Simple Syndication (RSS) update feeds provided by a server. The Mobile Host can eliminate the polling process, as the RSS feeds can now be directly sent to the Mobile Host when they are updated. From the commercial viewpoint, with the Mobile Host there can be a reversal of payment structures in the cellular world. Traditionally, the information-providing Web service client has to pay to upload his or her work results to a stationary server (where other clients may then have to pay again to access the information); in the Mobile Host scheme, responsibility for payment can be shifted to the actual clients—the users of the information/services provided

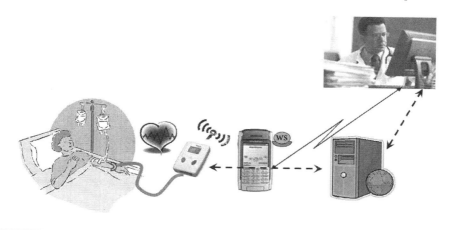

Figure 13.3   Mobile Host in a remote patient tele-monitoring scenario.

by the Mobile Host. Thus, the Mobile Host makes it possible for small mobile operators to set up their own mobile Web service businesses without resorting to stationary office structures.

Moreover, as a Mobile Host, the mobile terminal can provide access to information such as pictures, audios, videos, tags, documents, location details, and other information of individuals in a seamless, interoperable way. This sort of information can be used in building communities of practice (Wenger, 1998). Web 2.0 is the main driving force in this domain, with support for peer production, sharing, collaboration, distributed content, and decentralized authority. In communities of practice, some of the most innovative and valuable information is not made available online, but lies within groups of practice (e.g., m-learning communities, groups of specific expedition interests). Here, peers can browse through the information, add tags, and give their suggestions or comments. We have developed one such system, named MobileHost CoLearn System, that helps in the m-learning 2.0 domain, a conglomeration of Web 2.0, mobile and e-learning domains. The system is explained in detail in the following section.

Another important concept in Web 2.0 is the "End of the Software Release Cycle." Software has traditionally been released as a product. On the other hand, within the Web there is no need to install programs; software is delivered as a service. Software as a Service (SaaS) is the next generation of software. Traditional, standalone software will be replaced by open Web services that can be accessed online or reused and combined to create new applications and mashups. With the help of Mobile Hosts, mobile social software that foster community building in learning environments and collaborative learning applications (e.g., applications for collaborative learning, resource creation, and annotation, or applications for knowledge sharing) can be delivered as services via mobile phones.

Moreover, Mobile Hosts enable OTA and the dynamic deployment of new services to smart phones. By this means, services can automatically be downloaded and deployed to the Mobile Host. The services can be downloaded as individual classes or packaged jar files, and can be deployed dynamically to the Mobile Host. The feature was successfully tested with a PersonalJava-based Mobile Host. The OTA feature makes the Mobile Host provide a platform for smart phones, helping in seamless integration of new mobile applications and services. Examples of such applications can be P2P games or context-aware hotel bookings that can be discovered with the mobile Web service discovery mechanism. The applications or updates can also be automatically installed on all the subscribed Mobile Hosts. (Srirama, 2008a) Many interesting applications of the Mobile Host are also possible in the context-aware computing domain. For example, the mobile Web service provisioning concept can be used to realize context sources on the mobile device as services, and make this context information available to the Context Distribution Frameworks in the Internet. Any mobile application that is attuned to a user's context will leverage knowledge about *who* the user is, *what* the user is doing, *where* the user is, and *which* terminal is in use. Context-awareness helps in reaching the goal of a personalized experience when the user interacts with the mobile services (Pashtan, 2005).

13.6 MobileHost CoLearn System

Many m-learning application scenarios can be envisioned with the Mobile Host, such as podcasting, mobile blogging, mobile learning media sharing service, expertise finder service, and similar scenarios. In the mobile learning media sharing scenario, learners can share audio or video lecture recordings or go on a field study and take pictures of the location. Peers can then browse through the pictures taken, add tags, and give their suggestions or comments. Through an expertise finder feature, learners can look for reliable access to learning resources, persons who share the same interests, and experts with the required know-how who can help with achieving better results. In the e-learning aspect, these experts can share the information among other users. Examples of these use cases could be exchanging mathematical formulas with experts validating them or even correcting them. With in our research, we have developed one such system, MobileHost CoLearn System (Srirama, 2008a), helping in collaborative learning and community building. With the system, learners are no longer put under constraints of specific time and place. MobileHost CoLearn System also helps in locating the right person who can provide us with exactly the knowledge that we need and who can help us solve exactly the problems that we face. The collective intelligence from the participants of the system plays an essential role. The main idea is that, under the right circumstances, groups of people are occasionally more intelligent than their members are as individuals (Surowiecki, 2005).

13.6.1 Features of the MobileHost CoLearn System

The MobileHost CoLearn system presents a novel approach to expert finding within a truly mobile collaborative learning environment, targeted not only within the framework of the learner's social network, but also within the social networks of his or her acquaintances, and the social networks of the acquaintances of his or her acquaintances, and so forth. Such an expert finder flow usually leads to the discovery of more than one potential expert, and the learner's subjective decision about which of them is the most knowledgeable can be based either on the rating for the expert's level of expertise in the field, or on the path that the expert finder request has traveled before reaching the respective expert. After having found an expert, the learner is provided with all the necessary information in order to contact that expert for further assistance regarding specific issues.

Alongside the valuable knowledge that flows within the system from experts to non-experienced learners, the system supports the retrieval of a variety of literature resources, such as articles, proceedings, books, URLs, master and PhD theses, and unpublished resources that have been tagged by the learners. As tagging is something subjective, a three-level scale of relevance of a tag to a resource has been introduced. Apart from retrieving specific resources at the time when they are needed, the system maintains image and audio resources within photocasting and podcasting channels, through which they are automatically distributed to all subscribers as soon as they become available. This is one of the most suitable ways for sharing up-to-date knowledge, as the amount of information that can be conveyed through such broadcasting channels is enormous. Thus the system enables learners to

- Manage their personal and expertise data and make it available to other learners.
- Search for experts in a specific field within a truly collaborative mobile environment, and expand the expert finder search beyond the borders of the social network of the current user by allowing the forwarding of expert finder requests an arbitrary number of times until a real expert has been found.
- Manage the data of the experts that have been found, and to narrow down the list of experts by filtering by name and expertise field with a specific minimum level of expertise.
- Contact experts for further assistance and query them about problems faced.
- Manage their own databases of literature resources, organize them into categories of journal articles (e.g., conference proceedings, books, URLs, theses), tag these resources with specific keywords associated with a three-level scale of relevance to the resource (low, medium, high), narrow down the list of literature by filtering the resources (by title, author, tags), and make the literature databases accessible to other learners.
- Retrieve the different types of literature resources of other learners by specifying the tag of interest and its minimum relevance to the resource.
- Manage their photocasting and podcasting channels, create new photocast episodes by capturing images with the integrated camera device or browsing the file system of the mobile phone, and create new podcast episodes by capturing sound with the integrated audio recording device or browsing the file system of the mobile phone, manage the metadata of the channels and their episodes, and tag them with specific keywords.
- Subscribe to photocasting and podcasting channels of other learners, define preferences (such as whether to automatically download the content or just be notified of its availability, and the maximum size of the downloaded content), and unsubscribe from the photocasting and podcasting channels later on.
- Automatically receive new broadcasts as soon as they are available, in case their content complies with the user preferences, and narrow down the list of broadcast episodes by filtering by author and tags.
- Write comments to the received broadcasts, and retrieve the feedback that other learners have left.

13.6.2 Sample Scenarios of the MobileHost CoLearn System

An example expert finder scenario when using the MobileHost CoLearn System is illustrated in Figure 13.4. Anna encounters some difficulties while creating a Unified Modelling Language (UML) model of the system she is going to implement. She knows a couple of people, such as Bob and Brandon, who are either

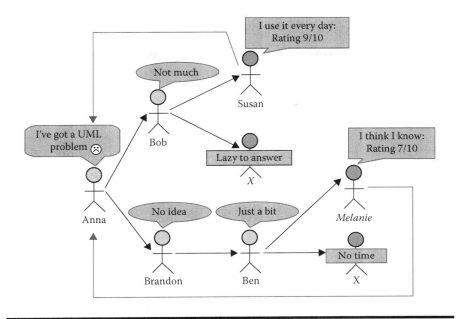

Figure 13.4 Expert finder scenario, with the help of Mobile Hosts. (From Ivanova, I., Mobile Web services for collaborative learning, Master's thesis, RWTH Aachen University, Aachen, Germany, 2007. With permission.)

likely to know something about her problem or are acquainted with other people with more extensive knowledge than themselves. She asks them to help her find the most knowledgeable person in the field by sending them the request. If they do not know other people with a more extensive knowledge in UML than themselves, they can return themselves as a result, stating their subjective assessment for their level of expertise in the field; if they know one or more people with better knowledge, they can forward Anna's request to them. As a result, Anna's request is spread through a network of people having a higher level of knowledge at each stage of the request path. She receives responses from Susan and Melanie, containing their subjective rating for expertise in the field (9 and 7 out of 10, respectively). Anna can make a subjective decision about which of them to approach for help, based either on their levels of expertise in the field or on the paths that the requests have passed before the responses are returned to her. In this way, a network of people, some of them not knowing each other at all, help Anna find the most suitable expert for her problem.

In the expert finder scenario of the MobileHost CoLearn System, every participant should act as both provider and client. Thus, their smart phones should act as both Mobile Hosts as well as mobile Web service clients, demonstrating the necessity for the existence of the Mobile Host in the system. Moreover, as the Mobile Host, the mobile terminal can provide access to information like pictures, audios, videos, tags, documents, location details, and other learning services. Thus, the system supports podcasting, mobile blogging, mobile learning, a media sharing service, etc. In the mobile learning media sharing scenario, learners can share audio or video lecture recordings or go on a field study and take pictures of the location. Peers can then browse through the pictures taken, add tags, and give their suggestions or comments. In an expertise finder scenario, learners can look for reliable access to learning resources from experts (Ivanova, 2007).

13.6.3 Usability Evaluation of the MobileHost CoLearn System

Once the MobileHost CoLearn System was realized, the usability of the developed system and the learner satisfaction with its information resources have been evaluated with the help of learner testing and a preliminary created questionnaire. Users had to perform certain tasks in order to evaluate the functionality of the system and its modules. The learner testing has been performed in two sessions with a duration of approximately 2 h. Altogether, seven pupils tested the system, most of them in the age range 26–31 years, and it was assumed that everyone knew the phone numbers of all the others. The tests were conducted in a T-Mobile GPRS environment. The operator provided public IP addresses for the participating Mobile Hosts. All learners had been using mobile phones in their everyday life, and most of them were feeling comfortable with the text input capabilities of their mobile device, the calendar and scheduling functionalities, the capturing of images with the phone camera, and the viewing of images, but only some of them had used audio recording before. None of them had used a mobile application for collaborative learning before.

The overall system evaluation section included 25 questions, which are a subset of the 50-question database of the Software Usability Measurement Inventory (SUMI)—a rigorously tested and proven method of measuring software quality from the end learner's point of view (University College Cork, 2007). The SUMI database embraces the learner's opinion of the usability of the system, including measures such as learnability and understandability, the reliability of the system (such as fault tolerance and recoverability), the maintainability of the system (such as stability), the efficiency of the system (such as time and resource behavior), and the functionality of the system (such as accuracy and suitability). For the evaluation of the results of the SUMI questionnaire, the System Usability Scale (SUS) (Brooke, 2007) was used, which is based on a 5-grade scale and yields a single number in the range from 0 to 100. The final scores for the overall satisfaction of each of the seven learners of the MobileHost CoLearn system ranged from 78 to 95 points. This results in an average learner satisfaction of 87.86 points out of 100 points, or approximately 88%.

Regarding the learner evaluation of the Expert Finder module, the learners were truly satisfied with the collaborative environment that the Expert Finder module creates. They were happy with the fact that they could not only see the path that the request has traveled before reaching them, but also read the comments that each forwarder has made to the next forwarder on the path. As a suggestion for further improvement, they mentioned that they would like to have the learner photos available, and to have not only a textual but also a graphical visualization of the path, as much as this is possible on the screen of the specific mobile device being used. All in all, the learners found the learner interface of the Expert Finder module to be simple and intuitive, and were very satisfied with the seamless coupling of the Expert Search and the Expert Rating Web Services with the learner interface. The learners said that they would also like to have a mechanism for automatic discovery of experts, so that if they do not know anyone who might be an expert in the field, the system could perform an automatic search. Further details of the questionnaire and the analysis are available at Srirama (2008a).

13.7 Mobile Enterprise

From the previous sections, we can conclude that service delivery and management can be done from the Mobile Host with reasonable performance latencies. We have also observed several applications with the Mobile Host, and even followed the usability evaluation of one such system. Mobile Hosts and their clients can together establish a *Mobile Enterprise* in a cellular network (Srirama and Jarke, 2009a). However, an established Mobile Enterprise poses several technical challenges, both to service providers as well as to the mobile operator. Some of the critical challenges and the associated research are addressed in the following sections.

Figure 13.5 shows the Mobile Enterprise and hints at the critical challenges posed to mobile phone users and operators. The challenges can be summarized as follows:

1. As the Mobile Host provides services to the Internet, the devices should be safe from malicious attacks. For this, the Mobile Host has to provide only secure and reliable communication in the vulnerable and volatile mobile ad hoc topologies.
2. In terms of scalability, the Mobile Host has to process a reasonable number of clients over long durations, without failure and without seriously impeding normal functioning of the smart phone for the user.
3. Since the Mobile Host follows the Web service communication that is very verbose, it poses serious issues with the energy efficiency of the devices. In the communication domain, the transmission is more of an energy consumer than CPU processing cycles (Raghunathan et al., 2002).
4. The huge number of Web services possible, with each Mobile Host providing some services in the wireless network, makes the discovery of the most relevant services quite complex. Proper discovery mechanisms are required for successful adoption of mobile Web services into commercial environments.

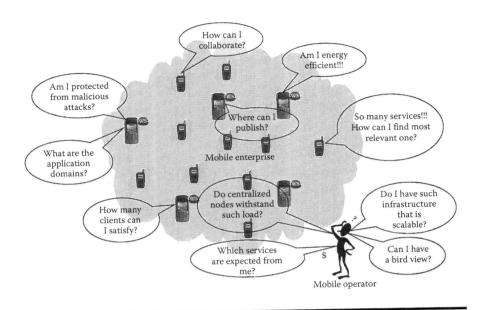

Figure 13.5 Mobile Enterprise and the critical challenges posed to mobile phone users and the operator.

5. The discovery poses some critical questions: Where to publish the services being provided by Mobile Hosts? Should they be published with the centralized UDDI (Bellwood, 2002) registries available on the Internet or is the operator going to offer some help? This also raises questions about whether centralized nodes can withstand such high loads or whether some alternatives are to be looked at.

6. From the mobile operator's perspective, the Mobile Enterprise poses a number of questions: What are the services expected by the mobile users from the operator? Can the operator monitor the communication and have a bird's-eye view of the complete network, so that business scenarios can be drawn out of it? Do the operators have the infrastructure that can scale to such requirements?

We tried to address these questions in our study, and the following sections discuss our QoS and discovery research of the Mobile Host in detail. Apart from the need for a central monitoring infrastructure for Mobile Enterprise, the QoS and discovery analyses also have identified the need for intermediary nodes helping in the successful deployment of smart-phone-based mobile Web service providers in the cellular networks. Based on these requirements, a *Mobile Web Services Mediation Framework* (MWSMF) (Srirama et al., 2007b) is designed and established as an intermediary between the Web service clients and Mobile Hosts in the mobile network, using ESB technology. The details are addressed in the following sections.

13.8 QoS Aspects of Mobile Web Service Provisioning

Providing proper QoS—especially appropriate security and reasonable scalability—for the mobile Web service provisioning domain was observed to be very critical. In terms of security, the Mobile Host has to provide secure and reliable communication in the vulnerable and volatile mobile ad hoc topologies. In terms of scalability, the layered model of Web service communication introduces a lot of message overhead to the exchanged verbose XML-based SOAP messages. This consumes a lot of resources, since all this additional information is to be exchanged over the radio link. Thus, to improve scalability, the messages are to be compressed without affecting the interoperability of the mobile Web services. The message compression also improves the energy efficiency of the devices, as there will be less data to transmit.

13.8.1 Security Analysis of the Mobile Host

Once the Web services are deployed with the Mobile Host, the services and the host are vulnerable to different sorts of security breaches such as denial-of-service attacks, man-in-the-middle attacks, intrusion and spoofing, etc. Moreover, with easily readable mobile Web services over the network, the complexity to realize security increases further. For traditional wired networks and Web services, a lot of standardized security specifications, protocols, and implementations exist, such as WS-Security (Atkinson et al., 2002) and Security Assertion Markup Language (SAML) (Cantor et al., 2005), among others, but not much has been explored and standardized in wireless environments. Some of the reasons for this poor state might be the lack of widely active commercial data applications, to date.

In our security analysis of the Mobile Host, we have analyzed the adaptability of WS-Security to the mobile Web service provisioning domain. WS-Security provides ways to add security headers to SOAP envelopes, attach security tokens and credentials to a message, insert a timestamp, sign the messages, and encrypt the message. We mainly observed the additional latency caused to performance of the Mobile Host with the introduction of security headers into the exchanged SOAP messages. The performance penalties of different encryption and signing algorithms were calculated, and the best possible scenario for securing mobile Web services communication was suggested.

From this analysis, we could conclude that because of resource limitations, not all of the WS-Security specification can be applied to the Mobile Host. The results of our security analysis suggest that the mobile Web service messages of reasonable size, approximately 2–5 kb, can be secured with Web service security standard specifications. The security delays caused are approximately 3–5 s. We could also conclude from the analysis that the best way of securing messages in mobile Web service provisioning is to use Advanced Encryption Standard (AES) symmetric encryption with 256 bit key, and to exchange the keys with RSA 1024 (Rivest et al., 1978) bit asymmetric key exchange mechanism and sign the messages with RSAwithSHA1. But there are still high performance penalties when the messages are both encrypted and signed. We therefore suggest encrypting only those parts of the message that are critical in terms of security, and signing the message. The signing on top of the encryption can completely be avoided in specific applications with lower security requirements (Srirama et al., 2007a).

For providing proper end-point security for mobile Web service provisioning, basic *service-level authentication* and *user-intervened authorization* were realized. The mechanisms were realized for the PersonalJava-based Mobile Host, using some of the features provided by the PersonalJava platform such as the TaskSwitch (Bloor, 2002). In the service-level authentication, an authentication service is provided at the Mobile Host which accepts a username and password and validates the client. Alternatives such as where the authentication service is handled by third-party servers and services were also considered. In the user-intervened authorization, each of the services provided at the Mobile Host can be configured to obtain the provider's (person using the Mobile Host) acceptance before providing the respective service to the Web service requestor. Critical issues such as disapprovals, user being busy, and timeouts were also considered. Realization of single sign-on, where identity and credentials can be maintained for multiple sessions and parties, has also been studied. Further details of our security analysis, such as the hardware support and adaptation of Semantics-Based Access Control (SBAC) (Naumenko, 2007) for mobile Web services, are beyond the scope of this chapter and are available at Srirama (2008a).

While the security analysis identified the best means of securing mobile Web services, a potential client of the Mobile Host from the Internet can follow full WS-Security standard. This pushes the necessity for some mediation framework as the legitimate intermediary in the mobile Web service invocation cycle, transforming the messages to supported standard. For clarity, if we consider the encryption scenario, the messages sent by the client can be encrypted with any other symmetric encryption algorithm such as TRIPLEDES, AES-128, AES-192 (RSALabs, 2008), etc. Since the Mobile Host can not implement the complete WS-Security, the security of the message is to be verified at the intermediary, and the message is to be encrypted again using AES-256 before sending the message across the radio link to the Mobile Host.

13.8.2 Scalability Aspects of the Mobile Host

Similar to providing secured communication for mobile Web services, attaining proper scalability is also crucial in achieving appropriate QoS from the Mobile Host. *Scalability* is defined as the Mobile Host's ability to process a reasonable number of clients over long durations, without failure and without seriously impeding normal functioning of the smart phone for the user. From the regression analysis of the Mobile Host for checking its scalability, the Mobile Host was successful in handling eight concurrent accesses for reasonable service like *location data provisioning service* with a response size of approximately 2 kb. The main observed reason for not being able to process more mobile Web service clients was due to the transmission delay, which constituted 90% of the mobile Web service invocation cycle time. The Mobile Host's scalability is thus inversely proportional to increased transmission delays. The transmission delays can be reduced in two ways: (1) by achieving higher data transmission rates with current generation telecommunication technologies and (2) by reducing the size of the message. In the scalability analysis of the Mobile Host, our study mainly concentrated on the second issue (i.e., reducing the size of the message being transmitted over the radio link). The approach also improves the energy efficiency of the devices, as the amount of energy consumed in the transmission of a single byte of data is approximately equivalent to the energy consumed for 8000 CPU cycles (Hollmen and Manner, 2008).

Web services communication is a layered communication across different protocols. Considering SOAP over HTTP binding, at the lowest level is the transportation protocol, TCP. On top of TCP lies the HTTP communication. Then SOAP communication is over the HTTP protocol. The application communication and protocols, for example WS-Security, lie on top of SOAP. Hence, any message exchanged over the Web service communication consists of some overhead across all the different layers. The size of the message is

$$B_{MSG} = B_{TP} + B_{MTP} + B_{SOAP} + B_{APP}$$

where B_{TP}, B_{MTP}, B_{SOAP}, and B_{APP} are the message overheads over transportation, message transportation, SOAP, and application protocols, respectively. Since we are considering message exchange over the radio link, B_{MSG} has to be reduced to the minimum possible level (Laukkanen and Helin, 2003). For this, the messages are to be compressed/encoded in the optimal way. The minimal encoding may not always be the best solution. The first reason for this is that the encoding should be efficient, both in terms of message size reduced and extra performance penalties added to the devices. Second, the encoding mechanism should not affect the interoperability. If an attempt is made to reduce the overload at B_{TP} or B_{MTP}, the interoperability of the Web services is seriously impeded. Therefore, the best position to target the encoding process is at the B_{SOAP} and upper levels, which means that the XML-based SOAP messages are to be compressed.

To obtain smaller message sizes, the messages can be compressed with standard compression techniques such as Gzip, or with XML-specific compression techniques such as XMill. Canonical XML (Boyer, 2001) standard targets the logical equivalence of these compressed XML messages. There is also an effort with the Fast Web Services (Sandoz et al., 2003), Fast Infoset standard draft (Sandoz et al., 2004), Efficient XML (AgileDelta, 2007), BinXML (Ericsson, 2003), etc., to specify a binary format for XML data that is an efficient alternative to XML in resource-constrained environments. Similarly, there is some effort with the Binary Format for

Metadata (BiM) (Heuer et al., 2002) standard for the binary encoding of Moving Picture Experts Group (MPEG-7) metadata. BiM is designed in a way that allows for fast parsing and filtering of the XML data at the binary level itself, without having to decompress again. (Ericsson and Levenshteyn, 2003) gives a comparison of different compression technologies for XML data, and specifies the best scenario for the Web service message exchange across smart phones. The analysis suggests that BinXML is the best option (BiM was not considered in this analysis) to compress Web service messages considering compression ratio, processing time, and resource usage.

Based on the analysis at (Ericsson and Levenshteyn, 2003), we have adapted BinXML for compressing the mobile Web service messages. BinXML replaces each tag and attribute with a unique byte value and replaces each end tag with 0xFF. By using a state machine and 6 special byte values including 0xFF, any XML data with circa 245 tags can be represented in this format. The approach is specifically designed to target SOAP messages across radio links. Hence, the mobile Web service messages are exchanged in the BinXML format, and this has reduced the message of some of the services by 30%, drastically reducing the transmission delays of mobile Web service invocation. The BinXML compression ratio is very significant where the SOAP message has repeated tags and deep structure. The binary encoding is also significant for the security analysis, as there was a linear increase in the size of the message with the security incorporation. The variation in the WS-Security encrypted message size for a typical 5 kb message is approximately 50% (Srirama et al., 2008a).

Similar to the security analysis, the scalability analysis also raised the necessity for an intermediary or proxy node. As BinXML compression adapted by the study is not an open standard, not all the messages transmitted over the radio link can be based on this standard. If a client sends an uncompressed message over the mobile network, the transmission is not very efficient, even though the Mobile Host can process such a request. In such a scenario, the middleware framework should encode/decode the mobile Web service messages to/from XML/BinXML formats in the mobile operator proprietary networks.

13.9 Mobile Web Service Discovery

In a commercial Mobile Enterprise with Mobile Hosts, and with each Mobile Host providing some services for the Internet, the number of services expected to be published could be quite high. Generally, Web services are published by advertising WSDL descriptions in a UDDI registry (Alonso et al., 2004). But with the huge number of services possible with Mobile Hosts, as each Mobile Host provides some personalized services to the network, a centralized solution for discovery is not the best idea. The centralized registry solutions can have bottlenecks and can make single points of failure. Besides, mobile networks are quite dynamic due to the node movement. Devices can join or leave the network at any time, and can switch from one operator to another operator. This makes the binding information in the WSDL documents inappropriate. Hence, the services are to be republished every time the Mobile Host changes the network.

Dynamic service discovery is one of the most extensively explored research topics in recent times. Most of these service discovery protocols are based on the announce-listen model, like the one in Jini (Sun Microsystems, 2001). In this model, a periodic multicast mechanism is used for service announcement and discovery. But all these mechanisms assume a service proxy object that acts as the registry and is always available. For dynamic ad hoc networks, assuming the existence of devices that are stable and powerful enough to play the role of the central service registries is inappropriate. Hence, services distributed in the ad hoc networks must be discovered without a centralized registry, and should be able to support spontaneous P2P connectivity. Dustdar and Treiber (2006) propose a distributed P2P Web service registry solution based on lightweight Web service profiles. They have developed View-based Integration of Web Service Registries (VISR) as a P2P architecture for distributed Web service registry. VISR's distributed registry model allows VISR peers to operate on a common data structure and provides a common vocabulary. The approach is context aware.

Similarly, Konark service discovery protocol (Lee et al., 2003) was designed for discovery and delivery of device-independent services in ad hoc networks. In this approach, a node multicasts the differences between services that a node knows and the ones others seem to know. These messages are called *delta messages*. In other words, each participating node advertises its knowledge of services minus the network's knowledge. Hence, the individual nodes complement one another's limited knowledge to attain the global view of the entire network. In line with these developments, Universal Plug and Play (UPnP) (UPnP Forum, 2003) discovery protocol allows devices to advertise their services to *control points* (i.e., clients) on the network. UPnP control points discover services of interest to them either by passively listening to these advertisements or by actively multicasting service discovery request messages.

Considering these developments and our need for distributed registry and dynamic discovery, we have studied alternative means of mobile Web service discovery and realized a discovery mechanism in the JXTA/JXME network (Srirama et al., 2008b). Before discussing the discovery issues, the following section introduces the Mobile Host's entry into P2P networks.

13.9.1 Mobile Web Service Provisioning in P2P Networks

P2P is a set of distributed computing model systems and applications used to perform a critical function in a decentralized manner. P2P takes advantage of the resources of individual peers (such as storage space, processing power, content), which are all critical for smart phones, and achieves scalability, cost sharing, and anonymity, thereby enabling ad hoc communication and collaboration. In order to reap the benefits of P2P by achieving increased application scope and targeting efficient utilization of resources of individual mobile peers, we tried to adapt the Mobile Host into P2P networks. For this, many of the current P2P technologies such as Gnutella, Napster, and Magi are studied in detail (Milojicic et al., 2003). Most of these technologies are proprietary, and generally target specific applications. Only project JXTA (Wilson, 2002) offers a language-agnostic and platform-neutral system for P2P computing.

JXTA technology, also known as *Juxtapose*, is a set of open protocols that allow any connected device on the network, ranging from cellular phones and wireless PDAs to PCs and servers, to communicate and collaborate in a P2P manner. JXTA enables these devices running on various platforms to not only share data with each other, but also to use the functions of their respective peers. JXTA technology is based on proven technologies and standards such as HTTP and TCP/IP, is independent of programming language, networking platform, or system platform, and can work with any combination of these. JXTA peers use XML as a standard message format, and create a virtual P2P network over the devices connected over different networks. Moreover the JXTA community has developed a light version of JXTA for mobile devices, called JXME (JXTA for J2ME). JXME eliminates many of the low level details of the P2P systems like the transportation details. Mobile peers can communicate with each other using the best of the many network interfaces supported by devices such as Bluetooth, Wi-Fi, and GPRS, among others. Considering these advantages and features of JXTA, the Mobile Host was adapted into the JXTA network, to check its feasibility in P2P networks.

Figure 13.6 shows the architecture of the eventual deployment scenario of Mobile Hosts in the JXME network. The *virtual P2P network*, also called the

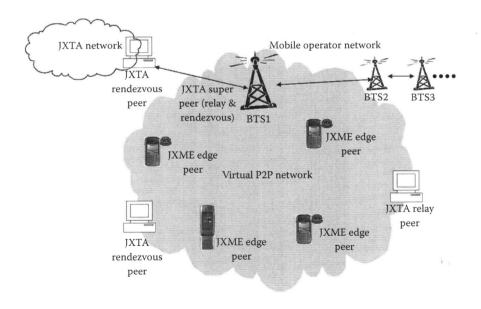

Figure 13.6   Virtual mobile P2P network with Mobile Hosts.

mobile P2P network, is established in the mobile operator network, with one of the nodes in the operator proprietary network acting as a JXTA super peer. JXTA network supports different types of peers to be connected to the network. The general peers are called *edge peers*. An edge peer registers itself with a *rendezvous peer* to connect to the JXTA network. Rendezvous peers cache and maintain an index of advertisements published by other peers in the P2P network. Rendezvous peers also participate in forwarding the discovery requests across the P2P network. A *relay peer* maintains route information and routes messages to peers behind the firewalls. A *super peer* has the functionality of both relay and rendezvous peers. In the mobile P2P network, the super peer can exist at Base Transceiver Station (BTS) and can be connected to other base stations, thus extending the JXTA network into the mobile operator network. Any Mobile Host or mobile Web service client in the wireless network can connect to the P2P network using the node at base station as the rendezvous peer. The super peer can also relay requests to and from the JXTA network to the smart phones. Standalone systems can also participate in such a network as both rendezvous and relay peers if the operator network allows such functionality, further extending the mobile P2P network. Within this mobile P2P network, the participating smart phones can be addressed with peer ID as addressed in Section 13.4, eliminating the need for a public IP, which is required to address and access services deployed with the Mobile Host (Srirama et al., 2008b).

13.9.2 Mobile Web Service Discovery Approach

Once the virtual P2P network is established, the network can help in P2P-based mobile Web service discovery. In JXTA/JXME, the decentralization is achieved with the advertisements. All resources such as peers, peer groups, and the services provided by peers in the JXTA network are described using advertisements. Advertisements are language-neutral metadata structures represented as XML documents. Peers discover each other, the resources available in the network, and the services provided by peers and peer groups by searching for their corresponding advertisements. Peers may cache any of the discovered advertisements locally. Every advertisement exists with a lifetime that specifies the availability of that resource. Lifetimes gives the opportunity to control out-of-date resources without the need for any centralized control mechanism. To extend the lifetime of an advertisement, the advertisements are to be republished.

Thus, to achieve alternate discovery mechanism for mobile Web services, the services deployed on the Mobile Host in the JXME virtual P2P network are to be published as JXTA advertisements, so that they can be sensed as JXTA services among other peers. JXTA specifies "modules" as a generic abstraction that allows peers to describe and instantiate any type of implementation of behavior representing any piece of "code" in the JXTA world. Thus, the mobile Web services are published as JXTA modules in the virtual P2P network. The module abstraction includes a module class, module specification, and module implementation. The module class

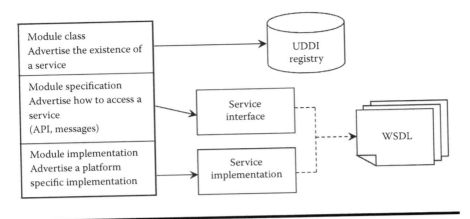

Figure 13.7   Mapping between JXTA modules and Web services.

is primarily used to advertise the existence of behavior. Each module class contains one or more module specifications, which contain all the information necessary to access or invoke the module. The module implementation is the implementation of a given specification across different platforms. Figure 13.7 shows the mapping between JXTA modules and Web services. The collection of module abstractions represents the UDDI in the sense of publishing and finding service description, and WSDL in the sense of defining transport binding to the service.

To publish the mobile Web services in the JXTA network, a standard *Module Class Advertisement* (*MCA*) is published into the P2P network, declaring the availability of a set of Web service definitions in that peer group. Once new Web services are developed for the Mobile Host, the WSDL descriptions of these services are incorporated into the <parm> element of the *Module Specification Advertisements* (*MSA*), and are published into the P2P network, by attaching themselves to the MCA. The MSA contains a unique identifier (MSID) that also includes the static module class ID (MCID), which identifies the MCA to which it belongs. Thus, MCA declares the existence of the Web service, and MSA provides metadata of the service. The MSAs are published into the JXME network with an approximate lifetime, which specifies the amount of time the Mobile Host wants to provide the service. The MSAs are cached at rendezvous peers or any other peers with sufficient resource capabilities. Once the lifetime expires, the MSAs are automatically deleted from the P2P network, thus avoiding stale advertisements. The MSA can be published to the network by a service developer or even by the Mobile Host.

Once published to the mobile P2P network, the services can later be discovered by using the keyword-based search provided by JXTA. The MSAs carrying the Web service descriptions can be searched by the name and description parameters. This basic search returns a large number of resulted services, returning every service that matches the keyword. Since the discovery client in mobile Web services scenario is a smart phone, the result set should be quite small, so that the user can scroll through the list and can select the intended services. Hence, the JXTA search resulted services are ordered according to their relevancy using Apache Lucene tool (Cutting, 2008). Lucene is an open source project that provides a Java-based high-performance, full-featured text search engine library. Lucene allows adding indexing and searching capabilities to user applications. Lucene can index and make searchable any data that can be converted to a textual format. Using the tool and its index mechanism, the search results were ordered/filtered, and the advanced matched services were returned to the discovery client (Srirama, 2008a).

13.9.3 Categorization Support of Mobile Web Service Discovery

In order to make the mobile Web service discovery process more efficient, we tried to adapt the categorization feature of the UDDI specification. All four main UDDI data structure types—Business Entity, Business Service, Binding Template, and Tmodel—provide *categoryBag* structure to support attaching categories to data. A categoryBag consists of one or more *keyedReferences*. Each keyedReference contains attributes to identify category information. When the relationship between different categories is needed to describe a more complex entity, *keyedReferenceGroup*, which contains a set of keyedReference, is used. With the intent of borrowing categorization to our P2P-based mobile Web service discovery mechanism, we incorporated the keyedReferenceGroup information into a *categoryPack* element and attached it to an MCA. When an MSA is to be published in the network, it should search for the most appropriate MCA to match by the category information provided by keyedReferenceGroup in MCA, and should attach itself to the respective MCA.

For further improving categorization of mobile Web services, we resorted to the peer groups feature of JXTA. With the reference to some popular industry categorization standards such as North American Industry Classification System (NAICS) and United Nations Standard Products and Services Code (UNSPSC), we designed a hierarchical categorization structure for mobile Web services. Altogether, it consists of four levels and 57 peer groups, with mobile Web Service Group (mWSGroup) on the top level. The group structure is a first draft to realize the idea of categorization. More details can be obtained from (Srirama et al., 2008b). With the peer group hierarchy in place, a potential service requester first joins his group of interest and then issues the query for the discovery. When a super peer that is in the peer group receives the request, it looks through the cache and sends the request to the still available peers in the group. Each peer that receives the request from its super peer conducts the search process in its local cache by searching for the MCAs according to the keyword given by the mobile requestor. An MCA maps the general description of Web services under the same group. Since more than one Web service could possess the same general description but different detailed description and content, one MCA could possibly match more than one MSA. After the discovery process, the matched MSAs are provided as the search results. If the number of found MSAs is small, then the searching results could already be accepted and sent back to the service requester. Otherwise, the Lucene-based Advanced Matching/Filtering of Service is performed on the result set, and the most relevant services are returned to the mobile client.

Detailed performance analysis of the discovery approach was conducted to verify the scalability of the approach. The details can be obtained from (Zhu, 2008). In summary, the mobile Web service discovery mechanism, by following the P2P, surely surpasses the problems of announce-listen mechanisms. The approach is conceptually similar to Konark, but is truly scalable by following the open standards of Web service technologies and P2P protocols of JXTA. Mobile Web service discovery still preserves the SOA, and the service descriptions are stored only at nodes with higher capabilities (super peers). The service descriptions cannot be maintained at all participating smart phones because of resource limitations. At the same time, discarding already discovered advertisements can lead to bottlenecks because of replica searches. Thus, the discovery approach significantly differs from VISR where all nodes are registries, and from UPnP where no node is a registry and all the search results are discarded. The service descriptions of mobile Web service discovery also benefit from well-known standards of Web services technologies (Srirama, 2008a).

Similar to the QoS analysis of the Mobile Host, the mobile Web service discovery also raised the necessity for intermediary nodes, acting as super peers and helping in the JXME publishing and discovery processes by hosting services like Lucene-based filtering mechanisms. Apart from this, mobile Web service clients generally prefer using services of the Mobile Host based on several context parameters such as location, time, device capabilities, profiles, and load on the Mobile Host. Most of these details cannot be provided just based on keywords. Semantic matching of services gives the most appropriate and relevant results for mobile Web service discovery. Thus, the intermediary should maintain the individual user profiles, personalization settings, and context-sensitive information of the participating mobile clients and devices for the *context aware service discovery* of mobile Web services. Moreover, the semantic discovery process is heavy in terms of resource consumption, and needs to be performed at a standalone intermediary or distributed middleware framework.

13.10 Enterprise Service Integration

Mobile Hosts with proper QoS and discovery mechanisms enable seamless integration of user-specific services to the Mobile Enterprise. Moreover, services provided by the Mobile Host can be integrated with larger enterprise services, bringing added value to these services. However, enterprise networks deploy disparate applications, platforms, and business processes that need to communicate or exchange data with each other or, in the specific scenario addressed by this chapter, with Mobile

Hosts. The applications, platforms, and processes of enterprise networks generally have non-compatible data formats and non-compatible communications protocols. Besides, within the domain of our research, the QoS and discovery study of the Mobile Host offered solutions in disparate technologies like JXTA. This leads to serious integration troubles within the networks. The integration problem extends further if two or more such enterprise networks have to communicate among themselves. We generally address this research scope and domain as *Enterprise Service Integration* (Srirama and Jarke, 2009a,b).

13.10.1 Deployment Details of the MWSMF

Based on the requirements for the distributed middleware framework from QoS and discovery analysis of mobile Web services, we tried to realize a Mobile Web Services Mediation Framework (MWSMF) for smart phones. The proposed deployment scenario of the MWSMF is shown in Figure 13.8. The mediation framework is established as an intermediary between the Web service clients and the Mobile Hosts in the JXTA network. The virtual P2P network is to be established in the mobile operator network, with one of the nodes in the operator proprietary network acting as a JXTA super peer, as explained in Section 13.9.1. Once the mobile P2P network is established, the Web service clients can discover the services using the mobile P2P discovery mechanism, and can access the deployed services across the MWSMF and JXTA network. The mediation framework ensures the QoS of the mobile Web service messages, transforms them as and when necessary, and routes the messages, based on their content, to the respective Mobile Hosts.

Apart from security and improvements to scalability, the QoS provisioning features of MWSMF also include message persistence, guaranteed delivery, failure handling, and transaction support. External Web service clients that do not

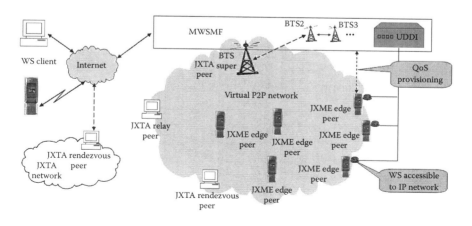

Figure 13.8   **The MWSMF deployment scenario.**

participate in the mobile P2P network can also directly access the services deployed on the Mobile Hosts via MWSMF, as long as the Web services are published with any public UDDI registry or the registry deployed at the mediation framework and the Mobile Hosts are provided with public IPs. This approach evades the JXME network completely. Thus, the mediation framework acts as an external gateway from the Internet to the Mobile Hosts and mobile P2P network.

13.10.2 Integration Options for the MWSMF

From the MWSMF deployment scenario shown in Figure 13.8, it can be derived that the MWSMF should have a distributed framework and should integrate multiple platforms with support for heterogeneous technologies such as JXTA, Web services, etc. However, realizing such a middleware is not a trivial task. As the integration troubles are well known and observed for quite some time, the research in the domain offers different solutions (Keen et al., 2004).

Point-To-Point: One of the first methods used to integrate applications has been by using point-to-point integration solutions. Under this scheme, a protocol or format transformer is built at one or either end between a pair of applications. One of the advantages of this scheme is that both applications have good knowledge about each other, thus getting to a tight coupling. The principal disadvantage of this scheme is the difficulty of integrating a new application to the system due to the high number of protocols or format transformers which have to be implemented. This architecture is very popular solution for small-scale integration problems.

Hub-And-Spoke: This architecture is also known as the message broker, and it provides a centralized point that all applications connect to by using lightweight connectors. This centralized point is called the hub or broker. The connectors act as adapters, and translate data and messages between different applications to canonical formats. Two of the advantages of the hub-and-spoke architecture are that all the message transformation and routing is done by the hub, and the number of connections for the integration is reduced with respect to the point-to-point architecture. Unfortunately, like any centralized architecture, the centralized hub becomes a single point of failure.

Enterprise Message Bus: An enterprise message bus is an infrastructure of communication where the integration between applications can be done in a platform- and language-independent way. It is composed of a message router and publishing (subscription) channels, where applications use request and response queues to interact with each other via a message. Consumers write a request to the request queue, and providers listen to the request queue waiting for a request for their services. The resultant messages are then added to the response message queues.

Enterprise Service Bus: By using ESB integration, applications do not communicate to each other directly, but communicate by using a SOA-based middleware backbone. ESB basically consists of a set of service containers that are interconnected with a reliable messaging bus. ESB supports multiple integration paradigms in order to fully support the variety of interaction patterns that are required in a comprehensive SOA between these service containers (Endrei et al., 2004). Thus, it has support for service-oriented architectures in which applications communicate through reusable services with well-defined and explicit interfaces, message-driven architectures in which applications send messages through the ESB to receiving applications, and event-driven architectures in which applications generate and consume messages independently of one another. Considering the benefits of ESBs over the other solutions, we proceeded with the technology for realizing the mediation framework.

13.10.3 Realization Details of the MWSMF

As discussed earlier, ESBs are recent developments in the enterprise integration domain, and a standards-based ESB solves the integration problems elevated by the MWSMF. Hence, we tried to realize the MWSMF based on ESB technology. For establishing the mediation framework, many of the current ESB products such as the Sonic Software, Artix, Cape Clear, etc., were studied. Of these tools, *Sonic SOA Suite* and *FioranoESB Suite* are two ESB products that are based on proprietary messaging middleware. Iona Technologies extends its legacy EAI architecture to achieve *Iona Artix ESB* (IONA Technologies, 2005). *PolarLake* and *FusionWare* take a server-centric, connector-based approach in their ESB product design. Only *Cape Clear* Software and *Cordys ESB* systems use a truly open and distributed SOA. (Borck, 2005) from InfoWorld provides a detailed survey of the products and their pros and cons, mainly focusing on the achieved interoperability, management, scalability, security, supported features, and their value for money. The survey recommends the Sonic SOA Suite among the contemporary ESB products. ESB products like *BEA AquaLogic Service Bus* (BEA, 2008) and *IBM WebSphere software* (IBM Corporation, 2008) were not considered in this survey.

A recent analysis of ESB products, provided at (MacVittie, 2006), suggests BEA AquaLogic Service Bus, considering the features provided by each product, without considering any performance measurements. But most of these ESB products are based on proprietary message middleware or extensions to EAI architectures. SUN has defined Java Specification Request (JSR) 208, *Java Business Integration* (*JBI*) specification (Ten-Hove and Walker, 2005), for enterprise application integration. JBI specification defines a platform for building enterprise-class ESBs using a pluggable and service-based design. The Java-based standard allows building enterprise integration systems by using plug-in components which interoperate through mediated normalized message exchanges. ESB products such as Service Mix and OpenESB are based on JBI specification. Observing these different ESB tools, standards, and surveys, we considered JBI and ServiceMix (Apache Software Foundation, 2008) for the realization of MWSMF.

13.10.3.1 Components of the Mediation Framework

Figure 13.9 shows the basic components of the mediation framework handling various tasks (Srirama et al., 2007b). ServiceMix, by following the JBI architecture, supports two types of components: *service engines* and *binding components*. Service engines are components responsible for implementing business logic, and they can be service providers/consumers. Service engine components support content-based routing, orchestration, rules, data, transformations, etc. Service engines communicate with the system by exchanging normalized messages across the *normalized message router* (NMR). Message normalization is the process of mapping context-specific data to a context-neutral abstraction to transport data in to a standard format. The normalized messaging model is based on the WSDL specification. The service engine components are shown as straight-lined rectangles in the figure.

Binding components are used to send and receive messages across specific protocols and transports. The binding components marshal and unmarshal messages to and from protocol-specific data formats to normalized messages. The *normalized message* consists of the message content, also called payload, message properties, or metadata, and optional message attachments referenced by the payload. The message attachments can be non-XML data. JBI uses the normalized messages for interaction between consumers and providers, whether service engines or binding components.

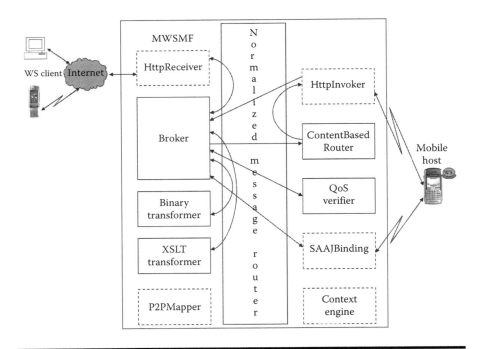

Figure 13.9 Basic components of the MWSMF.

The binding components are shown as dashed rectangles in the figure. Service engines and binding components interact with the NMR via a delivery channel that provides a bidirectional delivery mechanism for message packets, called *messages exchanges* (ME). The JBI system supports four message exchange patterns: *In-Only, Robust-In-Only, In-Out*, and *In-Optional-Out* (Ten-Hove and Walker, 2005).

The *HttpReceiver* component shown in Figure 13.9 receives the Web service requests (SOAP over HTTP) over a specific port and forwards them to the *Broker* component via NMR. The main integration logic of the mediation framework is maintained at the Broker component. For example, in case of the scalability maintenance, the messages received by the Broker component are verified for mobile Web service messages. If the messages are normal Http requests, they are handled by the *HttpInvoker* binding component. If the messages comprise Web service messages, the messages are transformed to BinXML format at the *BinaryTransformer* component. The request messages are then sent to the Mobile Host via the *SAAJBinding* component. The SAAJBinding component invokes Web services according to SOAP with Attachments specification. The BinXML adapter at the Mobile Host first decodes the binary messages back to the XML format, after which Web service requests are processed. The response messages follow the same track along the NMR, and are returned to the mobile Web service client via the HttpReceiver component. The mobile Web service message optimization scenario is explained in detail, in the next section.

The *QoSVerifier* component mainly helps in maintaining the security for the mobile Web service messages. The component verifies the security of the received messages, and transforms the message to the security level supported by the mobile Web services. The component uses the *XSLTTransformer* component in this process, basically to transform the mobile Web service messages using Extensible Stylesheet Language Transformations (XSLT). Currently, we are working with the *P2PMapper* binding component that transforms the JXTA messages to/from normalized messages, thus making the invocation of the mobile Web services feasible across the JXME network. As of now, the mobile Web service invocations are possible only through the SAAJBinding component. We are also trying to realize a binding component for the context engine that helps in context-aware service discovery of the mobile Web services. The details will be addressed by our future publications in this domain.

The components are deployed into the framework using a spring-based xml configuration file. Spring is a lightweight container, with wrappers that make it easy to use many different services and frameworks. Lightweight containers accept any Java Bean, instead of specific types of components (Tate and Gehtland, 2005). The configuration uses WS-Addressing to route the messages across the components via the normalized message router. WS-Addressing is a specification of transport-neutral mechanisms that allows Web services to communicate addressing information. It essentially consists of two parts: a structure for communicating a reference to a Web service endpoint, and a set of Message Addressing Properties, which associate addressing information with a particular message (Box et al., 2004).

Figure 13.9 shows only those components that are most relevant for the discussion in this chapter. Apart from these main components, the mediation framework also supports supplementary components that help in the successful deployment of Mobile Hosts in the mobile enterprise. For example, the mediation framework also supports automatic start-up of Mobile Hosts. Generally, hand-held devices have many resource limitations, such as low computational capacities, limited storage capacities, limited battery power, and similar limitations. To conserve these resources, the Mobile Host features of smart phones are to be turned on only when the provider is prepared to deliver and receives a request from the mobile Web service client. For help in this regard, the MWSMF identifies the contact details of the phone when the request is for a particular Mobile Host, and sends an SMS to the device. A generic program is run on the smart phone that starts the Mobile Host automatically and activates its services and features, when the SMS message is received. The application is developed using the *PushRegistry* feature of MIDP 2.0 from Wireless Messaging API (WMA) (Sun Microsystems, 2008). The SMS messages are sent to the smart phone following a specific application protocol. Currently, there is support only for the basic features of the Mobile Host such as starting the server and authenticating the client. The person with the Mobile Host can also opt to turn down this request from the client.

13.10.3.2 Mobile Web Service Message Optimization Scenario

Once the MWSMF is established in the Mobile Enterprise, the components discussed in the previous section are deployed with the mediation framework. To clarify, let us consider a scenario where the mobile Web service messages are exchanged across the MWSMF and its components. For example, in the case of improving the scalability of the Mobile Host, the messages received by the mediation framework from external clients are compressed using BinXML encoding, and the binary messages are sent over the radio link. The scenario is called *mobile Web service message optimization* (Srirama, 2008a). The message flow of the scenario is as follows:

- First, the HttpReceiver component receives the mobile Web service request or the HTTP request from the client.
- The HttpReceiver sends the message to the Broker component through the NMR.
- The Broker component examines the message for Web service request and, if it is a normal HTTP request, transfers the message to HttpInvoker. If the message comprises a mobile Web service request, the Broker component transfers the message to the BinaryTransformer component through the NMR.
- The BinaryTransformer component BinXML encodes the message and transfers the response message back to the Broker component.
- The Broker component sends the message to the HttpInvoker via the NMR.

- The HttpInvoker generates the request to the Mobile Host by setting the BinXML data to the body of the HTTP message.
- The Mobile Host processes the request using a BinXML adapter and sends the response message back to the HttpInvoker.
- The HttpInvoker sends the response back to the Broker component via the NMR.
- The Broker component transfers the response to the BinaryTransformer through the NMR.
- The BinaryTransformer decodes the BinXML data to the XML format and transfers the response message back to the Broker component.
- The Broker component returns the response to the HttpReceiver component through the NMR.
- The HttpReceiver component returns the response back to the client.

13.10.4 *Evaluation of the Mediation Framework*

Once realized, the MWSMF was tested extensively for its performance and scalability issues. The Mobile Web Service Message Optimization scenario was evaluated to verify the stability of the MWSMF on a laptop. The laptop has an Intel Pentium M Processor 2.00 GHz/1 GB of RAM. A Java-based server was developed and run on the same laptop on an arbitrary port (4444), mocking the Mobile Host. The server receives the service request from the client and populates a standard response. The response is then BinXML encoded, and the compressed response is sent back to the client in the HTTP response message format. By considering this simple server, we can eliminate the pure performance delays of the Mobile Host and the transmission delays of the radio link, thus getting the actual performance latencies of the MWSMF. For load generation, the study used a Java clone of the ApacheBench load generator from WSO2 ESB (WSO2, 2007; Apache Software Foundation, 2009b). The load generator can initiate a large number of concurrent Web service invocations simulating multiple parallel clients.

From this analysis, we have observed that the mediation framework was successful in handling up to 110 concurrent requests without any connection refusals. Higher numbers of concurrent requests were also possible (we have tested up to 250 concurrent requests), but some of the requests failed as the mediation framework generated connection refusal errors because of the load. It was also observed that the mean duration of handling a single request across several concurrent successful requests at the MWSMF is approximately 150 ms. The values can significantly grow when the deployment scenario is established on reasonable servers with high resource and performance capabilities. The results from this analysis show that the mediation framework has reasonable levels of performance, and the MWSMF can scale to handling large number of concurrent clients, possible in the deployment scenario shown in Figure 13.8. Further details of the analysis are available at (Srirama, 2008a).

13.11 Mobile Access to LAS Services—A Use Case of Mobile Enterprise

While the previous sections have discussed the details of the Mobile Host and the challenges and research associated with establishing the Mobile Enterprise, the rest of this chapter concludes the discussion with a case study, where the research has been applied. The Mobile Enterprise and MWSMF were employed for load balancing the Leight Application Server (LAS) in providing MPEG-7-based multimedia services to mobile users.

13.11.1 The Leight Application Server

LAS is a lightweight application server designed as a community middleware that is capable of managing users and multiple hierarchically structured communities along with their particular access rights, as well as a set of services accessible to users. LAS is a platform-independent Java implementation and mainly offers MPEG-7 multimedia services to users. A community application can make use of the offered services by simply connecting to the server and then remotely invoking service methods. A very prominent feature of LAS is that the server functionality is easily extensible by implementing and plugging in new services and respective components. LAS also has security management support to guarantee access controls. (Martinez et al., 2002; Spaniol et al., 2006)

MPEG-7-based multimedia services offered by LAS refer to services such as non-linear digital storytelling, multimedia upload, download, and tagging/annotation services. LAS offers a set of services, rather than one monolithic MPEG-7 service. Each of these services offers functionality for a given domain of the MPEG-7 standard (e.g., multimedia content, collections, classification schemes). Since MPEG-7 is XML-based, the basic functionality of an MPEG-7 LAS service (i.e., its methods) is to work on XML documents conforming to the MPEG-7 XML schema. MPEG-7 documents should be stored in an XML database that supports execution of XPath, Xquery, and XUpdate expressions. An MPEG-7 service should offer support for creation of new MPEG-7 descriptions, query for complete or partial MPEG-7 descriptions, and modification of MPEG-7 descriptions. Each MPEG-7 XML document must be valid against the MPEG-7 XML schema at all stages.

Even though, LAS is a reliable application server offering MPEG-7 multimedia services, it is not pure Web service architecture; it was not designed under the SOA paradigm and important aspects such as scalability, and distributed services were not taken into account. QoS and performance problems have been observed recently by LAS users. For many years, LAS has been used on top of traditional networks infrastructures to provide the services required by social software such as Virtual Campfire (Cao et al., 2007). Load balancing and cluster support were observed to be the immediate requirements for improving the performance of the LAS. Most recently, multimedia services are being offered to mobile phone users, and this requirement further enhances the problem.

13.11.2 Load Balancing for LAS

The QoS of the LAS can be improved either by changing the architecture of the LAS to have the cluster support or by employing a middleware framework taking care of the load balance issues. We went with the second option, and tried to adapt our knowledge from the mobile Web service provisioning domain to the LAS. Basically, we developed the components that provide a Web service interface to LAS services. We subsequently designed scenarios such that the requests are diverted to the less occupied server among a cluster of LASs (Srirama and Jarke, 2009a). The approach can be of three modes: (1) directly accessing the MPEG services through the Mobile Host; (2) accessing the service through the mediation framework of the Mobile Enterprise; and (3) indirect way of using the Mobile Host to connect to the mediation framework. All the three scenarios were tested for performance loads and can be seen in Figure 13.10.

13.11.2.1 Accessing LAS Services through the Mobile Host

Since LAS, as of now, supports the access of services only across the HTTP protocol, the invocation of these services from mobile phones becomes a little tricky. While studying mobile Web service clients and providers, we have taken the design decision that the exposure and access of services from/to mobiles would only be through Web services and WSDL. As mentioned previously, mobile Web services basically enable communication via open XML Web service interfaces and standardized protocols also on the radio link, where proprietary and application- and terminal-specific interfaces are still required today.

Based on our design decision, the Mobile Host provides a Web service interface of the LAS services for the external nodes. Under this architecture, only the Mobile

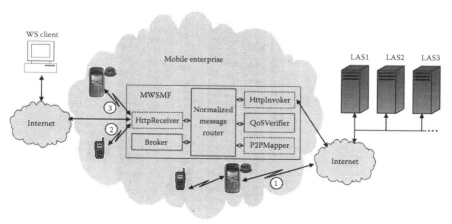

Figure 13.10   Provisioning LAS services to the Mobile Enterprise.

Host directly connects to the LAS, and it is the only access point to the LAS services from mobiles. The interaction is numbered 1 in Figure 13.10. Theoretically, the Mobile Host can be replaced by a standalone node that provides Web service interfaces for the LAS services. Since we are also interested in personalized services and user specific profiles, we were still proceeding with the Mobile Host. Services provided by the Mobile Host are published using WSDL in order to enable the consumer to find those services and invoke them. We were also successful in publishing and discovering these services using the mobile Web service discovery mechanism.

13.11.2.2 Mobile Access through the MWSMF

In this solution, the services are directly accessed from the MWSMF. The middleware takes care of all the load balancing issues. The main advantage of the middleware solution is its transparency to users and programmers, both on the client side as well as on the server side. Inside LAS, no necessary changes are required in order to couple it with the MWSMF, and users of LAS only need to connect to a single point, the MWSMF, in order to access any LAS server they are interested in. Imagine a set of LAS servers, which could be different instances of the same implementation or servers completely different from each other. Now, consider a group of Mobile Hosts or mobile Web service clients trying to access several services from those LAS servers. The current way to achieve this goal would be to configure each Mobile Host with the specific connection to the right LAS server, based on the services offered by it.

However, with the MWSMF solution, the middleware framework replaces all direct contact between Mobile Hosts and LAS servers, so that all communication takes place via the normalized message router. The MWSMF receives all the requests from mobile clients and applies a Java-based management process that selects the right LAS server to be invoked. Once the LAS server has been selected, the framework invokes the appropriate service. As soon as the response from the LAS server is received by the MWSMF, it is forwarded to the mobile requester. The interaction is numbered 2 in Figure 13.10.

13.11.2.3 Accessing the Middleware through the Mobile Host

In this solution, the services are directly accessed from the Mobile Host. Here, the Mobile Host contacts the middleware framework instead of contacting the LAS server directly. From the middleware, the invocation scenario is as explained in Section 13.11.2.2. The interaction is numbered 3 in Figure 13.10. The benefit with the scenario is that the mobile Web service client can be totally unaware of the middleware framework. This is the most realistic scenario in terms of the applications, as it is the Mobile Host that is providing some personalized solution. The Mobile Host is integrating its services with some services from the LAS to polish the result, and the mobile requester is generally unaware of the complete backend process.

This is the most realistic scenario, but the worst of the three solutions in terms of the performance. One invocation cycle of this scenario has two invocation cycles across the radio link and one across the Internet. Hence, a lot of latencies are added to the invocation cycle times. Moreover, the solution also adds a lot of network load. Assuming that the next generation mobile communication technologies achieve higher data transmission rates and that operators provide flat rates, this is the most feasible scenario for Mobile Enterprise Service Integration (Srirama and Jarke, 2009a).

13.12 Conclusions

The developments in the Web services domain, the improved device capabilities of smart phones, and the improved transmission capabilities of the cellular networks have lead to the mobile Web services domain. While mobile Web service clients are common these days, this chapter shows that it is now feasible to deliver basic Web services from smart phones. Mobile Hosts enable seamless integration of user-specific services to the enterprise, and this chapter mentioned some of the example applications in support of this argument. Based on these results, a Mobile Enterprise can be established in a cellular network with the participating Mobile Hosts and their clients. However, such a Mobile Enterprise established poses several technical challenges, both to the service providers and the mobile operator. The most significant of these aspects are providing proper QoS (especially in terms of security and reasonable scalability), discovery of a huge number of possible services, and integration issues of providing Web services from smart phones in a Mobile Enterprise.

The chapter tried to summarize the challenges and research associated in this domain with mobile Web service provisioning and establishing the Mobile Enterprise. The security study showed that mobile Web service messages of reasonable size, approximately 2–5 kb, can be secured with WS-Security specification. Scalability analysis showed that the performance gain to the Mobile Host with binary compression is quite significant, both in terms of improved scalability and battery life. The discovery analysis identified an alternative for mobile Web services discovery by using the features of P2P networks, especially those of the JXTA network. Later, the details of establishing the ESB technology-based MWSMF are addressed, along with the performance metrics. The research has significant application scope, and the discussed mobile access to the LAS services scenario explains where and how the study has been applied. In short, the study concludes that the Mobile Enterprise can be established with reasonable performance latencies. Moreover, the services provided by Mobile Hosts and the Mobile Enterprise offer significant help in integrating personalized mobile services to the regular enterprise and vice versa. Thus, the study and the results are of significant value to the enterprise service integration domain.

The study provides a lot of scope for future research in this domain. Currently, we are working with many auxiliary features for the mediation framework, such as context-aware service discovery. The security domain can be extended to provide proper end-point security and access control mechanisms. The scalability analysis can be extended to a better compression mechanism than the BinXML. We are also attentive of all the additional performance penalties added to the smart phone with these additional features. Another detailed performance analysis of the Mobile Host within the Mobile Enterprise is required, checking that these added features will not restrict the main mobile user from using his smart phone in general fashion, such as making phone calls.

Acknowledgments

This work is supported by the Ultra High-Speed Mobile Information and Communication (UMIC) research cluster (http://www.umic.rwth-aachen.de/) at RWTH Aachen University. The authors would also like to thank R. Levenshteyn and M. Gerdes of Ericsson Research for their help and support.

References

AgileDelta (2007) Efficient XML: Lightning-fast delivery of XML to more devices in more locations, Agile Delta, http://www.agiledelta.com/product_efx.html

Alonso, G., Casati, F., Kuno, H., and Machiraju, V. (2004) *Web Services Concepts, Architectures and Applications*, Springer Verlag, Berlin, Germany.

Apache Software Foundation (2008) Apache servicemix, http://incubator.apache.org/servicemix/home.html

Apache Software Foundation (2009a) Apache axis, Apache Software Foundation, http://ws.apache.org/axis/

Apache Software Foundation (2009b) Apache benchmark, Apache Software Foundation, http://svn.apache.org/repos/asf/jakarta/httpcomponents/httpcore/trunk/contrib/src/main/java/org/apache/http/contrib/benchmark/

Atkinson, B. et al. (April 2002) Web services security (WS-security), Technical Report, Microsoft, IBM and Verisign.

Balani, N. (2003) Deliver Web Services to mobile apps, IBM DeveloperWorks, http://www.ibm.com/developerworks/edu/wi-dw-wiwsvs-i.html

BEA AquaLogic (2008) BEA aqualogic service bus 2.6: The enterprise service bus for the agile business (Product data sheet), Technical Report, BEA Systems, Inc., San Jose, CA, http://www.bea.com/content/news_events/white_papers/BEA_AquaLogic_Service_Bus_ds.pdf

Bellwood, T. (2002) UDDI version 2.04 API specification, Technical Report, UDDI Committee Specification, http://uddi.org/pubs/ProgrammersAPI_v2.htm

Benatallah, B. and Maamar, Z. (2003) Introduction to the special issue on m-services, *IEEE Transactions on Systems, Man, and Sybernetics—Part A: Systems and Humans*, 33(6):665–666.

Bloor, R. (2002) PersonalJava training from SonyEricsson, Wireless Developer Network Symbian DevZone, http://www.wirelessdevnet.com/symbian/rb.18.html

Borck, J. R. (2005) Enterprise service buses hit the road, Infoworld, pp. 26–40, http://www.infoworld.com/article/05/07/22/30FEesb_1.html

Box, D. et al. (2000) Simple object access protocol (SOAP), version 1.1, W3C Note, W3C, http://www.w3.org/TR/soap/

Box, D. et al. (2004) Web services addressing (WS-addressing), Technical Report, W3C Member Submission, http://www.w3.org/Submission/2004/SUBM-ws-addressing-20040810/

Boyer, J. (March 2001) Canonical XML, W3C Recommendation, http://www.w3.org/TR/xml-c14n, http://www.ietf.org/rfc/rfc3076.txt

Brooke, J. (2007) SUS—A quick and dirty usability scale, http://www.usabilitynet.org/trump/documents/Suschapt.doc

Cantor, S., Kemp, J., Philpott, R., and Maler, E. (2005) Assertions and protocols for the OASIS security assertion markup language (SAML) V2.0, Technical Report, OASIS Standard, http://docs.oasis-open.org/security/saml/v2.0/saml-core-2.0-os.pdf

Cao, Y., Spaniol, M., Klamma, R., and Renzel, D. (2007) Virtual campfire—A mobile social software for cross-media communities, in *Proceedings of I-Media'07, International Conference on New Media Technology and Semantic Systems*, Graz, Austria.

Chappell, D. A. (2004) *Enterprise Service Bus*, O'Reilly Media, Inc., Sebastopol, CA.

Christensen, E., Curbera, F., Meredith, G., and Weerawarana, S. (2001) Web services description language (WSDL) 1.1., W3C Working Group Note, W3C, http://www.w3.org/TR/wsdl

Cutting, D. (2008) Apache lucene, http://lucene.apache.org/

Dustdar, S. and Treiber, M. (2006) Integration of transient Web services into a virtual peer to peer Web service registry, *Distributed and Parallel Databases*, 20:91–115.

Endrei, M., Ang, J., Arsanjani, A., Chua, S., Comte, P., Krogdahl, P., Luo, M., and Newling, T. (2004) Patterns: Service-oriented architecture and web services, IBM Redbooks, http://www.redbooks.ibm.com/abstracts/sg246303.html

Ericsson, M. (2003) A study of compression of XML-based messaging, Technical Report, Växjö University, Växjö, Sweden.

Ericsson, M. and Levenshteyn, R. (2003) On optimization of XML-based messaging, in *Second Nordic Conference on Web Services* (*NCWS 2003*), Växjö, Sweden, pp. 167–179.

ETC (2007) New media review, European Travel Commission, http://www.etcnewmedia.com/review/default.asp?sectionid=10&overviewid=6

Gehlen, G. (October 2007) Mobile web services—Concepts, prototype, and traffic performance analysis, PhD thesis, RWTH Aachen University, Aachen, Germany.

Gottschalk, K., Graham, S., Kreger, H., and Snell. J. (2002) Introduction to web services architecture, *IBM Systems Journal: New Developments in Web Services and E-commerce*, 41(2):178–198.

3GPP (2006) Third generation partnership project, 3GPP, http://www.3gpp.org/

Heuer, J., Thienot, C., and Wollborn, M. (2002) Binary format, in B. S. Manjunath, P. Salembier, T. Sikora (eds.), *Introduction to MPEG-7: Multimedia Content Description Interface*, John Wiley & Sons, Chichester, U.K., pp. 61–80.

Hollmen, J. and Manner, J. (July 2008) Sensor network routing requirements of structural health Monitoring, Network Working Group, Internet-Draft, http://tools.ietf.org/id/draft-manner-roll-shm-requirements-00.txt

IBM (2006) Websphere studio device developer, IBM, http://www-306.ibm.com/software/wireless/wsdd/ (August 16, 2006).

IBM Corporation (2008) WebSphere enterprise service bus, WebSphere Software, http://www-306.ibm.com/software/integration/wsesb/

Ichikawa, K. (2002) The view of NTT DoCoMo on the further development of wireless Internet, in *Tokyo Mobile Round Table Conference*, Tokyo, Japan.

IETF. (1999) Hypertext transfer protocol version 1.1., IETF RFC 2616, Internet Engineering Task Force, http://www.ietf.org/rfc/rfc2616.txt

IONA Technologies (2005) ESB: Evolving beyond EAI, Technical Report, IONA Technologies, Waltham, MA.

Ivanova, I. (2007) Mobile web services for collaborative learning, Master's thesis, RWTH Aachen University, Aachen, Germany.

Johnson, D., Perkins, C., and Arkko, J. (2004) Mobility support in IPv6, RFC 3775, Technical Report, Network Working Group, http://www.ietf.org/rfc/rfc3775.txt

Keen, M., Bishop, S., Hopkins, A., Milinski, S., Nott, C., Robinson, R., Adams, J., Verschueren, P., and Acharya, A. (2004) Patterns: Implementing an SOA using an enterprise service bus, IBM RedBooks, http://www.redbooks.ibm.com/redbooks/pdfs/sg246346.pdf

kSOAP2 (2008) kSOAP2—An efficient, lean, Java SOAP library for constrained devices, SourceForge.net, http://sourceforge.net/projects/ksoap2

Laukkanen, M. and Helin, H. (2003) Web services in wireless networks: What happened to the performance, in *Proceedings of the International Conference on Web Services (ICWS'03)*, Las Vegas, NV, pp. 278–284, CSREA Press, Las Vegas, NV.

Lee, C., Helal, A., Desai, N., Verma, V., and Arslan, B. (2003) Konark: A system and protocols for device independent, peer-to-peer discovery and delivery of mobile services, *IEEE Transactions on Systems, Man, and Cybernetics—Part A: Systems and Humans*, 33(6):682–696.

LibertyAlliance (2006) The liberty alliance project, LibertyAlliance, http://www.projectliberty.org/

MacVittie, L. (2006) Review: ESB suites—Make way for the enterprise service bus, Technical Report, Network Computing, http://www.networkcomputing.com/channels/appinfrastructure/showArticle.jhtml;jsessionid=CKJYQ2X3W1HY0QSNDBECKICCJUMEKJVN?articleID=181501276

Martinez, J. M., Gonzalez, C., Fernandez, O., Garcia, C., and de Ramon, J. (2002) Towards universal access to content using MPEG-7, in *Proceedings of the 10th ACM International Conference on Multimedia*, Juan-les-Pins, France, pp. 199–202, ACM Press, New York.

Microsoft Corporation. (November 1996) Distributed component object model protocol-DCOM/1.0, Draft, Microsoft Corporation, Redmond, WA.

Milojicic, D. S., Kalogeraki, V., Lukose, R., Nagaraja, K., Pruyne, J., Richard, B., Rollins, S., and Xu, Z. (2003) Peer-to-peer computing, Technical Report, HP Laboratories, Palo Alto, CA, http://www.hpl.hp.com/techreports/2002/HPL-2002–57R1.pdf

Naumenko, A. (2007) Semantic-based access control in business networks, PhD thesis, University of Jyväskylä, Jyväskylä, Finland.

NTTDoCoMo (2009) i-mode Technology, NTTDoCoMo, http://www.nttdocomo.com/technologies/present/imodetechnology/index.html

O'Reilly, T. (2005) What is Web 2.0, O'Reilly Network, http://www.oreillynet.com/pub/a/oreilly/tim/news/2005/09/30/what-is-web-20.html

OMA (July 2004) OMA web services enabler (OWSER): Overview, Open mobile alliance group, OMA, http://www.openmobilealliance.org/release_program/docs/OWSER/V1_0-20040715-A/OMA-OWSER-Overview-V1_0-20040715-A.pdf

OMG (2004) Common object request broker architecture: Core specification, object management group, http://www.omg.org/docs/formal/04-03-12.pdf

Pashtan, A. (2005) *Mobile Web Services*, Cambridge University Press, Cambridge, U.K.

Pawar, P., Srirama, S. N., Van Beijnum, B., and van Halteren, A. (2007) A comparative study of nomadic mobile service provisioning approaches, *International Conference and Exhibition on Next Generation Mobile Applications, Services and Technologies* (*NGMAST 2007*), September 12–14, 2007, pp. 277–286.

Raghunathan, V., Schurgers, C., Park, S., and Srivastava, M. B. (2002) Energy-aware wireless microsensor networks, *IEEE Signal Processing Magazine*, 19(2):40–50.

Rivest, R., Shamir, A., and Adleman, L. M. (February 1978) A method for obtaining digital signatures and public-key cryptosystems, *Communications of the ACM*, 21(2):120–126.

RSALabs (2008), Cryptographic technologies, RSALabs, http://www.rsa.com/rsalabs/node.asp?id=2212

Sandoz, P., Pericas-Geertsen, S., Kawaguchi, K., Hadley, M., and Pelegri-Llopart, E. (2003) Fast web services, http://java.sun.com/developer/technicalArticles/WebServices/fastWS/

Sandoz, P., Triglia, A., and Pericas-Geertsen, S. (2004) Fast infoset, http://java.sun.com/developer/technicalArticles/xml/fastinfoset/

Schulte, R. W. (2007) The enterprise service bus: Communication backbone for SOA, Gartner Inc., http://www.gartner.com/DisplayDocument?id=504645

Spaniol, M., Klamma, R., Janßen, H., and Renzel, D. (September 2006) LAS: A lightweight application server for MPEG-7 services in community engines, in *Proceedings of the I-KNOW'06, Sixth International Conference on Knowledge Management*, Graz, Austria, *Journal of Universal Computer Science Proceedings*, Springer, pp. 592–599.

Srirama, S. (2008a) Mobile hosts in enterprise service integration, PhD thesis, RWTH Aachen University, Aachen, Germany, October 2008.

Srirama, S. (2008b) Mobile web service provisioning project, URL http://www-i5.informatik.rwth-aachen.de/lehrstuhl/staff/srirama/MWSP.html

Srirama, S. N. and Jarke, M. (2009a) Mobile enterprise—A case study of enterprise service integration, in *Third International Conference and Exhibition on Next Generation Mobile Applications, Services and Technologies* (*NGMAST 2009*), Cardiff, U.K., IEEE Computer Society, Washington, DC (accepted for publication).

Srirama, S.N. and Jarke, M. (2009b) Mobile hosts in enterprise service integration, *International Journal of Web Engineering and Technology* (*IJWET*), Induscience Publishers, 5(2) (2009):187–213.

Srirama, S., Jarke, M., and Prinz, W. (2006a) Mobile web service provisioning, in *International Conference on Internet and Web Applications and Services* (*ICIW06*), Guadeloupe, French Caribbean, pp. 120–125, IEEE Computer Society, Washington, DC.

Srirama, S., Jarke, M., and Prinz, W. (2006b) Mobile Host: A feasibility analysis of mobile Web Service provisioning, in *Fourth International Workshop on Ubiquitous Mobile Information and Collaboration Systems, UMICS 2006, a CAiSE'06 Workshop*, Luxembourg, June 2006, pp. 942–953.

Srirama, S., Jarke, M., and Prinz, W. (2007a) Security analysis of mobile web service provisioning, *International Journal of Internet Technology and Secured Transactions* (*IJITST*), 1(1/2):151–171.

Srirama, S., Jarke, M., and Prinz, W. (2007b) Mobile web services mediation framwork, *Middleware for Service Oriented Computing* (*MW4SOC 2007*) *Workshop @ 8th International Middleware Conference 2007,* ACM Press, November 26–30, 2007.

Srirama, S. N., Jarke, M., and Prinz, W. (2008a) MWSMF: A mediation framework realizing scalable mobile web service provisioning, in *International Conference on MOBILe Wireless MiddleWARE, Operating Systems, and Applications* (*Mobilware 2008*), Innsbruck, Austria, *ACM International Conference Proceeding Series*; Vol. 278, Article No. 43.

Srirama, S. N., Jarke, M., Prinz, W., and Zhu, H. (2008b) Scalable mobile web service discovery in peer to peer networks, in *IEEE Third International Conference on Internet and Web Applications and Services* (*ICIW 2008*), Athens, Greece, pp. 668–674, IEEE Computer Society, Washington, DC.

Sun Microsystems (2001) Jini architecture specification—Version 1.2, Technical Report, Sun Microsystems, Inc. http://www.sun.com/software/jini/specs/jini1.2html/jini-title.html

Sun Microsystems (2006) Sun Java wireless toolkit, Sun Microsystems, http://java.sun.com/products/sjwtoolkit/

Sun Microsystems (2008) Wireless messaging API (WMA), JSR 120, JSR 205, Sun Developer Network, http://java.sun.com/products/wma/index.jsp

Surowiecki, J. (2005) *The Wisdom of Crowds*, Random House, Inc., New York.

Tate, B. A. and Gehtland, J. (2005) *Spring: A Developer's Notebook*, O'Reilly, Sebastopol, CA.

Ten-Hove, R. and Walker, P. (2005) Java TM business integration (JBI) 1.0-JSR 208, Final Release, Technical Report, Sun Microsystems, Inc., Santa Clara, CA.

Thomas, K. (1999) Fourth generation (4G) wireless communications, http://www.4g.co.uk/

University College Cork, (2007) SUMI- the de facto industry standard evaluation questionnaire for assessing quality of use of software by end learners, University College Cork, Ireland, http://sumi.ucc.ie/index.html

UPnP Forum (2003) Universal plug and play device architecture, Technical Report, UPnP Forum, http://www.upnp.org/resources/documents/CleanUPnPDA101–20031202s.pdf

van Halteren, A. and Pawar, P. (2006) Mobile service platform: A middleware for Nomadic mobile service provisioning, in *Second IEEE International Conference on Wireless and Mobile Computing, Networking and Communications* (*WiMob 2006*), Montreal, Canada.

Wenger, E. (1998) *Communities of Practice: Learning, Meaning, and Identity*, Cambridge University Press, Cambridge, U.K.

Wilson, B. J. (2002) *JXTA*, New Riders Publishing, Indianapolis, IN.

WSO2 (2007) WSO2 ESB benchmark, WSO2, https://www-lk.wso2.com/~asankha/benchmark/

Zhu, H. (2008) Scalability of P2P based mobile web services discovery, Master's thesis, RWTH Aachen University, Aachen, Germany.

Chapter 14

Real-Time Web for Mobile Devices

David Linner and Rafael Grote

Contents

14.1 Introduction

In contrast to classical means of communication like voice or video, the Web is based on the communication of changes to resource states. Adding a phrase to a wiki page, posting to a blog, or sending a comment to a chat all changes the corresponding resource on the Web server. Interactive communication for a user results from updating a resource on the Web server and receiving the latest updates done by other users. Real-time communication on the Web means communicating changes to the state of a resource without delaying user interactions represented by those state changes. The time limit for communicating a state change between two users is application dependent. For example, in online forums, several minutes up to an hour are acceptable between posting and reply, while for a chat application 10 s during delivery of a message is unacceptable. In relation to Web applications, the term *real-time* is mostly used to describe interactions at a speed comparable to natural communication. Achieving such application depends on the actual amount of data to be communicated, the available network protocols, the characteristics of sending and receiving devices, as well as on the characteristics of the underlying network.

Today, most real-time Web applications rely on Asynchronous JavaScript (Ajax; Kesteren, 2008) or Ajax-based workarounds (Russell, 2006) to enable server-push behavior. The actual interaction between multiple users (and thus Web browsers) in terms of synchronizing states is consequently not reflected. Under consideration of networks with high packet delay and packet loss, we decided to not simply add a layer on top, but rather develop a new application protocol complementing the Hypertext Transfer Protocol (HTTP) for real-time use cases. Realizing the protocol complementary to HTTP allows using network resources much more efficiently than possible when realized on top of HTTP. Our protocol bases on the concept of a distributed shared memory (Protic et al., 1996). Numerous clients of a Web application can simultaneously access and modify a shared space. This uses network resource more efficiently. Operations on the space are conveniently performed via a slim application programmer interface integrated with the Web-application runtime environment of the Web browser.

In the following section, we will discuss present approaches for bidirectional communication on the Web and point out inherent limits and drawbacks. We distinguish between bidirectional communication and real-time synchronization of states. In Section 14.3, we discuss how states are synchronized in distributed applications, assuming that messages are exchangeable among server and several clients. The discussion of approaches for bidirectional communication and real-time state synchronization is followed by the presentation of our real-time Web solution, which is also suited for mobile use cases.

14.2 Bidirectional Communication in the Web

Today, communication in the Web is unidirectional. The client always controls interaction between client and server. Besides replying requests, the server is not able to send data independently to the client. For real-time multi-user Web applications, it is crucial to have low-latency bidirectional client–server communication. The core issue is to let Web servers independently send data to browsers, which is called server-push. Even though communication is usually client and initiated, Web servers need the ability to push data proactively toward clients, to comply with requirements of real-time Web applications.

14.2.1 Reverse Ajax

The term Reverse Ajax summarizes a variety of techniques based on Asynchronous JavaScript, intended to realize a server-push mechanism, or at least to simulate it.

The traditional way to keep a Web application up to date is polling. That is, the client periodically requests the Web server for new data. The server cannot deliver emerging events on its own; rather, it has to wait for the next client request. Thus, the delay between the occurrence of an event and client notification depends on the polling rate, and increases with a shorter interval between requests. On the other hand, polling too frequently causes high server load.

As an enhancement to polling, so-called long-polling or slow load tries to reduce the number of requests. As usual, the client has to make periodic requests. The server delays the response until new data is available. Since occurring events can directly be delivered to clients through the waiting request, there is no longer a wasteful delay on client notifications. However, there is an overhead of connection establishment for each event.

The idea behind HTTP streaming, also known as Comet (Russell, 2006), is to keep a long-lived HTTP connection between Web server and client. The browser only establishes a connection once. Customized Web servers are able to keep the connection alive without running into timeouts. In case of an event, the server just has to stream the new data through the hold connection. Client side, the connection is observed for incoming data that is delivered in terms of an event.

At present, browsers do not yet have native support for Comet streaming technologies; therefore, workarounds based on Ajax have to be implemented. Due to the lack of standardization, these workarounds are mostly browser-specific. Sophisticated JavaScript frameworks have been developed to address this issue, and provide easy-to-use programming interfaces for Web applications. In future, Reverse Ajax might be superseded by upcoming HTML 5 communication practices.

14.2.2 Alternative Approaches

Reverse HTTP (Lentczner and Preston, 2009) is a very recent and yet experimental approach based on HTTP, in which connections are still client initiated, but communication is started by the server. It exploits the HTTP upgrade (Fielding et al., 1999) mechanism to turn a HTTP socket around, reversing client and server roles. After a client has established a HTTP connection, this client acts like a server and awaits a HTTP request. Using two client-initiated, persistent HTTP connections—a regular and a reversed one—bidirectional communication via HTTP is made possible.

The Bidirectional-streams Over Synchronous HTTP (BOSH) protocol (Paterson et al., 2008) is intended to tunnel XMPP (Saint-Andre, 2004) over HTTP. It emulates a bidirectional connection by using multiple HTTP connections. While Reverse HTTP and BOSH work with regular browsers, both require a customized server.

Several browser plug-ins, like Adobe Flash or Java Applets, provide proprietary programming interfaces allowing the creation of plain sockets and thus establishing connections to arbitrary remote servers. To avoid malicious scripts from abusing these, feature access is controlled by a security policy. Although being accessible from embedded scripts, plug-ins do not provide a standardized way to bidirectional Web communications.

14.2.3 HTML 5

Associated with the standardization of HTML 5 (Hickson, HTML 5, 2009), two approaches are currently developed: unidirectional Server-sent Events and full-duplex WebSockets. Both use long-lived TCP connections. The communication protocols are text and based and very similar to HTTP.

Server-sent Events (Hickson, Server-Sent Events, 2009) are used to supply Web browsers continuously with up-to-date information. Use cases would be, for example, news feeds or event notifications. On the client-side, Server-sent Events can easily be subscribed through a simple JavaScript interface. It connects to a server specified initially with a Uniform Resource Identifier (URI). Event handlers are called on incoming messages, on connection establishment, and whenever an error occurs or the connection is terminated. Besides using this simple interface, it is also possible to subscribe to Server-sent Events in HTML with the event-source element.

WebSockets (Hickson, The WebSocket protocol, 2009) allow full duplex communication. A WebSocket connection is established by a client with a HTTP request, which is followed by an initial handshake. Afterward, it behaves almost like a plain TCP connection. However, no raw binary data is delivered to JavaScript, since it does not support that and the usage should keep simple. Instead, data exchange is completely text and based.

WebSocket connections traverse firewalls, NAT-routers, and proxy servers seamlessly. Since the WebSocket protocol is mostly compatible with HTTP, existing solutions like HTTP load balancers can be utilized further on. Additionally, cross-domain communication and cookie-based authentication are supported.

The application programming interface is event and oriented, as is usual in the JavaScript world. Incoming data is buffered and delivered to a predefined callback. A send function allows data to be issued to the remote server.

WebSockets are nearly as versatile as ordinary TCP sockets, but incompatible to them for a variety of reasons. Using plain TCP sockets would involve serious security risks. Web sites would be able to initiate TCP connections to any desired service. Spammers could abuse computers of Web surfers by simply contacting any SMTP servers. Another reason not to use TCP sockets is to keep application programming interfaces simple. JavaScript developers should not be faced with complex to use, byte-orientated interfaces.

14.3 Real-Time Synchronization

Real-time synchronization is required in the disciplines of parallel and distributed simulation, distributed virtual environments (e.g., distributed interactive simulation, multi-player online games) and multi-user interactive applications. In contrast to concurrency control of distributed database systems, synchronization of the mentioned systems requires the correct ordering of events to keep them in a consistent state.

14.3.1 Time Constraint

Algorithms in the area of distributed virtual environments and interactive simulations mainly differ from classic synchronization algorithms in their additional requirement for keeping the simulation quick and responsive. Human factors studies showed that latencies above 100 ms are noticed by humans (Bailey, 1982; Woodson, 1987). Depending on the type of application, the maximum acceptable delay is supposed to be between 50 ms and at most 300 ms (Henderson, 2001; Pantel and Wolf, 2002; IEEE standard for distributed interactive simulation communication services and profiles, 1996).

Though the term "real-time" may indicate that a hard real-time constraint is specified, this does not apply to all algorithms introduced in the following sections. Indeed, some of them are rather best effort algorithms trying to minimize network latency.

14.3.2 Bucket Synchronization

The bucket synchronization mechanism was introduced with MiMaze, one of the first fully distributed multi-player games (Gautier and Diot, 1998). Its intention is to reorder data units in a distributed system by means of the time they were issued. Therefore, clocks of the participating clients are synchronized with NTP (Mills, 1992) and time is divided into fixed length sampling periods. Each of these time intervals is assigned to a bucket. Now data units can be sorted into buckets according to the time they were issued. After a certain delay, all data units in a bucket can be processed together. Data arriving after that delay are discarded. Thus, the delay should be substantially larger than the average network latency.

14.3.3 Round-Based Algorithms

Several round-based synchronization algorithms (GauthierDickey et al., 2004; Baughman et al., 2007) are based on a stop-and-wait mechanism; clients have to announce their state changes before the simulation time advances. These algorithms especially address cheating in peer-to-peer multi-player games.

The lockstep protocol (Baughman et al., 2007) breaks rounds into two steps:

1. Clients decide their turns, but do not announce them. Instead, a cryptographically secure one-way hash is announced.
2. After all clients announced their hashes, they reveal their unencrypted moves. Each client can validate the moves by comparing them to the respective hash values.

Improvements to this protocol facilitate acceptance of late and out-of-order announces. In addition, spheres of influence are introduced, which is an application-specific measure to determine currently relevant clients (i.e., spatial distance of player avatars in a virtual environment). Thereby, the number of remote hosts a client has to wait for before revealing its decision is limited. The greatest drawback of all lockstep variants is its performance: all clients have to wait for the slowest one.

The New-Event Ordering protocol (NEO) (GauthierDickey et al., 2004) solves this problem by bounding the maximum round length. Indeed, this restriction does no harm to real-time simulations, since they are already time constrained, as stated in Section 14.3.1. In real-time multi-player games, for example, player moves have to be announced within a specific time to keep the game playable.

NEO with pipelined rounds may handle multiple rounds. This works in a way similar to the technique of pipelining instructions in a processor (Flynn, 1995). In particular, pipelining enables NEO to process the next round before completing the current one, which is a further enhancement is dynamic adjustment of the round duration. Consequently, clients do not wait longer than necessary with regard to the actual network latencies.

14.3.4 Optimistic Synchronization

The previously mentioned algorithms follow a conservative synchronization strategy. They do not playout events before not being sure it is safe to do so with respect to consistency. Optimistic synchronization algorithms process incoming messages as soon as they are received. This is considered "optimistic," because it is assumed there are no causality errors (Fujimoto, 1999). Optimistic synchronization follows the ideas of optimistic concurrency control (Kung and Robinson, 1981), which is based on the assumption that most database transactions do not conflict with each other. Thus, optimistic concurrency control is as permissive as possible in allowing transactions to execute without using locking. However, it is referred to as optimistic locking. In optimistic concurrency control, transactions are subdivided into three phases:

1. Read: In the first phase, clients may read values from the database, including the ones intended to be modified. Copies of the retrieved values are stored to a local cache. Write operations are executed locally on the cached values during this phase.
2. Validation: After all operations of the first phase have been completed, the locally modified values are submitted to the database. There concurrent transactions are checked for possible conflicts. If a conflict is detected, it must be resolved using a conflict resolution algorithm. If the conflict cannot be resolved, a transaction has to be aborted as the last resort. For read-only queries, it must be determined that the result the query would return is actually correct. Validation ensures the integrity of the database.
3. Write: In the last phase, validated transactions are committed. Consequently, local modifications are made global. This phase is optional and applies only to successfully validated write operations.

Optimistic concurrency control is well suited for environments with a low contention for data. If there is only a small probability for conflicting transactions, it may lead to higher throughput than other concurrency control methods. However, highly conflicting environments may cause repeatedly restarting of transactions, reducing performance significantly.

14.3.4.1 Time Warp

The time warp algorithm (Jefferson, 1985; Fujimoto, 1999) optimistically processes events. In case of inconsistencies, it tries to repair them using the last state known to be consistent. Therefore, local and remote operations as well as snapshots of the application's state are stored, building a history.

The application state may become inconsistent as soon as an operation is received that should have been executed before (e.g., caused by high network latency of a client). In other words, the ordering of events is disturbed, and consistency cannot be warranted any longer.

Using the history, the state is rolled back to the last consistent state saved before the lately received operation was issued. After inserting the late operation, the remaining operations from the history are executed. This is done in a fast-forward mode until a consistent, present state is restored.

This approach has several shortcomings. First, it requires the clients to do many computations for rollbacks and to maintain the history. Second, it adds complexity to the application to support rollbacks. Finally, rollbacks and replaying of operations are time consuming, and may not be suitable for interactive applications. Consider a scenario in a multi-player ego-shooter game, where a rollback occurs after a player was shot down. In fact, this player may have shot the other player before firing on his part. After replaying this late event, both players may experience this as a strange situation.

14.3.4.2 Trailing State Synchronization

An improved time warp algorithm is proposed with trailing state synchronization (Cronin et al., 2004). Instead of storing a huge number of state snapshots in the history, it maintains merely a constant number of states. It works with only two state copies. The first one is the leading state, which is actually presented to users, processing recent events. A second copy of the state, called the trailing state, is running at a delay relative to the leading state.

If the leading state becomes inconsistent due to a late event, it is simply replaced with a copy of the (yet consistent) trailing state. After the late event is inserted into the history, the copied state is fast-forwarded until a recent leading state is restored. Thus, it is no longer necessary to rollback states in the case of late events, resulting in a saving of computational power. Moreover, memory capacity is saved due to the reduced number of states being stored.

14.3.5 Latency Hiding

Experienced, subjective delays of remotely triggered events can effectively be hidden by either introducing artificial delays on local operations or precalculating future values.

The first option is known as local lag (Mauve, 2000). Similar to a buffer in a streaming system, events are held back for a short period. Thus, messages can be delivered simultaneously on local and remote machines in a peer-to-peer architecture. A similar effect can be achieved for a centralized architecture with dumb clients that do not apply operations locally before not being acknowledged by the server. Hence, all clients, including the requesting one, experience a certain delay before the operation is applied.

The second technique, called dead reckoning (Gautier and Diot, 1998; Fujimoto, 1999; Bernier, 2001; IEEE standard for information technology—protocols for distributed interactive simulations applications. Entity information and interaction, 1993), is well suited to synchronize continuously changing objects like position vectors. Knowing the last values of an object, the next value can be calculated. Thus, lost updates are compensated by interpolating the missing value. Anyway, this technique may be used to determine values between updates, for example, to smooth a continuous movement.

14.3.6 Summary

Existing approaches for bidirectional communication in the Web are yet considered workarounds. Maybe this will change in future with WebSockets. However, none of the suggested concepts addresses the problem of data synchronization.

Originating from diverse disciplines, various techniques for real-time synchronization exist. The principal aim of data synchronization is to keep shared data in a consistent state.

Several synchronization algorithms use locking to avoid conflicting data modifications. However, this technique may unpredictably delay synchronization. Consequently, it is not suitable for real-time multi-user Web applications.

Alternatively, serializability of data access may be preserved by chronological ordering of synchronization events, based on timestamps and logical clocks. To avoid the overhead of synchronizing physical clocks via NTP, time may be partitioned into rounds.

Algorithms for optimistic synchronization of parallel simulations appear to be too complex for a generic concept. Rollback mechanisms have to be implemented, which are time consuming and may bother interactivity in real-time Web applications.

More promising are optimistic algorithms for concurrency control and data replication. Without locking, resources are accessible at any time. Optimistic access to resources may result in conflicting modifications, in which case conflict resolution is suggested. As a last resort, a modification request is rejected.

14.4 Real-Time State Synchronization for Web Applications

Based on our findings, we designed Pulsar, a synchronization system for the state of Web resources. The system is not intended to replace the existing Web application model. In fact, it realizes perfectly the idea of Representational State Transfer while complementing the hypertext stack to synchronize states with low delays. As shown in Figure 14.1, Pulsar is divided into three main components: a server, a client component embedded into Web browsers, and a communication protocol used for synchronization.

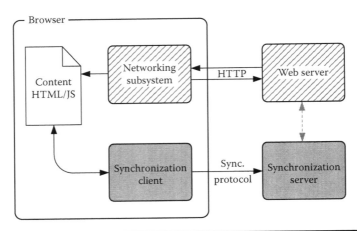

Figure 14.1   Architecture overview.

While the protocol and the synchronization mechanism as a whole are explained in detail throughout this section, reference architectures of the client and server components are briefly suggested in Section 14.5. First, it is explained why a centralized architecture was chosen.

14.4.1 Rationale for a Centralized Architecture

The existing Web architecture is client–server-based. This simple architecture model may have drawbacks with regard to scalability, but is best manageable from the business perspective. Indeed, there are various reasons for using client–server architecture in our case:

- Centralized architectures are better suited for synchronization, because there is a single instance to solve conflicts and care for consistency. Causal and even total order of events is guaranteed implicitly (Guo et al., 2003).
- A server constitutes a central point to supervise restrictions as well as security policies and to guarantee fairness. The additional overhead of consistency checking on each machine, as needed in peer-to-peer systems, is avoided.
- Though a server may be considered a single point of failure, this problem exists a priori in a client–server-based Web. Probably, a distributed synchronization system may be more robust. Nevertheless, it becomes useless without the Web server, whose Web applications it synchronizes.
- Web browsers may run on machines located in private networks or hidden behind firewalls. The chance of a centralized system working properly (without further configuration) in such an environment is substantially higher than for a peer-to-peer system.
- Scalable peer-to-peer architectures are either based on multicast networks, which are not yet widely available, or rely on routing mechanisms, resulting in additional network latency.
- Many existing (massive) multi-player online games are implemented on centralized architectures (Bernier, 2001; Färber, 2002; Guo et al., 2003).

14.4.2 Communication Protocol

The communication between the Pulsar server and browsers is organized in rounds with a fixed order of events. Each round is organized as shown in Figure 14.2. From the server's point of view, rounds are subdivided into three phases:

1. **Wait for incoming requests**. The Pulsar server waits until a request-message from each client has been received. The requests are instantly parsed but not yet applied, since they may interfere with each other. The waiting time is bounded by the maximum round time, preventing clients with very high network latencies from slowing down the entire system.

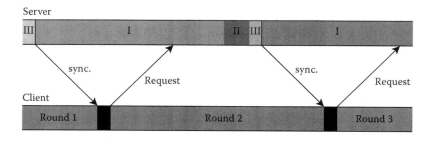

Figure 14.2   Round-based interaction between server and a single client.

2. **Resolve conflicts and apply requests**. Without reference to their accurate arrival times, all requests issued in the same round are considered as simultaneous. Consequently, concurrent changes of an object value are conflicting, and have to be detected and resolved first. Thereafter remaining requests are applied.

3. **Distribute updates**. Clients are notified about the resulting updates of this round. This phase is divided into two parts, which are executed sequentially:
   a. For each client, an individual synchronization message is constructed, containing the relevant updates of this round as well as status codes indicating whether their requests were successful.
   b. The synchronization messages are sent to the clients, initializing the next round.

14.4.2.1 Client Behavior

Most often, clients behave passively. Applications may access a local copy of the distributed state. While reading the state happens instantly without restrictions, modifications are held back first and are locally cached as a request. During a round, a client collects requests and subsumes them locally. Whenever a new synchronization message is received, the client is synchronized in an atomic procedure by applying the following steps:

1. Advance to the next round. The actual round number has been submitted with the sync-message.
2. Reply with a request-message, which acknowledges successful receipt and contains the requests collected during the last round.
3. Apply incoming updates, which may be contained in the synchronization message.
4. Optionally notify client applications about possible updates.

14.4.2.2 Message Types

As indicated in the previous sections, communication is message based. There are four different types of messages used for communication:

Sync-message: At the beginning of a round, the Pulsar server sends a synchronization (sync) message to each client. It announces the actual round, identified through its round number. Additionally, the request-message received at last from the client is acknowledged. However, the main purpose is to distribute recent updates of the shared state, which are contained in the message body. Besides, it contains status codes indicating whether a client's requests for the last round were successful. Table 14.1 shows the format of a synchronization message.

Request-message: The collected requests of each round are subsumed in a request-message, as shown in Table 14.2. After a round ends, it is sent by a client to the Pulsar server. Requests are modifying operations, including create for creating new objects, update for changing a value, and delete for removing objects again. Request-messages are sent whenever a new sync-message is received, thus acknowledging the reception. In addition, the last sync-message successfully received before this round is acknowledged, so that the server is notified about intermediate message loss.

Join-message: Clients willing to join a synchronization session indicate this by sending a join-message to the server. This message contains an URI identifying the session to join. Optionally, this URI may point to a document describing rules and restrictions for this session. If the desired session does not yet exist, it will be started on the fly. Table 14.3 shows the structure of a join-message.

Table 14.1   Format of Sync-Messages

Protocol Version	Client ID	Round No.	Last Received
Request status codes			
State updates			

Table 14.2   Format of Request-Messages

Protocol Version	Client ID	Round No.	Last Received
Requests			

Table 14.3   Format of Join-Messages

Protocol Version	Client ID	Round No.	Last Received
Session identifier (URI)			

Table 14.4 Format of Init-Messages

Protocol Version	Client ID	Round No.	Last Received
Session configuration parameters			
Initial state			

Init-message: The first message sent to a client is the initialization (init) message. It contains all necessary information to bootstrap a newly participating client to the synchronization process, as shown in Table 14.4. Therefore, mainly the shared objects have to be submitted to the client. Particularly, the relevant data of each shared object is included. Thus, the init-message is very similar to a sync-message, with the distinction that the init-message only contains instructions to create objects. However, init-messages are sent at the same time as sync-message and also indicate the beginning of a round. Consequently, the client continues with the regular synchronization procedure.

14.4.2.3 Adaptive Round Time

Using fixed round times preserves slow clients from delaying the whole synchronization process. On the other hand, doing so excludes clients with high network latency, as well as wasting time if all clients reply quickly. Adaptive round times do not pinpoint on specific network latency, allowing the Pulsar server to wait no longer than needed. Additionally, defining a maximum round time prevents round times from being determined by the slowest client. Using adaptive round time, a round finishes as soon as the minimum round time elapses and all clients have replied with a request-message, but at the latest after the predefined maximum round time has been reached. Thus, waiting for lost messages or clients with poor network latency is avoided.

14.4.2.4 Join Procedure

The process of joining a Pulsar session consists of two steps:

1. The client sends a join-message to the Pulsar server. This message contains an URI identifying the session to join. Optionally, this URI may point to a document describing rules and restrictions for this session. If the desired session does not yet exist, it will be started on the fly.
2. The server replies with an init-message as soon as the next round begins. It contains all necessary information to bootstrap a newly participating client to the synchronization process. In particular, it contains a recent copy of the shared state.

After having received and processed the init-message, the client's state should be consistent with the shared state, and the client is ready to participate in the regular procedure of round-based updates and requests. Figure 14.3 illustrates the join procedure.

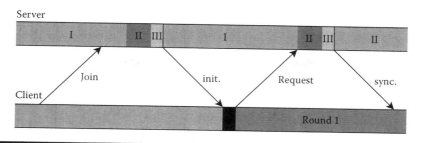

Figure 14.3   A client joins a synchronization session.

Since the complete state is stored at the central server, clients are able to join at any point in time. The initial state submitted to a client is guaranteed to be up to date. Other clients of the same session are notified about new clients joining as well as about leaving clients.

There is no special procedure intended for leaving a synchronization session again. A client willing to quit a session just stops sending requests and ignores incoming sync-messages. After a dedicated number of rounds (depending on server configuration), a client not replying to the Pulsar server is eliminated.

14.4.2.5 Event Serializability

Many synchronization algorithms suggested in the Section 14.3 depend on synchronization of physical clocks. Accurately adjusted system times facilitate the ordering of remotely received events by means of their time stamps. Pulsar abstains from synchronizing system times, since it would be too expensive for a lightweight synchronization mechanism. Moreover, browser-embedded clients are usually not able to modify the system time.

However, serializing incoming messages is necessary, and cannot be determined by the order of reception, which would be an illegal disregarding of network latency. It is required to identify late messages and allow reordering of incoming messages. Therefore, messages are assigned to rounds and labeled with their corresponding round numbers. Each client sends exactly one message per round, making the round number a unique identifier. Concurrent requests of the same round are considered simultaneous, and are handled equally in case of a conflict.

14.4.2.6 Simultaneity

To guarantee fairness among clients, it must be ensured that rounds are simultaneous. Thus, synchronization updates must be delivered at the same time for each client. Consider, for example, a multi-player game, in which a movement of an object must be presented to all players at the same moment, and each player must be granted the same amount of time to react to this event.

Client 1, with high latency

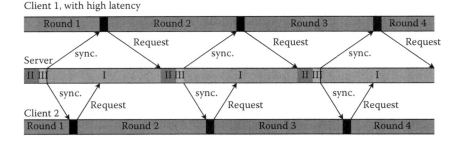

Figure 14.4 Suspension of rounds, caused by different network latencies.

In a round-based system, simultaneity has a special meaning. It is not essential that state updates arrive at exactly the same point in time at each client machine; rather, rounds are considered synchronous. A client with high network latency may receive update messages a little bit later than other clients by means of physical time. As a result, the beginning of each round is slightly delayed for this client. Since all messages are affected by this network latency, this delay has no effect on round times. Apart from a certain jitter, at least the average round time is constant and equal for all clients. Of course, it is crucial to select a round time superior to the estimated network latency.

This approach does not allow determining the chronology of events occurred in the same round on distinct clients. These events are considered simultaneous. Thus, the timeline is not entirely continuous, but until a certain scale, which equals the predefined round time. The server treats simultaneous synchronization requests as a conflict. How such a conflict should be resolved depends on the specific application, and is decided based on a preset policy.

Strictly speaking, simultaneous events do not need to happen at the same time but in the same round. Figure 14.4 shows clients with varying network latencies experiencing rounds at different physical times.

14.4.3 State Synchronization

The proposed synchronization mechanism is not thought of as a messaging system. It is a distributed-shared-memory-architecture for fast synchronization of shared states between browsers. Thus, Web applications should not care about message exchange. Rather, they should work with objects, which are synchronized behind the scene. Pulsar is responsible for managing synchronization and resolving conflicts. The browser's embedded synchronization client maintains a copy of the shared state, which has to be kept consistent with the Pulsar server. Furthermore, shared data should be presented to client applications in a clear fashion. In this section, the synchronization functionality of Pulsar is discussed.

14.4.3.1 Data Model

The shared data is organized into objects that are identified by unique names. Objects are tree-like hierarchical structures. They are composed of a set of properties, which are name-value pairs containing either a primitive value or again a nested object structure. Alternatively, properties may contain a collection of values (including nested objects) subsumed in an array. Primitive values particularly may have one of the types: boolean, number, or string.

The suggested object structure is compatible to Web objects of existing Web technologies such as the document object model, which is used in hypertext markup and scripting languages. This is essential, since Pulsar should integrate well into the Web. Besides, the primarily targeted programming language on the client side is JavaScript.

14.4.3.2 Groups, Scopes, and Access Restrictions

In many use cases, applications need to implement some kind of subdivision concept (e.g., channels, rooms, spheres of influence, separate levels, or teams). Therefore, clients are able to assign themselves to a group or create new groups.

Being divided into groups also effects which objects are visible to clients. Whenever an object is created (no matter if predefined or ad hoc), the creator is assigned as its owner and the object is associated with a scope. There are three diverse kinds of scopes:

- Global scope: The global scope is unique for the entire synchronization session. Global objects are generally visible from clients of all groups.
- Group scopes: Since a session may contain multiple groups, each group constitutes a distinct scope. Group objects are dedicated to exactly one group, as are clients. Clients can access objects belonging to their group, but not the ones of other groups.
- Client scopes: Each client is exclusively associated with an individual client scope. Client objects are always created in the client scope of its owner. Clients may access objects of foreign client scopes as long as the accessing and the owning client belong to the same group. Thus, client objects are weakly associated with a group, but change their affiliation as soon as the owning client does.

Each scope constitutes a separate namespace, which allows creating multiple objects of the same fashion (and name) without naming conflicts—for example, a "position" object for each client in its individual scope.

The concept of groups and scopes should not be confused with also supported sessions, which are strictly separated from each other, allowing multiple instances of an application (or different applications) running concurrently on the same Pulsar server.

To restrict access on objects further, they can be defined with different access modes (no access, read only, and fully accessible). These modes are applied for remote clients, while the owner of an object always has full access.

14.4.3.3 Conflict Resolution

If in the same round two or more clients are requesting to modify the same property of an object, their requests are conflicting. In that case, a conflict manager has to resolve the conflict by deciding which request prevails. Since conflict resolution is highly application specific, depending on the synchronization-server configuration, one of different conflict resolving policies is chosen. The following conflict resolving strategies exist:

Random: One of the conflicting requests is randomly chosen, with an evenly distributed probability. The actual reception time of the requests during the round does not matter. Due to its simplicity and fairness, this is the default strategy.

Consensus: All requests are compared with each other. They will succeed only if all of them are equal; otherwise, all requests are declined. Two requests are equal if they want to write exactly the same value to the same resource.

Majority: If some but not all of the conflicting requests are equal, it may be possible to find a majority. In that case, the request of the majority will succeed. Otherwise, no request—or alternatively, a random one—is chosen.

Priority: The priority strategy prefers a single client. In particular, this may be the owner of the object in conflict. The request of the preferred client always succeeds. Consequently, this strategy is not considered fair. In fact, it must be combined with a second resolution strategy, which is applied if none of the requesting clients is prioritized (i.e., the owner is not part of the conflict).

Merging: In some cases, conflicting values may be merged. For example, two changes on different positions of a string may be applied concurrently without being in conflict. In some use cases, it may even be applicable to merge a number by using its mean value. However, this strategy may not be used in a large number of applications. Additionally, string merging may require other changes to the synchronization concept. Thus, this strategy is considered more theoretical.

However, at best, conflicts should be avoided, which is done with several preventive measures. First, read operations are local, and do not conflict with modification requests. Second, concurrent write-accesses to different properties of the same object do not conflict with each other if the accessed properties are addressed directly via their property path. Finally, keeping round times as short as possible statistically reduces the probability of conflicting requests in the same round.

Indirect or application-level conflicts result from the application logic. This type of conflict is not narrowed to a single object, but rather to multiple objects being modified in the same round. While the individual modifications are not conflicting, the sum of modifications breaks the application rules. Consider, for example, a virtual environment, where two objects move to the same place, although they should collide.

Application-level conflicts are not really conflicts in terms of synchronization, and do not harm consistency of the shared state. Thus, it is not a primary goal to avoid or solve indirect conflicts, and this should be addressed on the application level in the first place. However, it may be an option to integrate rule-checking logic into Pulsar, warranting a set of application-defined rules.

14.4.3.4 Fair Handling of Clients

Fairness is of vital importance for the synchronization mechanism. Being used in multi-player games or even commercial applications like a platform for live auctions, fairness is required for user acceptance. To be considered fair, client requests must be handled equally, and independent of network latencies. Furthermore, updates should be delivered simultaneously to all clients to avoid advantages for clients with low network delays.

Since Pulsar is round based, fast requests are not privileged, and processed together with later received requests within the same round. Thus, clients with high network latency normally have no disadvantage as long as their requests are received before a round has elapsed. In the case of a conflict, a fair resolution is found depending on the configured conflict policy.

Instead of using artificial delays for fast clients, it is assumed that the actual time of event delivery does not matter as long as the (average) round time is equal for all clients. Thus, each client—or rather, each user in the case of an interactive application—has the same time to react to incoming updates. The event of receiving a synchronization message—and with it the beginning, respectively the ending, of a round—is considered simultaneous for each client independent from the absolute time, as stated in Section 14.4.2.6.

14.4.3.5 Consistency

Since the proposed architecture is centralized and all requests are processed by the Pulsar server, we have implicitly guaranteed causal and total order (Guo et al., 2003). Thus, consistency is guaranteed as long as messages are delivered reliably and in order. Actually, it must be able to distinguish between lost synchronization messages (sent by the server) and request-messages (sent by clients).

Loss of a client request-message is undesirable, but does not affect consistency. Since a lost message is never received by the Pulsar server, it is also not propagated to the remaining clients, and the shared state remains equal for all participants. If local lag is enabled—which is the default setting—even the requesting client remains in the origin state.

More important are synchronization messages that contain updates to the actual state. A client missing one or more of these updates is left in an outdated state. Since the server detects any lost updates, the next successfully received synchronization message contains all missed updates, bringing the client back to the current state.

However, there is a chance that a client being temporarily in an outdated state creates inconsistency by manipulating an outdated value. For example, a client may increase an integer value, which in fact was changed by another client previously, but the corresponding update was lost for the first client. To prevent such inconsistencies, clients should send outstanding requests on their own in good time before the end of a round. This solution would require clients to maintain their own timers and guess the actual round time. Thus, an alternative way is actually preferred: the Pulsar server handles such request-messages, which are sent after missing a round and potentially contain late requests, such as a late message.

Duplicates and late messages (requests as well as synchronizations) are problematic for consistency in a similar fashion. Since values addressed in a late message may have changed intermediately, straightforward applying of late messages possibly leads to inconsistency. In fact, the correct order of messages is warranted for server and clients by using round numbers, which allows identifying late messages and handling them in a special way.

It is necessary to detect whether a property being accessed in a late request has been modified since the request's round. Therefore, each property is tagged with a time stamp of last access. Every time a property is modified, the current round number is set as a time stamp. By means of this time stamp, the Pulsar server individually decides whether to apply or discard each late request.

14.4.4 Web Application Programming Interface

The synchronization client, embedded into browsers, provides an application programming interface for Web developers. This ought to be available in the browser's regular JavaScript environment, integrated with the browser-supplied set of default JavaScript libraries.

The interface has a high level of abstraction for accessing shared objects as if they were local. Web applications barely have to care for synchronization issues, because synchronization is mostly hidden behind the scenes.

Multiple event handlers notify applications about initialization of a connection, incoming synchronization updates, and errors. Listing 14.1 shows an extract of the application programming interface in WebIDL (McCormack, 2008). In particular, the interface to a single connection is shown. Multiple connections from different windows, different tabs, or even from the same Web application can exist concurrently. Incoming messages share the same port and are internally multiplexed by means of an application ID.

Integrated dead-reckoning features allow easy integration of dead-reckoned objects in real-time multi-user Web applications. This technique is especially suitable for frequently changing objects controlled by a single client, for example, moving objects in a multi-player game.

Listing 14.1   Extract of the Connection Interface in WebIDL

```
interface PulsarConnection
{
        // event handlers
        attribute EventListener oninit;
        attribute EventListener onsync;
        attribute EventListener onfail;

        // connection
        void join(in DOMString host,
        in DOMString port,
        in DOMString session);
        void disconnect();

        // access shared objects
        Object getObject(in DOMString scope,
              in DOMString objId);
        void create(in DOMString scope,
              in DOMString objId,
              in Object object,
              in DOMString mode);
        void update(in DOMString scope,
              in DOMString property,
              in any newValue);
        void remove(in DOMString scope,
              in DOMString objId);

        // dead reckoned objects
        Object getDeadReckonedObject(
              in DOMString scope,
              in DOMString objId,
              [Optional] in Function adjFunc);
        void createDeadReckonedObject(
              in DOMString scope,
              in DOMString objId,
              in Object object);
};
```

14.4.4.1 Working with Shared Objects

The interface involves methods to create, remove, and update objects. Updating objects is only for optional use, since modifications are usually detected automatically. There are several ways of sharing an object:

■ Create a shared object from a local one using the create function. The object can be deleted again with the remove function.

- Retrieve a remotely shared object using the getObject function.
- For dead-reckoned objects, the two functions mentioned above are replaced with its dead-reckoning-enabled counterparts.

Once an object is shared, it is used like a local object. It can be read and modified without using special functions. The local copy is directly accessed. It is not necessary to use the update function. Local modifications on objects are automatically propagated toward the Pulsar server. Therefore, shared objects are observed by the synchronization system. This may be realized by using special getters and setters in a proxy object. Indeed, the implementation of an observation mechanism remains a concern of the implementation.

14.4.4.2 Integrated Dead-Reckoning System

For application developers, it is not necessary to implement their own dead-reckoning algorithms, because a generic one is already included in Pulsar. Given an approximation function, each client can calculate local values for a dead-reckoned object. Incoming updates for such objects are merged with the local values to avoid unnatural, choppy behavior. Therefore, a custom adjustment function is used. To save network bandwidth and computational power on clients and the server, it is possible to update dead-reckoned objects not in each round but, for example, only in every tenth round.

Dead-reckoned objects are naturally written only by one entity. This could be a client that owns that object, such as a vehicle or avatar in a game. Nevertheless, it could as well be the server, which controls a global object like a bullet. Remote entities (the clients not controlling the object) use the same algorithm to calculate continuous values. This actual computation of values for a dead-reckoned object is usually done by the application itself.

In the course of time, the values calculated individually drift apart. The dead-reckoning system corrects the values periodically, using the predefined adjustment function.

Imagine a distributed Web application synchronizing a ball moving at constant speed. One of the clients calculates its position and propagates it once a round to the other clients. The other clients, of course, make use of dead reckoning to be able to redraw the ball multiple times in each round. Since all clients use the same algorithm to calculate the ball's position, the divergence between the clients—and thus, the correction—is very small. It is obvious that there must be a correction from time to time. Otherwise, the ball's positions would drift apart after a few seconds. However, for many applications, it may be sufficient to synchronize, for example, once a second. For a supposed round time of 50 ms, this would be merely every 20th round. Round-based updating of values led to a delayed synchronization in the first instance. Consequently, flooding server and clients with exorbitant numbers of updates for continuously changing objects was limited. Restricting dead-reckoning correction extends this idea and avoids unnecessary data exchange.

14.5 Implementation

In this section, design details of the communication protocol as well as the architecture and implementation of the client and server components are presented.

14.5.1 Real-Time Synchronization Protocol

For communication between the Pulsar server and browsers, a lightweight, text-based protocol on top of UDP is used. To satisfy requirements of state synchronization, it is designed to distribute small amounts of data quickly to a large number of clients. Thus, the protocol focuses mainly on low-latency communication as well as data consistency, and less on reliability or throughput.

14.5.1.1 Message Format

Generally, each message consists of a header and a body. The header format is equal for all types of messages. As illustrated in Table 14.5, the header contains the following elements:

- Protocol: Name and version of the communication protocol used.
- Client ID: Identifier of the client application, to allow multiple clients with the same host address and port (e.g., multiple browser windows or tabs).
- Round number: Number of the current round. Rounds are numbered sequentially. The round sequence is linear, and allows the receiver to order messages and recognize message losses, and is also used to identify messages.
- Last received: Indicates the last message received before the current round. This is used to acknowledge received messages, thus being able to detect message loss.

While the header is equal for all types of messages, they differ in their body. There are four different types of messages used for communication, as explained in Section 14.4.2.2. Table 14.1 (sync-message), Table 14.2 (request-message), Table 14.3 (join-message), and Table 14.4 (init-message) show the exact format of the different message types.

14.5.1.2 Data Exchange Format

For exchange of Web objects with a text-based protocol, a text-based representation of data would be suitable as well. In the Web, two text-based representation formats

Table 14.5 Generic Message Format

Protocol Version	Client ID	Round No.	Last Received
Body			

are frequently used: the Extensible Markup Language (XML) (Sperberg-McQueen et al., 2008) and, recently becoming popular, the JavaScript Object Notation (JSON) (Crockford, 2006). Compared to XML, JSON provides several advantages such as its better compatibility with JavaScript as well as the more compact string representation, which saves bandwidth during network communication.

Consequently, using JSON as data exchange format is an appropriate choice. It is a powerful, Web-inherent, and approved technique for serialization of simple object structures to a compact plain text format, which is easily readable for humans. Objects are well represented, due to the hierarchic structure of JSON and its support of basic primitive data types. JSON is supported by a huge number of programming languages and may efficiently be parsed, due to its simple structure. It is well integrated with JavaScript, which is supported by all major browsers and frequently used in Web applications.

Optionally, modifications on objects can be restricted by defining a schema (Zyp, 2008) for received JSON strings. Thus, incoming data can be validated even before they are applied. Among others, this allows predefined object structures, fixed property types, and verification of values by using enumerations or ranges for numbers. Schemes can be individually defined for each object or a set of objects.

It would obviously be a bad idea to serialize and transmit the entire object every time a single property changes, especially for large objects. There are two different strategies to transfer object changes, described in the following.

The object difference strategy detects distinctions between two objects and subsumes them into a new object representing the difference. The greatest advantage of transmitting object differences is their atomic character. Numerous properties can be changed in a single request with transactional behavior. If one property update fails (e.g., due to a conflict), the whole request is abandoned.

To determine the changes made to an object, the original and the changed object are examined. Both object versions are compared recursively, also descending into nested objects. Changed as well as newly added property values are copied to a separate object structure, which represents the difference. This difference is serialized to a JSON string.

With this technique, nearly all kinds of changes can be modeled, regardless of whether just a single property or the whole object has changed. Even new properties may be inserted, if this is permitted. Only removing of properties is not possible, since JSON does not support an undefined value, which would be a suitable representation for a property to be deleted. Furthermore, manipulating arrays is limited to either replacing the whole array or using an array-like object (numbers as property names addressing indices), which can change, as well as adding particular entries. However, with this approach, it is impossible to insert or remove individual elements or to shift the array.

Another approach directly addresses a changed property's path. This path identifies a single, possibly nested property using its and its parent's names, each separated with a dot. Thus, in each request, only one property can be updated. Apart from losing the transaction character, this technique has several advantages over object differences: needless conflicts may be avoided due to more subtle addressing, properties to be deleted can be individually identified, and distinct array elements may be modified.

To take advantage of both methods, a combined approach is actually used. Complete objects as well as single properties can be addressed by their path. Addressed properties may also be nested objects. A transmitted update value is allowed to be an object difference (represented through a valid JSON string) or a replacement for a primitive value.

Table 14.6 shows an example of serialization with all three suggested approaches. It relates to the example object serialization in Listing 14.2 and the modification of it in Listing 14.3.

14.5.1.3 Reliability

The primary goal was to create a fast and lightweight synchronization protocol. Consequently, UDP was preferred over TCP, abdicating on transmission control. Nevertheless, some of the TCP core features were implemented with lightweight replacements.

Table 14.6 Example of Data Exchange Representations with Different Strategies

Strategy	Addressing	New Value

Object difference	obj	{
		    sub: {
		       a: changed!,"
		       c: true
		    },
		    items: [1, 777, 3]
		 }

Property addressing	obj.sub.a	"changed!"
	obj.sub.c	true
	obj.items[1]	777

Combined approach	obj.sub	{
		       a: "changed!,"
		       c: true
		 }

	obj.items[1]	777

Listing 14.2   JSON Serialization of Example Object

```
var obj = {
      prop: "Hello world!",
      sub: {
              a: "test",
              b: 123.45,
              c: false
      },
      items: [ 1, 2, 3 ]
}
```

Listing 14.3   JSON Serialization of Modified Object

```
var obj = {
      prop: "Hello world!",
      sub: {
              a: "changed!",
              b: 123.45,
              c: true
      },
      items: [ 1, 777, 3 ]
}
```

Ascending round numbers allow detection of packet loss, duplicates, and out-of-order packets. Additionally, message headers contain a field specifying the last message received. Knowledge of the application allows smart solutions instead of the generic TCP strategies like retransmission of lost packets.

The Pulsar server keeps track of updates sent to any client within the last round. These updates may differ from client to client because of various sessions, different client scopes, and group affiliations. Lost messages may be detected by means of ascending round numbers. For the Pulsar server, it is yet sufficient to monitor whether clients reply within a round. Instead of resending the last message, contained updates will be merged into the next round's synchronization message. Consequently, a client receiving the latest synchronization message is definitely up to date and in a consistent state after processing contained updates.

Clients implement a different strategy: copies of sent request-messages are stored (if not empty), and may be resent in the case of not being acknowledged by the Pulsar server.

Duplicate and out-of-order messages are recognized by their round numbers and handled as late messages. Clients can just ignore late synchronization messages, since the contained updates were already merged into a more recent message, which apparently has been processed before.

In contrast, the Pulsar server accepts late messages. As far as possible, requests are applied, but conflicts with requests that are more recent are detected and the conflicting late request is discarded. Accordingly, clients with very poor or temporarily impaired network connections are disadvantaged. However, every attempt is made to keep the connection and accept requests.

14.5.2 Server Component

The Pulsar server is a separate entity, even though it is able to interact with a Web server. It maintains a shared state, processes modification requests from clients, and distributes resulting updates. As shown in Figure 14.5, the server component is subdivided into two mainly independent subsystems:

1. The communication subsystem receives and parses incoming request-messages, takes care of clients, and controls the round duration. It constructs outgoing synchronization messages from current state updates. Additionally, this subsystem manages reliable communication.
2. The synchronization subsystem maintains the shared state. It processes modification requests and is responsible for resolving conflicts to keep the shared state consistent. Applied requests are summarized to a list of resulting state updates, ready to be dispatched to clients.

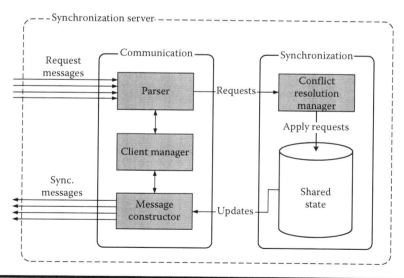

Figure 14.5 Reference architecture of the server component.

14.5.3 Client Component

The client component is embedded with Web browsers. This may be realized in the form of a browser extension or plug-in, as well as an incorporated implementation delivered together with the browser. On the one hand, this component interacts with the Pulsar server and is responsible for maintaining a replicated copy of the distributed shared state. On the other hand, synchronization features are offered to local Web applications. Figure 14.6 shows the reference architecture of the client component, including the three major subsystems:

1. Synchronization Service: A central synchronization service communicates with the server. It receives incoming messages, parses them partially, and delivers them to the corresponding endpoints.
2. Synchronization endpoint: For each synchronization connection, a separate endpoint object is created. Endpoints maintain a copy of the shared state. Incoming updates are parsed and applied to that state as well as delivered to Web applications. Modifications on the state are collected during a round. At the end of a round, requests are summarized and issued to the server.
3. Application programming interface: Web applications activate state synchronization through embedded scripts. Access to shared objects is synchronized automatically, and event handlers notify about incoming updates. Therefore, a sophisticated application programming interface is provided, which is

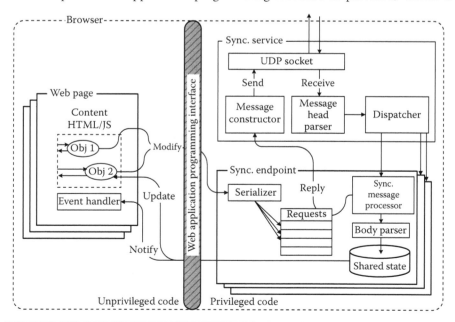

Figure 14.6 Reference architecture of the client component.

embedded directly into the browser's JavaScript libraries. Interaction with untrustworthy content code is a potential security risk and must be handled carefully. This interface is particularly described in Section 14.4.4.

14.5.3.1 Reference Implementation for Mozilla Firefox

A synchronization system for the Web requires a browser-embedded client for Web browsers, which provides a suitable application programming interface. On the other side, this cannot be achieved with a simple JavaScript library, for the very reason that access to sockets on a low level is needed.

There are multiple ways to embed such a system into existing browsers. The obvious way is to take an open source browser and modify its source code. However, this would involve several disadvantages. In the first place, it would break compatibility with future releases, and needs all modifications to be ported to new versions. Secondly, it requires recompilation and reinstallation of the entire browser. Thereby, the possibility to simply distribute the software would be lost.

A cleaner approach is offered with browser plug-ins, extensions, and other add-ons. Most major browsers provide interfaces to add-on components. For example, Mozilla's Web site (at http://addons.mozilla.org) hosts more than 6000 Firefox extensions. In contrast to plug-ins, Firefox extensions are much more integrated to the browser. Moreover, an extension is installed with just a few mouse clicks. The installation process may be started directly from a Web site. Thus, a sample implementation is developed for Mozilla Firefox, which is one of the most prevalent Web browsers.

Following the specification defined above, the implementation comprises three major subsystems: the synchronization service interacting with the server, the endpoint component maintaining the shared state, and the Web application programming interface. Figure 14.7 shows an overview to the client architecture in a component diagram.

Using the XPCOM model, the subsystems of the client architecture are realized as components. Thus, utilization of different programming languages, namely JavaScript and C++, for implementation is possible. Although the C++ components generally ought to be compilable on diverse platforms, this is out of scope. Hence, development was delimited to the Windows systems in the first place.

14.6 Experimental Validation

In this section, the concept of Pulsar is evaluated regarding performance, fairness, and scalability. In particular, measurements are supposed to determine synchronization delays, analyze fairness of conflict resolution, and examine the scalability of Pulsar. Therefore, the client and server implementations presented in Section 14.5 are used.

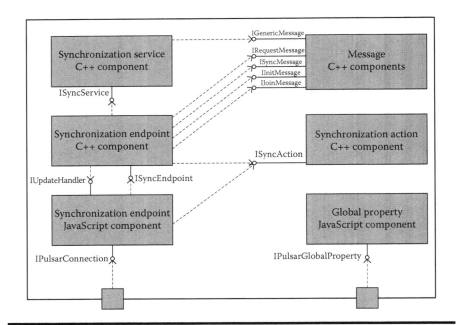

Figure 14.7   Overview of the client architecture.

Usually, it would be appropriate to compare a concept to state-of-the-art technologies. However, the presented approaches in Section 14.2 differ too much from the proposed concept for a meaningful comparison. On the one hand, we have technologies for bidirectional client–server communication in the Web; on the other hand, we have a client-to-client synchronization system. Implementation of a similar system with state-of-the-art server-push Web technologies for the purpose of comparison would evidently be biased, is very complex to realize, and is definitely not within the scope of this thesis.

Consequently, this evaluation does not comprise comparative measurements. Rather, measured values of delays and fairness are analyzed with regard to the expected results.

14.6.1 Setup of a Test Environment

The test bed consists of a server, a software router, and multiple client machines. Figure 14.8 shows its network architecture. The actual hardware and software configurations of each machine are specified below:

■ Adenin: Hosting the Pulsar server and a Web server. Running on Windows Server 2008 64 bit with 8 GB of RAM, Intel Xeon 2 GHz quad core CPU, and two network adapters.

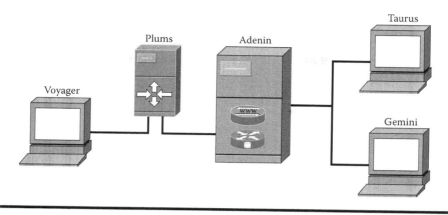

Figure 14.8 **Network architecture of the test bed.**

- Plums: Routing PC used for simulating artificial network effects (iptables in combination with a proprietary program). Running on Ubuntu Linux 9.04 32 bit with 1 GB of RAM, Pentium 4 3,2 GHz CPU, and two network adapters.
- Taurus: Client machine. Running on Windows XP 32 bit with 2 GB of RAM, and Intel Pentium 4 2.8 GHz CPU.
- Voyager: Client machine. Running on Windows XP 32 bit with 2 GB of RAM, and AMD Athlon 64 2.2 GHz CPU.
- Gemini: Client machine. Running on Windows Vista 32 bit with 2 GB of RAM, and Intel Core 2 Duo 2.4 GHz dual core CPU.

14.6.1.1 Simulated Network Effects

The test bed is subdivided into two networks, which are connected through a router. The router (Plums) is running on a Linux operating system and routes traffic between the two networks. Network packets are routed with iptables to a special user space program. This program simulates diverse network effects:

- **Latency**: Each packet is delayed for a certain amount of time, simulating network latency.
- **Jitter**: Network latency varies randomly within a specified range, simulating jitter.
- **Packet loss**: Random packets are dropped with a specified probability, simulating loss of data.

One of the client machines (Voyager) is separated from the Pulsar server by this router (Plums), so that measurements can be influenced by artificial network effects. Since the values of these effects are adjustable, series of measurements with increasing values are possible. Specified values are always related to a complete server–client–server round trip.

14.6.2 Test Procedure

For evaluation of the system, a distributed Web application was implemented using Pulsar. It consists of an arbitrary number of client instances (slaves) as well as a controlling instance (master). From Pulsar's point of view, all of them are clients. To avoid confusion, the terms "slaves" and "master" are used. Figure 14.9 shows the Web application deployed to the test bed.

The Web application instances share an object holding a single value. A simple procedure is used for evaluation, consisting of the following events:

1. The master client starts the procedure by writing a specific value (token) to the shared object. The actual time is measured (T_0).
2. All clients receive the updated value. The master takes the time (T_1). As soon as a slave is notified about the update containing the token value, it replies by writing its unique ID to the shared object.
3. Since all slaves receive the token in the same round, they are modifying the shared object concurrently and their requests are conflicting. Only one of the slave requests will succeed.
4. All clients again receive the updated value. The master logs which slave request has been successful and takes the time (T_2).

A sequence diagram visualizing the procedure is shown in Figure 14.10. The arrows symbolize sync-messages and request-messages. They are labeled with the value changes of the shared object. For simplicity, only two clients—the master and a slave—are shown. In reality, at least two slave clients are participating, so that replies are conflicting.

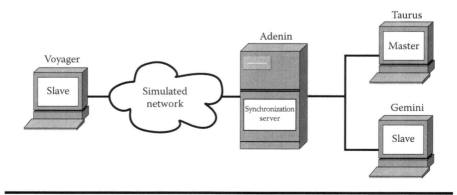

Figure 14.9   Evaluation Web application deployed to the test bed.

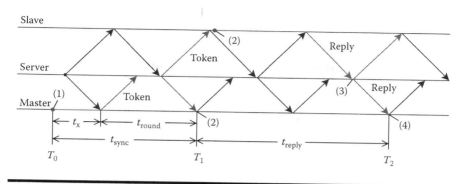

Figure 14.10   Sequence diagram of the test procedure.

14.6.2.1 System Configuration

The Pulsar server is configured to use the random conflict resolution policy. Thus, an even distribution of success among the clients is expected. The minimum round duration is set to 10 ms to avoid too much load on the client machines tainting the results. The maximum round duration is set to 125 ms.

The evaluation Web application is configured to repeat each test 1000 times. Hence, the average values of the results ought to be statistical meaningful.

In the first place, the impacts of the simulated network effects on the measurements are determined. Additionally, scalability tests are performed. In particular, the following parameters are tuned in test runs:

■ Latency is varied within the range of 10–120 ms.
■ For a fixed latency of 50 ms,
  – jitter is varied between ±0 and ±40 ms
  – packet loss rates are varied between 0% and 50%.
■ For scalability tests, the number of clients is increased in steps of 500 clients until the system becomes too slow.
■ A second scalability test is performed with varying request rates, increasing the system load.

14.6.3 Measurements

As mentioned in the description of the test procedure, the master client logs the time and also slave updates at various points in time. Furthermore, the traffic is logged with Wireshark on each machine, allowing additional analysis. For detailed examination of the tasks done by the Pulsar server, values logged by the server can be used. Thus, the duration of the individual server phases can be determined. By means of the measured values, the following aspects are examined.

14.6.3.1 Synchronization Delay

The synchronization delay denotes the time between triggering a synchronization request (i.e., writing to a shared object) and notification about the corresponding update. Figure 14.10 illustrates the synchronization delay as

$$t_{sync} = t_{round} + t_x + t_c$$

where

t_{round} is the round duration
t_x is the remaining time of the actual round after a request was triggered

This value is variable limited by the round duration, so that $0 \leq t_x \leq t_r$ is valid.

The last value in the equation above, t_c, is a small constant, specifying the delay caused by the client implementation. It is omitted in Figure 14.10, because it is very small compared to the other delays. The client delay mainly comprises the duration of processing a request-message, including parsing and applying contained updates as well as notifying the client Web application. The round duration is not affected by this delay, since synchronization messages are replied to immediately before the message is processed, as stated in Section 14.4.2.

The time elapsed until another client has replied to a request, t_{reply}, is also illustrated in Figure 14.10. It is a special case of the synchronization delay, where a request is triggered by an update notification in the beginning of a round. Thus, it is exactly the double round time:

$$t_{reply} = 2 \cdot t_{round}$$

The sum of the both delays defined above is the time needed for two-way client-to-client communication:

$$t_{echo} = t_{sync} + t_{reply}$$

By means of measured time values (T_0, T_1, and T_2), the periods described above are determined, as shown in Figure 14.10. The synchronization delay is the measured duration between issuing the token (event 1) and notification about the corresponding update (event 2):

$$t_{sync} = \Delta(T_1, T_0) = T_1 - T_0$$

The two-way communication delay is the duration between issuing the token (event 1) and notification about the responding slave update (event 4):

$$t_{echo} = \Delta(T_2, T_0) = T_2 - T_0$$

14.6.3.1.1 Hypothesis

The results of the measurements are average values based on a large number of indicated values. Hence, it is assumed that t_x approximates to its probabilistically expected value $E(t_x) = t_{round}/2$. Thus, a synchronization delay of

$$t_{sync} = 1.5 \cdot t_{round} + t_c$$

is expected. Consequently, the time for client-to-client communication can be calculated with

$$t_{sync} = 3.5 \cdot t_{round} + t_c$$

The current implementation is estimated to cause a roughly constant delay t_c of less than 5 ms, depending on hardware configuration and actual load. The round duration t_{round} depends on the variable network parameters as described in the following section.

14.6.3.2 Round Duration

The concept of Pulsar involves adaptive round times. Therefore, the round duration is highly dependent on network latency. The server generally waits for slow clients to reply if a maximum delay is not exceeded. Consequently, jitter and message loss influence the round duration as well. The latter one may even cause the server to wait until the waiting time limit is reached. However, clients known to be too slow (respectively losing to many messages) are ignored.

The round duration is constituted from several components: the server processing delay, the client processing delay, and the network latency. The last of these comprises the entire round trip time between client and server, because a message is sent within a round in both directions.

14.6.3.2.1 Hypothesis

Since all measurements (not including the scalability test) involve only a small number of clients (exactly three for the described setup), the client and server delays can be neglected. Thus, the round duration is mainly determined by network latency. Since jitter is randomly positive or negative signed, it increases or decreases latency to the same extent. Hence, jitter is not discoverable in the resulting average values and it is expected that

$$t_{round} = t_{latency}$$

is valid. If additional message loss appears, it may increase the duration of individual rounds. However, this effect is hardly discoverable for average values. As soon as a single client loses too many messages, the server stops waiting for it. Thus, the impact on round durations is capped.

14.6.3.3 Fairness of Conflict Resolution

As described in Section 14.6.2, a master client provokes synchronization conflicts. Two slave clients concurrently request the same resource. The master client logs the distribution of prevailing requests. A statistic of success in conflicts can be derived for each client. Evenly distributed success indicates fair behavior of the conflict manager. With two participating slave clients, a resulting success of 50% is expected for each client. A deviation is assumed to become statistical significant when it differs by more than 5%. Hence, resulting values ought to be in the interval between 47.5% and 52.5%.

14.6.3.3.1 Hypothesis

Latency does not affect fairness. The Pulsar server waits until all request-messages have arrived. Afterward, the conflict manager resolves conflicts without consideration of arrival times. However, late messages are ignored in conflict resolution. Consequently, a client can no longer win conflicts as soon as its network delay exceeds the wait limit. An even distribution is expected for latency values below the wait limit.

The same applies to jitter. As long as requests arrive in time, no effect should be discoverable. The server is expected to behave fairly.

The behavior is entirely different when messages are lost. Though a lost request-message is detected and resent, it will arrive too late, and is ignored in conflict resolution. Consequently, the percentage of successful requests should decrease to the same extent as message loss increases for a client.

14.6.3.4 System Scalability

Scalability tests in a centralized system mainly bear on the server component. Though the client implementation is concerned about the number of incoming updates, it does not care for the total number of clients connected to the system. The server is influenced by both aspects: the rate of incoming requests and the number of connected clients.

A special setup is needed to run the Pulsar server with a huge number of clients. It is not possible to run 100 or even 1000 browser instances on a few machines. Even though it may be possible to open a large number of synchronization connections from

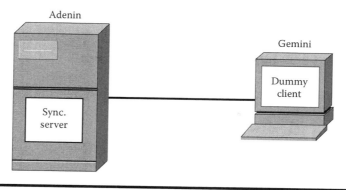

Figure 14.11   Changed setup of the test bed for scalability measurements.

a single Web application, parallel notifications on all connections would dramatically slow down the browser. This setup would not comply with a normal usage of Pulsar.

Indeed, a simple dummy client was implemented for test runs with a large number of participating clients. It is a lightweight console application, missing many features. However, it acts like a normal client from the server's point of view. A single instance is able to open and concurrently maintain a large number of synchronization connections. As a special feature, it allows extensive usage of protocol built-in operations, such as updating objects or changing groups, to stress-test the Pulsar server.

The test setup is slightly changed for this measurement, as shown in Figure 14.11. The regular clients are replaced by a single dummy client instance. It runs on Gemini—the most powerful client machine. For detailed examination of the tasks done by the Pulsar server, values logged by the server can be used. Thus, the duration of individual server phases (see Section 14.4.2) is determined.

14.6.3.4.1 Hypothesis

Phase I is completely determined by the delay of client responses. This phase is thoroughly examined with the previous measurements and will therefore not be considered here. More interesting are phases II and III, which involve server activity.

Phase II comprises conflict detection and resolution as well as application of synchronization requests to the shared state. The duration of phase II is expected to grow slightly with an increasing number of clients. The number of synchronization requests should have a greater impact on the duration of that phase. It is expected to grow linearly with an increasing number of requests.

Phase III is influenced by both effects as well. In phase III.1, updates are summarized to individual synchronization messages. It is expected to last longer with more clients as well as with an increasing number of requests. Phase III.2 comprises only sending of the synchronization messages. Its duration is expected to grow to the same extent as the number of clients increases. The number of requests may increase the duration of phase III.2 slightly, since more requests imply more data to be sent. However, sending of messages should not take as much time as the preceding phases.

14.6.4 Analysis of Results

In the following sections, the measured values are analyzed and compared to the expected values, stated in the hypotheses above.

14.6.4.1 Synchronization Delay

The measured delays for synchronization (t_{sync}) and two-way communication (t_{echo}) between master and slave clients with increasing latency are illustrated in Figure 14.12. Additionally, the measured values are supplemented with error bars, indicating the difference of individual values to the average. On the x-axis, the amount of latency added to the round trip time between client (Voyager) and server (Adenin) is specified. In fact, that value is approximately equal to the overall round trip time, due to the very low latency in the local area network.

Obviously, the variance of measured values is notably great, as indicated by the error bars. This can be explained by considering the varying t_x values mentioned above. Since token requests are triggered randomly during a round, t_x fluctuates

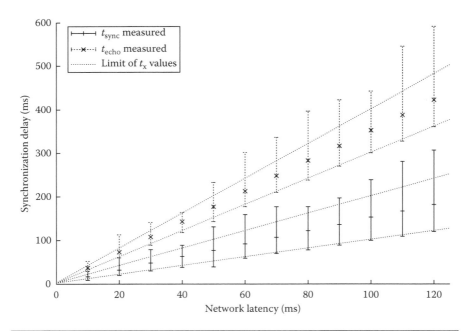

Figure 14.12   Average values and variance of measured synchronization delays.

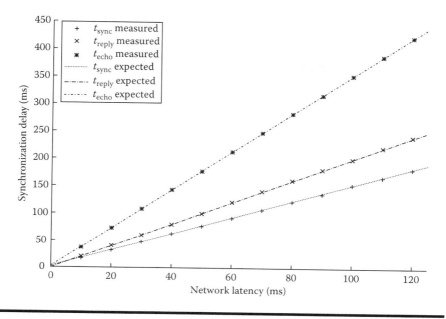

Figure 14.13 **Synchronization delay of individual intervals.**

between 0 and the actual round duration. With the dashed lines, the limits of t_x dispersion are visualized. The lower edge of that dispersion appears to conform to the variance of measured values. Hence, the negative variance is entirely explainable with this effect.

On the upper edge, the variance is exceeding the t_x dispersion. However, this additional variance does not appear to influence the average values. Consequently, it is considered to be induced by a couple of outliers. A reason for that may be high computational load on a client machine or a temporary congestion on the network.

Figure 14.13 shows the measured delays overlaid with the respective theoretically expected values. Besides the measured delays for synchronization and two-way communication, additionally the reply synchronization delay (t_{reply}) is illustrated. It is calculated out of the first two values. Evidently, the measured values perfectly match expectations.

14.6.4.2 Client Delay

Apparently, the graphs of the expected t_{sync} and t_{echo} functions are slightly displaced in the y-dimension. This is not surprising with regard to the function definitions in Section 14.6.4.1, both containing a constant value t_c. This value denotes the client delay. It comprises the processing of a synchronization message, including the notification of client Web applications.

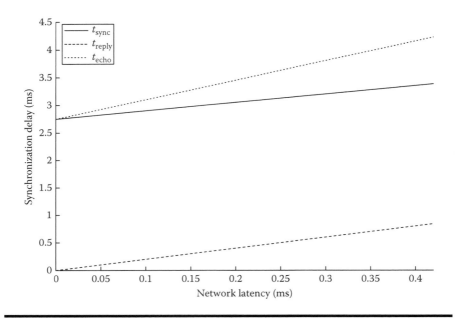

Figure 14.14 Zoom to intersections with *y*-axis showing suspension.

The actual value of t_c was determined by fitting the functions independently to their corresponding measured values. For both functions a value of

$$t_c = 2.75 \text{ ms}$$

has been achieved. Zooming in to the intersection between the functions and the *y*-axis illustrates this result, as done in Figure 14.14.

14.6.4.3 Round Duration

Out of the delays, the average round duration experienced on a client machine has been calculated. Additionally, the round duration on the server was measured independently. Figure 14.15 shows the results of both measurements mapped onto the expected results, with increasing network latency. Obviously, the measured values on client and server match the expected outcome well. Thus, the round duration for a small number of clients is mainly determined by the round trip time between the slowest client and server, as stated in Section 14.6.4.3.

The impact of jitter and packet loss to round durations is shown in Figure 14.16. While latency remained constant at a value of 50 ms, jitter and packet loss were varied. As expected, jitter has no discoverable impact on the average round duration. Since jitter alters network latency upward as well as downward to the

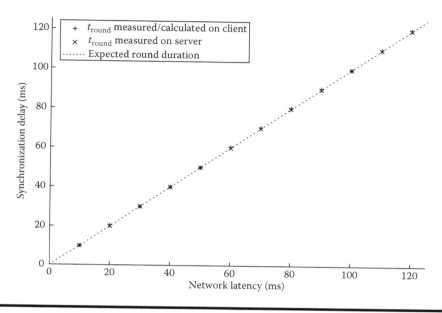

Figure 14.15   Measured and expected round duration.

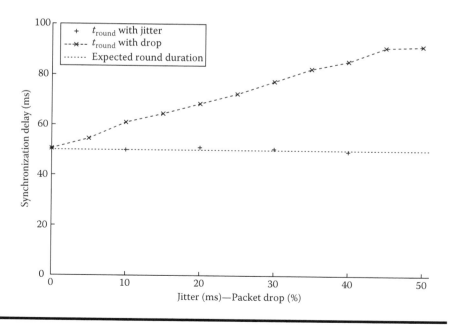

Figure 14.16   Impact of jitter and packet loss to round durations.

same extent, its average is equal to zero. Consequently, jitter is not discoverable in the average values.

Contrary to the hypothesis, packet loss has substantial influence on the average round duration. That may be explained by the fact that each lost message causes individual rounds to last until the maximum round duration is reached. Besides, the applied probabilities for packet loss go up to 50%, which is very high and more unlikely to appear in the real world.

14.6.4.4 Fairness of Conflict Resolution

In this measurement, the individual success of clients on conflicting requests was compared. Figures 14.17 through 14.19 show the percentage of success of the client Voyager in conflicts. This client was exposed, one after another, to an artificially raised network latency, varying jitter values, and packet loss, reflected on the x-axis of the respective figures.

Figure 14.19 shows the measured results for different latency values. All of them are within the expected interval. Thus, the hypothesis turned out to be true. The fact that one of the clients was exposed to higher network latency did not influence the results. As soon as the latency grew higher than the specified maximum round duration of 125 ms, the messages of the delayed client arrived late and underlied in conflicts.

As seen in Figure 14.18, jitter also turned out to have no impact on fairness of the synchronization. As expected, the success on conflicting requests stayed at approximately 50% for the client Voyager.

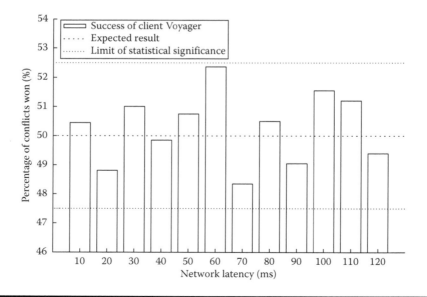

Figure 14.17   Distribution of success on conflicting requests.

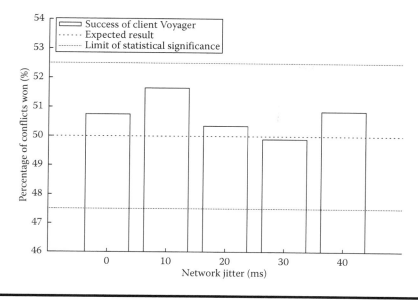

Figure 14.18   Impact of jitter on conflict resolution.

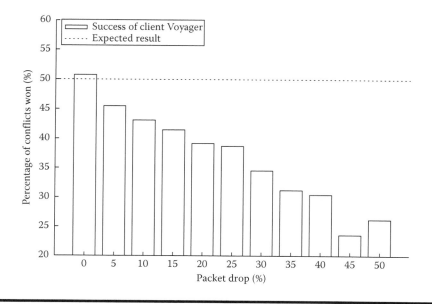

Figure 14.19   Impact of packet loss on conflict resolution.

In contrast to that, packet loss clearly influences the success of an affected client. Figure 14.19 shows the decreasing number of won conflicts with a growing packet loss rate. Since a lost request-message means a sure loss of a test run, the rate of lost messages is directly reflected in the success rate. The remaining test runs are still equally decided.

The values exceeding a 30% drop rate appear to be slightly irregular. For those values, the number of test runs was partly smaller than intended, caused by repeated suspending of client Voyager due to the huge number of lost messages.

In conclusion, Pulsar and its conflict-resolving strategy can be considered fair, as long as latency remains in a reasonable range and messages are not lost.

14.6.4.5 Scalability against Number of Clients

The first scalability measurement was intended to determine the maximum number of clients concurrently connected to the Pulsar server. The load caused by synchronization requests was limited to a minimum. Hence, the server was mainly engaged in sending and receiving empty messages. Actually, clients did not trigger any requests at all. However, a single client would be able to synchronize data continuously without changing the results significantly. This behavior would apply to a use case where a great majority of clients is just listening to updates.

Figure 14.20 shows the effective round durations broken down to processing phases with an increasing number of clients. The number of clients significantly

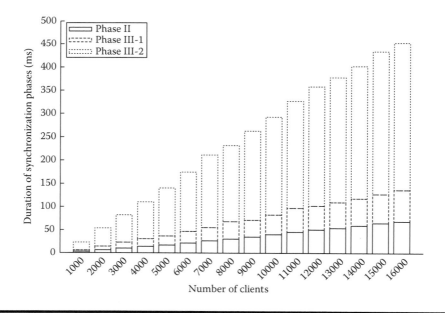

Figure 14.20   Duration of processing phases without load.

influences the effective round times. The diagram omits phase I (waiting for replies), since it only reflects performance of the dummy client and is not meaningful for the scalability test. The duration of phase II grows linearly, but remains reasonable low. This matches the expectation, since there were no requests to process. However, this phase needs still a notable amount of time. Examination of empty messages takes some time as well.

Most of the time is consumed in phase III. In the first part of this phase, individual synchronization messages are constructed for each client. Thus, a linear growth of phase III.1 with an increasing number of clients is comprehensible. The same argument applies to the second part of that phase, where the constructed messages are conveyed. Nonetheless, it is astonishing that phase III.2 lasts as long as the entire other phases. Message sending consumes roughly half of the server processing time.

In contrast to the first test run, a second one was done with a medium load. In every round, each client triggered a request with a probability of 1%. Even though that may not appear to be much in the first place, a huge number of requests may result in total, depending on the number of clients. Moreover, rounds are quite short. Consequently, for a round duration of 50 ms the resulting probability to trigger a request within one second would be 20%.

Figure 14.21 shows the results of this repeated test run. Evidently, phase III.2 is no longer prevailing. Both phase II and III.1 increased. Contrary to expectation, phase II is still not dominating. Conflict resolution and applying of requests seems to proceed rapidly, compared to message construction and sending. In

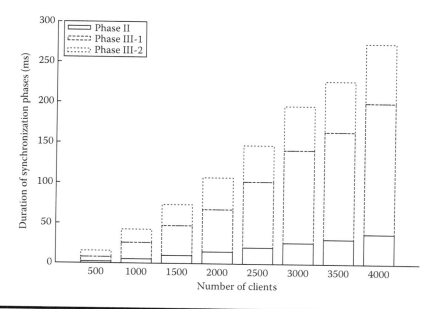

Figure 14.21   Duration of processing phases with load.

particular, phase III.1 (message construction) increased in this run, and now consumes the largest amount of time.

14.6.4.6 Scalability against Number of Requests

In a further scalability measurement, the time consumed by synchronization phases was determined depending on the number of requests per round. In contrast to the previous measurement, the actual incoming requests are now counted. However, the number of clients is still relevant, since each incoming request has to be delivered to all participating clients.

Figure 14.22 shows the duration of synchronization phases with an increasing request rate and a constant number of 500 clients. As measured values are continuous, the results are rendered as curves instead of showing a histogram. As done before in the histograms, the values of preceding phases are subsumed so that the curves are stacked.

In this measurement, phase III.1 (construction of messages) clearly lasts longer than phases II and III.2. Actually, it grows exponentially with an increasing number of requests. Even though the complexity of message construction depends on both the number of clients awaiting a message and the number of contained updates, this behavior may be caused by an implementation problem. Anyway, a linear growth has been expected. Nevertheless, the results are reasonable for a huge number of connected clients.

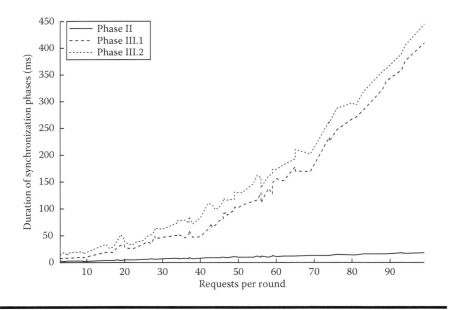

Figure 14.22 Duration of processing phases for 500 clients with increasing request rate.

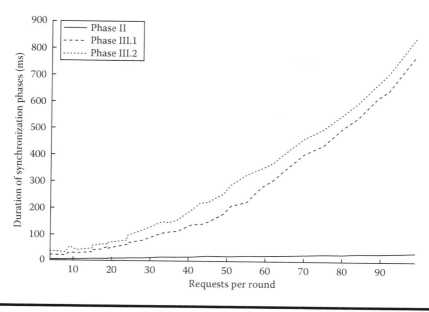

Figure 14.23 **Duration of processing phases for 1000 clients with increasing request rate.**

Increasing the number of clients to 1000 shows similar results, displayed in Figure 14.23. As expected, the measured values are twice as large as previously.

14.6.5 Summary

The evaluation omitted a comparison to state-of-the-art approaches of bidirectional Web communication, which are limited in plain client–server messaging. The difference to Pulsar is too large for a meaningful comparison, as explained at the beginning of this section.

Several measurements with a reasonable small number of clients showed that the effective round durations stayed within the expected range. Round duration appeared to be approximately equal to the round trip time between server and slowest client. If clients exceed a specified latency limit, they are suspended and no longer affect round duration. Furthermore, it turned out that jitter may delay individual rounds, but does not affect the average round duration. On the contrary, packet loss affects the round duration.

Pulsar proved to be fair. Regardless of network latency and jitter, clients had the same chance to be the first one reacting on delivered updates. Conflicting requests were resolved in a fair manner. As expected, packet loss appeared to have a negative effect on success in conflicts. This behavior is intentional, avoiding late (or resent) messages to change shared values afterward.

Measurements with a large number of clients showed that server-side processing time depends on the number of clients and the rate of issued requests. The 4000 concurrently participating clients are processed in 120 ms without load, or respectively 280 ms with a reasonable request rate of 1%. The Pulsar server handles 16,000 clients in a single session within 500 ms.

14.7 Mobile Port to Android

The concept of synchronization followed the goals of low traffic and lightweight implementation. Consequently, Pulsar already satisfies the main requirements for running on a mobile device. Nevertheless, a new client implementation is necessary to run on Android, since neither Firefox nor the Pulsar extension is suitable.

The Android browser does not provide any extension or plug-in model available in the standard Android Software Development Kit. We derive our own Web browser, based on the Web runtime component of Android. Through the Java programming interface, the runtime environment is extended with the Pulsar client interface. For the Web application itself, there is no difference between the Android implementation and the Firefox reference implementation.

14.8 Application Examples

To demonstrate the power of Pulsar, we implemented several sample applications. This section introduces the most interesting applications in brief.

14.8.1 Pong

A simple demo application for Pulsar is the well-known game, Pong. We implemented a browser version of this game using SVG and JavaScript. Data exchange between clients for synchronizing the game state is realized thoroughly with Pulsar. Communication with a Web server is not necessary. In detail, each client (respective player) owns shared objects representing information such as the player's name and current score. The other clients are able to get this data by accessing these shared objects. Additionally, Pulsar synchronizes the ball and the paddles of each client as separate objects. Since those objects move continuously, their synchronization utilizes the built-in dead-reckoning features. More precisely, the paddle and ball positions are propagated only occasionally. Each client interpolates the intermediate positions independently.

14.8.2 MovieStar

Combining mobile and fixed devices, MovieStar is a great example for the diversity of Pulsar. Similar to karaoke in music, MovieStar gives the player a role in his favorite movies. Mobile devices of the players are used as cameras and for displaying storyboards. Players become actors or camera operators. As a result, they create

Figure 14.24   Recording scene with Android mobile phone while playing MovieStar.

new interpretations of famous movie scenes. A desktop computer or Web-enabled television serves as screen for presentation of the produced short films. Figure 14.24 shows two players recording a scene with an Android mobile phone.

Pulsar synchronizes all participating devices. Shared objects hold the game state to control the application flow. All devices display the same (or a suitable) view at the same time. For example, the camera operator's mobile uploads movies to YouTube and publishes its URL through Pulsar.

14.9 Conclusion

We presented a concept for low-latency synchronization of real-time multi-user Web applications. A server component was specified, which maintains the shared state and resolves conflicts in a fair manner. The suggested communication protocol is round based, using adaptive round times. Although it mainly focuses on fast data exchange, basic reliability features are included.

Shared application states are quickly synchronized between browsers. Conflicts are resolved automatically, while clients are treated in a fair manner. Pulsar is Web-inherent, embedding seamlessly into browsers. It provides a sophisticated application programming interface on a high level of abstraction, enabling Web developers to create distributed Web applications easily.

Embedded client software was implemented for the Firefox browser, in the form of a browser extension. The implementation strictly followed the suggested specification, including all features of the proposed concept. Several prototype applications proved stable and fluid to run.

The evaluation showed that the proposed concept of a synchronization system works as expected. Though the synchronization is not done in real time, it is performed with a low latency. The delay of synchronization remained within the expected range. The round duration mostly depends on the round trip time between clients and server, determined by network latency. Jitter has no effect on the average round duration, but message loss may lengthen rounds.

The system turned out to be fairly independent of network latency or jitter. During conflict resolution, no particular client was preferred. Furthermore, it proved to be scalable within a reasonable range. The 4000 concurrently participating clients are processed in 120 ms without load, or respectively in 280 ms with a medium load. A total number of 16,000 clients were reached during testing, consuming less than 500 ms of processing time.

Pulsar integrates seamlessly with browsers as well as Web servers. As proved in the evaluation, the data transmission is reasonable fast, clients are handled fairly, and the shared data remained in a consistent state during all tests.

References

Bailey, R. W. *Human Performance Engineering: A Guide for System Designers*. Englewood Cliffs, NJ: Prentice Hall Professional Technical Reference, 1982.

Baughman, N. E., M. Liberatore, and B. N. Levine. Cheat-proof playout for centralized and peer-to-peer gaming. *IEEE/ACM Transactions on Networking*, 15: 1–13, 2007.

Bernier, Y. W. Latency compensating methods in client/server in-game protocol design and optimization. In *Proceedings of the 15th Game Developers Conference*, San Jose, CA, 2001.

Crockford, D. The application/json Media Type for JavaScript Object Notation (JSON). IETF, 2006.

Cronin, E., A. R. Kurc, B. Filstrup, and S. Jamin. An efficient synchronization mechanism for mirrored game architectures. *Multimedia Tools and Applications*, 23: 7–30, 2004.

Färber, J. Network game traffic modelling. In *Proceedings of the 1st Workshop on Network and System Support for Games*, ACM, Braunschweig, Germany, 2002, pp. 53–57.

Fielding, R., J. Gettys, J. Mogul, H. Frystyk, L. Masinter, P. Leach, and T. Berners-Lee. Hypertext transfer protocol—HTTP/1.1. RFC 2616 (Draft Standard), June 1999. Updated by RFC 2817.

Flynn, M. J. *Computer Architecture: Pipelined and Parallel Processor Design*, 1st edn. Sudbury, MA: Jones & Bartlett Publishers, Inc., 1995.

Fujimoto, R. M. *Parallel and Distributed Simulation Systems*. New York: John Wiley & Sons, Inc., 1999.

GauthierDickey, C., D. Zappala, V. Lo, and J. Marr. Low latency and cheat-proof event ordering for peer-to-peer games. In *Proceedings of the 14th International Workshop on Network and Operating Systems Support for Digital Audio and Video*, ACM, New York, 2004, pp. 134–139.

Gautier, L. and C. Diot. Design and evaluation of MiMaze, a multi-player game on the Internet. In *Proceedings of the IEEE International Conference on Multimedia Computing and Systems*, IEEE Computer Society, Washington, DC, 1998, p. 233.

Grosskurth, A. and M. W. Godfrey. A reference architecture for Web browsers. In *Proceedings of the 21st IEEE International Conference on Software Maintenance*, IEEE Computer Society, Washington, DC, 2005, pp. 661–664.

Guo, K., S. Mukherjee, S. Rangarajan, and S. Paul. A fair message exchange framework for distributed multi-player games. In *Proceedings of the 2nd Workshop on Network and System Support for Games*, ACM, New York, 2003, pp. 29–41.

Henderson, T. Latency and user behaviour on a multiplayer game server. In *Proceedings of the Third International COST264 Workshop on Networked Group Communication*, Springer-Verlag, London, U.K., 2001, pp. 1–13.

Hickson, I. HTML 5. W3C Working Draft, W3C, 2009.

Hickson, I. Server-sent events. W3C Working Draft, W3C, 2009.

Hickson, I. The WebSocket protocol. Standards Track, IETF, 2009.

IEEE standard for distributed interactive simulation communication services and profiles. IEEE Std 1278.2-1995, 1996.

IEEE standard for information technology—Protocols for distributed interactive simulations applications. Entity information and interaction. IEEE Std 1278-1993, 1993.

Jefferson, D. R. Virtual time. *ACM Transactions on Programming Language and System*, 7: 404–425, 1985.

Kesteren, A. V. The XMLHttpRequest Object. Working Draft, W3C, 2008.

Kung, H. T. and J. T. Robinson. On optimistic methods for concurrency control. *ACM Transactions on Database System*, 6: 213–226, 1981.

Lentczner, M. and D. Preston. Reverse HTTP. Internet-Draft, IETF, 2009.

Mauve, M. Consistency in replicated continuous interactive media. In *Proceedings of the 2000 ACM Conference on Computer Supported Cooperative Work*, ACM, New York, 2000, pp. 181–190.

McCormack, C. Web IDL. Working Draft, W3C, 2008.

Mills, D. Network time protocol (Version 3) specification, implementation and analysis. IETF, 1992.

Pantel, L. and L. C. Wolf. On the impact of delay on real-time multiplayer games. In *Proceedings of the 12th International Workshop on Network and Operating Systems Support for Digital Audio and Video*, ACM, New York, 2002, pp. 23–29.

Paterson, I., D. Smith, and P. Saint-Andre. XEP-0124: Bidirectional-streams Over Synchronous HTTP (BOSH). Standards Track, XMPP Standards Foundation, 2008.

Protic, J., M. Tomasevic, and V. Milutinovic. Distributed shared memory: Concepts and systems. *IEEE Parallel and Distributed Technology: Systems and Technology*, 4: 63–79, 1996.

Russell, A. Comet: Low latency data for the browser, 2006, http://alex.dojotoolkit.org/2006/03/comet-low-latency-data-for-the-browser, (last accessed June 2009).

Saint-Andre, P. Extensible messaging and presence protocol (XMPP): Core. IETF, 2004.

Sperberg-McQueen, C. M., F. Yergeau, E. Maler, J. Paoli, and T. Bray. Extensible markup language (XML) 1.0, 5th edn. W3C Proposed Edited Recommendation, W3C, 2008.

Woodson, W. E. *Human Factors Reference Guide for Electronics and Computer Professionals*. New York: McGraw-Hill, Inc., 1987.

Zyp, K. JSON schema proposal. Working Draft, 2008.

Chapter 15

Super-Peer-Based Mobile Peer-to-Peer Networks

Sathish Rajasekhar, Ibrahim Khalil, and Zahir Tari

Contents

A peer-to-peer (P2P) system has an overlay network built on the physical network, that is, the Internet. As the number and variety of P2P overlay networks increases, resource sharing between heterogeneous P2P overlay networks becomes more important. For example, wireless devices capable of holding gigabytes of data, reasonable processing power with moderate costs can be integrated into the existing P2P systems. Since the file-sharing applications are increasing rapidly, we need a way to collaborate and communicate among them. Hence, an architecture for resource sharing among heterogeneous P2P overlay network for fixed and mobile peer devices is required.

P2P architectures that help in cost sharing and reduction, resource aggregation, and interoperability are classified as structured, unstructured, and super-peer (SP)-based architectures. In unstructured architecture, there is no predefined topology for linking the peers to each other. Routing is done by flooding the network, that is, each peer sends the query to its neighboring peers, which then send it to their neighboring peers and so on. In the SP-based architecture, some peers are responsible for indexing and locating the shared data. The structured architecture associates each data item with a key and each peer is responsible for storing a range of keys and associated data. These existing architectures have several disadvantages. Napster uses a centralized directory approach, which makes it hard to scale and is vulnerable to service failure. Gnutella employs the flooding-based techniques, thereby limiting search. The work in [7,12,27] proposes P2P query protocols; however, they fail to efficiently service requests under the environment of unstable connectivity and unpredictable network congestion. Most of the prominent versions of the structured P2P systems [32] do not consider providing Quality of Service (QoS) at all. Some of them consider network proximity, but they come at the cost of expensive overlay maintenance protocol. Also, proximity-based routing can compromise load sharing in P2P overlay networks. Therefore, a file-sharing mechanism that can retrieve resources in a P2P overlay network while providing QoS efficiently is necessary. This chapter proposes and details such an architecture, which is an extension of SP-based concept. Basic experimental results show that the proposed architecture is robust and efficient in terms of average search response time and Query Hit Ratio (QHR).

15.1 Motivation: The Need for a New Architecture

The prevalent structured, unstructured, and SP-based architectures lack the following qualities:

- *Lack of mechanisms for transparent integration of mobile devices*: The present day computer users share content not only on PCs but also on devices such as laptops and PDAs. The storage capacity of devices such as PDAs have increased multifold to hold gigabytes of data. These small devices usually get connected to the network via Wi-Fi, mostly while they are in a Wi-Fi hotspot. Therefore, any new file-sharing architecture should have mechanisms to integrate both fixed and mobile devices. The existing architectures do not support the seamless integration of mobile devices in P2P wireless networks. For seamless integration and increased efficiency of information exchange operation, it is important for the SP to transparently authenticate a peer device in a fixed or wireless network. That is, it automatically senses the mobile peer device if the MAC address of the peer device is present in its repository, and also tracks the location of each peer device.

■ *Lack of QoS support*: Since there will be enormous amount of data transfer such as DVD movie clips, music file downloads, etc., between peers or SPs, QoS routing, faster search mechanisms, and caching of popular files and their locations at SP nodes are necessary to speed up data delivery. Unfortunately, these qualities are lacking in the existing structured, unstructured, or SP-based architectures. Most of the work done on P2P systems is to effectively search for data using Distributed Hash Table (DHT) mechanisms [4,32]. The prominent versions of the structured P2P systems [32] do not consider providing QoS at all. Some of them consider network proximity, but they come at the cost of expensive overlay maintenance protocol. Also, proximity-based routing can compromise load sharing in P2P overlay networks. In SP-based approaches, most peer devices exhibit different characteristics such as updating their location and state information to their respective SPs, and they are not properly exploited. An SP gathers information (e.g., what files a peer can share) from peer devices within its cluster, but does not use QoS to search or route a query.

■ *Lack of overall reliability*: Peers possess different characteristics with respect to their capabilities, such as bandwidth, available storage, processing power, etc. Exploiting these different capabilities in a P2P network can be done by SPs to provide better overall reliability. The current architectures provide better scalability than broadcast networks [22], but are not reliable. We also observe that an SP in existing architectures suffers from a single point of failure as it manages most of the activities of a large number of peers within its cluster. The replication of SP node database is also lacking in the existing architectures.

The lack of the above-mentioned qualities in the existing architectures is the main motivation behind proposing an enhanced SP-based architecture that supports QoS, takes care of reliability in the P2P overlay, and allows integration of mobile devices. To illustrate how such architecture could work, we explain with an example. Let us consider an airport with Wi-Fi connectivity, as shown in Figure 15.1. The passengers in an airport want to share information and files between fixed and mobile devices such as PDAs or laptops. Let us assume that there are a large

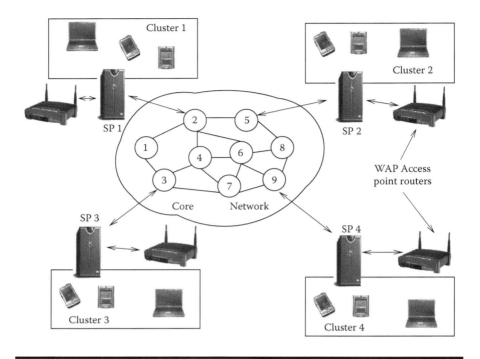

Figure 15.1   Wi-Fi hotspots connectivity to the Internet.

number of peer devices within a cluster. A cluster consists of peers in the same neighborhood where peer devices are able to communicate with each other using Wi-Fi or by wired networks. There might be several clusters under the supervision of an SP. The SP manages intelligently and collectively the entire operations of these clusters and might as well communicate with other SPs to search and obtain data.

For example, if a passenger at the airport wants to download a music file, say *Toxic*, then the passenger's peer device generates a query to its SP. The SP will return information as to which peer device has the music file *Toxic*, whether the service provider's device is in the same cluster or in a different network belonging to other SPs. Such seamless authentication (such as popping up in a Wi-Fi environment and getting connected automatically to the P2P network if the user is authorized) and search mechanisms with QoS support do not exist in any of the existing architectures [7,26,32,35].

Therefore, this chapter proposes a generalized, scalable, and robust extended SP-based file-sharing architecture (ESPA) to offer reliable, fast, and optimal downloads by applying QoS path selection between SPs. Since several paths exist between source and destination SPs, QoS path-selection mechanism helps us choose the one that offers the best possible path. Our architecture also supports load sharing and caching mechanisms for effective and optimal file sharing. We also propose a protocol (Mobile Authentication and Resource eXchange (MARX)), which can use the standard look-up algorithms such as chord [32]. The registration process of peer devices in our extended SP-based architecture is simple and automated. The active replication of the SP node database in our architecture eliminates single point of failure, thus making it robust and reliable even during adverse conditions. The features in our scheme are essential for effective file sharing, but are not prevalent in any of the existing architectures.

15.2 Limitations of Existing P2P Architectures

P2P computing has attracted a lot of attention in the file-sharing and data management community. Many systems have been developed for managing shared data in P2P networks. A centralized communication model in P2P system holding an index of files that peers can share is Napster [3]. In Napster, peers connect to a central database, identifying themselves and requesting or sharing files with other peers. Since the advent of Napster, several other P2P architectures and applications have been proposed. They are typically unstructured, structured, or SP-based architectures and usually do not integrate with each other. The P2P architecture proposed in this chapter is an extension of the existing SP-based file-sharing system. In this section, we discuss the limitations of several existing P2P file-sharing architectures including SP-based systems, and compare them with our proposed architecture to highlight the key differences.

15.2.1 Unstructured

Currently, there are many unstructured P2P file-sharing systems. One of the most common unstructured P2P system is Gnutella [26], which has no central server, and is the first of its kind in P2P to implement a fully distributed file search by flooding the P2P network while searching files. LimeWire [34] is an open source program written in Java. It allows users to share any type of files. LimeWire supports Gnutella's open-protocol, prejudice-free development environment. Since nobody owns the Gnutella protocol, anyone can use it to send or respond to queries, and no entity will have an hold on the P2P network or over the information flowing through it. All computers running a program utilizing the Gnutella protocol are said to be on the Gnutella Network (gNet). This free environment allows malicious operations from peers, which is a major drawback of Limewire and Gnutella.

Freenet [36] system architecture supports replication and retrieval of data, and protects the anonymity of authors and readers. It is infeasible to determine the origin or destination of data. There is no central server like in Napster nor a constrained broadcast as in Gnutella. The requests are routed to the most likely physical location. The peers store keys, data, and addresses. When a search request arrives, the data may be in the table or the request is forwarded to the addresses with the most similar keys till an answer is found. The key and address of the answer are inserted into the table, with the least recently used key evicted to make way for the new key. Freenet is completely decentralized, robust, and scalable, and adapts dynamically to changing peer population.

PeerDB [31] is a P2P system designed with the objective of high level data management in unstructured P2P networks. It exploits mobile agents for flooding the query to the peers such that their hop-distance from the query originator is less than a specified value, that is, TTL (time-to-live). Then, the query answers are gathered by the mobile agents and returned back to the query originator. The architecture of PeerDB consists of three layers, namely, the P2P layer that provides P2P capabilities (e.g., facilitates exchange of data and resource discovery), the agent layer that exploits agents as the workhorse, and the object management layer (which is also the application layer) that provides the data storage and processing capabilities.

Although, most unstructured P2P systems are fault tolerant, some of them are prone to a single point of failure. For example, a dynamic file-sharing approach using ORION protocol is proposed in [16]. Figure 15.2 illustrates the ORION protocol with a four-node scenario. Node *A* issues a query for four files 1, 2, 3, and 4. The results are found in nodes *B* (files 1 and 2), *C* (files 1, 2, and 3), and *D* (files 2, 3, and 4), respectively. Node *B* sends files 1 and 2 to node *A*. Node *B* receives files from nodes *C* and *D*, filters the redundant files, and sends the remaining two files,

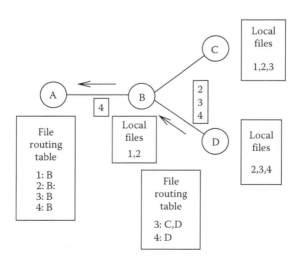

Figure 15.2   File sharing using ORION.

as shown in Figure 15.2. Here, node *B* is a single point of contact for nodes *C* and *D*, which could possibly lead to single node failure.

The unstructured architectures use the flooding model for the decentralized P2P overlay systems. Each peer device keeps a user-driven neighbor table to locate data items, typically files. This leads to excessive P2P network bandwidth consumption and remote or unpopular data files may not be found due to the limit of lookup typically imposed by the TTL. For example, if a search takes $O(N)$ times in the unstructured network, it will take $O(N/M)$ times in our proposed SP network architecture (where M is the average number of peer devices connected to a single SP). This nearly eliminates the problem of P2P network flooding.

15.2.2 Structured

Several systems like Content Addressable Network (CAN) [25], Chord [32], Pastry [28], and Tapestry [37] implement distributed hash tables (DHTs) in structured systems. In such systems, a query is associated with a key and is routed to a peer holding the key value. Each peer has to maintain certain neighbors and connections, thereby providing static connections between participating peers.

P2P Information Exchange and Retrieval (PIER) [15] is a massively distributed query engine built on top of a DHT. It aims to bring database query processing facilities to widely distributed environments. PIER is a three-tier system organized as follows: applications (at the higher level) interact with the PIERs query processor (at the middle level), which utilizes an underlying DHT (at the lower level) for data storage and retrieval. An instance of each DHT and PIER query processor component runs on every participating node. PIER currently implements a particular kind of DHT, called Content Addressable Network (CAN) [25]. The difference between PIER and our architecture is that, basic and advanced services run on top of trusted static SPs in our proposed system by protecting data through redundant techniques, whereas PIER runs on untrusted servers. To improve performance, our architecture allows data to be cached in SPs anywhere, anytime. Additionally, monitoring of usage patterns allows adaptation to regional outages and denial of service attacks; monitoring also enhances performance through proactive movement of data for load sharing.

P-Grid [2] is a P2P lookup system based on a virtual distributed search tree, similarly structured as standard DHTs. In P-Grid, each peer holds part of the overall tree depending on its path, that is, the binary bit string representing the subset of the trees information that the peer is responsible for. A decentralized and self-organizing process builds the P-Grids routing infrastructure, which is adapted to a given distribution of data keys stored by peers. This process also addresses uniform load distribution of data storage and uniform replication of data to support uniform availability. On top of P-Grids lookup system, other self-organizing services may be implemented (e.g., identity, adaptive media dissemination, trust management). Unlike our proposed architecture, which is independent of the overlay network, P-Grid relies on a specific virtual distributed search tree. Like P-Grid, other structured P2P systems usually provide a basic lookup infrastructure on top of which other services and applications may be deployed. For instance, over Chords lookup system, we find services such as i3 [18], a large-scale, reliable multicast, and applications such as CFS (Cooperative File System) [9], a P2P read-only storage system that provides provable guarantees for the efficiency, robustness, and load balancing of file storage and retrieval. Likewise, on top of the Pastry object location and routing substrate [28], we find PAST [29], a large-scale P2P persistent storage utility, that manages data storage and caching, and SCRIBE [6], an application-level implementation of anycast for highly dynamic groups, dependent on DHTs.

OceanStore [17] is a utility infrastructure designed to span the Internet and provide continuous access to persistent information. It relies on Tapestry [37], an overlay location and routing infrastructure, such as DHTs, that provides location-independent routing of messages directly to the closest copy of an object or service using only point-to-point links and without centralized resources. It envisions a cooperative utility model in which consumers pay a fee in exchange for access to persistent storage.

The main difference between our SP-based architecture and the structured ones is that the entire system based on our proposal is virtually broken into a search of information from a smaller set of SPs, each containing indexed information from its sets of peers. Therefore, the search time is much faster in our architecture as compared to the structured P2P systems. In the structured P2P networks, every peer device is given equal responsibility irrespective of a peer's computing and storage capabilities. This leads to the deterioration of system performance as less capable peer devices are involved in searching and routing operations. This problem is alleviated in our SP-based approach. The SPs are very powerful devices with heavy processing capability and more storage capacity. This ensures that the SPs share the load according to the capability of their peer devices leading to overall better performance.

15.2.3 Hybrid

In hybrid P2P systems, the control information is exchanged through a central server, while data flow takes place in a P2P manner. The eDonkey [1], a popular P2P file-sharing system, does not rely on a single central server. A file can be downloaded from several different peers at once, and can be shared by a peer before it is completely obtained. The eDonkey file-sharing service belongs to the class of hybrid P2P architectures [14]. Its architecture comprises of two applications, which form the eDonkey network: the eDonkey clients and the eDonkey servers. The eDonkey client is used to share and download files. The eDonkey server operates as an index server for file locations and distributes addresses of other servers to clients. In the eDonkey network, files are not transmitted through the server. Every eDonkey user is eligible to set up a server. When a client connects to the eDonkey service, it logs onto one of the servers (using a TCP connection) and registers all files it is willing to share. Each server keeps a list of all files shared by the clients connected to it. When a client searches a file, it sends the query to its main server. The server returns a list of matching files and their locations. The client may resubmit the query to another server, if no matches have been returned. The major communication between client and server is typically implemented by TCP connections on port "4661." Additional communication between clients and servers, for example, further queries and their results, are transmitted via UDP on port 4665. When an eDonkey client decides to download a file, it first gathers a list of all potential file providers and then asks the providing peers for an upload slot. Upon reception of a download request, the providing client places the request in its upload queue. A download request is served as soon as it obtains an upload slot. eDonkey clients may restrict their total upload bandwidth to a given limit. An upload slot becomes available when a minimum fair share of the upload limit is possible. When an upload slot is available, the providing client initiates a TCP connection to the requesting client, negotiates which chunk of the file is exchanged, and transmits the data. The eDonkey protocols split the file into separate pieces, denoted as chunks. A chunk has typically a size of 10 MB. The consuming client can reassemble the file using the chunks or parts of chunks. A client can share a file as soon as it a has received a complete chunk. A major feature of eDonkey is that the consuming client may operate in the multiple source download mode. In this mode, the downloading client issues in parallel two or more requests to different providing clients and retrieves data in parallel from the providers. Since an eDonkey client may leave the eDonkey service at any time, the requesting client has to renew its download request periodically otherwise the requests are dropped. In order to reliably check the availability of a client, the eDonkey protocol uses TCP connections on port 4662 for the communication between the clients. A client-to-client connection is terminated by the eDonkey application after an idle period of 40 s (TTL). The communication between eDonkey servers is very limited, which is a major drawback. Also, the servers contact each other periodically but with small frequency in order to announce themselves and to send back a list of other servers. In this way, the servers maintain an updated list of working servers and affirm the search efficiency of the eDonkey service.

The differences in routing capability allow hybrid P2P systems to offer better scalability than centralized models. But the hybrid systems suffer from scalability problems for control information that flows through a single peer. Replication and redundancy through SP overlay in our architecture eliminates this problem and provides high reliability. Also, hybrid systems are used for mission critical applications, but their solutions are limited to solve relatively small-scale problems, unlike our proposed SP-based architecture. In our approach, every SP in the P2P overlay contains large index information of peers under its control. When total overlay is taken into consideration with many

SPs, the searchable information becomes voluminous, which is handled effi-ciently in our SP-based approach, but is absent in the hybrid models.

15.2.4 Super-Peer

A new wave of P2P systems is advancing an architecture of centralized topology embedded in decentralized systems, forming an SP-based network. For example, Edutella [21] is a P2P system for data management in SP networks. In Edutella, a small percentage of SP nodes are responsible for indexing the shared data and rout-ing the queries. The SPs are assumed to be highly available with very good comput-ing capacity. The available SPs are arranged in a hypercube topology, according to the HyperCuP protocol [30]. When a peer connects to Edutella, it should reg-ister at one of the SPs. Upon registration, the peer provides to the SP its Resource Description Framework (RDF)-based metadata. The initial Edutella services are to process the queries based on RDF metadata, replicate service that provides data availability and workload balancing, perform mapping between the metadata of different peers to enable interoperability between them, and lastly, service anno-tated materials stored anywhere within the Edutella network. Edutella is imple-mented on top of an SP network, where SPs are dynamically selected.

The existing typical SP-based approaches are efficient, scalable, and manageable. However, one of the limitations of these approaches is the way an SP is selected. They dynamically select an SP from a set of peers in the P2P network. The selected SP may have limitations in terms of processing power, storage capacity, and also may leave the P2P network within a short notice. Thus, SPs become a potential single point of failure for its peers quite often. This problem is overcome in our architecture by actively monitoring the status of static SPs and replicating the operational index of the SP database to keep the P2P network alive in case of failures. Most of the existing SP-based solutions, except ours, do not support simple seamless authentication for wireless and fixed peer devices. As significant portion of P2P traffic are delay-sensitive applications, such as music file downloads, movie clips, and DVD downloads ensuring QoS to such applications becomes necessary. The existing SP-based approaches do not consider QoS, but our proposal is built on a stable and static SP-based model that sup-ports QoS by efficient path selection among SPs. It also caches popular files and their locations to significantly reduce search and download time. The replication of popular files can reduce the load on SPs and enhances the overall reliability of the P2P overlay in our proposed architecture, which is not supported by other SP-based architectures.

15.3 Proposed Wi-Fi ESPA

This section introduces the concept and architectural design of our proposed P2P file-sharing approach. Users on the move use devices such as a laptop or a PDA that is capable of exchanging information using Wi-Fi. Wi-Fi is used to define the wireless technology in the IEEE 802.11 specification including the wireless protocols 802.11a, 802.11b, and 802.11g. A Wi-Fi hotspot is defined as any location in which 802.11 wireless technology both exists and is available for use to consumers [13]. Wi-Fi hotspots are locations that provide public network access to wireless users carrying laptops and PDAs. These hotspots are generally located at train stations, airports, coffee lounges, hotels, universities, and even on the streets, making wireless network accessible to mass public. A peer device such as laptop or a PDA communicates with other such peer devices using Wi-Fi hotspots after proper authentication. For effective QoS data transfer, a generic three-layered, SP-based architecture is proposed to support information exchange and file management in wired and wireless networks. The proposed architecture consists of application/interface layer, the QoS layer and the search layer. The design objectives are not just to search data faster, but also to provide QoS support to data delivery and search mechanism using suitable techniques.

15.3.1 Application/Interface Layer

The objective of the application/interface layer is to provide a suitable interface for peer devices to generate a search query and respond with possible search locations. It interacts with other layers to provide and establish optimal locations for file requests from peers. For example, when a query is generated from a peer, the search layer is invoked. The search will generate many possible locations for a file to be fetched from. However, the file is essentially requested from the location that provides the best QoS enabled path (i.e., faster download). The following functional components facilitate the query search and presentation of such results:

- *Application interface*: The peers communicate through WAP to be authenticated, generate the query, and provide information to the service requester.
- *Query information module*: The functions of this module are twofold. First, it checks for authentication of the peer device in its cluster. This is done by checking the local information source SP database (LISSP). Second, the query that is generated by the service requester for file sharing is processed.
- *Update query module*: Updates the requester with the availability of the best device for file transfer based on the information provided by the QoS module. File statistics are collected, which trigger updates in the cache to store the most popular files and their locations, based on the frequency of downloads and other storage constraints.

15.3.2 QoS Layer

When a search is invoked, multiple possible file locations are returned. For every location, there are many possible paths to reach the destination peer. Different paths have different link characteristics in terms of hop count, delay, and bandwidth. QoS layer helps in selecting the best path between peers for efficient file transfer using the QoS Broker and Internet distance map service (IDMaps) [10]. It also caches the search results and saves the popular files on SPs, which can significantly reduce the network load and improve response time. The QoS layer supports the functional blocks as described below:

■ *QoS Broker*: It provides QoS between the requesting peer device and the device that is offering the service. It establishes a reliable connection between the requester and the service provider in an optimal manner by using a suitable QoS routing algorithm. Since the search results return multiple possible paths, the QoS routing algorithm will select the most suitable path that is likely to provide the least download time with maximum available bandwidth on the selected path. In order to select the most suitable path, the link bandwidth between the source-destination peers has to be known. This information can be obtained by lookup service such as IDMaps [10].

■ IDMaps: IDMaps is a global architecture for Internet host distance estimation and distribution [10]. IDMaps provides network distances quickly and efficiently in terms of metrics such as latency or bandwidth between Internet hosts. Higher level services collect distance and bandwidth information to build a virtual distance map of the Internet and estimate the distance between any pair of IP addresses. Internet addresses are grouped by address prefix (AP) to reduce the number of measurements and number of tracers required to provide useful distance estimation. An AP is a consecutive address range within which all assigned addresses are equidistant to the rest of the Internet. Experimentation on a 4200 node network on the Internet was conducted and the following results were observed:

– Mirror selection using IDMaps gives considerable improvement over random selection.
– Network topology affects IDMaps performance.
– Tracer placement algorithms that do not rely on knowing the network topology can perform as well as or better than algorithms that require a priori knowledge of the topology.
– Adding more tracers after a certain threshold (over 2%) diminishes improvement.

Overall, use of IDMaps provides network distance in terms of latency or bandwidth and is scalable. This information is fed to the QoS Broker, which then efficiently computes the best path between any two peers.

■ *Caching*: It is a feature that is added into our system architecture. Let us consider the example of a recent popular music file or a video clip on the P2P network. Peer devices may always request the most popular music files or video clips that are available from some other SP nodes. Every time a request for a popular music file or a video clip is issued, instead of searching the same file every time from other peers or SPs, these files, their locations or both can be cached. Therefore, we classify our cache module as file cache (F-cache) and location cache (L-cache):

- *F-cache*: The F-cache stores the most frequently accessed files for peers that belong to an SP to save valuable network resources such as bandwidth and reduce response time (minimizing delay) by providing the file from the local storage. Since the F-cache has finite capacity, the files consume the available storage space rapidly. The F-cache is not able to hold all files due to storage constraints and large peer requests for distinct files. Hence, caching policy is used to select the most suitable files to be stored.
- *L-cache*: This stores the most frequently accessed locations based on statistics maintained in the update information module. The SP receives large number of search results for a search request, which are then processed by the QoS module to provide the best location for that request. This consumes the computing and processing power of the SP. However, storing the location of the best search result does not consume disk space even though there may be large number of such refined results. Doing so reduces processing, computing, disk space, and query overheads. Signaling and overhead traffic in the P2P network are also reduced.

15.3.3 Search Layer

As a large amount of information is available within SPs, searching data in a scalable and efficient way is important. The search layer deploys existing search tools such as CAN and CHORD, which provide hash-table-like functionality on an Internet-like scale. Even though existing search techniques are used, the generated search information is filtered through the QoS module to refine selection and generate the best available path. Hence, search combined with QoS plays a pivotal role in our proposed P2P file-sharing architecture.

- *Lookup module*: This assists in finding the requested file or resources by using any of the existing search modules [19] such as CAN, Pastry, Chord, and Tapestry. The requested resource may be present in the same cluster or a different cluster under the same SP. Resources can also be found in SPs that are located in clusters other than the source cluster. In a P2P overlay network, peer nodes are connected to each other logically and each peer's information is available in its LISSP database.
- *LISSP*: This is a centralized information database that authenticates every peer device in the network. Any service request by a peer in the P2P network is processed through LISSP. It contains a list of active devices, **Act. Dv.**, and registered devices, **Rgd. dv.**, which are used for authentication process.

Figure 15.3 shows our proposed architecture. A device willing to exchange information with another device can exist in the same cluster, different cluster, or in a different network. A peer that requests a service is termed as a *service receiver*,

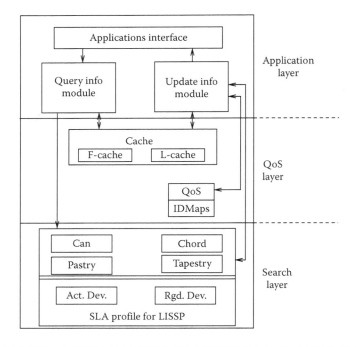

Figure 15.3   Architecture overview of the proposed ESPA.

and a peer that offers the requested service is called a *service provider*. Both service receiver and service provider are authenticated through WAP. If the device's MAC address is not found via WAP, the LISSP is checked for the peer device's profile. Once admitted after authentication, the query generated by the peer device is serviced. First, the F-cache and L-cache are checked for the requested resource. If it is found, the update info module will process the query and the file transfer takes place. Otherwise, the query is sent to the lookup module, which in turn will find the available resources. IDMaps will gather the available bandwidth of different resources and will send information to the QoS Broker, which in turn will select the best available peer. The QoS Broker will inform the update information module and the communication between the service provider and the requester takes place. In the next section, we discuss this in detail.

15.4 Mobile Authentication and Resource eXchange Protocol

A protocol is an agreed-upon format for transmitting data between two entities. In other words, it is a set of rules that govern communication between sending and receiving entities. The proposed architecture for file transfer and transparent authentication has been outlined in the previous section. To achieve efficient file transfer over the best QoS path, we propose simple rules, which can define communication among peers using the functional blocks defined in our architecture. Hence, a Mobile Authentication and Resource eXchange P2P protocol (MARX) for communication between the service provider (e.g., a peer device or an SP holding the requested content) and the service receiver (peer device receiving file) becomes necessary. For example, if a peer device x, wants to download a file from an SP, the steps involved are:

- Mobile device x having a unique MAC address is registered and authenticated in a cluster.
- Mobile device x sends a request to the SP for a file.
- Mobile device x gets a response regarding the availability of the requested file through the QoS Broker. Several possible locations of the file are returned, which could be in the same cluster or in a different cluster.
- Since there are many possible locations, the QoS Broker computes routing paths to select the best possible location to retrieve the file for the service requesting peer.
- Mobile device x requests connection establishment to the selected peer (i.e., service provider peers) and receives data from it.
- After fetching the file, mobile device x sends a disconnection message to the service provider by sending the "transfer complete" signal and then the service provider device disconnects from device x.

Here, as stated earlier, we make the assumptions that mobile devices in the clusters are registered peer devices known to WAP and inform the SP of the cluster regarding their sharable information. These peer devices will be present in the same cluster and do not change clusters during file transfer operation. We also assume that clusters are independent with no overlapping of clusters.

15.4.1 Protocol Overview and Semantics

The MARX protocol is a dynamic protocol used for communication and file transfer between peer devices. The protocol defined is used for file discovery, connection setup, maintenance, and termination of connection. When the protocol returns several possible file locations, the QoS Broker of the SP must decide the best possible location to fetch the requested file. Now, every single path has several links between the service providers and the service receiver. These link capacities may vary depending on the traffic load. However, the available capacities on these links can be obtained using IDMaps. Once, the distance information is made available to the SP system, the QoS Broker deploys efficient routing algorithms to select the best QoS path. To facilitate such efficient selection and implementation of MARX, we define the following protocol semantics that are used by peer devices to join an SP-based P2P network and to share information with other peers.

- *JOIN*: To indicate a peer's arrival in a cluster to its SP. This also establishes the authentication of the peer device. This device may be registered in the same cluster or in a different cluster. The peer device will send its unique ID (MAC address) and its original SP (i.e., where it was registered first) for faster authentication.
- *INVTE*: This is confirming the peer's authentication from its SP and at the same time requesting the peer to send its list of sharable files.
- *RQUERY*: A file request can be made by a peer to its own cluster SP or by the SP to other SPs for file location. The request query message is forwarded to the query information module to find suitable peer device for data transfer. The request query contains the service receiver ID, and the file it is requesting for.
- *AVLBE*: An SP indicates the availability of resources from its peer by invoking the LISSP database. The service provider device informs its SP of its availability.
- *ESTBLMNT*: It establishes a session between a sender device and a receiver device indicating the file transfer after computing and selecting the best QoS path.
- *SEND*: The data is sent from the service provider to the service receiver. The service receiver peer receives the requested information.
- *RECV*: This indicates that the service receiver peer received the data from the service provider.
- *CLOSE*: This terminates the connection between communicating peers safely by discarding the established connection.

15.4.2 MARX Protocol Operations

While there are several steps involved during a successful file transfer between two peer devices, the three key operations that actually enable us to retrieve a file from a suitable location are device and information registration, peer authentication, and efficient data retrieval. Based on our architecture, when a mobile device enters the periphery of a cluster, it is automatically sensed by a WAP and is admitted to the P2P system if it is already found to have been registered in the system. The authentication process is very crucial for the transparent and seamless connectivity of peer devices for file transfer operations. Subsequent operations involve query submission, searching possible data locations, and selecting the best possible QoS path for retrieval of data once the location is determined. The information retrieval can be from the SP cache, different clusters, or from a different P2P network depending on the availability of data. In this section, we discuss the registration, authentication, and data retrieval operations in detail.

15.4.2.1 Registration

Any new device can be registered in any cluster of a given SP. When a new device communicates with the WAP database of a cluster, the details and the identity of the peer device is matched against the LISSP database. If the device is a new device, which is not registered to any cluster, then the device must send its unique MAC address to the WAP database. This in turn updates the LISSP database with the new device information. The new device must also inform the LISSP database of the files it wishes to share. This information is updated from time-to-time for all the peer devices connected to the network. The new device will carry its unique ID (MAC address) and its SP ID where it was registered for the first time.

When a device from another cluster enters a new cluster, information regarding its original SP is transmitted to the new SP. The new SP retrieves the registration details of the peer from the original LISSP database (on the original SP) and endorses the peer device to use and share information from the present cluster. Every cluster contains an LISSP database. An LISSP database is very much required because a cluster can avoid threats or attacks from hostile devices by ensuring that only registered devices are allowed to participate in file transactions. Figure 15.4 depicts an LISSP database, which consists of the authentication and the mobile access device (MAD) (i.e., the peer devices) information of the peers in a cluster.

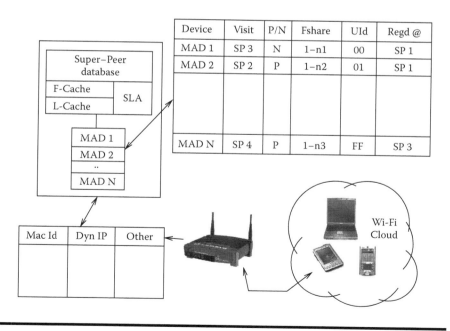

Figure 15.4 **Entries for LISSP database at Wi-Fi hotspot for peer devices.**

Each peer device will register the following information in the LISSP database:

- A general device identity as to what kind of peer device it is. This identity is used to list all the peer devices in the SP database. This data can be used for the purpose of load sharing. For example, laptops are more powerful when compared to PDA devices. Hence, SP may proportionally allocate resources based on the type and the capacity of the peer device.
- Last "Visit" helps to identify where the peer device was last active. This information also helps the current SP to pull the peer's latest list of sharable items. For example peer a was originally registered at SP1 and the files that the peer was sharing were x and y at that time. When peer a moved to SP5, the files peer a shared were files x, y, and z. This updated information is not available from the original SP, nor is known to the current SP. But, the current SP can pull the latest list of sharable files from the last "visited" SP.
- P/N status indicates the availability of the peer in the cluster. The status value "P" indicates that the peer is present in the cluster and its sharable information is valid. Therefore, the SP will not direct any requesting peer to fetch information from such a peer that is not active.
- FShare lists all the files that the peer device is willing to share with other devices and its SP. This information is updated frequently and periodically. The LISSP database gets updates as and when the device updates itself with information. For example, if the peer wants to share a new file, it will immediately update the LISSP database.
- UnqeID is the unique ID of the peer device. This is a very important feature, for security reasons. By having this unique address of a device, malicious attacks and threats can be avoided by allowing only registered devices. We obtain the MAC address of a device for the purpose of authentication.
- Regd @ provides information as to where this peer device was first registered. Using this information, the SP node of the current P2P network can reconfirm and obtain details of the peer device with the LISSP database for the peer's identity.

15.4.2.2 Peer Authentication

The peer device that enters the P2P network should not be a malicious or unknown peer. Our architecture takes due and diligent care to ensure that such malicious or unknown devices are not allowed to share files in the P2P environment. For this purpose, a peer device's authenticity is verified at multiple check points. Figures 15.5 and 15.6 show the logical view of cluster organization as to how the devices are connected to the SP through a WAP and the internal organization of the WAP database connected to LISSP database.

Given the above scenario, three possibilities arise related to peer device authentication. First, the device information can be found in the WAP database.

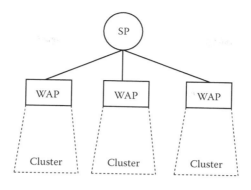

Figure 15.5   Logical view of cluster organization.

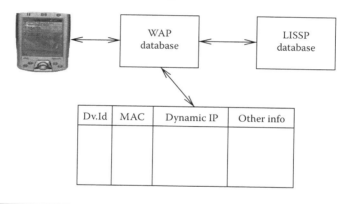

Figure 15.6   Mobile device connectivity.

The peer device sends its unique MAC address, dynamic IP address, and its device ID. As the device enters the cluster, it senses the WAP. The WAP database checks the MAC address of the device. If the MAC address is present, then it authenticates the device. Steps 1 and 2 in Figure 15.7 illustrate that the device senses the WAP (step 1) and the WAP database authenticates the device (step 2).

Second, if the device information is not found in the WAP database, a search is initiated in the LISSP database. This is shown in Figure 15.8. Step 1 checks the WAP database. If the information of the peer device is not found, then WAP probes the LISSP database for the authenticity of the device, which is shown in step 2 of Figure 15.8. If there is a match, steps 3 and 4 confirm that the peer device is a registered user under the SP.

Finally, at times, the device information may not be available either in the WAP database or in the LISSP database. A global search (search other SPs) is undertaken to find the peer device's registration, as shown in Figure 15.9. The search initiated by WAP and LISSP databases does not produce a match, indicated in steps 1 and 2

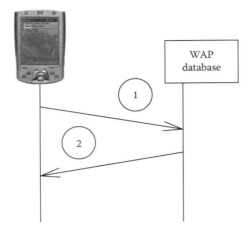

Figure 15.7   Mobile device authenticated via WAP database.

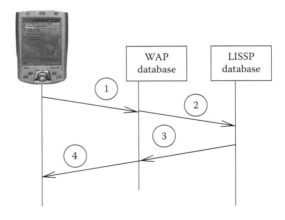

Figure 15.8   Mobile device authenticated via LISSP database.

of Figure 15.9. Efforts to search for other SPs (peer indicates which SP it was regis-tered originally) are initiated as shown in step 3. Steps 4, 5, and 6 confirm that the peer device is indeed a registered device although in a different SP. It can now pro-ceed with file-sharing and retrieval operations, since authentication is confirmed.

15.4.2.3 Data Retrieval

Distributed search tools such as CAN and CHORD in the search layer can return multiple data locations. These possible data locations can be within the SP cache, or peer(s) of the same cluster, or peers/super-peers in a different P2P network. Our

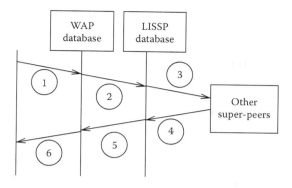

Figure 15.9 Mobile device authenticated via global search.

protocol addresses three different modes of communication. They are: data in cache, data within a cluster, and data in different networks.

■ *Data in cache*: Once the device is authenticated, the peer device may get involved in file transfer operations. When the peer device requests a file, a query is received by its SP. If the requested data file is present in the SP cache, it is returned to the service receiver peer quickly, thereby saving bandwidth and processing time. The operational steps for the service receiver peer are shown in Figure 15.10. After receiving the query, it is forwarded to the cache of the SP as shown in steps 1 and 2. If the cache contains the requested file, then it notifies the service receiver that the requested file is available in cache as indicated

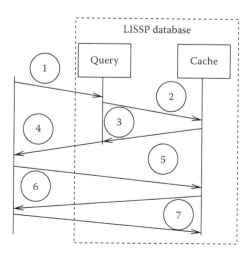

Figure 15.10 Requested file in cache.

in steps 3 and 4. The peer device (service receiver) will then establish a connection requesting the file (step 5), which is then delivered (step 6) to the requesting peer and finally disconnected (step 7) after the file transfer is complete.

■ *Data within a cluster*: The query generated is processed through the query information module. If the cache does not contain the requested file, it is then passed on to the lookup module. The lookup module searches the requested file in the hash table of the LISSP. If one or more peer devices within the cluster have the requested file, then the QoS Broker identifies the optimal peer, based on the QoS path. However, if the peers are within the clusters of the SP, there will not be much of an impact on QoS path selection in terms of performance.

Figure 15.11 represents the communication tasks within a cluster. When a query is initiated from the service receiver after authentication, an available service provider is identified through the QoS Broker of the LISSP, as shown in steps 1, 2, 3, and 4 of Figure 15.11. Once the service receiver knows the service provider, the session establishment (step 5) is made, data transfer takes place (step 6) and terminated (step 7) when transfer is complete.

■ *Data in different networks*: The service provider at times cannot be found in the same vicinity or in the same cluster. When the desired information is not available within its cluster, then the QoS Broker makes a global request. The global search request for the file is issued to neighbor SPs in different networks. Even though the requested information is present in other SPs, some SPs may not respond. This may be due to the SPs being very busy processing information within their own clusters or because of capacity constraints. The QoS Broker gets an updated list of SPs willing to communicate and transmit information to the requesting device. The QoS Broker then decides the optimal SP based on certain parameters such as delay,

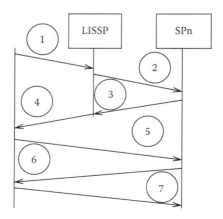

Figure 15.11   Requested file in same cluster.

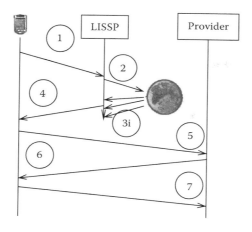

Figure 15.12 Requested file in different network.

available bandwidth, and the number of hops in terms of capacity-to-hop count ratio [23] from the requested device. The selected SP service provider is then passed on to the service requester via the update query module. The data transfer between the service provider and the service requester now takes place.

Figure 15.12 details the steps for a case when data can only be found outside the SP clusters. The LISSP will issue a global search to find the requested file as shown in steps 1 and 2. Many search results are returned to the QoS Broker of LISSP. To determine the best possible location, the QoS Broker uses QoS routing algorithm. This routing algorithm selects the path that has the maximal path capacity [23]. Once a QoS path is identified and established, the QoS Broker informs the query update module of the new service provider to be made known to the service receiver (steps 3 and 4). Steps 5, 6, and 7 are very much similar to file transfer between devices in the same cluster.

15.5 Replication of LISSP Data

Replication [8] of LISSP data is used to increase data availability at low communication cost: if an SP is temporarily inaccessible, then transactions can continue using different SPs while attaining the minimum communication cost thereby eliminating the risk of single point failure. By storing multiple copies of LISSP data at several locations in different SPs, there is an increased data availability and accessibility to peers despite the source SP being down. The distributed SP-based P2P system proposed in this chapter consists of a set of distinct SP sites that communicate with each other by sending messages over a P2P communication network. An SP site may become inaccessible due to site or network failure thereby causing service disruption to peers in the P2P network. The LISSP database contains valuable authentication information used to admit valid peers into the P2P network. If such data becomes unavailable, peers will not be able to participate in P2P transactions even if they connect to other SPs. Also, an SP caches popular files in F-cache, popular file location information in L-cache, and other statistical data that can improve the performance of the proposed SP-based P2P system. While the size of F-cache could be large, L-cache (with hashed index) is not huge and is required to speed up search for peers in P2P network. Therefore, to keep the system functional, this issue is dealt by partially replicating partial LISSP database to neighboring SPs.

Ideally, replicated data management in SPs is transparent. Transparency implies that peers have the illusion of using single copy of the data or the object, even if multiple copies of a shared data exist or replicated SPs provide a specific service, the peers should not have any knowledge of which SP replica is supplying with the data or which SP is providing the service. It is conceptually reassuring for the peers to believe there is a single source SP that is highly available and trustworthy, although different replicas create that illusion. Ideal replication transparency can rarely be achieved; approximating replication transparency is one of the architectural goals of replicated data management. For maintaining replication transparency, two different architectural models [8] are widely used.

In passive replication, every peer communicates with the single source SP called the primary source. In addition to the primary SP, one or more neighboring SPs are used to backup the partial LISSP data, as illustrated in Figure 15.13. A peer requesting service always communicates with its SP in the cluster. If the SP is up and running, then it provides the desired service of authentication and file transfer

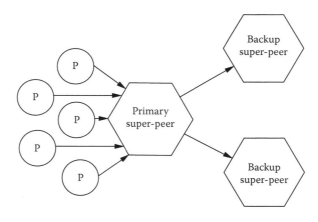

Figure 15.13   Passive replication of partial LISSP data in primary SP to neighboring SPs.

for the peer that enters the cluster. To keep the states of the backup neighboring SPs consistent, the SP performs an automatic multicast [20,33] of the updates to the backup neighboring SPs. If the SP fails, then one of the neighboring SPs is elected as the SP for the failed SP's clusters, handling the peer requests and sending responses to the peer. The SP backup protocol has the following specifications. At most one super-peer, is the primary SP for any peer at any given time. Each peer maintains a variable L (leader) that specifies the replica to which it will send requests. Requests are queued at the primary SP.

There may be periods when there is no primary SP. This happens during a changeover, and the period is called the failover time. When repairs are ignored, the primary SP backup protocol implements a service that can tolerate a bounded number of faults over the lifetime of the service. Here, unless specified otherwise, a fault implies an SP crash. Since the primary SP returns a response to the peer after completing the automatic multicast, when a peer receives a response, it is assured that each non-faulty replica has received the update. The primary SP is also required to periodically broadcast heartbeat messages. If a backup neighboring SP fails to receive this message within a specific window of time, then it concludes that the primary SP has crashed and initiates an election. The new leader neighboring SP takes over as the new primary SP. Figure 15.14 illustrates the steps of the primary backup protocol. If the response is not received due to an SP crash, then the request for service is retransmitted. Multiple retransmissions may be necessary until a backup neighboring SP becomes the new primary SP. This happens during the failover time.

Lemma 15.1   *At least (m + 1) SP replicas are sufficient to tolerate the crash of m SPs.*

Proof: Here, we are considering that essential data in LISSP databases are exchanged among SPs on regular basis. Therefore, SPs are almost perfect replicas of each other, and to keep the system up and running, only one SP needs to be operational out of m SPs.

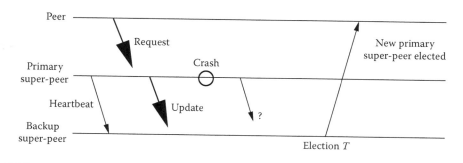

Figure 15.14   Illustration of primary backup protocol. (Note: Here '?' indicates that the backup SP fails to receive heartbeat message.)

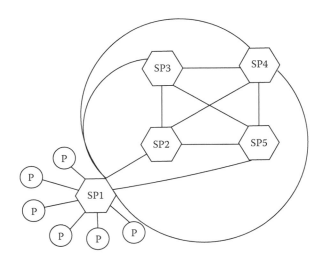

Figure 15.15   Active replication of SPs.

Lemma 15.2   *The smallest failover time for neighboring SP to become a primary SP in our primary backup protocol as illustrated in Figure 15.14 is* $\tau + 2\delta + T$.

Proof: For the primary backup protocol as illustrated in Figure 15.14 τ is the interval between the reception of two consecutive heartbeat messages, T is the election time, and δ is the maximum message propagation delay from the primary server to a backup server. It is obvious from Figure 15.14 that the two-way propagation delay is 2δ. Therefore, total time, that is, the smallest failover time, is $\tau + 2\delta + T$.

An SP-based P2P system shared by a group of peers is illustrated in Figure 15.15. Each peer uses LISSP database of its associated SP. Whenever a peer transaction causes an update, for example, a new peer joins, an existing peer leaves, a new set of files are added into the SP index, LISSP database of its associated SP propagates the update to each of the neighboring SPs using total order multicast (so that all copies of the LISSP become identical). As seen in Figure 15.15, peers come and join SP1. If SP1 gets updated due to P2P transactions, its partial LISSP database is replicated at SP2, SP3, SP4, and SP5. If SP1 fails, any of the neighboring SPs can act as a primary backup.

15.6 Path Selection

There are several algorithms to compute paths from a given source to a destination. The classical shortest path routing algorithms are Dijkstra's algorithm and Bellman–Ford algorithm. We select the best available path between SPs based on a routing algorithm, which maximizes the available path capacity-to-hop count ratio [23]. In fact, the algorithm obtains the paths with maximal ratio of available capacity to hop count from the source SP node to all the other SP nodes. The complexity of the algorithm is similar to that of the Bellman–Ford algorithm. In structured overlays, peers are satisfied by merely locating requested files in a probabilistically bounded, small number of network hops. Thus, QoS in terms of bandwidth-rich paths are actually ignored. QoS path selection algorithms deployed in the proposed ESPA architecture overcome these shortcomings by allowing SPs to choose best possible paths. This not only maximizes network utilization, but also helps peers receive an enhanced file download service.

15.7 Security

Security is one of the important aspects of any architecture. We only consider basic seamless authentication of peer devices. We discuss in brief the challenges of security. Securing data-sharing applications is a very challenging issue due to the open and autonomous nature of P2P networks. The structured P2P overlay systems provide a substrate to facilitate construction of large-scale, decentralized applications for content distribution. Despite being robust and fault tolerant, these overlays are not secure, and even a small fraction of malicious peers can disrupt message delivery throughout the overlay. The diversity and openness of P2P systems can actually be harmful since autonomous parties without having prior trust relationships can easily pool resources.

In Pastry or Chord overlay, an attacker may target a particular victim node/peer and manipulate its entries and neighbors set to redirect it to a hostile node [5]. This may prevent the victim's access to the P2P systems and its resources. In other words the victim's access to the overlay network is completely taken over by the attacker. The attacker can gain access to objects and can delete, deny, or corrupt them. By delegating the nodeID (i.e., unique IDs for peer devices) assignment job to a centralized, trusted SP, such problems can be prevented. In fact, the SP will not allow any peer to join and participate in resource sharing without valid authentication.

Pastry and Chord use proximity-based routing algorithms. If attackers gain access to routing tables, they can fake proximity to increase bad entries in the tables, and make nodes appear close even though they are far away [5]. The SP's restrictive communication model does not allow a malicious peer to join a network without having its MAC address validated.

The routing primitives implemented by current structured P2P overlays provide a best-effort service [24] to deliver a message from source to destination peer. Routing will fail if any of the peers between source and destination becomes faulty. To make routing robust and secure, at least one copy of the message should reach the destination. This is achievable in our SP-based architecture as path selection is performed dynamically between SPs and the paths of the overlays are constantly monitored. Therefore, the chance of selecting a faulty path is relatively small.

15.8 Experimental Results

In this chapter, we have proposed an extended SP-based P2P file-sharing architecture, which supports QoS-enhanced search and caching of popular files and their locations. In this section, we present results of basic experiments to show the advantages of this architecture in terms of average response time, QHR, the effect of varying number of search requests for different network size, and routing power. We compare them to existing file-sharing architectures such as eDonkey and Limewire. The objectives of these experiments are to demonstrate the strength and impact of our architecture in terms of scalability and performance.

Figure 15.16 shows the effect on the number of visited SPs and the efficiency of search results when varying the number of SPs. The results are averaged over 5000 search requests. As the number of search requests increases, the number of visited SPs grows quickly. Although the number of visited SP increases with the number of search requests, on average, fewer SPs need to be visited to find the requested file. When the user requests 15 files, on average, 3 SPs need to be searched to find the requested file. When the number of search requests increases to 960, on average, only 0.3 SPs need to be searched to find the requested file. This is due to the fact

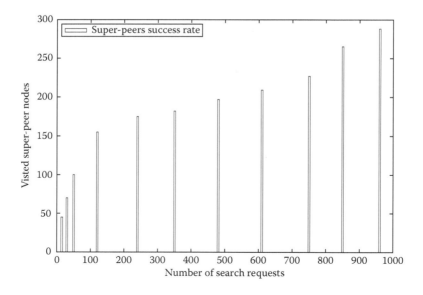

Figure 15.16   The effect of varying number of search requests.

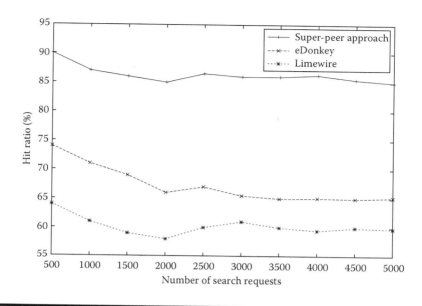

Figure 15.17 Query hit ratio.

that the LISSP not only caches popular files, but also caches the locations of other files that have been searched by the SP but not cached yet. It is likely that many of the files or their locations can be found in the LISSP cache.

Figure 15.17 shows the QHR in the simulation trace results for our SP-based architecture and two other unstructured P2P systems, the Limewire and eDonkey. QHR is defined as the ratio of the number of responses that the resources are found to the number of queries that are sent to find the desired resources. This ratio is a good indicator of the possibility of finding a resource among large P2P file-sharing networks. Experiments show that our SP-based P2P system has better QHR when compared to Limewire and eDonkey. This is because it is likely that data or its location is found in the LISSP's cache for a requested search. Even if the file is not found in the LISSP cache, the SP initiates a search to other neighboring SPs that have large number of peers in their clusters. In fact, when an SP is connected to neighboring SPs, it is virtually connected to more number of P2P file-sharing devices. When large numbers of peers are active in such clusters of the searched SPs the availability of resources such as files is high. Limewire and eDonkey users can only search resources within their limited-sized networks. Therefore, we see that the possibility of finding the desired resources is better in an SP-based architecture when compared to Limewire and eDonkey. Figure 15.17 also shows that eDonkey has better QHR than Limewire. This is because eDonkey has more shared files when compared to Limewire.

The average response time for search requests is shown in Figure 15.18. The proposed ESPA architecture has a better response time when compared to eDonkey

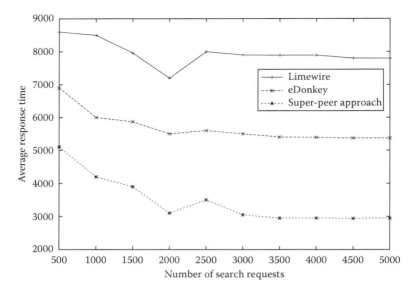

Figure 15.18   Average response time.

and Limewire. This is because our proposed architecture uses the SP, which minimizes the search and response time. The response time for a search from the file or location cache of the LISSP is obviously very fast. Since the queries are flooded in Limewire, the forwarding operations increase the response time of queries. Hence, Limewire has the worst average response time.

A search is successful if the SP discovers at least one peer (or a neighboring SP) containing the requested file. The most frequently accessed files by many peers are the files that are most popular. These popular files and their locations are cached in the LISSP. The ratio of successful searches to total searches made is termed as success rate (or accuracy). A search can result in multiple discoveries (hits) of a requested file. Figure 15.19 shows how object popularity affects the success rate of different search schemes in a highly dynamic environment. In Figure 15.19, popularity decreases as we move to the right along the *x* axis. The first data point represents the accuracy of the search schemes for files 1–10, the second for files 11–20, etc. Here, file 1 is the most popular file with file 100 being the least popular one. Our SP-based search method exhibits high accuracy as compared to modified breath first search (BFS) and Gnutella-type flooding. The modified BFS and flooding show decreasing accuracy as popularity drops. This difference becomes large for medium popular files. Although Gnutella and modified BFS are able to discover maximum number of files, these approaches do not scale as they have huge overheads. To minimize the overheads, we have used time to live (TTL) values, which restricts the search region of these two methods. That is why the success rate is higher in our scheme compared to the other two methods. This also indicates that less popular objects receive less search queries.

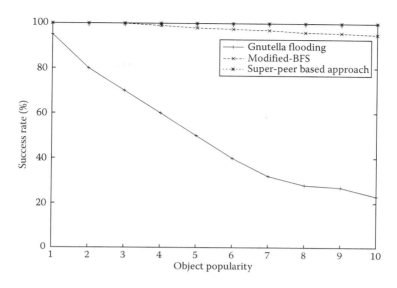

Figure 15.19   Decreasing popularity from left to right.

Existing P2P search algorithms generally target either the performance objective of improving search quality from a peer's perspective, or the objective of reducing the search cost from an Internet management perspective. Our proposed architecture aims to reduce the search cost by building the LISSP database in the SPs, which manages number of clusters and contains the peer resource information thereby providing a pool of popular objects frequently accessed by the peer community. A query is first routed to the SP and most likely to be satisfied there because of the location and file cache in the LISSP of the SP. This will significantly minimize the amount of network traffic and the search scope. The metrics we use for search in P2P system are average response time per query and query success rate. What we are really concerned about is not the absolute success rate of the queries but the cluster relative success rate for a number of clusters under the management of an SP. This is defined as the number of queries that can be satisfied in the cluster over the number of queries that can be satisfied by a flooding search. Figure 15.20 shows the cluster relative success rate in clusters of different sizes for different search satisfactions. The cluster relative success rate increases with cluster size and decreases as the query satisfaction value increases. The cluster size increases as the number of peers that join the clusters increase. The probability of finding a requested search file is quite high when more peers are active in the cluster and they communicate to the LISSP database about their resource availability. Hence, from Figure 15.20, it is seen that cluster-relative success rate for 1 search query satisfaction is much higher than 10 and 50 search query satisfaction.

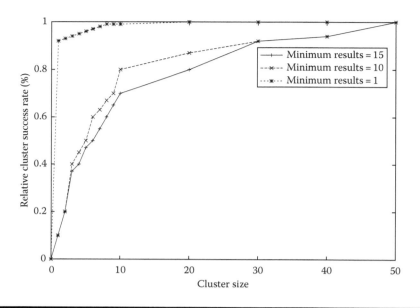

Figure 15.20   Cluster's relative success rate.

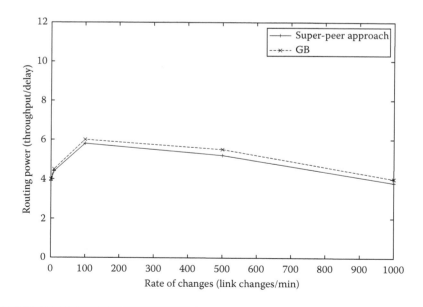

Figure 15.21   Routing power.

In our next simulation experiment, the performance measure we consider is routing power, which is computed as the average search throughput over the average search delay. The average search delay is calculated as the average time spent by all search messages in the P2P network. For search replies delivered to the destination, the delay contribution is the complete end-to-end delay. For those search replies not reaching the destination, the contribution is the amount of time from generation to the end of the simulation. The average search throughput is computed as the percentage of generated search replies delivered to their destination during the simulation time. We clearly observe in Figure 15.21, when MARX is deployed in SPs, the routing power is comparable to that of the Gafni–Bertsekas (GB) [11] protocol. GB is a distributed routing protocol for generating loop free routes in networks with frequently changing topology. Like GB, our protocol performs well at high rate of capacity changes and quickly adapts to topological variations (addition of new SPs in a P2P network).

15.9 Conclusion

This chapter described a scalable, robust, and generalized architecture ESPA for providing good file transfer between devices in a network by authenticating and accounting for the device. We automate the process of registration, and describe a P2P protocol and an SP node database to effectively carry out communication. It is envisaged that there will be enormous amounts of data transfer such as DVD movie clips and music file downloads between SPs. To support enhanced services for such traffic, the proposed ESPA integrates features such as QoS routing, faster search mechanisms and caching of popular files and their locations at SP nodes to speed up data delivery. However, the ESPA may be prone to a single point of failure when traffic load is high, as the SP manages most of the activities of large number of peers within its cluster. Therefore, ESPA proposes to eliminate this single point of failure by replicating the SP's node database, essential to carry out file transactions to neighbor SPs. Basic experiments showed the strength and impact of ESPA in terms of scalability and performance.

References

1. A measurement-based traffic profile of the edonkey filesharing service. Lecture Notes in Computer Science, Vol. 3015, Springer, Berlin, Germany, pp. 12–21, 2004.
2. K. Aberer, A. Datta, and M. Hauswirth. P-Grid: Dynamics of self-organising processes in structured peer-to-peer systems. In R. Steinmetz and K. Wehrle (eds.), *Peer-Peer Systems and Application, Lecture Notes in Computer Science*, Vol. 3485, Springer, Berlin, Germany, pp. 137–153, 2005.
3. K. Abeer and M. Hauswirth. Peer-to-peer information systems: Concepts and models, state of the art, and future systems. *ACM SIGSOFT Software Engineering Notes*, 26(5):326–327, 2001.
4. H. Balakrishnan, M. F. Kaashoek, D. Karger, R. Morris, and I. Stoica. Looking up data in P2P systems. *Communications of the ACM*, 46(2):43–48, 2003.

5. M. Castro, P. Druschel, A. Ganesh, A. Rowstron, and D. Wallach. Secure routing for structured P2P overlay networks. In *Proceedings of the Fifth Symposium on Operating Systems Design and Implementation*, Boston, MA, pp. 299–314, Dec. 2002.

6. M. Castro, P. Druschel, A. Kermarrec, and A. Rowstron. SCRIBE: A large-scale and decentralized application-level multicast infrastructure. *IEEE Journal on Selected Areas in Communications*, 20(8):1489–1499, 2002.

7. I. Clarke, O. Sandberg, B. Wiley, and T. W. Hong. Freenet: A distributed anonymous information storage and retrieval system. In *Proceedings of the ICSI Workshop on Design Issues in Anonymity and Unobservability*, Berkeley, CA, pp. 311–320, 2000.

8. E. Cohen and S. Shenker. Replication strategies in unstructured P2P networks. In *Proceedings of the ACM SIGCOMM*, Pittsburgh, PA, pp. 177–190, Aug. 2002.

9. F. Dabek, M. F. Kaashoek, D. Karger, R. Morris, and I. Stoica. Wide-area cooperative storage with CFS. In *Proceedings of the 18th ACM Symposium on Operating Systems Principles*, Banff, Canada, pp. 202–215, 2001.

10. P. Francis, S. Jamin, C. Jin, D. Raz, Y. Shavitt, and L. Zhang. IDMPAS: A global Internet host distance estimation service. *IEEE/ACM Transactions on Networking*, 9(5):525–540, 2001.

11. E. Gafni and D. Bertsekas. Distributed algorithms for generating loop-free routes in networks with frequently changing topology. *Proceedings of IEEE Transactions on Communications*, 29(1):11–18, 1981.

12. L. Gong. JXTA: A network programming environment. *IEEE Internet Computing*, 5(3):88–95, 2001.

13. C. Hesselman, H. Eetink, I. Widya, and E. Huizer. A Mobility-aware broadcasting infrastructure for a wireless internet with hotspots. In *Proceedings of the First ACM International Workshop on WMASH*, San Diego, CA, pp. 103–112, 2003.

14. T. Hossfeld, K. Tutschku, and F. U. Andersen. Mapping of file-sharing onto mobile environments: Feasibility and performance of eDonkey with GPRS. In *Proceedings of the IEEE Wireless Communications and Networking Conference*, Vol. 4, New Orleans, LA, pp. 2453–2458, Mar. 2005.

15. R. Huebsch, J. M. Hellerstein, N. L. Boon, T. Loo, S. Shenker, and I. Stoica. Querying the Internet with PIER. In *Proceedings of the 19th International Conference on Very Large Databases*, Berlin, Germany, pp. 321–332, 2003.

16. A. Klemm, C. Lindemann, and O. P. Waldhorst (eds). *P2P Computing in Mobile Ad Hoc Networks*, Lecture Notes in Computer Science, Vol. 2965/2004, Springer, Berlin, Germany, 2004.

17. J. Kubiatowicz, D. Bindel, Y. Chen, S. Czerwinski, P. Eaton, D. Geels, R. Gummadi et al. Oceanstore: An architecture for global-scale persistent storage. In *Proceedings of Ninth International Conference on Architectural Support for Programming Languages and Operating Systems*, Cambridge, MA, pp. 190–201, 2002.

18. K. Lakshminarayanan, A. Rao, I. Stoica, and S. Shenker. Flexible and robust large scale multicast using i3. Technical Report UCB/CSD-02-1187, UC Berkeley, Berkeley, CA, 2002.

19. K. Keong Lua, J. Crowcroft, M. Pias, R. Sharma, and S. Lim. A survey and comparision of P2P overlay networks. *Proceedings of the IEEE Communications*, 7(2), 72–94, 2005.

20. C. K. Miller. *Multicast Networking and Applications*, 1st edn., Addison Wesley, Reading, MA, 1998.

21. W. Nejdl, B. Wolf, C. Qu, S. Decker, M. Sintek, A. Naeve, M. Nilsson, M. Palmer, and T. Risch. EDUTELLA: A P2P networking infrastructure based on RDF. In *Proceedings of the 11th International World Wide Web Conference*, Honolulu, HI, pp. 604–615, 2002.

22. W. Nejdl, M. Wolpers, W. Siberski, C. Schmitz, M. Schlosser, I. Brunkhorst, and A. Loser. Super-peer based routing strategies for RDF-based P2P networks. In *Proceedings of the Web Semantics*, pp. 536–543, 2004.

23. S. Rajasekhar, B. L. Smith, and Z. Tari. QoS path routing based on capacity to link ratio in networks. In *Proceedings of the Networks Parallel and Distributed Processing and Applications NPDPA*, Tsukuba, Japan, pp. 138–142, 2002.

24. S. Rajasekhar, Z. Tari, and R. J. Harris. Quality of service in large networks. In *Proceedings of the ATCRC Telecommunications and Networking Conference*, Perth, Australia, 2001.

25. S. Ratnasamy, P. Francis, M. Handley, R. Karp, and S. Shenker. A scalable content-addressable network. In *Proceedings of the ACM SIGCOMM*, San Diego, CA, pp. 161–172, 2001.

26. M. Ripeanu. P2P architecture case study: Gnutella network. Technical Report TR-2001-26, University of Chicago, Chicago, IL, 2001.

27. M. Ripeanu, I. Foster, and A. Iamnitchi. Mapping the gnutella network: Properties of large-scale P2P systems and implications for system design. *IEEE Internet Computing*, 6(1):50–57, 2002.

28. A. Rowstron and P. Druschel. Pastry: Scalable, distributed object location and routing for large-scale P2P systems. In *Proceedings of the IFIP/ACM International Conference on Distributed Systems Platforms*, Heidelberg, Germany, pp. 32–35, 2001.

29. A. Rowstron and P. Druschel. Storage management and caching in PAST, a large-scale P2P storage utility. In *Proceedings of the 18th ACM Symposium on Operating Systems Principles*, Banff, Canada, pp. 188–201, 2001.

30. M. Schlosser, M. Sinetek, S. Decker, and W. Nejdl. HyperCup-Hypercubes, ontologies and efficient search on P2P networks. In *Proceedings of the International Workshop on Agents and P2P Computing*, Bologna, Italy, 2002.

31. N. W. Siong, B. C. Ooi, K. L. Tan, and A. Y. Zhou. PeerDB: A P2P based system for distributed data sharing. In *Proceedings of the 19th International Conference on Data Engineering*, Bangalore, India, pp. 633–644, 2003.

32. I. Stoica, R. Morris, D. Karger, M. F. Kaashoek, and H. Balakrishnan. Chord: A scalable P2P lookup service for internet applications. In *Proceedings of the ACM SIGCOMM*, San Diego, CA, pp. 149–160, 2001.

33. A. Striegel and G. Manimaran. A survey of QoS multicasting issues. *IEEE Communications Magazine*, 40(6):82–87, 2002.

34. D. Stutzbach, R. Rejaie, N. Duffield, S. Sen, and W. Willinger. On unbiased sampling for unstructured P2P networks. In *Proceedings of Sixth ACM SIGCOMM on Internet Measurement*, Rio de Janeriro, Brazil, pp. 27–40, 2006.

35. C. Tang, Z. Xu, and S. Dwarkadas. P2P information retrieval using self-organizing semantic overlay networks. In *Proceedings of the ACM SIGCOMM*, Karlsruhe, Germany, pp. 175–186, 2003.

36. H. Zhang, A. Goel, and R. Govindan. Using the small-world model to improve freenet performance. *The Journal of Computer and Telecommunications Networking*, 46(4):555–574, June 2002.

37. B. Y. Zhao, L. Huang, J. Stribling, S. C. Rhea, A. D. Joseph, and J. D. Kubiatowicz. Tapestry: A resilient global-scale overlay for service deployment. *IEEE Journal on Selected Areas in Communications*, 22(1):41–53, 2004.

Chapter 16

Woodapples: A New Approach for Context-Aware Mobile Marketing

Stephan Haslinger, Florian Skopik,
Daniel Schall, and Martin Treiber

Contents

16.1 Introduction

In this chapter, we present our project Woodapples [WOOD]—information people love. Within the last year, we (Tisco, a software company based in Vienna) and the Distributed Systems Group worked on several projects. Tisco was involved in the project MyMobileWeb [MYMW], where it was the work package leader for implementing location-based services within the framework. Section 16.2 discusses the middleware layer, which was implemented within MyMobileWeb [MYMW] to gather contextual information such as location from the mobile handset. Section 16.3 focuses on Woodapples [WOOD], a new mobile marketing platform. Woodapples [WOOD] offers a new way of dealing with mobile information, mainly marketing information. The middleware layer, which is explained in Section 16.2, plays a central role within Woodapples [WOOD].

16.2 Context Information for Mobile Services: Part 1

Location-based services are a key driver in today's telecom market, even if the power of location-based services is not fully exhausted in current telecom systems. To build intuitive location-based services for mobile handsets, one success factor is to cover a broad range of mobile handsets available on the market and to make the services context-aware. Within the EUREKA project MyMobileWeb [MYMW], we implemented a framework to obtain any contextual information from handsets using various capabilities of the mobiles. Contextual information is any information we can obtain from the handset and that can be used for any kind of service. The most obvious information is location information. Within our framework, we built an architecture that can obtain location information from various sources and is not bound to any special handset capability. Furthermore, the architecture can be used to obtain various other context information such as battery level. This information in addition is then used to offer special services to the customer. For this, a correlation of the context information has to be done, which is based on a correlation engine for contextual information.

In the last years, the technology world has witnessed a very powerful new trend: more and more people have started using their handsets when accessing the Internet. The main driving force behind this trend is the improved versatility, power, and usability of mobile handsets, which in turn enabled the market to invest more in building mobile-phone-friendly Web services. One new revenue stream for telecom operators in the next years is definitely location-based services on mobile handsets. Analysts at Gartner expect location to become a mainstream mobile application within 2–5 years. They see the market growing from 16 million users in 2007, to 43.2 million users in 2008, to 300 million users by 2011 [PMF08]. To launch successful location-based services, there is a need for a proven development framework for those services. It has to be reliable, create a community of developers that are focusing on it, and, of course, it should try to cover nearly all mobile handsets on the market. The current situation, with big suppliers (Google, Apple, Nokia, RIM, etc.) developing their own frameworks and ecosystems, does not encourage interoperability. Location information is nothing else than contextual information that can be retrieved from the mobile handset and can be used to build special services for the user. Besides location there is a lot of other contextual information that is interesting to obtain, such as battery level of the handset and weather forecast for a special location. All this information correlated can be used to build new mobile services for mobile handsets with a very high flexibility. Imagine a mobile service you can access from your Web browser to get tourist information, for example, Barcelona. You are a tourist on a trip to Barcelona. In the airport, you get an e-mail from your hotel reservation system informing you that they had to cancel your booked hotel room. We assume that you installed the framework from MyMobileWeb [MYMW] on your handset, which has the possibility to retrieve context information about your environment. You browse to your hotel information site, which shows you all four- and five-star hotels in Barcelona, as this is predefined in your private context information stored on your handset. You book the hotel. Afterward, the site leads you to another site offering you a plan on how to use public transport to get to your hotel. Once you come to your hotel and enjoy some hours of rest, you are in the mood for some sightseeing and enquire about places of interest. The Web site offering you this information gives you a list of interesting places. As the service is aware that it is one of those rare rainy days in Barcelona, indoor places, such as museums, are ranked higher than outdoor places. You just click on the sights of interest and another service starts where a navigation system for walkers comes up and guides you through Barcelona. Services like this rarely exist and are always bound to a certain handset or to a certain technology your mobile handset has to support. We are proposing a framework that can assist developers to build such services with a very broad range of mobile handsets available on the market. To build such a framework, it is crucial to build data sets of context information that are well known by the services on the back-end side as well as on the mobile side.

16.2.1 Architecture of the Framework

Figure 16.1 shows the architectural overview of the framework, which is built within the project MyMobileWeb [MYMW]. Input channels are channels that can retrieve contextual data from the handset. An input channel can be any sort

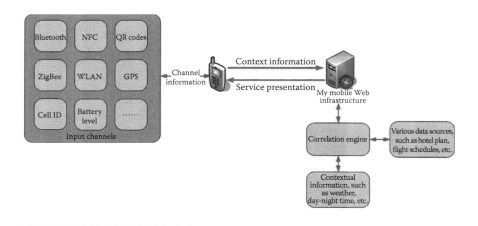

Figure 16.1   Architecture of the framework.

of hardware capability of the handset that can deliver context data. In Figure 16.1, we assume as examples the GPS sensor, QR codes, battery level, etc. This contextual data alone would not allow a mobile service to give any valuable information to the user. Therefore, this data needs to be correlated with suitable information to build mobile services described in the introduction. Location-based services, just as any other context-aware service, are aimed at customizing their contents, appearance, and behavior to optimize their usefulness. Pure location information is central to such services, but it is not enough to make the most with respect to usability. Imagine a recommender system that offers you nearby restaurants on a map, not only will it have to adapt such map to your screen size, but also it will have to consider the resized result and know whether icons on the map are visible or not, seamlessly adapting to small devices, and it may filter vegetarian restaurants as you love meat. So there is some context information, not only the location that can enhance LBS. Furthermore, since application domains are unpredictable, establishing a boundary on what may be useful or not in a given domain will constrain the developers trying to offer innovative LBS. The presence of such, potentially huge, amount of factors and their influence in the usefulness of the offered service makes it necessary to have an automatic correlation process. Correlation in then crucial in LBSs as a mechanism for selecting context-dependent contents based on location and non-location information.

16.2.2 Implementation of the Framework

This section covers the concept of the middleware layer in detail, which is used on mobile handsets to fetch context information, which is, in our case, nowadays mainly location information. It discusses how a middleware layer would have to be set up to be reused within the project MyMobileWeb [MYMW]. The chapter discusses different possibilities how a middleware layer could be implemented. We believe that following our ideas helps other developers to choose an appropriate architecture for their specific problems.

One of the main steps in development of context-aware services is acquisition of location information from mobile device. Since there are no standardized APIs or libraries available at the moment, this step involves development of device specific code targeting particular device. The resulting code is usually not portable to other devices since it is dependent on underlying device hardware, operating system, and device driver libraries provided by device vendor. Because of this reason, the resulting solutions are usually targeting only restricted and small set of devices available on the market. In order to extend the set of devices that support such service, same functionality must be rewritten multiple times for different devices from different device vendors. All of this makes development of location-based solutions time-consuming process, which slows down development of new services. So, the first problem that has to be addressed by the framework design and architecture is covering as many mobile device types with as much reusable code as possible. Second, since described framework is targeting development of Web-based services and it is intended as extension to MyMobileWeb [MYMW] solution, communication channel between location information acquisition code running on the device and back-end server must be established. Through this communication channel, back-end server side must be able to trigger acquisition of location information on the mobile device, and on another side, through the same communication channel, code running on the device must deliver the location information back to the server side.

16.2.2.1 Requirements

On one side, MyMobileWeb [MYMW] is based on Web server side technology for which the client side software is the Web browser installed on the user mobile device. On the other hand, in order to acquire context information from a mobile device, a dedicated software must be run on the device itself. The requirements imposed by this on the overall framework architecture are

Server side triggering of location information retrieval from the mobile device must be done through a Web browser as a client-side software.

Communication channel must be established between the MyMobileWeb [MYMW] server side and the location information acquisition framework running on the mobile device for delivery of location information. Additionally, there are different sorts of location information types (GPS position, Cell ID, NFC ID, etc.). Therefore, the server side must have a possibility to specify which location information type is required for a particular service. Since there are many different mobile device types on the market with different capabilities for acquisition of location information, there also must be a possibility of detecting which types of location information can be acquired on a particular mobile device. This information must be delivered to the server-side application prior to actual acquisition of location information. Therefore, the following requirements must be satisfied by the architecture as well:

■ Detection of device capabilities regarding location information types on mobile device itself. Notifying server side about device capabilities.
■ Possibility of specifying required context information type by the server side. Further on, there are many different types of location information and also different ways of collecting it from different devices. It is obvious that it is not possible to cover all of it with one monolithic application. This leads to the conclusion that the architecture must be designed in a way that makes it easily extendible with additional modules. Intention of additional modules is to easily add support for additional location information types and additional location information acquisition methods, as well as providing the possibility to reimplement already existing modules for different device types. This means that an additional architectural requirement is
  – Modular design that is easily extendible by additional modules. Finally, there are some cases in which the user should be able to acquire location information on his own (without this action being triggered from server side) and then use it (deliver it to the server side) later on.
  – Possibility to trigger context information acquisition not only by the server side but by the user as well.

16.2.2.2 Triggering Context Information Acquisition from the Server-Side Application

This section covers various methods of triggering the context information acquisition. For example, if the user is browsing Web pages that are not offering any context information services, it does not make any sense that the acquisition framework is running all the time on his mobile device and blocking its resources. The acquisition framework should be started only when the user browses a Web page that really offers some location-based services. Furthermore, the context information should be acquired exactly at the moment when the server side is ready to process the received information. To achieve this, a mechanism that enables the server side to trigger context information acquisition on the user's mobile phone must be designed. An additional requirement is that, since the client software on the mobile device is a Web browser, the triggering mechanism must be implemented through means provided by the Web browser's client-side application.

16.2.2.3 Channel for Delivery of Context Information to the Server-Side Application

Once the location information acquisition has been triggered and location information has been gathered on the mobile device, a communication channel between the acquisition framework on the mobile device and the server side application must be established in order to deliver acquired data. This, once again, can be implemented through the means provided by the Web browser installed on the user's mobile device, or by establishing a direct connection between the acquisition framework on a mobile device and the server side.

16.2.2.4 Detection of Device Capabilities

Detection of device capabilities does not only consist of finding out what sort of device it is and deciding which context information can be retrieved from this device. It also involves the detection of the version of the acquisition framework that has been installed on this device and informing the server side about it. Based on this information, the server side can then trigger the desired context information acquisition method or it can inform the user that in order to use the particular service he must install additional software on his device (or simply informing him that the service cannot be used).

16.2.2.5 Specification of Desired Context Information by the Server

Once information about the device capabilities has been delivered to the server side, the server must be able to inform the acquisition framework installed on the mobile device about the type of context information it would like to receive (Cell ID, GPS position, NFC ID, etc.) and the acquisition framework must be able to gather the desired location information accordingly.

16.2.2.6 Modular Design

From the previous requirements, it can be seen that there are some common tasks that must be performed by the acquisition framework regardless of the context information type that is used. Those common tasks are

- Detection of device capabilities.
- Communication with the server (reporting device capabilities and delivering acquired data). Those functions can be grouped together in one common module. On the other hand, the current implementation of the context information acquisition is not only dependent on the context information type,

but also on the device type, the underlying operating system, and hardware. Therefore, the implementation of this functionality should be divided and separated into different modules (plug-ins) for each location information type and, in some cases, even for different device types.

16.2.2.7 Triggering Context Information Acquisition by the User

In some cases, it can be useful to give the user the possibility to acquire the context information on his own without this being triggered from the server side (off-line mode). It also must be possible to deliver the acquired data to the server side later on. For this purpose, the acquisition framework must provide some sort of user interface through which the user can see what type of context information he can get from his mobile device, select the desired type, and trigger the acquisition. The collected data must be stored locally on the mobile device until the user visits the Web site that will then request the stored information from the acquisition framework.

16.2.2.8 Possible Solutions

During design phase, two main implementation possibilities were considered and both of them, together with their advantages and disadvantages, are presented in the following sections. Also, a third possibility, which is a hybrid combination of the first two, will be shown.

16.2.2.8.1 Web Browser Extensions–Based Architecture

This architecture is based on extending the mobile Web browser with a set of plug-ins, through which location information would be acquired from the mobile device and delivered to the server side. The concept of this architecture is presented in Figure 16.2.

The mobile device is extended with an additional plug-in that is capable of communicating with the device drivers of the handset through native APIs. The native APIs are used to fetch context information from the device, which is then sent back to the server side.

Advantages of this approach

■ Since MyMobileWeb [MYMW] is a Web-based technology, meaning that the primary client side software is a Web browser, this architecture simplifies integration of location-based services into the rest of the MyMobileWeb [MYMW] framework. Since the main client software on the mobile device for MyMobileWeb [MYMW] is a Web browser, the easiest way to deliver context information from a mobile device to the server side is to extend the browser capabilities in a way that it is able to acquire context information

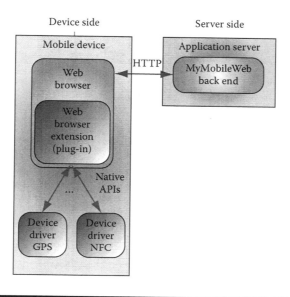

Figure 16.2   Web browser extension.

from the mobile device. Doing this through a plug-in mechanism (provided by most mobile Web browsers on the market) is an obvious and straightforward way.

■ As the plug-in is a part of the Web browser, this architecture reduces problems with synchronization of the browser and the back-end server, since the browser itself is responsible for context information acquisition (synchronization problems between Web browser and the acquisition framework are more noticeable in the Java2ME-based architecture, which is explained in detail in the following chapter).

■ Since MyMobileWeb [MYMW] also incorporates AJAX capabilities through usage of JavaScript, this architectural approach makes it easier to implement a JavaScript event mechanism (creating additional events when context information is acquired and available for usage on the server side). This approach eases the usage of location information during the Web site development phase for developers that use the MyMobileWeb [MYMW] framework.

Disadvantages of this approach

■ The first big disadvantage is the fact that the source code is Web browser dependent. There are a couple of different mobile Web browsers on the market today, and in order to cover most of the market, totally different extensions would be required for each of the existing Web browsers (since all of them have different plug-in writing principles and requirements). Furthermore, whenever a new Web browser appears on the market, a completely new implementation from scratch would be required.

■ Since most of the Web browsers offered on the market today are implemented as mobile device native applications (unlike Java2ME applications, and OperaMini, which is one exception to this rule), plug-in development is also done for the native platform or through ActiveX controls in the case of Windows-based mobile devices. This implies dependencies upon the underlying operating system as well.

■ Because of the reasons mentioned above, this architecture is not time resilient. Whenever some Web browser or operating system disappears from the market, the existing source code for this browser or system would become obsolete.

16.2.2.8.2 Java2ME-Based Architecture

This architecture is based on Java2ME technology. In this solution, a standalone program is installed on a mobile device and runs as a server that provides context information whenever it receives requests through a standard localhost socket connection. Figure 16.3 presents this concept.

Although Figure 16.2 looks similar to the concept described in the previous chapter, there is one major difference. In this case, the Web browser does not need to be extended. In this case, the Web browser does not have the capability to acquire the context information from a mobile device on its own. Instead, there is a separate, standalone, and totally independent software (Java2ME Server Midlet)

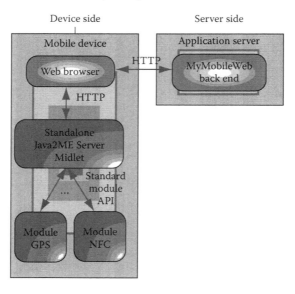

Figure 16.3   Java2ME architecture.

installed on the mobile device that is in charge for context information acquisition from the device. Whenever the MyMobileWeb [MYMW] server side requests context information, the Web browser just forwards the request to the ServerMidlet, which in turn collects the location information from the Server Midlet and forwards it back to the MyMobileWeb [MYMW] server side.

Differences to previous approach

■ The implementation is completely independent of the Web browser used on the mobile device. The only interaction between the Web browser and the standalone Server Midlet is done through classical TCP/IP socket connection through the usage of HTTP protocol. This is, by definition, supported by all Web browsers and removes necessity to support different Web browsers individually and thus provides better market coverage with less time and effort.
■ Java2ME as technology for the implementation of the acquisition framework ensures operating system independence and source code that is interoperable between different devices. Java2ME can be seen as mobile device vendor independent since it is being widely accepted and supported by all major mobile device vendors. The usage of this technology ensures very high market coverage.
■ Because of the reasons mentioned above, this approach is more time resilient than the first one, in a way that it is not dependent on the currently existing Web browsers. It will integrate seamlessly with every new Web browser that could appear on the market. Besides this, this approach is much more device and vendor independent since it is based on Java2ME standard.
■ An additional advantage when using a standalone (Web browser independent) concept is that the architecture can be extended easily for an off-line mode of work, when fetching of context information is not triggered by the server side but by the user instead. The off-line mode can be supported by adding GUI that is presented to the user in the case that the Server Midlet is started manually (and not triggered by the server side). This provides the possibility to the user of collecting context information at any time. Collected information can then be used with any Web site providing context information services that is built with the usage of MyMobileWeb [MYMW] technology.

Disadvantages of the approach

Although this architecture has some obvious advantages over the previous one, it also introduces some disadvantages and problems that are not presented in the first approach:

■ First of all, this approach complicates data (context information) exchange with the MyMobileWeb [MYMW] back-end server. Since the client software for MyMobileWeb [MYMW] is a Web browser and the acquisition framework runs as a standalone application on the mobile device, there must be some mechanism implemented to establish a communication channel between the server side and the acquisition framework through a Web browser (this communication channel will be used for triggering a context information acquisition from the server side and for delivery of the context information back to the MyMobileWeb [MYMW] server).

■ Second, in order to request and receive context information, the Web browser must connect to the acquisition framework running on the device through a standard TCP/IP socket connection with usage of HTTP protocol for data exchange. This imposes possible time-out problems, since in some cases, location acquisition can be a time-consuming process. If the communication between the Web browser and acquisition framework is synchronous, which means that the Web browser sends a request to the acquisition framework and then waits for the response, it can happen in certain cases that the session times out (Web browser concludes that server is not responding) if context acquisition process takes too long.

■ Furthermore, if the communication between the Web browser and the acquisition framework on the device is asynchronous, which means that the Web browser sends a request to the acquisition framework on the device and gets redirected back to the server-side Web page before the location information has been acquired, there is a problem of how and when to deliver the location information to the server side.

■ Besides this, in the case of an asynchronous communication, the architecture should be designed in a way that it enables the implementation of some sort of event-based mechanisms on the server side in order to make usage of context information easier during the Web site development process.

■ This architectural approach, although providing big overall benefits, introduces some additional and new problems. Although none of those problems is technically unsolvable, and none of them represents major drawbacks, they should be considered during implementation phase and solved accordingly.

16.2.2.8.3 Hybrid Java2ME and Web Browser Extensions Architecture

This architectural solution is a combination of the architectures described previously. The idea of this architecture is to combine advantages of both approaches into a single solution in order to minimize problems that both architectures have inherently. The high-level idea of this approach is presented in Figure 16.4. As it can be seen from Figure 16.4, this architecture is just a hybrid composition of the previous two. The idea is to try to take the best from both and combine it into a single solution.

Advantages of this solution

■ Easier implementation of data exchange and synchronization between server and client side through Web browser extensions (plug-ins).
■ Easier implementation of event-based development mechanisms through Web browser extensions.

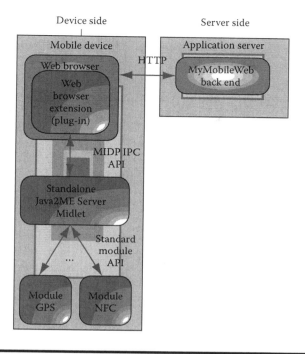

Figure 16.4   Hybrid architecture.

■ Retaining most of the Web browser and underlying operating system independence through implementing most of the functionality in Java2ME technology and keeping only a small functional part as a Web browser extensions
(implementing "thin" Web browser extensions with a small codebase and
using them only as a "glue" between MyMobileWeb [MYMW] server side
and the acquisition framework implemented in Java2ME technology).
■ Retaining the off-line mode of work through the location information acquisition Server Midlet.

Disadvantages that still remain in this approach

■ Although "thin," Web browser extensions are still used. This implies a certain
amount of dependency to both the Web browser and the underlying operating system. Certainly not the kind presented in the pure Web browser extensions architecture, but nevertheless a noticeable one.
■ Heavy prototypes. A lot of time and effort is needed in order to develop a
prototype, since it requires the implementation of the whole Java2ME-based
architecture in addition to the "thin" Web browser extensions for a couple
of browsers.

16.2.2.9 The Chosen Architecture

As a result of different architectural considerations, the conclusion was made that Java2ME-based architecture is the most appropriate one for mobile device context acquisition framework implementation. This decision was made because of the following reasons:

■ Although introducing some additional problems, the Java2ME-based architecture has big advantages over the Web browser extensions–based architecture in terms of independence. It does not rely on specific browser or underlying operating system features, thus providing better market coverage with less time and effort.
■ The implementation of hybrid Java2ME and Web browser extensions architecture, although having certain advantages, is time and effort consuming for initial framework implementation.
■ After Java2ME-based architecture implementation is finished, it can be easily upgraded later to the hybrid version. Because of this, during the implementation phase, special attention was given in order to make this upgrade as easy as possible. Upgrading the architecture will be possible with only minor changes done to Java2ME application code and only by adding implementations of "thin" Web browser extensions on top of the already existing architecture.

16.3 Woodapples: Part 2

The Web 2.0 paradigm encourages users to collaborate and share knowledge and information, so the Web is no longer a "read-only" information repository. However, the collaboration is usually neither structured—there is no interaction link between the users—nor can the users in general apply the same procedure in other collaborations. This makes it difficult—if not impossible—to manage interactions that might span multiple users and services not to mention over a certain period of time. With Woodapples [WOOD], we are employing a platform for mobile marketing, where humans and services comprise a collaborative system not only incidentally but for a certain time span.

Woodapples [WOOD] is a mobile marketing platform, which reuses the work from the middleware layer presented in Section 16.2. On top of the middleware layer, a new midlet got implemented, which we call the notification engine. With the notification engine, it is possible to retrieve information when entering a special location area.

16.3.1 Mobile Information/Marketing Platform—Woodapples: Bringing the Pieces Together

Tisco provides a platform for mobile marketing, which is named Woodapples [WOOD]. A movie illustrating how Woodapples [WOOD] is working can be found under http://www.youtube.com/watch?v=wq7F8gwIa1A.

This section gives an overview how Woodapples [WOOD] works. The idea is to provide a new mobile marketing platform, which takes into account human preferences.

The idea behind this platform is to build an advertisement platform such as Google ads for mobile handsets. Service providers can set up campaigns with a specific budget, a time span, and a location radius. The marketing info will be then published to people within a certain location area in a specified time span if they are interested in this information. Potential customers interact with the platform through dedicated interfaces (see Human-Provided Services [SGD+07]), and set up their preferences so that the platform really provides the information they want.

A generic scenario of an interaction flow will typically assume an advertiser (service provider in general), who wants to place an advertisements. The campaign should be of some exclusivity and thus narrowed to only a small range of so-called experts. An expert will be somebody whose reputation [SSD09a] is well known in the community, and preferably being a longtime customer. The reason for the exclusivity could be a voucher included in the advertisement.

The mobile marketing concept includes that basically (but not exclusively) only people, who enter the dedicated vicinity, will receive the notification. Receiving a location-based advertisement will not be considered a spam message, since the person is registered for the notification service and willing to receive information concerning specific topics.

If the advertisement includes a voucher, the addressee will be attracted to directly interact with the advertiser (e.g., enter the shop or restaurant, etc.). Redeemed vouchers can be recorded as success. The ratio between redeemed and not redeemed vouchers will be a measurement for the success of the campaign. The advertiser will be able to refine the client's profile if the ratio tends to one of both extremes (0 or 1).

For the expert, it will strengthen his reputation (increase his expertise [SSD09a]) if he embraces the offer to interact, since this interaction clearly distinguishes him from so-called "only said to be experts." If he rejects too many times, he will lose even his reputation as expert ("How else should he know?").

If the expert is not able to get in contact with the advertiser currently, he might want to forward the advertisement to friends. The reason for forwarding the info could be an incentive clause in the advertisement (e.g., 10 new customers implying one meal for free).

From case to case a "newcomer" will be among the friends. Why does he get notified? The reason for that might be, that he recently dropped the wish to get in touch with this specific topic (e.g., "never ate Asian food, want to try it the first time"). In that case, the forward interaction of the expert represents a recommendation to a potential new customer. Since the expert is well known for good recommendations, the newcomer will take his chance. But, for some reason, he declines to act alone. He invites his friends or even spontaneously similar minded people (see [SSD09b] for interest similarity models) in the vicinity to join in (technically this includes a trick: the profile of the newcomer needs to be temporarily updated with the new interest). The message to the invited people could include a motivating comment on this event being the "very first time" ("teach me how to eat with chop sticks"). Later, the newcomer could rate the recommendation and thus update his own profile (being interested in the topic), and indirectly the reputation of the expert.

The generic scenario can be applied to any kind of advertisement with products or services, but is not limited to marketing examples. The principle can be advantageous for a wide set of interactions, for instance:

■ Education platform: Offering (advertising) a new course → contacting experts (e.g., former attendees) → recommendation finds new clients → inviting buddies to join in → a group attends the course
■ Social games: A spontaneous idea to do an activity in a group → expert joins in but recommends also to friends → group gathers and conducts the game

Most of the technical parts of Woodapples [WOOD] are already implemented.

16.3.2 Woodapples [WOOD]: A Use Case

This section describes a common use case that can be provided with Woodapples [WOOD]. We assume a use case where a vegetarian restaurant is setting up a marketing campaign. Figures 16.5 and 16.6 show the architectural principles behind this use case.

■ 09:00 A new vegetarian restaurant has opened in down-town Vienna. It wants to attract people from the surrounding area and offers a limited special menu only available for those contacted via Woodapples [WOOD].

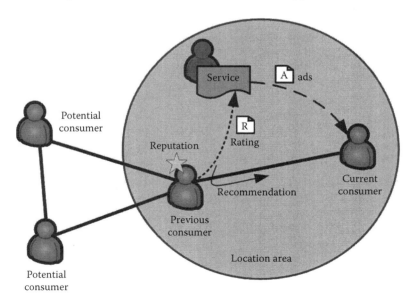

Figure 16.5   Use case description.

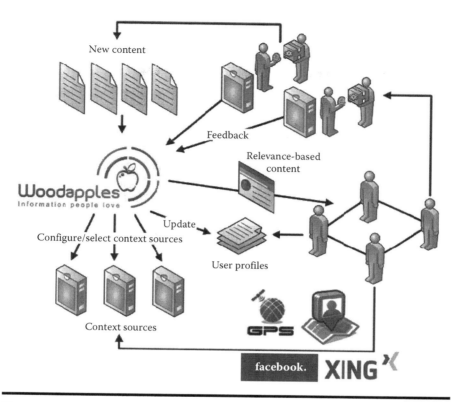

Figure 16.6   Woodapples-architectural overview.

- 09:30 It sets up the marketing campaign in the Woodapples [WOOD] platform having to specify only that it is offering vegetarian food, and its opening
  hours. Internally, Woodapples [WOOD] constantly monitors and reconfig-
  ures the context and preference sources that provide the core data to determine the relevance of an individual context element for a particular user.
- 09:45 As Woodapples [WOOD] is a self-aware system, it knows about all
  involved system elements, i.e., the set of current content for delivery, the online
  user base, granularity of their preferences, and available context sensors. Bob,
  one of the content consumers, has subscribed to food-related information. Bob
  likes to dine out, but does not really have any preferences for a certain cuisine.
- 11:00 Woodapples [WOOD] switches from coarse-grained to fine-grained
  location information as lunch time approaches. Doing so preserves the user's
  privacy as context information is only retrieved when deemed relevant.
- 11:15 Woodapples [WOOD] is aware of interactions between users. For
  doing so, it can select between various social networking platforms, selecting
  the ones that deliver the best information for the current user base. Coupling
  social structure with location, Woodapples [WOOD] becomes aware that Bob

and his girlfriend are within the same area. She is a vegetarian. Subsequently, Woodapples [WOOD] sends out the content about the new restaurant to both of them, reasoning that they would like to have lunch together.

■ 12:00 Bob displays the personalized code for special offer at the vegetarian restaurant. Subsequently, Woodapples [WOOD] receives feedback via the restaurant that the content delivery was relevant indeed.

16.4 Summary and Outlook

In this chapter, we presented an approach of how a middleware layer for mobile handsets, which is capable of gathering context information from the handset, could look like. We believe that our detailed discussion that led us to a final architecture can help other developers when facing a similar problem.

On top of the framework we implemented a new product, called Woodapples [WOOD]—Information people love. The product is a mobile marketing platform, which delivers information to mobile users, taking into account their preferences and location. Within the next months and years, we will refine Woodapples [WOOD].

References

[WOOD] Online at: http://www.woodapples.com

[MYMW] Online at: http://mymobileweb.morfeo-project.org/

[PMF08] M. Palmer. Handset makers find a route map to the future. *Financial Times*, February 2008, online at: http://www.ft.com/cms/s/0/186df4b2-db69–11dc-9fdd-0000779fd2ac.html?nclick_check=

[SGD+07] D. Schall, R. Gombotz, C. Dorn, and S. Dustdar. Human interactions in dynamic environments through mobile Web services. In *IEEE 2007 International Conference on Web Services* (*ICWS*), July 9–13, 2007, Salt Lake City, UT.

[SSD09a] F. Skopik, D. Schall, and S. Dustdar. The cycle of trust in mixed service-oriented systems. In *Proceedings of the 2009 35th Euromicro Conference on Software Engineering and Advanced Applications* (*SEAA*), August 27–29, 2009, Patras, Greece.

[SSD09b] F. Skopik, D. Schall, and S. Dustdar. Start trusting strangers? Bootstrapping and prediction of trust. In *Proceedings of the 10th International Conference on Web Information Systems Engineering* (*WISE*), October 05–07, 2009, Poznan, Poland.

Chapter 17

JAVA IMS Mobile Application Development

Francisco Javier Díaz, Claudia Alejandra
Queiruga, and Jorge Horacio Rosso

Contents

17.1 Motivation

During the year 2008, the Computer Science School of the UNLP (National University of La Plata) signed a cooperation agreement with Ericsson Company in Argentina with the goal of undertaking activities to promote innovation, training, and research in the area of NGN (New Generation Networking) and IMS (IP Multimedia Subsystem), primarily addressing issues related to the development of Java applications on this new vision of converged networks.

Students from the "Software Lab" subject, which belongs to the fourth year of the Computer Science and Systems degrees, learn to develop Java client applications, both for desktop and mobile devices. We consider that having a high-quality converged network infrastructure like IMS, which makes it possible to integrate services that are usually provided by heterogeneous networks, is a challenge for the construction of new applications on mobile devices, where communication and sharing play a central role.

According to Patrik Heldmund, innovation area manager in Ericsson-Argentina, the expectations in the development of Java applications on IMS are to find innovative ideas on how to use the technology and create services that create real value for the consumer. Ultimately, the driving force for network transformation will be related to applications and end user services. The cooperation between Ericsson and UNLP create interesting synergy effects for the students, the university, and Ericsson, and is a successful example of the need to create more industry–university projects in this field.

To fully understand the services provided by the IMS architecture, we illustrate an example of daily life according to current communication behavior, where mobility and device diversity are already in common use. IMS encourages new experiences for users, promoting the use of a convergent network which guarantees a comfortable bandwidth, quality of service, and device neutrality. When the odds of communication increase, physical distances become shorter, and media simultaneity and device diversity further enrich the experience.

Let us imagine a situation in which a group of friends share the experience of a rock concert by means of their devices connected to an IMS network. Following, we provide a possible IMS use case by means of a simple story.

David goes to the first show of his friend Charlie's band in the most important venue of the city and remembers that his friends Fabiana and Pedro had to leave town unexpectedly. David decides to share with them the moment in which Charlie does his drum solo by sending them a live video which he captures with his cellphone camera.

"I cannot beleive it!! Charlie is playing incredibly well!!" Fabiana writes David, as she shows the video to Pedro. David and Fabiana exchange opinions while they enjoy the fine music.

Pedro surprises his father by sending him an invitation to his TV, and while they share the video they talk about Charlie's new songs. Fabiana also contacts Charlie's brothers, who enjoy their artist brother's music miles away from their birth city, in their PCs at their office.

17.2 What Is IMS?

The major evolution in fixed and mobile telephony networks occurred in the last 20 years. Today, we live in a world in which digital communication has changed the way people communicate. Mobile telephony has a central role in this change, as it is not only used for "talking," but also for capturing and reproducing videos, taking pictures, consulting work schedules, news pages, using dynamic maps, etc.

Historically, voice, data, and video services have been provided on dedicated network infrastructures, even by different telecommunication providers. In turn, it took multiple types of devices exclusively designed to access the services provided by each type of network (cell phone, notebook, fixed telephone, TV, etc.) using different identities and without the possibility of "moving" (roaming) in a transparent way among those devices. These different networks have offered us multiple, powerful, and attractive ways of communication; nevertheless, the services offered behave as islands (video, voice, data, email, etc.) with no synergy between them.

Figure 17.1 shows the services offered by each access technology: cell phone, cable modem (TV and Internet), digital subscriber line (DSL) (fixed telephony and Internet) and Wi-Fi. The limitations given by the base technology hinder agility in broadening the range of services offered as well as their integration by means of different channels.

Classic telephony employs a scheme based on an end device connected to a circuit-switched network with services supplied only by the provider of telephony or specific services. This scheme tends to disappear in the evolution of "networks."

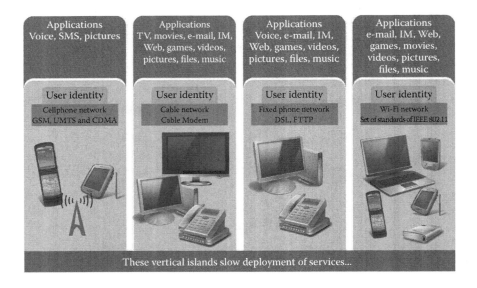

Figure 17.1   Without IMS, network functionality exists in silos, slowing deployment of services.

It is a fact that, in IP networks, any workstation (host) may act as a consumer or a provider of services, giving rise to P2P networks (peer-to-peer). At a time, the most widespread service was NAPSTER [L1]. Other P2P networks emerged later, such as Kazaa [L2] and eMule [L3], which were also used to share files. Other P2P networks promote instant messaging services, from the precursor ICQ [L4] to the popular SKYPE [L5] and MSN [L6], which permit connections between peers without the need of a central Internet service provider server.

IMS is a network architecture, which converges multiple services offered over a common infrastructure for IP data transmission. IMS is a set of specifications that describes the NGN architecture for implementing IP-based telephony and multimedia services. IMS defines a complete architecture and framework that enables the convergence of voice, video, data, and mobile networking technology over an IP-based infrastructure. It fills the gap between the two most successful communication paradigms: cellular and Internet technology. Did you ever imagine that you could surf the Web, play an online game, or join a videoconference no matter where you are using your cell phone from? This is the initial vision for IMS: to provide cellular access to all the services that the Internet provides. Another example of the new services IMS provides within the world of applications is thinking of systems such as TWITTER [L7], which currently enables people to find out what their friends are doing; IMS-TWITTER would also allow us to "see" what is happening, vote online, etc. This new form of communication allows any mobile telephony device to consume and/or produce multimedia services, which combine voice on IP calls, teleconferences, file transference (music, video, general documents, etc.), Web navigation, instant messaging, etc., opening the doors to attractive multimedia applications [R1]. Also, communication companies can provide users their service packages in an increasingly attractive way, combining, for example, a flat rate for voice calls over the fixed and cellular lines, broadband Internet, and IP television [R1].

The 3GPP consortium (Third Generation Partnership Project) developed the first standard for delivering "Internet services" over GPRS (General Packet Radio Service) [L8]. This specification is called R5 (Release 5) 3GPP (3GPP R5). IMS is part of that specification. Then, the vision evolved into a standard that contemplated other networks apart from GPRS, such as wireless networks, fixed telephoning, and CDMA2000 [L9]. Thus arises a new specification called R7 (Release 7) 3GPP (3GPP R7) in which the consortiums 3GPP2 [L10] and TISPAN [L11] also participated.

The IMS architecture over network telecommunications offers a complete convergence of standard services (fixed and mobile telephony, cable and satellite TV, Internet). With IMS, telecommunication operators can integrate fixed calls with mobile ones and voice with broadband Internet access, TV and video on demand, opening doors to new multimedia residential and business applications [R1].

Figure 17.2 shows how IMS provides a common consolidated communications network. This way, the user has access to a multiplicity of different services, integrated through a single identity with the capacity to change media nimbly and improve its experience.

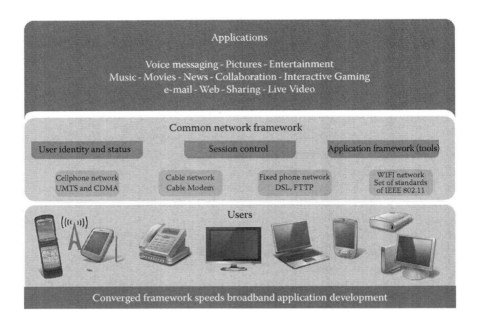

Figure 17.2 With IMS, the networks converge in a common network framework.

17.3 IMS Architecture

The IMS architecture design proposed by the 3GPP is based on functional layers. This solution facilitates the services offered through the IMS network independent of the access network. Thus, any IMS network subscriber that communicates throughout their mobile telephone or PC uses the same presence and group management service, to name a few, regardless of the access technology used. However, the potentials for bandwidth and latency of connections could be different, like the processing speed in the IMS device [R2].

In turn, layer-based design minimizes the dependency between them, facilitating the addition of new access networks to the IMS network. For example, WLAN (Wireless Local Area Network) technology was added to the IMS architecture in Release 6 and fixed broadband access in Release 7 [R2,R3].

Figure 17.3 illustrates the layered modular design of the IMS architecture.

17.3.1 The Role of the SIP Protocol

Internet applications typically need to create and maintain sessions between participants of a communication (client-server or peer-to-peer) to facilitate the data exchange between them. In turn, the exchange of multimedia information is key in applications over an all-IP network such as IMS, and session management becomes an essential issue. This led to the adoption of a protocol for session management

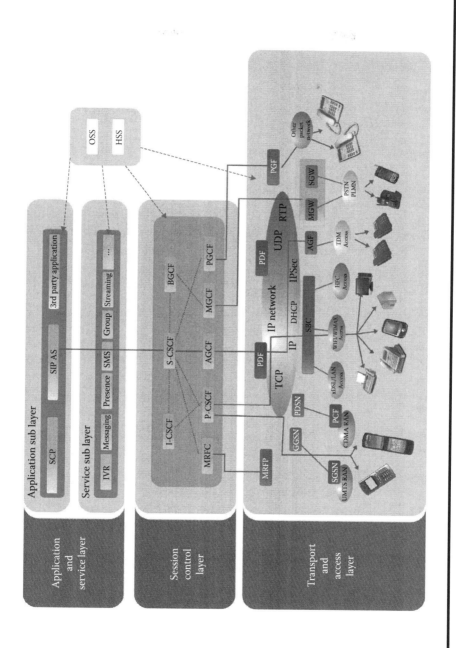

Figure 17.3   IMS architecture–layered design.

characterized by its simplicity, maturity, extensibility, and flexibility regarding mobility such as SIP. SIP (Session Initiation Protocol) is a protocol developed by IETF belonging to the application layer of TCP/IP model and is used to establish, modify, and terminate multimedia sessions in an IP network. The complete specification is available on the IETF RFC 3261 [L12]. Typical applications are related to video, voice, gaming, messaging, call control, and presence. An SIP session could be a phone call between two or more people or a videoconference. The role of SIP in these applications is, during the establishment of the session, the negotiation of the communication parameters between its participants. This negotiation involves the medium (text, voice, video, or other) transport protocol, typically RTP (Real-time Transport Protocol) for streaming and encoding technology (codec). Once these parameters are agreed upon, participants communicate using the selected method. At this point, SIP delegates the transmission in an appropriate protocol. Once the communication in completed, SIP is used again to end the session.

One of the design features of SIP is its ability to operate in collaboration with other protocols used for specific tasks, such as SDP (Session Description Protocol) to describe the parameters previously mentioned in the negotiation phase and RTP for multimedia data transport.

In a word, the role of SIP in relation to these applications is session management and consensus of its attributes, delegating transport of voice, video, and streaming to other protocols such as RTP (Real-time Transport Protocol)/RTCP (RTP Control Protocol) and SCTP (Stream Control Transmission Protocol). All this is accomplished by exchanging text messages.

User identification in IMS is based on the addressing scheme proposed by SIP. SIP addresses are similar to e-mail, prepending prefix *sip*: or *sips*: (secure SIP) and can also contain additional parameters that may indicate preferences. Some examples of URIs are

sip:tom@domain.com

sips:jerry.brown@example.com

sip:mafalda@linti.unlp.edu.ar; transport=tcp

SIP is a request-response type protocol and its messages have a format, which is similar to HTTP (Hypertext Transfer Protocol) and SMTP (Simple Mail Transfer Protocol), used in Web pages and e-mail distribution, respectively. One of the objectives of SIP design was to transform telephony into another Internet service, which is why it was based on two of the most disseminated protocols. The SIP message format consists of three parts: start line, header, and body. The start line enables to distinguish between the SIP messages that represent requests and those that represent responses. In the request SIP messages, the beginning line contains the method name, the request SIP URI, and the protocol version (SIP/2.0 now). The heading is composed of multiple fields that contain information related with the request, i.e., who initiates it, who is the receptor, and the CALL-ID (identifies a SIP dialogue). The information in these fields is the name-value type, enabling the

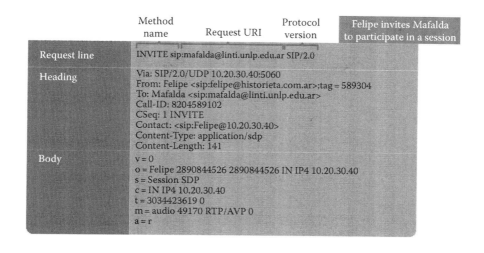

	Method name	Request URI	Protocol version	Felipe invites Mafalda to participate in a session
Request line	INVITE sip:mafalda@linti.unlp.edu.ar SIP/2.0			
Heading	Via: SIP/2.0/UDP 10.20.30.40:5060 From: Felipe <sip:felipe@historieta.com.ar>;tag = 589304 To: Mafalda <sip:mafalda@linti.unlp.edu.ar> Call-ID: 8204589102 CSeq: 1 INVITE Contact: <sip:Felipe@10.20.30.40> Content-Type: application/sdp Content-Length: 141			
Body	v = 0 o = Felipe 2890844526 2890844526 IN IP4 10.20.30.40 s = Session SDP c = IN IP4 10.20.30.40 t = 3034423619 0 m = audio 49170 RTP/AVP 0 a = r			

Figure 17.4   SIP Invite message.

existence of multi-value fields, like in the case of the field *Via* and *From*. Figure 17.4 presents an example of SIP INVITE request message, which in this case expresses the case of a user named Felipe who wishes to initiate a conversation with Mafalda.

The request type messages provided by the SIP protocol are described in Table 17.1 detailing the IETF RFCs in which they were defined. The main specification of the SIP RFC 3261 [L12] defined the six basic methods for session management. However, these were not sufficient to provide support services that are widely accepted today such as instant messaging and presence, which is why SIP was extended in the RFC 3428 [L17], 3265 [L15], 3856 [L20], 3857 [L14], and 3903 [L13] to include four more methods. In addition, other request type messages compete the range of messages provided by the SIP protocol defined in the RFC 3311 [L16], 2976 [L18], and 3262 [L19].

SIP response messages contain a starting line called state line conformed by the protocol version, state code, and a description of the error code (a reason phrase). The state codes are grouped in six categories also established in the RFC 3261 [L12] and summarized in Table 17.2.

Figure 17.5 shows an SIP OK response message example, which expresses that Mafalda accepts Felipe's invitation.

Next, we see the SIP dialogue between Mafalda and Felipe, held for a voice communication. The diagram in Figure 17.6 presents the sequence of SIP messages sent to establish the session, delegate the voice transport on the RTP protocol, and finally end the session.

Table 17.3 describes in detail the sequence of SIP messages exchanged between Felipe and Mafalda to establish the session, which will enable them to "speak," the delegation on the specific protocol that will transport (RTP) the voice and finally, the end of the session.

Table 17.1 SIP Request Messages

Name	Meaning
Specified in RFC 3261	
Invite	Establishes a session
Cancel	Cancels a pending session
Bye	Ends a session
Register	Maps a public URI with the current location of the user
Ack	Acknowledges the reception of an end response originated by INVITE
Options	Consults a server about its capabilities
Specified in RFC 3265, RFC 3856 Y RFC 3903	
Publish	Updates information on a server
Subscribe	Requests notifications about particular events
Notify	Notifies the User Agent about the occurrence of a particular event
Specified in RFC 3311	
Update	Modifies some session characteristics
Specified in RFC 3428	
Message	Transport PSNT telephone signaling
Specified in RFC 2976	
Info	Transport PSNT telephone signaling
Specified in RFC 3262	
Confirms the reception of a provisional response	

Components or elements of the SIP architecture:

User Agent (UA): A logical entity that can act as both a user agent client and user agent server.

User Agent Client (UAC): A user agent client is a logical entity that creates a new request, and then uses the client transaction state machinery to send it. The role of UAC lasts only for the duration of that transaction. In other words, if a piece of software initiates a request, it acts as a UAC for the duration of that transaction.

Table 17.2   Family of State Codes in the SIP Response Messages

Code	Class	Functions	Examples
1XX	Provisional informational	Request has been received and is being processed	100 trying 180 ringing 183 session in progress
2XX	OK	Successful, understood, accepted	20 OK 202 accepted
3XX	Redirect	Redirect or revise request	300 moved 305 use proxy
4XX	Client error	Error detected by the client, syntax error	401 unauthorized 404 not found 415 unsupported media type
5XX	Server error	Error detected by the server, cannot fulfill a valid request	500 not implemented 501 server timeout
6XX	Global network error	Request cannot be fulfilled by any server	600 busy everywhere 603 decline

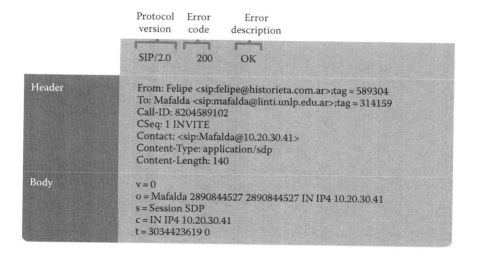

Figure 17.5   SIP OK message.

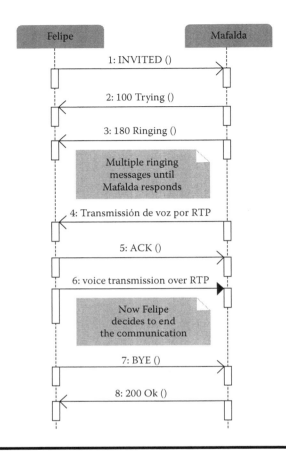

Figure 17.6   SIP dialogue between Felipe and Mafalda.

If it receives a request later, it assumes the role of a user agent server for the processing of that transaction.

User Agent Server (UAS): A user agent server is a logical entity that generates a response to a SIP request. The response accepts, rejects, or redirects the request. This role lasts only for the duration of that transaction. In other words, if a piece of software responds to a request, it acts as a UAS for the duration of that transaction. If it generates a request later, it assumes the role of a user agent client for the processing of that transaction.

The role of UAC and UAS, as well as proxy and redirect servers, are defined on a transaction-by-transaction basis. For example, the user agent initiating a call acts as a UAC when sending the initial INVITE request and as a UAS when receiving a BYE request from the callee.

Similarly, the same software can act as a proxy server for one request and as a redirect server for the next request.

The most relevant functions of each layer of the IMS network are detailed below.

Table 17.3   Sequence of SIP Messages between Felipe and Mafalda

INVITE sip:mafalda@linti.unlp.edu.arSIP/2.0 Via: SIP/2.0/UDP 10.20.30.40:5060 From: Felipe <sip:felipe@historieta.com.ar>;tag=589304 To: Mafalda <sip:mafalda@linti.unlp.edu.ar> Call-ID: 8204589102 CSeq: 1 INVITE Contact: <sip:Felipe@10.20.30.40> Content-Type: application/sdp Content-Length: 141 $v = 0$ 0 = Felipe2890844526 2890844526 IN IP410.20.30.40 s = Session SDP c = IN IP410.20.30.40 t = 3034423619 0 M = audio 49170 RTP/AVP 0 a = rtpmap:0 PCMU/8000	Felipe sends an INVITE message to Mafalda telling her that he wishes to communicate with her. The body of the message contains the negotiation parameters for voice transmission in SDP format.
SIP/2.0 100 Trying From: Felipe <sip:felipe@historieta.com.ar>;tag=589304 To: Mafalda <sip:mafalda@linti.unlp.edu.ar> Call-ID: 8204589102 CSeq: 1 INVITE Content-Length: 0	Mafalda's device responds automatically with a Trying message to signal that the INVITE has been received.
SIP/2.0180 Ringing From: Felipe <sip:felipe@historieta.com.ar>;tag=589304 To: Mafalda <sip:mafalda@linti.unlp.edu.ar>;tag=314159 Call-ID: 8204589102 CSeq: 1 INVITE Content Length: 0	While Mafalda's device rings, it automatically sends ringing messages to inform of this situation.

(continued)

Table 17.3 (continued) Sequence of SIP Messages between Felipe and Mafalda

SIP/2.0 200 OK From: Felipe <sip:felipe@historieta.com.ar>;tag=589304 To: Mafalda <sip:mafalda@linti.unlp.edu.ar>;tag=314159 Call-ID: 8204589102 CSeq: 1 INVITE Contact: <sip:Mafalda@10.20.30.41> Content-Type: application/sdp Content-Length: 140 v = 0 o = Mafalda 2890844527 2890844527 IN IP410.20.30.41 s = Session SDP c = IN IP410.20.30.41 t = 30344236190 m = audio 3456 RTP/AVP 0 a = rtpmap:0 PCMU/8000	Mafalda takes Felipe's call sending the OK response message. The body of the message contains the agreed parameters for voice transmission in SDP format.
ACK sip:mafalda@linti.unlp.edu.ar SIP/2.0 Via: SIP/2.0/UDP 10.20.30.41:5060 Route: <sip:Mafalda@10.20.30.41> From: Felipe <sip:felipe@historieta.com.ar>;tag=589304 To: Mafalda <sip:mafalda@linti.unlp.edu.ar>;tag=314159 Call-ID: 8204589102 CSeq: 1 ACK Content-Length: 0	Felipe confirms with an ACK message. This is the only request message that cannot be replied and is only sent for INVITE.
Now Mafalda and Felipe are talking through a voice channel established with the SDP parameters which were agreed upon by means of the OK method. RTP packets of audio data are going in both directions over ports 49170 and 3456 using PCMU/8000 encoding.	

Table 17.3 (continued)   Sequence of SIP Messages between Felipe and Mafalda

BYE sip:mafalda@linti.unlp.edu.ar SIP/2.0 Via: SIP/2.0/UDP 10.20.30.41:5060 To: Mafalda <sip:mafalda@linti.unlp.edu.ar>;tag=314159 From: Felipe <sip:felipe@historieta.com.ar>;tag=589304 Call-ID: 8204589102 CSeq: 1 BYE Content-Length: 0	To end the communication, any of the users, either Mafalda or Felipe, can send a BYE message to finalize the SIP session. In this case, Felipe decides to end the session.
SIP/2.0 200 OK To: Mafalda <sip:mafalda@linti.unlp.edu.ar>;tag=314159 From: Felipe <sip:felipe@historieta.com.ar>;tag=589304 Call-ID: 8204589102 CSeq: 1 BYE Content-Length: 0	Mafalda receives the BYE request and responds with an OK message, terminating the communication.

17.3.2 Application and Service Layer

In this layer, the services offered by the IMS network are available, whether provided in a standard way or as the new services that arise from the adoption of this new converging network. Typically, all the applications and services over the IMS network are running on SIP application services (SIP AS) located in the *Application* sub-layer. Its important to highlight the flexibility of the IMS architecture for the deployment of new services through third-party application servers, available as well through a peer-to-peer architecture based on SIP that do not require a central server. The Open Mobile Alliance (OMA) comprised by the main mobile operators and equipment manufacturers promotes the use of interoperable services among different devices, geographic locations, service providers, operators, and networks, named by OMA as *IMS enablers*. They provide specific Internet services with rich capabilities related to communities, such as presence information, group management, location, instant messaging (IM), push-to-talk over cellular, and IP conferencing. The IMS enablers collaborate on the development of multimedia applications from the *Services* sub-layer.

One of the key features of the IMS architecture is its flexibility to incorporate new services to end users. It is possible to use network resources (such as caller ID, user location service) and provide reliable services with feature-rich Web 2.0-type, among other things for entertainment and games. Currently, these collaboration capabilities and real-time communications provided by the IMS enablers are being incorporated into Internet services and applications. The generation of this new ecosystem of new applications is the great challenge of IMS. Multimedia push, real-time video sharing, real-time peer-to-peer multimedia streaming service, interactive gaming, and videoconferencing can be considered in the category of services.

17.3.3 Session Control Layer

This layer basically implements the control session. This aspect covers from the user registration, the SIP session routing, to the maintenance and management of the user data and policies that ensure service quality. The SIP session routing could consist on requests from a device, toward specific services hosted in an AS or toward another user in the network (peer-to-peer) of the same provider or doing roaming from another network or toward a predefined component which meets calls in eventual situations.

The core of the IMS architecture is the CSCF (call session control function) and its main functions include registration, session establishment, and SIP routing strategies, which it carries out through the following components: P-CSCF (proxy-call session control function), I-CSCF (interrogating-call session control function), and S-CSCF (serving-call session control function) [R2,R3]. In turn, the IMS architecture specifies a main database server called HSS (home subscriber server) which stores information of all the subscribers and data related to the services. P-CSCF is the first contact point of the end user with the IMS network; this implies that all SIP signaling traffic between the user device and the network passes through the P-CSCF.

The main function of I-CSCF is the recovery of the S-CSCF or the AS name that will attend the SIP request. It is carried out using the subscriber's profile information stored in the HSS. Another function of I-CSCF is to provide concealment of the topology among networks from different operators. S-CSCF is responsible for the registration, the SIP routing decisions, and the maintenance of the session states.

17.3.4 Transport and Access Layer

The most important goal of IMS is the convergence between fixed and mobile networks by creating a new paradigm in telecommunications services, where the system focuses on the user as opposed to the current paradigm focused on devices [R2]. This convergence was designed to give end users new communication experiences provided across multiple geographic locations, devices, access technologies, and services. The integration of fixed and mobile world presents the user the best of each, offering the convenience and availability of mobile services, and the reliability and quality of fixed [R2].

As shown in Figure 17.3, one of the most prominent features of IMS is the capacity to separate services from transport technology; thus a 3G mobile telephone connects to IMS network using protocols from the IP and SIP family in the same way that a PC does through the DSL. More significantly, in a mobile environment in which the user is able to move geographically, the IMS independent access not only enables the user to roam between different providers but the device could allow the user to change the connection method between different access technologies, making better use of available types of connections. For example, a telephone with Wi-Fi technology could change in a transparent way between the 3G and Wi-Fi access, and the users could also change the device, i.e., between a cell phone and the PC, maintaining the same session and user [R4]. In this layer, we can find the traditional Internet protocols and devices.

With a look into the OSI model, we have in the physical and data link layers the access technologies DSL, 3G, Cable Modem, Wi-Fi, Ethernet, etc., in the network layer the IP protocol, essential for the interconnection of all these underlying technologies, and in the transport layer the protocols TCP, UDP, RTP/RCTP, etc.

The core of the IMS network is based on the same pillars as the Internet, as illustrated in Figure 17.3. IP is the core of this converging network architecture. In turn, the IMS design was conceived as well taking into account the service security and quality as essential elements. Security in IMS networks is applied in three different scenarios: in the first contact point of the user with the IMS network, which is between the user device and the IMS network, among the different devices of the diverse layers of the operator's core network, and among the core networks of the different operators for the case of the use of *roaming*. Originally, IMS security support was provided by the IPSec protocol (RFC 2401, RFC 2406, RFC 2407, and RFC 2409) for access from both the user's device and for routing between different networks, then the TLS protocol (Transport Layer Security—RFC 2246) was added also for the same scenarios. Moreover, the protocol annexed AKA (Authentication and Key Agreement) for user authentication when accessing the network.

The service quality (QoS) is a key component of IMS; for a particular session, it could be determined by multiple factors, among them, the maximum bandwidth assigned to a user based on his subscription or the current state of the network. IMS supports multiple models of point-to-point quality service, the user's devices can use specific protocols of a link layer for resource reservation, i.e., PDP Context Activation for parameter reservation of QoS in mobile networks, RSVP (Resource ReSerVation Protocol, RFC 2205) or DiffServ (Differentiated Services, RFC 2475 and RFC 3260).

17.4 The Challenge Proposed by IMS for New Applications

The advantage of having a converging network of a higher bandwidth and service quality support enables us to build new multimedia applications and services which enhance user experience.

The possibility of using multimedia services in a simultaneous way encourages the construction of content rich applications, interactive and, in general terms, more suited to people's new communication behaviors that only consume what they are interested in (iTunes in an example), that register themselves in virtual communities, etc. The IMS network user can manage their own contents, share them with other users, such as "live video." Thus, IMS facilitates "user–user" and "user–content" communication.

Cell phones bring more and more benefits, their screens are more accurate and larger, they have photograph and video cameras incorporated, they reproduce music, and have GPS navigators. They are mobile devices with the capacity of being always on, representing a challenge for new applications that stop being isolated entities that exchange information through the user interface. The new generation of applications will be peer-to-peer, enabling to share music, games, boards, live video, etc.

17.5 Java Support for IMS and SIP

The neutrality of the Java platform, the ability to be available "everywhere" from mobile devices with limited hardware resources including servers to desktop PCs, laptops, etc., and the standardization of APIs to access the incorporated function-alities and build new ones without depending on manufacturer's particular imple-mentations, positions Java as an ideal and convenient development platform for building new converging applications through IMS. Another important Java fea-ture is its availability in different IDEs (integrated development environment) of the free software community, like the Eclipse and NetBeans initiatives.

JCP (Java Community Process) has various activities focused on the definition of multiple JSRs (Java Specification Requests) that provide support for the develop-ment of applications based on SIP in the different Java technology *flavors*.

In turn, the JAIN (Java APIs for Integrated Networks) initiative of JCP defines APIs (application programming interfaces) that enable the use of JAVA technologies to develop telecommunication services in converging networks [L21].

The JAIN initiative extends Java technology to provide service portability (write once, run anywhere), network independence (any network), and open development (by anyone) promoting a chain of open value ranging from the network equipment, computer and device manufacturers to the outsource service providers, having impact in the technological structure and telecommunication company businesses.

The goal of this initiative is to change from closed and proprietary systems to open environments, having an influence in the same way JEE has done in the IT industry. This way, the communication operators can extend their portfolio services by getting faster, simpler, and less expensive applications [L22].

Within the JCP, multiple APIs JAVA evolve in relation to SIP and IMS [L23–L30]; they are described in Table 17.4:

Java is the standard platform for both server-side and client-side IMS application development. Server-side applications require an SIP servlet container compatible with JSR 289 [L25] (or the earlier JSR 116 [L31]). Client applications are implemented using the JSR 281 [L29] and JSR 325 [L30] specifications.

The most significant initiative in relation to IMS client development is the ICP (IMS Client Platform) [R5], included in Ericsson SDS (Service Development Studio) [L32]. The ICP was proposed as a standard Java in 2005 under JSR 281, led by Ericsson and BenQ. In its final state, the JSR 281 standard was divided into two specifications: JSR 281 and JSR 325. The first one, called *IMS Service API*, approved as standard in July 2008, contains basic IMS features like logging and setting up audio and video sessions, as well as a generic framework to access IMS services. JSR 325, called *IMS Communication Enablers* and currently in progress, provides an interface to access specific IMS service enablers such as presence, group management, and instant messaging. This part was removed from the original version of JSR 281. The ICP, together with the related JSRs, simplifies the development of IMS client-side applications.

As for the development of IMS server-side applications, JCP developed the JSR 116 specifications called SIP Servlet API 1.0 and the JSR 289 SIP Servlet API 1.1. Both specifications are an abstraction of the SIP protocol based on the Java Servlet API. SIP Servlets are the Java components that run in a Servlet container within a server. The SIP container simplifies the creation of SIP applications managing the life cycle of the SIP servlets and providing support for the interaction between SIP servlets and SIP clients (User Agent) through SIP request and reply message exchange. A container can enqueue messages, manage states in the server, and transfer control to the components or the SIP servlets, which are responsible for addressing messages. As the SIP protocol is based on the popular HTTP Web protocol, Java SIP Servlets share a common nature with Java HTTP Servlets. Thus, the programming interface available in the SIP Servlet API will be familiar to developers of Web applications in Java, particularly for those who make use of the Servlet API.

In connection with the implementation of the JSR 289 and JSR 116, the SailFin [L33] project is the most important initiative within the free software community. SailFin is a java.net project contributed by Ericsson, which extends the JEE application server, GlassFish [L34], SIP Servlet technology, included in the Ericsson SDS (Service Development Studio) [L32]. GlassFish is an open-source server implemented by Sun Microsystems and, like SailFin, has dual license—GNU-GPL Version 2 and CDDL Version 1.0. There are other proprietary products with similar capabilities as

Table 17.4   JCP Java APIs for Handling SIP and IMS

API	*JSR*	*Java Platform*	*Description*
JAIN SIP	JSR32	JSE	Low-level API for SIP. Requires extensive knowledge of the SPI protocol. Provides call control management.
JAIN SIP Lite	JSR 125	JSE adaptable a JME	Abstraction of the SIP protocol which does not require a comprehensive understanding of the protocol. Being a lightweight API, it adapts to devices with scarce computing and memory capabilities.
SIP Servlet	JSR 289	JEE	Abstraction of the SPI protocol based on Java servlets.
			The SIP servlets are run and managed by a servlet container that hides the complexity of the protocol.
SIP for JME	JSR 180	JME	SIP for mobile devices with limited resources.
JAIN SIMPLE Presence	JSR 164	JME y JSE	API that allows clients and SIMPLE/SIP servers to exchange presence information between a client and a server SIMPLE/SIP.
JAIN SIMPLE IM	JSR 165	JME y JSE	API that allows instant messaging between clients SIMPLE/SIP. Generally applications use JSR 164 and JSR 165 Implementations together.
IMS services	JSR 281	JME	High-level abstraction of IMS technology and protocols to facilitate JME applications development.
IMS communication enablers	JSR 325	JME	High-level abstraction of IMS technology and protocols to facilitate application development JME. Provides access to the IMS enablers (presence, group management, Instant Messaging, etc.).

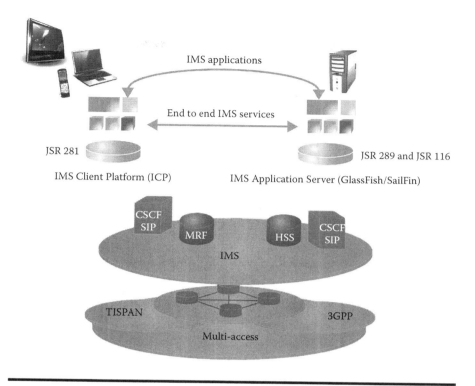

Figure 17.7   Implementations for Java IMS application development.

SailFin, among them, OCMS Oracle, WLSS BEA, and IBM Websphere; however, SailFin is the first one contributed by the free software community.

Figure 17.7 illustrates the two most relevant implementations described above that facilitate the development of client-side and server-side Java applications in IMS. They both provide high-level abstraction for the developer, hiding the particularities of IMS technology.

17.6 Java Development Environments for IMS Applications

Our school has over 10 years of experience training highly qualified students in development on Java technology through various graduate and postgraduate courses. The agreement of the Computer Science School with Ericsson-Argentina enabled us to leverage the strengths of both teams. Ericsson Company is a world leader in 2G and 3G mobile technologies and is now betting heavily on the convergence of multiservice networks to promote IMS technology and foster the development of innovative applications that exploit the facilities provided by this technology, being one of the leaders in the standardization of Java APIs for application development on IMS-capable mobile devices. Ericsson has established the SDS [L32], which is an integrated development environment based on Eclipse open access [L35] that facilitates the construction of IMS services and applications. We believe that Ericsson SDS is the best IDE that accompanies the evolution of the different standardizations promoted by the JCP in relation to IMS; also, being based on Eclipse turns it into a popularly adopted tool by the Java developer community.

SDS is a comprehensive tool for development, testing, and deployment of IMS services and applications, both on the client side and the server side. It is possible to develop client-side applications using ICP, Ericsson's implementation of JSR 281, and on the other end, SDS integrates and supports the GlassFish/SailFin server that implements the JSR 289 for development of IMS services. To facilitate testing of mobile applications, SDS also includes a set of mobile device emulators.

Ericsson SDS simulates a complete IMS infrastructure, enabling the developers to test their software through an execution environment that emulates all the real IMS network components in an only PC.

17.7 IMS Java for Mobile 2.0

17.7.1 Java Initiatives for Mobile Applications

Currently, JME is the most accepted technology among cell phone manufacturers, wireless connection carriers, and mobile application developers. It has been adopted as a standard development and application execution platform in most cell phones. The JME platform has evolved in response to the growing capabilities of the devices, which have been introduced in the mobile world, based on the MIDP (Mobile Information Device Profile) specification. Today, the tendency in the software industry is toward standardizing the access to these new hardware capabilities in mobile devices and to the new services provided through the Internet. JCP supports this tendency with the elaboration of a specification built over MIDP called MSA (Mobile Service Architecture).

The MIDP 2 specification defined by JCP is the JSR 118 [L36] is designed to operate on CLDC (Connected, Limited Device Configuration) 1.0 (JSR 30 [L37]), 1.1 (JSR 139 [L38]) and subsequent versions. This last specification defines some basic APIs and a virtual machine for mobile devices with limited resources, such as the cell phone case. MIDP is the foundational piece of JME technology; it defines a platform to deploy in a dynamic and secure way for optimized applications, with graphic capacities and with connectivity possibilities through different technologies (Wi-Fi, Bluetooth, Infrared, USB, etc.).

Today, the mobile world is not limited to cell phones; it also includes devices such as netbooks, smartbooks, Amazon Kindle, PlayStation 3, etc. In response

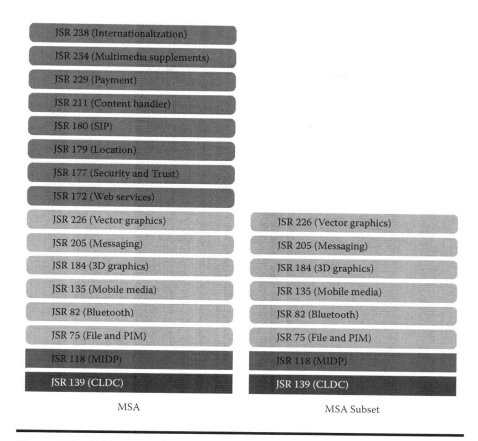

Figure 17.8   MSA 1.0 stack.

to this evolution, the Java ME platform was expanded to consider the new hardware and software capabilities of these devices, giving way to the JTWI (Java Technology for the Wireless Industry) platforms [L39] and later to MSA [L40], which are now a standardized environment for building mobile applications. These two new JME platforms include MIDP 2. Figure 17.8 shows the specification stack of the MSA and MSA Subset platform. The difference between both specifications is related with the option of including conditionally mandatory APIs that depend on the hardware capacity of the device, as is the case of the Location API (JSR 179) that depends on the existence of a GPS incorporated to the device.

17.7.2 Java Initiatives for IMS in Mobile Applications

As described, the MSA 1 specification [L40] for JME defines a set of standard functions for mobile devices. JCP continued to work on a new version of MSA called

MSA Advanced or MSA 2 (JSR 249) [L41], which improved the standard functionality by incorporating current technologies with a good future prospect, especially in the multimedia area such as the case of JSR 281 called IMS Services. Figure 17.9 shows the new APIs that were added to MSA to version 2 in relation with version 1. The incorporation of JSR 281 IMS Services API [L29] to MSA 2 standard shows that IMS technology will be available in the JME platform, promoting its adoption by mobile device manufacturing firms and with a favorable impact in the development of applications for these devices. Mobile application developers can take advantage of IMS multimedia capabilities such as QoS, single login to access multiple services, with the confidence that their applications will run on any device compatible with the MSA 2 platform.

17.7.3 Mobile Development JME: MIDlets

As already mentioned, MSA is based on MIDP and from its origins MIDP applications are called MIDlets. These applications are written with JME APIs that comprise the MSA standard and run on a mobile computing environment. MIDlets require a special execution environment given by a specific piece of software in the device called *Application Management System* (AMS) that controls the installation, execution, and life cycle of the MIDlet.

To build a MIDlet it is necessary to define a class that extends *javax.microedition.midlet.MIDlet* and overwrites at least the following three methods:

1. *startApp()*: It is invoked by the AMS to initiate the MIDlet or to resume the MIDlet execution in pause state.
2. *pauseApp()*: It is invoked by the AMS when certain events take place, for example, an incoming call.
3. *destroyApp()*: It is invoked by the AMS to release the resources allocated by the application.

A MIDlet has three different states that determine its functioning: Paused, Active, and Destroyed. These states correspond with the three methods earlier described called collectively as MIDlet life cycle methods. The AMS is responsible for controlling the MIDlet life cycle by invoking the life cycle methods. Figure 17.10 describes the MIDlet states and the methods invoked by the AMS that cause the passage between the different states.

The AMS decides when to invoke the life cycle methods: beginning the execution, the MIDlet passes to *Active* state, but what happens if it receives a call or a message while running? The AMS is in charge of changing the MIDlet state according to the external events that are produced. In this case, the AMS temporarily stops the MIDlet execution to attend the call or read the message, passing it to a *Paused* state.

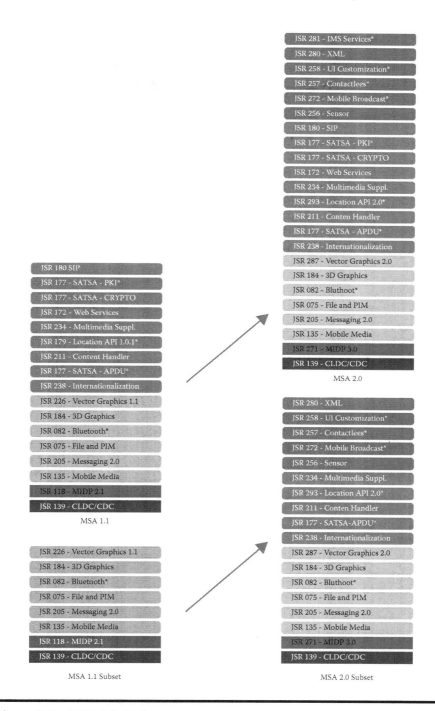

Figure 17.9 MSA 2.0 stack.

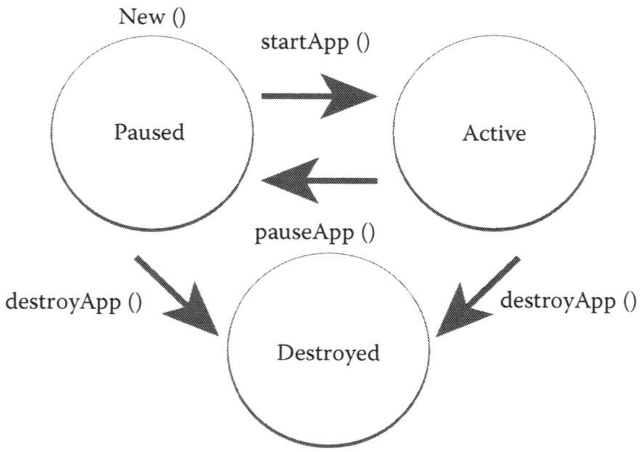

Figure 17.10   MIDlet life cycle.

17.8 Development of a Simple Chatroom on IMS Using Ericsson SDS

17.8.1 Introduction

Below, we describe an example of a simple chat application, *IjcuChat*, taken from the *Ericsson Developer Connection** Web site. Despite continuous growth, interest, and expectations regarding the development of Java applications on IMS networks, we must consider that this is a software development technology that is not yet mature and there is no consolidated documentation on the issues to take into account for the development of applications. While the IMS architecture design promotes the abstraction of the network architecture for application development, it currently requires the programmer to be aware of some key elements and some basic configurations of the components that constitute it. Today, programming an IMS Java application using the SDS requires management of the specific APIs to access the IMS network and services and the assembly of a simulated execution context of an IMS network by setting its key elements (CSCF, HSS, DNS, etc.). That is why in this description we also consider the commonly requested configuration aspects.

17.8.2 General Architecture of the IjcuChat Application

IjcuChat is a JME (CLDC 1.1/MIDP2.0) client implemented through a Java Midlet that connects to the application server called *Twitty* by means of the IJCU platform. The basic functionality of this application is the registration of users, the

---

* http://www.ericsson.com/developer/sub/open/technologies/ims_poc/tools/sds_40

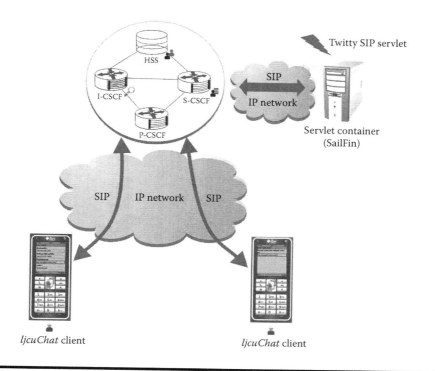

Figure 17.11   Conceptual *IjcuChat* architecture.

invitation from one user to another to start a conversation, and exchange messages between them. Figure 17.11 shows a conceptual scheme of the main components and the communication protocols used in the IMS *IjcuChat* Java application.

The user interface is a simple form that allows users to complete text fields required for the registration in the IMS network, the invitation and exchange of messages between two users. Figure 17.12 shows screenshots of the screens that enable the user to interact with *IjcuChat application; in this case, we will call the users of this application Bob and Alice.*

The Twitty server is a Java SIP server-side application implemented through a SIP Servlet (JSR 116) that runs in a SailFil application server. It provides a simple functionality that consists of sending the invitation message from one user to another to establish a conversation and forward the exchange of messages.

17.8.3 IjcuChat Functioning in a Simulated IMS Network in the Ericsson SDS

The DSD has a feature that facilitates the view of the exchange of SIP messages that pass through the CSCF through a flowchart. This feature is available in two ways, one is online and shows the exchange of SIP message during the application

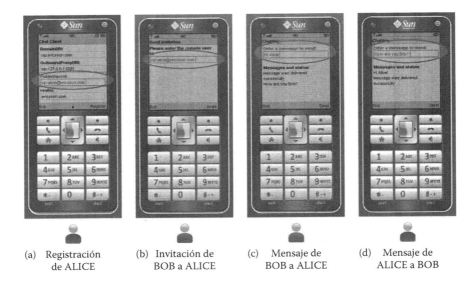

(a) Registración
de ALICE

(b) Invitación de
BOB a ALICE

(c) Mensaje de
BOB a ALICE

(d) Mensaje de
ALICE a BOB

Figure 17.12   *IjcuChat* **screenshots.**

execution, and the other is offline and shows the exchange recorded in a CSCF log file. This diagram is very educational to understand how the different actors in the IMS network interact; it is also a very useful tool when testing the application behavior in the different user interactions. Figure 17.13 shows an image capture of the SDS during the execution of the *IjcuChat* application and the *Twitty* server.

The flowchart in Figure 17.13 is described in detail in relation to the main actors of an IMS network and this way we can understand what happens in the *IjcuChat* application "backstage." Figure 17.13 shows the SIP methods used for the *IjcuChat application.*

SIP REGISTER: when the application starts, the first screen prompts the user registration in the IMS network (Figure 17.12), which results in the sending of an SIP REGISTER message to the CSCF component; this is the Alice and Bob case. Considering that the CSCF has three main components, P-CSCF, I-CSCF, and S-CSCF, in this interaction, the S-CSCF plays a main role. The functionality provided by the S-CSCF is to facilitate the establishment and ending of the SIP session with help from the HSS. The latter is the one that contains the subscriber's permanent data and the most relevant temporal data related with the users that are connected. Figures 17.14 and 17.15 show the screenshots from the *Provisioning* perspective of the SDS where it is possible to view the IMS network subscriber's permanent information and the temporary information, respectively; both aspects of the user management are stored in the HSS. In particular, the information presented is related with the registration of Alice and Bob in the IMS network. In conclusion, when the S-CSCF receives the request, SIP REGISTER delegates the

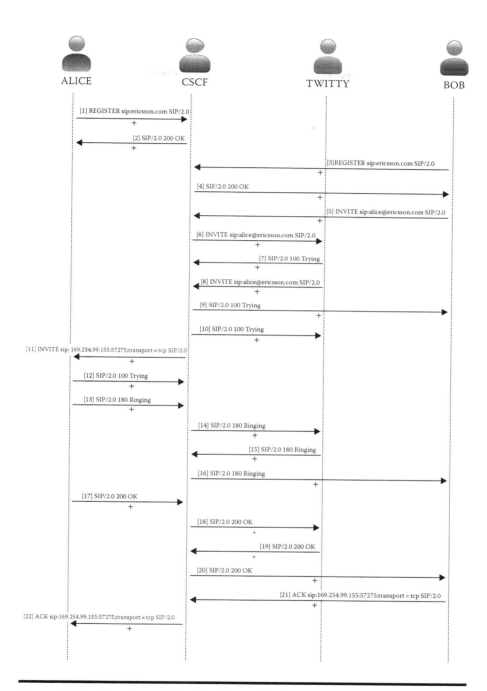

Figure 17.13   Flowchart of the *IjcuChat* application and the Twitty server.

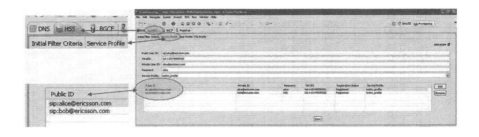

Figure 17.14 **Provisioning perspective that shows the subscribers loaded in the HSS.**

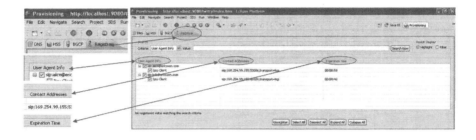

Figure 17.15 **Provisioning perspective showing the online registered users in the HSS.**

specific function of registration of the SIP session to later build and send the SIP answer to the user. In the flowchart in Figure 17.13, these interactions are summarized in the messages exchanged between the users Bob and Alice in the CSCF.

SIP INVITE and SIP MESSAGE: Once Bob and Alice are registered, Bob decides to invite Alice to a chat session. For this purpose, in the invitation screen (Figure 17.12), he indicates Alice's SIP address. With this user request, the application generates a message and sends a SIP INVITE message whose final receptor is Alice. In the IMS networks, this message in canalized through the CSCF that again plays a central role. The S-CSCF component is the most intensive processing node of the IMS nucleolus because it is responsible for determining what services will the users have available according to their profiles and which will be the application servers in charge of attending those requests. Figures 17.16 through 17.19 show the HSS configuration screen that relates the user's profiles with the services that will attend their requests. Particularly, each time a user interacts with an application that generates a SIP INVITE or SIP MESSAGE request message, they will be attended by the service identified as *twitty_profile*. Figures 17.18 and 17.19 show these configurations, called iFC (Initial Filter Criteria) in the IMS vocabulary. In our example,

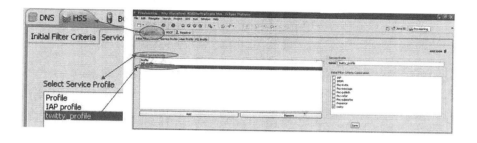

Figure 17.16   HSS configuration: Definition of the service profile twitty_profile.

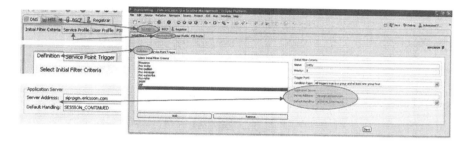

Figure 17.17   HSS configuration: Definition of the Initial Filter Criteria twitty.

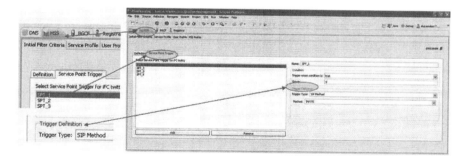

Figure 17.18   HSS configuration: iFC twitty INVITE criteria.

the established iFC is called *twitty*. In the flowchart in Figure 17.13, we can see that the SIP INVITE message sent from Bob to Alice is headed toward the CSCF who internally, as we explained previously, uses the S-CSCF component to determine through the iFC what server will attend the request; in this particular case, *Twitty*. Figure 17.17 shows the relationship between the iFC *twitty* and the server sip:pgm. ericsson.com where the *Twitty* service is hosted, and Figure 17.20 shows the reso-lution of the IP address in the DNS. As described before, *Twitty* provides a very

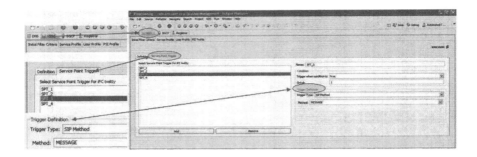

Figure 17.19 **HSS configuration: iFC twitty MESSAGE criteria.**

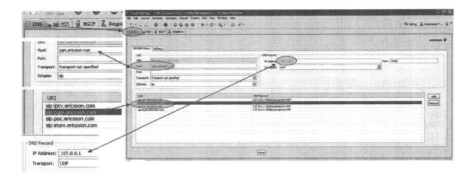

Figure 17.20 **DNS configuration in the provisioning perspective.**

simple functionality that will forward the SIP messages received addressed to Alice, once again through the CSCF. In relation to the SIP MESSAGE, the message flow between Bob and Alice is similar to the one described previously on the SIP INVITE message. The difference between them is that both Bob and Alice generate this type of message each time they chat using the *IjcuChat* client.

17.8.4 Analyzing the Use of JSR 281 in the IjcuChat

As mentioned before, *IjcuChat* is an application for mobile devices implemented by a MIDlet that uses the IMS network to communicate with the server and its set up in the chat session. Figures 17.21 through 17.28 show the most relevant code segments in which the API from JSR 281 is used to access the communication services provided by the IMS network. MIDP defines a mechanism called Push Registration that enables, among other things, to register MIDlets that correspond to network events such as incoming connections. IMS applications must declare the capacities it can manage though Push Registration. After an incoming call, the

```
public class ljcuChat extends MIDlet {

    private StringItem chatArea;

    private TextField domainURIField;
    private TextField outboundProxyURIField;
    private TextField publicUserIdField;
    private TextField realmField;
    private TextField privateUserIdField;
    private TextField mobilePhoneNumberField;
    private TextField passwordField;

    private Session session;
    private FramedMedia framedMedia;

    private String scheme = "imscore://";
    private String iari = "com.ericsson.client.ljcuChat";
    private String serviceId = "chat";
    private String aliceUserId = "sip:alice@ericsson.com";
    private String IARI_PREFIX = "urn%3Aurn-xxx%3A";
    private String transport = "tcp";
    private static String contentType = "text/plain";

    private CoreService coreService;

    public void startApp() throws MIDletStateChangeException {
        showRegistrationForm();
    }
```

Registration form fields in the IMS network

Application capability declaration fields

The application initiates with the display of the network registration form

Figure 17.21   Declaration of registration fields in the IMS network and of application capabilities.

```
private void showRegistrationForm(){

    Form registerForm = new Form("Register");
    Command registerCommand = new Command("Register", Command.OK, 1);
    Command exitCommand = new Command("Exit", Command.EXIT, 1);
    registerForm.addCommand(registerCommand);
    registerForm.addCommand(exitCommand);
    getDisplay().setCurrent(registerForm);
    domainURIField = new TextField("DomainURI:", "sip:ericsson.com", 40, TextField.ANY);
    outboundProxyURIField = new TextField("OutboundProxyURI:", "sip: 127.0.01:5081", 40, TextField.ANY);
    publicUserIdField = new TextField("PublicUserId:", aliceUserId, 40, TextField.ANY);
    realmField = new TextField("realm:", ericsson.com", 40, TextField.ANY);
    privateUserIdField = new TextField("PrivateUserId:", "alice@ericsson.com", 40, TextField.ANY);
    mobilePhoneNumberField = new TextField("MobilePhoneNumber:", "5143457900", 40, TextField.DECIMAL);
    passwordField = new TextField("Password", "alice", 10, TextField.PASSWORD);
    passwordField.setString("alice");
    registerForm.append(domainURIField);
    registerForm.append(outboundProxyURIField);
    registerForm.append(publicUserIdField);
    registerForm.append(realmField);
    registerForm.append(privateUserIdField);
    registerForm.append(mobilePhoneNumberField);
    registerForm.append(passwordField);
    registerForm.setTitle("Chat Client");
```
Creation and display of IMS network registration form

```
    registerForm.setCommandListener(new CommandListener() {
        public void commandAction(Command command, Displayable displayable) {
            if(command.getCommandType() == Command.OK) {

                RegisterThread registerThread = new RegisterThread();
                Thread thread = new Thread(registerThread);
                thread.start();
```
Initiate of the IMS network registration process

```
            }
            else if(command.getCommandType() == Command.EXIT)
                exit();
        }
    });
}
```

Figure 17.22 Registration form display.

```
protected class RegisterThread implements Runnable {
  public void run() {
    register();
  }

  public void register() {
    try {
      ljcuUserManagement ijcuUserManagement = ljcuUserManagement.getInstance();
      ijcuUserManagement.setDomainURI(domainURIField.getString());
      ijcuUserManagement.setOutboundProxyURI(outboundProxyURIField.getString());
      ijcuUserManagement.setPublicUserId(publicUserIdField.getString());
      ijcuUserManagement.setRealm(realmField.getString());
      ijcuUserManagement.setPrivateUserId(privateUserIdField.getString());
      ijcuUserManagement.setPassword(passwordField.getString());
      ijcuUserManagement.setTransport(transport);
      ijcuUserManagement.setPlatform(ljcuChat.this);

      Configuration myConfiguration = Configuration.getConfiguration();
      String [][] properties = new String [][] {{"Framed", contentTyper},
                     {"CoreService", serviceId, IARI_PREFIX + iari, "",""}};
      myConfiguration.setRegistry(iari, ljcuChat.class.getName(), properties);

      ConnectionState connectionState = ConnectionState.getConnectionState();
      ConnectionState.setListener(new ConnectionStateAdapter());

      String url = scheme + iari + ";serviceId=" + serviceId;
      coreService = (CoreService) Connector.open(url);
      CoreServiceClientAdapter coreServiceAdapter = new CoreServiceClientAdapter();
      coreService.setListener(coreServiceAdapter);

      initializeSession();
    }
    catch (IllegalArgumentException e3) {
      alert("Can not register the phone: receive IllegalArgumentException");
    }
    catch (ConnectionNotFoundException e2) {
      alert("Can not register the phone: receive ConnectionNotFoundException");
    }
    catch (IOException e1) {
      alert("Can not register the phone: receive IOException");
    }
    catch (Exception e) {
      alert("Can not register the phone: receive Exception");
    }

  }
}
```

Annotations:
- The application capabilities are declared using Push Registration
- IMS network registration transaction triggers sending a SIP Register to CSCF
- Inicio de formulario de invitación

Figure 17.23   Registration of the application capabilities and network registration.

AMS is responsible for launching the MIDlet whose declared capabilities match the ones required by the incoming call.

The Push Registration can be carried out statically or dynamically. Static registration consists on declaring the capabilities through properties in the MIDlet's (JAD) description file. In our example, we use dynamic registration that consists of declaring the capabilities through an array of strings, which will be established as a parameter to the method *Configuration.setRegistry()*, as shown in Figure 17.23.

After declaring the application capabilities, the MIDlet registers the user in the IMS network invoking the *Connector.open()* method, and thus the application can initiate and receive invitations for chat and text message exchange. The *Connector. open()* method performs the registration through the message exchange of the SIP REGISTER transaction displayed graphically in Figure 17.13. This method may

```
private void initializeSession()
{
    Form form = new Form("Initialize Session");
    form.setTitle("Send invitation");
    final TextField remoteUserId = new TextField("RemoteUserId", "sip:alice@ericsson.com", 40, TextField.ANY);
    remoteUserId.setLabel("Please enter the remote user Id:");
    form.append(remoteUserId);
    Command inviteCommand = new Command("Invite", Command.OK, 1);
    Command exitCommand = new Command("Exit", Command.EXIT, 1);
    form.addCommand(inviteCommand);
    form.addCommand(exitCommand);

    form.setCommandListener(new CommandListener()
    {
        public void commandAction(Command command, Displayable displayable)
        {
            if (command.getCommandType() == Command.OK)
            {
                StartSessionThread startSessionThread = new StartSessionThread(remoteUserId.getString());
                Thread thread = new Thread(startSessionThread);
                thread.start();
            }
            else if (command.getCommandType() == Command.EXIT)
            {
                exit();
            }
        }
    });

    getDisplay().setCurrent(form);
}
```

Creation and display
of the chat incitation form

Initiation of the chat session
invitation process

Figure 17.24 Invitation form display.

```
private class StartSessionThread implements Runnable {

    private String remoteUserId;
    private StartSessionThread(String remoteUserId) {
        this.remoteUserId = remoteUserId;
    }

    public void run() {

        invite(remoteUserId);
    }

    private void invite(String remoteUserId) {
        if(!isUriValid(remoteUserId)) {
            displayAlert("Please enter a valid URI", true);
        }
        try {

            session = coreService.createSession(publicUserIdField.getString(), remoteUserId);
            SessionAdapter sessionAdapter = new SessionAdapter();
            Session.setListener(sessionAdapter);

            framedMedia = (FrameMedia) session.createMedia("FrameMedia", Media.DIRECTION_SEND_RECEIVE);
            framedMedia.setAcceptedContentTypes(new String[] {"text/plain"});

            FramedMediaAdapter outgoingFramedAdapter = new FramedMediaAdapter();
            framedMedia.setListener(outgoingFramedAdapter);

            session.start();

        } catch (IllegalStateException e) {
            displayAlert("Problem starting a session-IllegalStateException", true);
        }
        catch (ImsException e1) {
            displayAlert("Problem starting a session-ImsException", true);
        }
        catch (ServiceClosedException e2) {
            displayAlert("Problem starting a session-ServiceClosedException", true);
        }
    }
}
```

Creation of an object that represents a session for media exchange and their configuration

Chat session invitation transaction that triggers a SIP INVITE

Figure 17.25   Session setting and media configuration.

```
private class SessionAdapter implements SessionListener {
  public void sessionAlerting(Session arg0) {}
  public void sessionReferenceReceived(Session arg0, Reference arg1) {}
  public void sessionStartFailed(Session arg0) {
    displayAlert("Session can't be started", true);
    terminateSession();
  }
  public void sessionStarted(Session arg0) {
    startSession();
  }
  public void session terminated(Session arg0) {
    initializeSession();
  }
  public void sessionUpdateFailed(Session arg0) {}
  public void sessionUpdateReceived(Session arg0) {}
  public void sessionUpdated(Session arg0) {}
}
```

Established session alert
that triggers the message exchange

Figure 17.26   Session events listener.

end successfully with the user registered in the network, or with an encapsulated failure as an exception. A code of nonsatisfactory answer as a result of the SIP REGISTER request is abstracted in an IOException.

Once the users are registered in the network, the MIDlet presents the form which allows the user to identify another user to start a chat session with. Figure 17.24 shows the code for the creation and display of this form.

Figure 17.25 shows the code related with the session establishment between two connection ends. The *CoreService.createSession()* method enables the creation of the session through the IMS network for the media exchange. In our example, we use the *FramedMedia* type, which represents a media transference connection in which the content is sent in packets and may be used for instant messaging as well as object serialization or file transference. Through the constant *Media. DIRECTION_SEND_RECEIVE*, it specifies that the media exchange will take place in both directions.

The object that represents the session is associated with a *listener* in charge of listening and processing the events it generates. In our example, the class *SessionAdapter* in Figure 17.26 implements this *listener*.

When the user at the other end of the session accepts the invitation, the implementation of the API JSR 281 generates an event that represents this situation and results in the invocation to the *sessionStarted()* method, which in our implementation invokes the MIDlet *startSession()* method that displays the form for entering and sending messages during the chat session, as shown in Figure 17.27.

Finally, Figure 17.28 shows how the message sending is done using the *FramedMedia* object created in the *StartSessionThread* class of Figure 17.25.

```
private void startSession() {
    ┌─────────────────────────────────────────────────────────────────────────┐
    │ Form form = new Form("Chatting");                                         │
    │ final TextField message = new TextField("Enter a message to sent:", "", 40, TextField.ANY);
    │ chatArea = new StringItem("\r\nMessages and status:\r\n","");             │
    │ chatArea.setText("");                                                     │
    │ form.append(message);                                                     │
    │ form.append(chatArea);                                                    │
    │ Command byeCommand = new Command("End", Command.EXIT, 1);   Creation and display
    │ Command sendCommand = new Command("Send", Command.OK, 1);   of the incoming messages form
    │ form.addCommand(sendCommand);                                             │
    │ form.addCommand(byeCommand);                                              │
    └─────────────────────────────────────────────────────────────────────────┘

    form.setCommandListener(new CommandListener() {
        public void commandAction(Command command, Displayable displayable) {
            if(command.getCommandType() == Command.OK) {
                ┌──────────────────────────────────────────────────────────────┐
                │ SendThread sendThread = new SendThread(message.getString());   │   Start of the message sending
                │ Thread thread = new Thread(sendThread);                        │   process to the other
                │ thread.start();                                                │   end of the session
                └──────────────────────────────────────────────────────────────┘
            } else if (command.getCommandType() == Command.EXIT) {
                EndSessionThread endSessionThread = new EndSessionThread();
                Thread thread = new Thread(endSessionThread);
                thread.start();
            }
        }
    });
    Display.getDisplay(this).setCurrent(form);
}
```

Figure 17.27 Display of the message sending form.

```
private class SendThread implements Runnable {
    private String message;
    public SendThread(String message) {
        this.message = message;
    }

    public void run() {
        send(message);
    }
    private void send(String message) {
        try {
            framedMedia.sendBytes(message.getBytes(), "text/plain", null);
        } catch (Exception e) {
            displayAlert("Problem sending message!", true);
        }
    }
}
```

Message sending through FrameMedia object

Figure 17.28   Message sending.

17.9 Conclusions

IMS (IP Multimedia Subsystem) is an intelligent network architecture standardized by 3GPP, with multiple converging services such as mobile and fixed telephony, cable and satellite TV, and access to the Internet (usually provided by different networks), through an IP common data transmission infrastructure. Since it is based on standard interfaces and protocols (i.e., SIP for session management), IMS favors an "open" chain of values that includes network, computer and device manufacturers, as well as third-party service providers promoting its adoption by telecommunication companies.

The IMS layer design facilitates, on one hand, the availability of those services offered independently in the access network and, on the other hand, their integration, making the creation of new services and application more agile. Also, the growth of mobile devices for their processing capabilities, connectivity, and functionality and its broad adoption by users will promote an interesting market for the adoption of a network with the characteristics of IMS. IMS fills the gap between the two most successful communication paradigms, cellular and Internet technology.

The vision of IMS of "any service, any screen, anywhere," enables users to share contents "live" with multifunctional devices through multiple virtual systems, promoting the creation of enriched Web 2.0 applications.

References

[R1] R.J.M. Tejedor, Convergencia total en IMS: IP multimedia subsystem. *Revista Comunicaciones World*, 214, 50–53, 2006, ISSN 1139-0867, Versión digital: http://www.idg.es/comunicaciones/articulo.asp?id=178173.

[R2] M. Poikselkä, G. Mayer, H. Khartabil, and A. Niemi, *The IMS: IP Multimedia Concepts and Services*, 2nd edn., John Wiley & Sons, Chichester, U.K., ISBN 0-470-01906-9, 2006.

[R3] R. Copeland, *Converging NGN Wireline and Mobile 3G Networks with IMS*, CRC Press, Boca Raton, FL, ISBN 978-08493-9250-4, 2009.

[R4] O. Rashid, P. Coulton, and R. Edwarts, Implications of IMS and SIP on the evolution of mobile applications, *2006 IEEE Tenth International Symposium on Consumer Electronics, 2006 (ISCE'06)*, Petersburg, Russia, ISBN 1-4244-0216-6, 2006.

[R5] P. Kessler, IMS client platform, *Ericsson Review*, No. 2. Año 2007.

Web Sites

[L1] NAPSTER: http://free.napster.com/
[L2] KaZaA: http://www.kazaa.com/
[L3] eMule: http://www.emule-project.net/
[L4] ICQ: http://www.icq.com/

[L5] Skype: http://www.skype.com

[L6] MSN: http://www.msn.com

[L7] Twitter: http://twitter.com/

[L8] GPRS: General Packet Radio Service. http://www.3gpp.org/article/gprs-edge

[L9] Code Division Multiple Access. http://www.tiaonline.org/standards/technology/cdma2000

[L10] 3GPP2: 3rd Generation Partnership Project 2. http://www.3gpp2.org/public_html/specs/index.cfm

[L11] TISPAN: Telecoms & Internet converged Services & Protocols for Advanced Networks. http://www.etsi.net/tispan/

[L12] J. Rosenberg, H. Schulzrinne, G. Camarillo, A. Johnston, J. Peterson, R. Sparks, M. Handley, and E. Schooler. RFC 3261: SIP: Session initiation protocol. June 2002. http://www.ietf.org/rfc/rfc3261.txt

[L13] A. Niemi. RFC 3903: Session initiation protocol (SIP) extension for event state publication. October 2004. http://www.ietf.org/rfc/rfc3903.txt

[L14] J. Rosenberg. RFC 3857: A watcher information event template-package for the session initiation protocol (SIP). August 2004. http://www.ietf.org/rfc/rfc3857.txt

[L15] A. B. Roach. RFC 3265: Session initiation protocol (SIP)-specific event notification. June 2002. http://www.ietf.org/rfc/rfc3265.txt

[L16] J. Rosenberg. RFC 3311: The session initiation protocol (SIP) UPDATE method. September 2002. http://www.ietf.org/rfc/rfc3311.txt

[L17] B. Campbell, J. Rosenberg, H. Schulzrinne, C. Huitema, and D. Gurle. RFC 3428: Session initiation protocol (SIP) extension for instant messaging. December 2002. http://rfc.dotsrc.org/rfc/rfc3428.html

[L18] S. Donovan. RFC 2976: The SIP INFO method. October 2000. http://www.ietf.org/rfc/rfc2976.txt

[L19] J. Rosenberg and H. Schulzrinne. RFC 3262: Reliability of provisional responses in the session initiation protocol (SIP). June 2002. http://www.ietf.org/rfc/rfc3262.txt

[L20] J. Rosenberg. RFC 3856: A presence event package for the session initiation protocol (SIP). August 2004. http://www.ietf.org/rfc/rfc3856.txt

[L21] Java Community Process. JSRs de la iniciativa JAIN. http://jcp.org/en/jsr/summary?id=jain

[L22] JAIN and Java in Communications. Sun White Paper. March 2004. http://java.sun.com/products/jain/reference/docs/Jain_and_Java_in_Communications-1_0.pdf

[L23] JSR 32: JAIN SIP API. http://jcp.org/en/jsr/detail?id=32

[L24] JSR 125: JAIN SIP Lite. http://jcp.org/en/jsr/detail?id=125

[L25] JSR 289: SIP Servlet v1.1. http://jcp.org/en/jsr/detail?id=289

[L26] JSR 180: SIP API for J2ME. http://jcp.org/en/jsr/detail?id=180

[L27] JSR 164: SIMPLE Presence. http://jcp.org/en/jsr/detail?id=164

[L28] JSR 165: SIMPLE Instant Messaging. http://jcp.org/en/jsr/detail?id=165

[L29] JSR 281: IMS Services API. http://jcp.org/en/jsr/detail?id=281

[L30] JSR 325: IMS Communication Enablers (ICE). http://jcp.org/en/jsr/detail?id=325

[L31] JSR 116: SIP Servlet API v1.0. http://jcp.org/en/jsr/detail?id=116

[L32] SDS (Service Development Studio) de Ericsson. http://www.ericsson.com/developer/sub/open/technologies/ims_poc/tools/sds_40

[L33] Proyecto SailFin. http://sailfin.dev.java.net

[L34] GlassFish. https://glassfish.dev.java.net

[L35] Eclipse. http://www.eclipse.org

[L36] JSR 118: Mobile Information Device Profile 2. http://jcp.org/en/jsr/detail?id=118

[L37] JSR 30: Connected, Limited Device Configuration. http://jcp.org/en/jsr/detail?id=30

[L38] JSR 139: Connected Limited Device Configuration 1.1. http://jcp.org/en/jsr/detail?id=139

[L39] JSR 185: JavaTM Technology for the Wireless Industry. http://jcp.org/en/jsr/detail?id=185

[L40] JSR 248: Mobile Service Architecture. http://jcp.org/en/jsr/detail?id=248

[L41] JSR 249: Mobile Service Architecture 2. http://jcp.org/en/jsr/detail?id=249

Chapter 18

User Evaluation of Mobile Phone as a Platform for Healthcare Applications

Javier Sierra, Miguel Angel Santiago, Iiro Jantunen, Eija Kaasinen, Harald Kaaja, Matthias Müllenborn, Nicolas Tille, and Juha Virtanen

Contents

18.1 Introduction

Health is one of the six future application fields for ambient intelligence (AmI) in everyday life that IPTS/ESTO (European Science and Technology Observatory network) names in their foresight report (IPTS/ESTO, 2003). Health applications include prevention, cure, and care.

Mobile phone–based AmI is among the first and most realizable steps toward the world of AmI (Roussos et al., 2005). Mobile phones can be used as secure personal devices to disclose personal information when needed. Personal mobile phones as the interaction medium also facilitate the user's keeping of ambient services in control. Using the mobile phone as the service platform facilitates the provision of services to users without the need to purchase additional equipment. Moreover, the familiarity of the platform eases the process of putting novel services into use. These characteristics make mobile phone–based, AmI solutions promising in the fields of healthcare and assistive services, especially those that require continuous monitoring of health parameters and timely feedback to the patient and/or the patient's caregiver(s).

Iso-Ketola et al. (2005) have highlighted the problem of interfacing wearable devices and suggested a mobile phone (with wireless technology such as Bluetooth or IrDa) as the interface device between the user and the body area network (BAN). Niemelä et al. (2007) have highlighted the possibilities of mobile-centric, AmI for independently living elderly people. Jung et al. (2008) showed how physiological (wearable) signal devices communicate in wireless body area network (WBAN) to the mobile system in ubiquitous healthcare systems. The problem with wearable devices is the small size of the battery, which leads to short operational time before recharging.

Mobile phones have been proposed as the platform for a wide range of health-related applications, including diabetes management (Lopez et al., 2009), mental health intervention (Matthews et al., 2008), weight management, obesity therapy (Morak et al., 2008), cancer assistance (Islam et al., 2007), and asthma management (Jeong and Arriaga, 2009). AmI is applicable in healthcare, but there are also some acceptance problems studied, for example, in hospitals.

The ethical challenges of AmI applications are multifaceted, for patients and healthcare professionals. AmI can make information and communication technology automatic and invisible. Some information about the individual patient and his or her behavior can be collected without the patient even noticing it. An illustrative example is the implanted cardiac defibrillator, which automatically applies electrical stimuli to the patient's heart when the defibrillator's microcircuitry recognizes abnormal cardiac electrical activity that is predictive of an imminent, potentially fatal ventricular fibrillation. The wide adoption of implanted cardiac defibrillators speaks to a strong consensus among regulators, physicians, and patients as to the favorable risk-benefit and cost-benefit aspects of this microcircuitry-based device which made its initial appearance about two decades ago. This example illustrates that new products based on advanced technologies should provide favorable risk- and cost-benefit attributes, along with human values such as privacy, self-control, and trust. These ethical issues are frequently raised as ethical concerns of ubiquitous computing (Bohn et al., 2005; Friedewald et al., 2006; Kosta et al., 2008).

The personal mobile terminal is a trusted device for personal data, providing facilities to ensure the user having control of different actions. This fact provides a good basis for ethically acceptable solutions (Kaasinen et al., 2006). Still, especially in healthcare, there are several ethical concerns, which is why we studied ethical issues parallel to the technical development of the mobile platform and the demonstrator applications. We defined ethical guidelines for the platform design and for the forthcoming applications. The ethical guidelines were based on six ethical principles: privacy, autonomy, integrity and dignity, reliability, e-inclusion, as well as the benefit for the society (Ikonen et al., 2008).

One of the main concerns in this medical technology scenario is to allow a proper interaction between the users (patients and doctors) and the devices. Usability of this system becomes an essential condition in order, not just to improve medical aid, but also to avoid any harm may be caused (Nielsen, 2005). Usability is also an issue for homecare professionals who travel and meet patients at their homes. Usability problems can lead to low productivity or low effectiveness of homecare professionals, for example, when using mobile homecare application with a PDA (personal digital assistant) or a computer. These tasks cover, for example, finding specific information, finding healthcare-related information, documenting new information, and switching to another patient (Scandurra et al, 2008). User acceptance of mobile applications targeted to hospital and professional use has also been studied by other researchers. Scandurra et al. (2008) have evaluated a mobile homecare application with home help service staff and they found that contextual differences such as terminology use and documentation practices hindered performance more than uncovered usability problems. Wu et al. (2005) have carried out a survey with healthcare professionals in Taiwan. They found that compatibility with work processes is the most important factor of information technology acceptance in the context of mobile healthcare. Kummer et al. (2009) have studied user acceptance of AmI and mobile technologies in hospitals in India and in Germany.

Their interview study revealed that in Germany fears related to new technologies, fears of surveillance, and fears of changing working conditions affected intentions to use new technology. In India, the fears were mainly related to fears of changing working conditions.

User acceptance of mobile healthcare applications targeted to personal use has been studied less. Beaudin et al. (2006) have studied user reactions to concepts in longitudinal health monitoring. They propose that mobile monitoring systems may be more readily adopted if they are developed as tools for personalized, longitudinal self-investigation that help end users learn about the conditions and variables that impact their social, cognitive, and physical health. In the study by Beaudin et al., the interviewed users envisioned conducting customized, short-term health investigations. Jeong and Arriaga (2009) propose heuristics based on an ecological framework to understand user needs. They have used asthma management as a case study and proposed for that application field five application concepts: asthma monitoring and information transfer, information transfer of personal health record, outdoor activity support (e.g., air quality warning messages), and indoor air quality management (e.g., detecting poor air quality).

In the course of our researches, we have been designing future healthcare applications based on a mobile architecture and platform. In this chapter, our aim has been to illustrate usage possibilities and evaluate them with users to get early feedback from the design. During the first phases of our project, we were describing application possibilities as scenarios. The main health application scenarios Smart Pillbox, Smart Listening Device, and Sleep Quality Logger were illustrated as animated movies (Figure 18.1).

The main concepts presented in the three scenarios were the following:

■ The Smart Pillbox allows dosage to be monitored based on recognizing when the pillbox is opened. The mobile phone keeps track of the openings, and afterward gathers the information to a data center accessed by doctors.
■ The Sleep Quality Logger includes a plaster equipped with sensors. The plaster is put on the forehead of the user where it monitors EEG overnight. In the morning, the user transfers the measurements to his mobile phone and further to the healthcare personnel.
■ The Smart Listening Device embedded in a mobile phone utilizes directional microphones with which the user can select the sound sources to be amplified and to be faded out.

Smart Pillbox and Sleep Quality Logger scenarios acceptance was first approached by focus groups in Spain and in Finland and its results were presented in Niemelä et al. (2007). In this chapter, we present the results of a survey and a series of user tests in laboratory involving our three scenarios.

Next, we give a deeper description of our mobile-centric platform characteristics. Then, we briefly describe the scenarios and the role played by the mobile

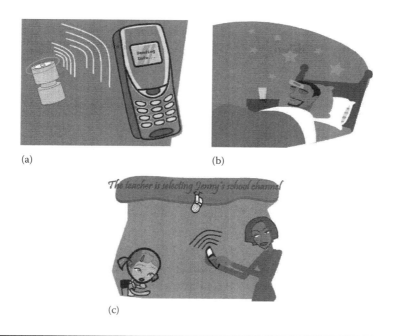

Figure 18.1   Illustrations of the animated scenarios: (a) Smart Pillbox, (b) Sleep Quality Logger, and (c) Smart Listening Device.

phone and the healthcare actors. In the "User Evaluation" section, methodology and results regarding scenario surveys and laboratory test with real users are presented. Finally, we discuss the evaluation results data and present a characterization of end user requirements, conclusions, and recommendations for future work.

18.2 Mobile Phone–Centric Architecture

Our vision is based on a mobile-centric approach to AmI for healthcare applications. In this vision, personal mobile devices provide trusted intelligent user interfaces and wireless gateways to sensors, networks of sensors, local networks, and the Internet (Broens et al., 2005; Jantunen et al., 2008a; Korhonen et al., 2003). We are developing a mobile phone–based architecture, which includes the personal mobile terminal with access to wireless sensors and tags in near proximity (Figure 18.2). Our architecture defines five different hardware entity types: terminal devices (mobile phones), active wireless sensor devices, RFID sensor tags, new RF memory tags, and back-end servers in the Internet.

The same architecture is able to support both existing sensors types and new RF memory tags. In this architecture, functionality offered by different sensors and tags can be integrated into independent modular subsystems that the mobile phone

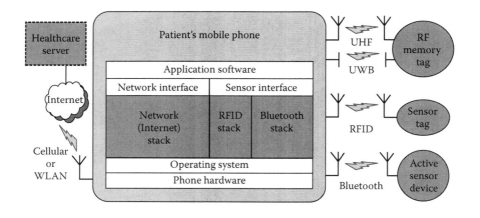

Figure 18.2   Architecture for mobile-centric approach to healthcare AmI.

is interacting with in its near proximity. Subsystem is a physical implementation of a certain functionality. It is an entity that includes all the necessary hardware and software resources to provide that functionality, which it offers as a "service" for other entities. For healthcare it could mean, for example, display services, measuring services, storage services, or networking services. By following that definition, in healthcare, such sensors and tags form measurement and storage functionality, respectively. Mobile phone is, for example, able to access available healthcare-specific local services by utilizing subsystems that provide reading patient healthcare data from measuring sensor devices and tags, writing the data on tags, displaying the data, and storing the data into secure RF memory tags. Another (network) subsystem provides access to back-end healthcare services in the Internet, i.e., healthcare servers and service providers (healthcare operators). Uploaded measured data can be processed by analysis and monitoring software and be made available for the relevant healthcare professionals for further analysis or diagnosis.

Available subsystems that enable AmI for healthcare applications provide such services either through embedded or separate devices. Clear benefit from the modular approach is a flexible architecture supporting different specialized subsystems, making possible fast and flexible development of new services. Subsystems can offer common services or more specialized ones, for example, tuned to target healthcare application. Each subsystem is attached to the mobile phone–centric architecture that takes it into use when needed.

Reading active (battery-powered) *wireless sensor devices* is based on Bluetooth, which is the standard BAN technology. The fact that the devices are active increases considerably the distance allowed between the reader and the sensor during the communications, but requires the sensors to have a power source of their own (e.g., battery). Autonomous sensor devices can, for example, periodically collect and send sensor measurements, or the sensor data can be streamed to the phone with a predefined rate (Quero et al., 2007), or the sensor devices can be polled at freely configurable intervals. For applications where classic Bluetooth is too power-hungry, for example, due to restrictions on device (and thus battery) size, a low-energy mode (Honkanen et al., 2004) has been developed for Bluetooth and is becoming a part of the Bluetooth standard in 2010.

The mobile phone also provides a wireless access to the Internet, which makes it possible to send sensor data to external servers for extra services or applications, and receive analysis results from healthcare professionals; diagnosis, feedback results, prescriptions, and further healthcare management based on sensor data.

The scenarios herein analyzed are supported by this part of the architecture: active wireless sensors and wireless access to the Internet. However, we aim to extend the future applications of our platform by including tags on AmI devices in the field of healthcare.

Reading sensor tags is based on RFID (radio frequency identification) communication, which is powered by an external reader integrated into the mobile phone. Tags can be passive, not having any battery of their own. Also active tags, providing their own power source for more complicated functions or longer communications range are possible, along with semi-passive tags which use their own power source for sensing and data logging, while communications can be powered by the reading device (mobile phone). Normally, RFID sensor tags are limited in memory size and in how often the measurements are available, unless there are regular readings performed by the reader device. Reading a sensor tag releases the tag's memory for new measurements.

There are two types of RFID radios available, one low-data-rate single channel (such as NFC which is already available in some commercial mobile phones, for example, Nokia 6131 NFC, Nokia 6212 Classic, and Nokia 6216 Classic models) and one high-data-rate dual channel (Jantunen et al., 2008b) for moving large amounts of data (in gigabit range). The latter one is interesting for RF memory tags, which combine both memory capacity and high data rate requirements. RF memory tags can also be used as an information media, for example, to provide medical information, user instructions, and containing also multimedia material (Kaasinen et al., 2008). In some applications, RF memory tags could even include medical history data, but this has many sensitive viewpoints and needs built-in security features.

The user would have direct benefit from the RF memory tags, from high-speed and high storage capabilities. Mobile phone is an appealing and personal device for secure readings, and this fits well for mobile-terminal-centric architecture. One possible use case is the Sleep Quality Logger presented later in this chapter. In that case, sensor system would log data to a memory readable through RF memory tag interface. This would make the device cheaper, as it would need no internal power scheme for wireless communication.

The operating system of the mobile phone provides application protocol interfaces (APIs) to application software for using integrated or wirelessly connected sensors and memory. The data can be processed on the phone, with information provided to the user on the phone user interface (display or audio output). Processed or unprocessed data or context information can be also sent to healthcare servers in the Internet, where the data can be logged for long-term monitoring of health status, and healthcare personnel, as well as the patient, can be alerted if acute symptoms are detected.

18.3 Scenarios

18.3.1 Smart Pillbox Scenario

The MEMS® (Medication Event Monitoring Systems) platform transfers dosing history data from the Smart Pillbox to the patient's mobile phone via a wireless Bluetooth communication and, from there, to data servers. The dosing history data enter a database, which is analyzed by appropriate statistical means, and the resulting adherence analyses are displayed in a variety of easily interpreted graphic and tabular formats. The comparison of the patient's drug dosing history with the prescribed drug dosing regimen leads to quantitative results of the patient's adherence to the prescribed medicines.

18.3.1.1 Typical Scenario

A doctor is unsure if his or her patient takes the prescribed medication(s) correctly. The doctor prescribed the patient's medication to be dispensed in Smart Pillbox. The pharmacist informs the patient how to use the Smart Pillbox via its own mobile phone or network. The patient uses his own mobile phone to transmit dosing history data at intervals such that they arrive prior to a scheduled visit, thus giving to the doctor the key data on the patient's drug intake between the visits. The patient has the opportunity to see some information about the quality of his drug intake in graphical or textual form on the display of his mobile phone. Those results are generated locally by the application running on the mobile phone. He can enter, if necessary, some additional information which helps the pharmacist and the doctor to understand and analyze the transferred dosing history data. Furthermore, even if the patient cannot attend the scheduled visit, at least the data are available to the pharmacist and physician, one of whom may opt to call or visit the patient, depending on the nature of the data and the risks. The management system analyses the patient's dosing history, compares it to the prescription, and generates a report, which is automatically forwarded to the pharmacist and the doctor, one of whom (depending on the local organization of caregiving) can then decide what is best to do next.

18.3.2 Sleep Quality Logger Scenario

The Sleep Quality Logger is a tool for sleep disorder screening at the patient's home. The patient's brain wave and movement data are collected overnight with an active tag called the Logger Patch. Mobile phone is used for transmitting measurement data to healthcare provider's Data Server, and for receiving instructions. The mobile phone also serves as user interface for the patient. Automatic Sleep Analysis application provides the attending physician with familiar numbers describing the sleep quality with no need for interpreting raw EEG-signals or questionnaire data.

18.3.2.1 Typical Scenario

Having felt extremely tired lately, Tom goes to the healthcare center at his working place. The attending physician decides that in addition to standard laboratory test results, she needs objective information about Tom's sleep. During the same visit, Tom receives a "Sleep Quality Logger" package, which comprises a Logger Patch, disposable adhesive electrodes, and a stand for battery recharging and data downloading from the logger. The stand also communicates wirelessly with mobile phone.

That night Tom snaps the electrodes to the Logger Patch and applies it on his forehead. The device starts logging brain waves and head movement automatically. Next morning, Tom detaches the logger patch and places it on the stand, which initiates data transmission automatically. After some 5 min, the mobile phone informs Tom that the data has been successfully transmitted to the healthcare provider's data server. Tom leaves for work. The data is automatically analyzed in the server and reviewed by a physician. A message is sent that the measurement should be replicated in order to confirm the results. The following night Tom repeats the measurement. The next day, he receives a message advising him to make an appointment with the physician.

During her next session with Tom, the physician logs into the Sleep Analysis application and reviews the automatic analysis results from Tom's data. Taking into account the laboratory test results, as well as other clinical information, she makes her diagnosis.

18.3.3 Assistive Listening Device Scenario

The solution investigated in the platform is a wireless smart microphone array with beam steering and other signal processing capabilities to enhance speech in noise. The array transfers the audio information wirelessly to the hearing instruments and can be controlled from a mobile phone through a Bluetooth link. Challenges are typically encountered in noisy environments, for example, when many speakers are present in the same room such as a classroom or conference room.

18.3.3.1 Typical Scenario

Jenny loves to sleep in. So as usual, she is a little late for school. She does not always pay attention in class, either. Jenny might seem like a typical student, but she is not. Part of the reason she finds it hard to concentrate is because she has a hearing loss.

One day, Jenny receives a new set of hearing instruments. The audiologist also gives her a small white box, which looks like an MP3 player. With this, Jenny can listen to music stored on her phone and talk to friends over the phone with her hearing instruments.

At school, the teacher asks if she can borrow Jenny's mobile phone. At first, she thinks the teacher is searching for music. Suddenly, Jenny hears a bip-bip-bip in her ears and Jenny knows that something else is up. The teacher explains that when the check mark is set next to Jenny's classroom on her mobile phone display, Jenny can hear the teacher. A smart microphone that sits on the teacher's desk picks up sound and transmits it to Jenny's hearing instruments. "Wow," Jenny thinks, "this thing is pretty cool."

The teacher tells Jenny that this new smart microphone has something called voice priority. This means it is smart enough to focus in on what Jenny needs to hear, like the voice of her teacher. And then it turns down other distracting sounds like that old fan that hums all day so she can hear and understand every word. The teacher does not have to wear the microphone anymore like in the old days and Jenny can even hear her classmates talking now. The sound of their voices always comes in loud and clear!

Jenny thinks about what an amazing difference the new smart microphone has made. "Before, I had to keep raising my hand to ask, 'What did you say?' Now, I can hear much more clearly." Jenny is convinced that she will get better grades. The only problem is that she still sleeps in. Oh well, there are some problems even the smart microphone cannot solve!

18.4 User Evaluation

We conducted Web surveys in Finland, France, and Spain, involving altogether 500 respondents who assessed all scenarios except for the assistive listening device that was considered too specific to be included in the survey targeted to general public. Parallel to the survey, we built proofs of concept that facilitated studying and evaluating user experience of the Smart Pillbox scenario and the assistive listening device scenario. The proofs of concept were evaluated with eight users in laboratory conditions.

18.4.1 Web Survey

Internet-mediated, Web-based questionnaires were developed to gather quantitative user feedback on ethical, acceptance, usage, and consumption topics. The survey looked forward to making a deep approach of user background in terms of scenarios and obtaining user assessments of the scenarios concept. Our main objective was obtaining data that allowed us to make a prediction of impact of the devices and getting cues regarding end user requirements.

Table 18.1   Number of Participants and Geographical Distribution per Country

Country	Number of Participants	Geographical Distribution
Finland	240 (120 per scenario)	Helsinki 10%
		Espoo/Vantaa/Kauniainen 9%
		Other metropolitan area 5%
		Turku/Tampere 8%
		Other town with over 50,000 inhabitants 15%
		Other town 29%
		Other municipality 24%
Spain	200 (100 per scenario)	Madrid 34%
		Barcelona 33%
		Other town with over 50,000 inhabitants 33%
France	60 (30 per scenario)	Paris 34%
		Marseille 33%
		Toulouse 33%

Survey sample was picked from the main cities of three countries: Finland, Spain, and France. The targets of the survey were people older than 18 years old who owned a mobile phone.

The sample consisted of a total of 500 individuals between 18 and 60 years of age (48.5% males, 51.5% females; quotas: 30.3% 18–30 YO, 26.1% 31–45 YO, and 43.7% 46–60 YO). The total of participants per scenario was 250 (Table 18.1).

18.4.2 User Tests in Laboratory

Our evaluation was based on a user-centered design perspective. Right now, UCD is the main method about how to develop usable systems (Karat and Karat, 2003). Thus, proper comprehension and evaluation of usability is required.

Usability is a multidimensional construct that can be examined from various perspectives. We chose ISO 9241, a suite of international standards on ergonomics requirements for office work carried out on visual display terminals, as our framework. Part 11 of ISO 9241 (1998) defines usability as the sum of three factors: effectiveness, efficiency, and satisfaction. Nowadays, it is commonly accepted that this standard offers a broader perspective of usability than other models (Abran et al., 2003).

18.4.2.1 Test Characteristics

Our usability evaluation sessions consisted of five empirical methods: scenario-based task performance (formal laboratory usability testing), user observation for critical incidents, concurrent think-aloud, questionnaires, and post-task interview. First, we showed participants the same flash animation regarding the proof of concept of the device that was presented to survey respondents. Then, they were asked to perform benchmarking tasks in the laboratory setting, and their performance was measured and recorded. At the end of each session, users filled in a satisfaction questionnaire and a brief wrap-up interview regarding user's final impressions, and conclusions closed the evaluation. Usability evaluation was conducted only to the pillbox application and the listening device application. The sleep logger was evaluated only as a scenario as the application did not include much user interaction.

18.4.2.2 Sample

Tests goals focused especially on conceptual issues, look and feel, and basic functionalities of the scenario devices. Then, we looked for participants using a mobile phone on a regular basis with no more specific characteristics (Table 18.2).

The Smart Pillbox group consisted of four participants, three males and one female, with ages between 26 and 30. Assistive listening device group consisted of four users, two females and two males, with ages between 21 and 45.

Table 18.2   Participant's Profiles in Laboratory Tests

	Participant	*Age*	*Gender*
Smart Pillbox	1	30	Male
	2	26	Male
	3	27	Male
	4	28	Female
Assistive Listening Device	1	45	Female
	2	21	Male
	3	27	Female
	4	27	Male

18.4.2.3 Processing of the Data

Both qualitative and quantitative data were obtained. Regarding qualitative data, we analyzed the natural behavior and interaction of the users with the application (what do they look for? in which way?), conceptual aspects (do they like it? would they use it?), and perceptions and impressions held by users during the interaction. Paying attention to quantitative metrics, efficacy was measured by scenario success ratios and number (and criticism) of errors during system use. For efficiency, there was not specific time-out for any of the scenarios; hence the facilitator judged if the time spent by the user was a usability standard or if the user was taking too long. Satisfaction was measured by a Likert scale 1–7 questionnaire. Deficiencies in the scenarios tests were categorized according to heuristic evaluation principles (Nielsen, 1994b). The issues were prioritized according to severity.

18.5 Evaluation Results

18.5.1 Smart Pillbox Scenario

18.5.1.1 Web Survey

Two hundred and fifty three of the respondents of this scenario described themselves as patients. Pill intake showed to be a rather frequent way of medication with 30% of participants using pills on a regular basis. This group of usual pill takers was characterized by three features: mainly taking just one type of medication, reporting some degree of oversights on intakes, and whose doctors not checking their intakes in an extended way. In our sample, females and people belonging to the 46–60 YO group tended to follow more often medical treatments.

Acceptance of the Smart Pillbox seemed to be really good since the great majority of the respondents (84%) would consent a record of their medicine intake with a device like the Smart Pillbox, if sending to their doctor or a medical center in a confidential and secure way. Those who rejected (16%) claimed the lack of control over who could read their health data and issues regarding security over transmissions.

In terms of controlling doses, 66% of the participants chose mobile phone as an appropriate medium of reminder. SMSs and sound alerts were the favorite ways of warning.

Some concerns expressed by the patients about the pillbox were

■ Possible failures between pillbox and mobile phone interaction (i.e., loss of data or system failures).
■ Reliability of the intake control method because "the fact that the pillbox gets opened doesn't mean you have actually taken your medication or the correct dose." Some respondents argued that for certain kind of pills consumers (with mental illness), it would not be of any use because sometimes "they don't want to take the medication."

- Easiness of use and comfort.
- Confidentiality.
- Price of the device and the possible added cost of the warning method (i.e., SMS).

Finally, the electronic pillbox had an overall assess of 3.2 out of 5 by patients respondents. Surprisingly, when asked about if they would expend money in the device, the group of people that claimed to usually take pills reported less interest than the non-medicated group (just 49% of pill takers would pay for it, instead of 68% of non-pill-takers).

The rest of participants in the Smart Pillbox Scenario—seven—were doctors. Their specialties were family medicine, geriatrics, neurology, palliative cares, and pediatrics. All of them except one used computers and network databases for carrying out certain tasks on their jobs such as keeping records of their patients' stories and processing medical data. Two of the doctors even used to give remote attendance to their patients. Therefore, the scenario was considered very useful for all of them. In fact, the possibility of carrying remote patient revision was seen as a key feature for integration in their workplaces. In relation to intake oversights control, five of the seven doctors reported giving a follow-up to their patients' medication.

After watching the scenario animation, six of the doctors were interested in having objective data on their patients' intake of the medicines they have prescribed. The one who answered not to be interested in it claimed it was because it seemed very difficult to integrate such a thing in his current medical system.

Disadvantages proposed by the doctors for such a system were

- Lack of resources in healthcare.
- Difficulty of suiting it for the elderly.
- Failures when not carrying both the mobile phone and the pillbox together.
- Possible high cost of the system.

The mean assessment the Smart Pillbox proof of concept by the doctors was 3.7 out of 5.

18.5.1.2 User Tests in Laboratory

In the usability tests, all participants managed to achieve the entire tasks with the Smart Pillbox. Figure 18.3 illustrates the ease-of-use assessment by the participants, showing a good interaction experience as every average rate goes over five (one being very difficult and seven very easy).

However, when comparing animated scenario assessment with after-test proof of concept valuation, we can see in Figure 18.4 that scores are generally lower. That reveals different critical points and usability issues found out along some of the tasks.

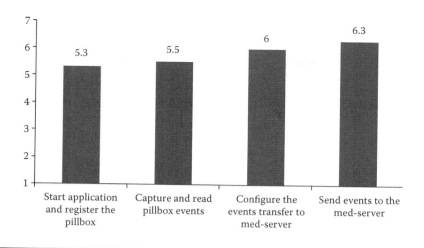

Figure 18.3 **Likert scale 1–7 for ease-of-use perception for each of the defined tasks in the test.**

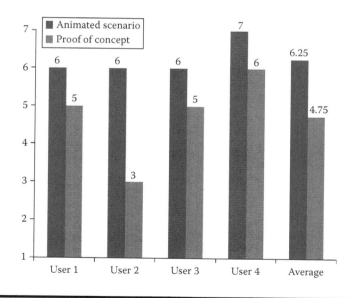

Figure 18.4 **Likert scale 1–7 comparison between assessments based on animated scenario and based on interaction with the devices.**

Usability tests identified five critical issues in the Smart Pillbox scenario:

■ The process of capturing data from the pillbox and displaying its events in the mobile phone was confusing. First, the user had to download the data from the pillbox through a main menu option, and then he should select another option to display intake events. This logic of mobile phone–pillbox interaction was felt unnatural and relied on too many steps.

■ Mobile phone interface had too many confirmation boxes even on noncritical actions.

■ Information of intakes was not presented in an appropriate way with adequate metaphors. Users barely understood the meaning of the information given to them.

■ Certain critical options of the configuration (i.e., the medical server setting) relied on the user or were easily accessible.

The terms and expressions used in the options texts are rather far from what would be a user valid description of action results. For example, "Upload data" instead of "Send your intake data to your doctor" or "Display Events" instead "Show my intake data." The same happened with most of the confirmation boxes' texts (Figure 18.5).

These usability issues mainly imply the transgression of three heuristics: a poor match between the system and the real world, shoving user to recall rather than recognition and failing to accomplish flexibility and efficiency of use.

Observing satisfaction questionnaire metrics, those items related to information being sufficient, menus and options adequate definition, appealing look and feel, and estimation of use were below the mean. Users found some menus and options to be not logically connected and to be redundant. Help menus or tips

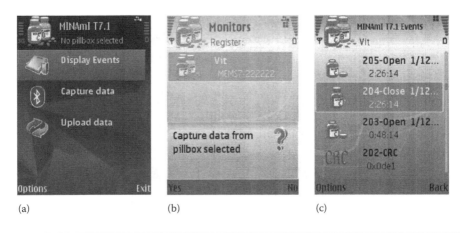

(a) (b) (c)

Figure 18.5 Smart Pillbox interface with main menu: (a) Capture, (b) Intake Data, and (c) Screens.

were unavailable and option texts were rather basic. Recommendation to family and friends got a low rate, which could be explained because none of them used to take pills and they thought that the device target would be only certain kind of public (elderly, diabetics, etc.). About interface design, participants found lack of consistency along the different menus.

At the final interview, users reported that the simplicity and the small number of options in the interface were good features and seemed to prefer complete automation between mobile phone and pillbox data transfer, being transparent to them. Implementation of alarms about intake oversights in the mobile phone application was also commented. It could be integrated on a "calendar-like" layout, where information about medication process (date of beginning and end, statistics about doses consumed and left, etc.) could be obtained.

18.5.2 Sleep Quality Logger Scenario

18.5.2.1 Web Survey

In this scenario, 245 participants selected the patient profile. From this sample, 59% reported some degree of tiredness because of not always sleeping well and 17% seldom or never had a proper sleep. It seems that, even having problems regarding sleep quality, it is very rare to find people that visit the doctor for that reason. Only 12% of our participants went to a professional asking for advice about it but negatively correlated to the age. That is to say, despite of the fact that women and people between 18 and 30 YO reported higher levels of tiredness for bad sleep, people of 46–65 were the most regular visitors.

After watching the proof-of-concept animation, participants showed a relatively high interest in the device, but only a 37% considered they would use it. Easiness of use and comfort were emphasized as attractive features mainly on account of the measurement at the patient's home. Most of respondents seemed to prefer automation in the process of measurement but 74% of them would like to take control of the data transmission to the medical center.

Those of the few respondents that actually suffered from apnea reported really good comments and looked forward to testing a device with these characteristics. The main profile willing to test it were middle (31–45 YO) and old (46–65 YO) people with sleep difficulties.

Some of the participant's impressions about the Sleep Quality Logger were

- Concerns about the comfort of sleeping with such a device: "For me the possibility to change posture during the night is essential. I'm suspicious how the sensor works if you move or if it only works when sleeping on your back…".
- The respondents wondered what kind of phone would be necessary to use the device and claimed that the system should be compatible with average mobile phones.

■ Easiness of use and availability of needed devices were repeatedly demanded.
■ Using the Logger to improve research on sleep disorders was considerably marked, just like one participant states: "poor quality of sleep is much more common than people think and sometimes it is the precursor of some other diseases."
■ In terms of security, patients wondered if the necessary electrical transfer for achieving the measurement will be safe in a home-based way.

The mean assessment by patients for this scenario was 3.74 out of 5.

In the doctor profile, we had five respondents, four of them in public practice, being their specialties palliative cares, rehabilitation, rural health, sleep unit study, and emergency. All of them used the computer regularly at their workplace for managing patient histories through a networked database. Regarding remote assistance, doctors considered it very useful overall for emergencies triggered by patients and monitoring certain vital constants, but none of them used to offer this service. Comfort and saving of time were obvious benefits of remote assistance.

The doctor in the field of apnea treatment was skeptical about achieving a good diagnosis at patient's home since the hospital has more resources for carrying out a proper evaluation.

Even though three of the doctors were interested in the Sleep Quality Logger, they were unsure of the accuracy that such an EEG-based assessment could have of patient's sleep quality and patients sleep stages as part of sleep apnea.

Doctors' mean assessment for the EEG Logger was 3.2 out of 5.

18.5.3 Assistive Listening Device Scenario

18.5.3.1 User Tests in Laboratory

In the laboratory tests of the Assistive Listening Device, all participants succeeded in achieving 100% of tasks goals. Figure 18.6 illustrates the ease-of-use assessment by the participants whose scores pointed out high values (all over five out of seven points).

There was almost no change between values given to the scenario animation and after-use proof-of-concept assessment (see Figure 18.7). Furthermore, the only user that differed in his assessment scored slightly higher the device after the proof of concept. Therefore, expectations of use after watching flash animation and experience after using the device seemed to correlate each other at a certain level.

However, participants faced some difficulties when interacting with the mobile interface. Usability testing identified four critical issues with the Assistive Listening Device:

■ Application first screen did not clarify that there were two profiles available with different interaction possibilities. All this information should be accessible from the main menu screen.

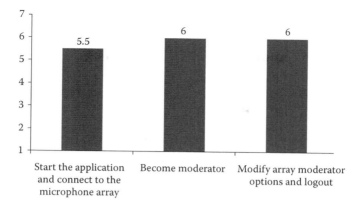

Figure 18.6 Likert scale 1–7 results for ease-of-use perception for each of the defined tasks in the test.

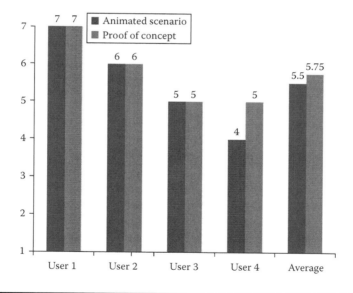

Figure 18.7 Likert scale 1–7 comparison between assessments based on animated scenario and based on interaction with the device.

- There was a mixture of concepts: "connect to an array" and "add a new array." It entails two steps for achieving just one objective: getting connected to an array.
- Lack of information: "Array status" showed that there was a connection established but not with which device.

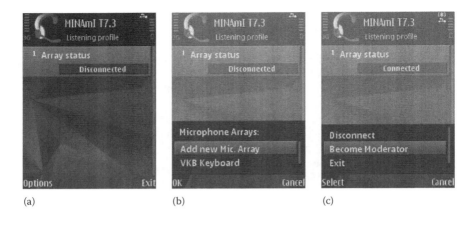

(a)   (b)   (c)

Figure 18.8   Assistive Listening Device interface with main menu: (a) Array Connection/Addition, (b) Profile Swap, and (c) Screens.

Critical information as current profile and array status cannot be rapidly distinguished from other information displayed. Users missed icons and color codes to improve connection and profile status recognition (Figure 18.8).

With regard to heuristics, these problems rely mainly on a poor visibility of system status, user having to depend on recognition rather than recall and not allowing a flexible and efficient use of the system.

In the satisfaction questionnaire analysis, we observed that users scored lower those items concerning the correct definition and integration of options, look and feel, and offered information. The logic of the array connection process and the fact that there was a "hidden" profile distinction may be the reason for these scores. There was also a lack of contextual help or tips, which may explain the low assessment. In terms of interface design, lack of icons and graphic aids and a bad use of screen space equally manifested a low average assessment.

In the final interview, our participants seemed to be really interested in such a device. All of them commented the huge possibilities that it could have for hearing-impaired people. Furthermore, some of them thought that this system could also be implemented for regular people to use in conferences, museums, etc.

18.6 Summary and Discussion

Under the paradigm of a mobile-centric approach, we have defined a platform suited in the field of healthcare. The mobile phone becomes not only a gateway between small sensors and healthcare servers, but also a trusted device to manage AmI services. Three scenarios have been proposed according to the platform and their related applications have been developed in a first stage.

The survey and laboratory tests have helped us to characterize the role of the mobile phone as a platform for developing health-related applications. We have obtained data involving not just the context of use but also a first interaction with devices. From these results, both strengths and weaknesses have been outlined along with implications regarding design and development of future systems.

The Smart Pillbox scenario showed a good acceptance between survey respondents. From our data, it seems that the current context of people who take pills approves of the introduction of a system like this. Not only patients reported medication oversights and being not checked by doctors, but also they confirmed the mobile phone as an appropriate way of interacting with the pillbox and managing its data.

From the doctors' side, the context for a Smart Pillbox was appreciated too. Doctors were immersed on a regular use of computers and networks at their work for processing medical data and patient's stories. On the other side, remote assistance seemed not to be very usual and patients and doctors contradicted themselves about the follow-up in the patients' medication intake. While doctors argued to control this factor, patients denied it.

User tests of the pillbox proof of concept showed that our mobile phone application was still far from the animated scenario idea. The higher scores in the assessment of the latter confirmed it. However, as a result of the total success in achieving the tasks and the good scores for ease of use, we can expect that the identified problems (mostly about allowing proper interaction metaphors and enabling a refined interface) could be easily improved in further iterations.

An obstacle for integrating a system like this in everyday life of possible users would be to allow an errorless and natural intake control and data management. User must never feel controlled by the device, but we must deal with his possible oversights in a friendly way. Another implication of system design is the possible limitations that healthcare systems could present to integrate it. Even doctors who liked the idea very much claimed that it should make their job more fluent and do not imply high costs.

The Sleep Quality Logger seemed to have a moderate acceptance. Our sample didn't have very high quality of sleep; the percentage of respondents that actually sleep well was very low. Consulting the doctor about this problem is even stranger. However, we must not be impressed by these results. People with sleep apnea are rarely aware of having difficulty breathing, even upon awakening (Slovis et al., 2001), which lead us to one of the advantages of the Logger. Being comfortable (nonintrusive to sleep) could allow home-based tests, which surely would improve the acceptance of those patients reluctant to sleep in a hospital. On the other side, the ecological validity that offers can be used in sleep research as an added value.

The skepticism of the doctor devoted to diagnosis of the apnea about accurate analysis at patient's home should be a priority in the Logger. System reliability and accuracy need to be granted for the experts.

Users managed to succeed in all the tasks of the Assistive Listening Device laboratory test. For this device, assessment of animated scenario and proof of concept correlated. One of the participants even assessed the application with a higher score.

Participants enjoyed interacting with the environment through the phone. Ease-of-use perception for each of the defined tasks got also high scores. However, some problems regarding interface design and logic were found. Users were frustrated because of the lack of system status visibility and were tired of having to go back and forth in order to recall critical information not accessible to them. Although the application was rather simple, the system lacks certain flexibility and needs to be refined for more efficient use.

With all the scenarios, survey respondents and test users were concerned about ethical issues. Medical experts were concerned about the security of health data and they also highlighted reliability, accuracy, technology replacing face-to-face examinations, and unintentional misuse as possible ethical problems.

In the Web survey, the respondents were concerned about information confidentiality, system reliability, and ease of use. In the proof-of-concept evaluations, the users emphasized the need to stay in control and predictability.

18.7 Conclusion and Further Work

The results of our studies give evidence that mobile phone can become an adequate device for implementing healthcare applications and assistive technologies. We have presented in this chapter three scenarios which differed highly in their main purpose and target. However, all scenarios are based on using the user's personal mobile phone as the platform for the healthcare applications.

Pill fatigue is a well-known effect that implies decrease in treatment fulfillment with time even in those responsible patients. It is proven that the degree of fulfillment of patients' treatments (called "medication adherence") is one of the main factors to achieve a positive therapeutic outcome and that poor adherence leads up to substantial worsening of disease, death, and increased healthcare costs (Osterberg and Blaschke, 2005).

Ecological data gathering and patient comfort are described by doctors and patients as characteristic barriers of sleep studies. The attitudes toward preventive behavior and screening tests regarding sleep apnea disease seem to be almost null; this can be a big problem since sleep apnea tends to occur without the patient being conscious of it. Fast home-based screening tests would be highly recommended for those persons belonging to risky groups, according to the *European Respiratory Journal* (2005).

Hearing-impaired children can have difficulties in improving their language in written, read, spoken, and signed forms. This poor language acquisition reverberates in dragging cognitive and social competences emergence. It has been found that the quality of the hearing aid (multichannel cochlear implants instead of single-channel, for example) improves meaningful use of sound by children in everyday situations (Robbins et al., 1991).

Our research verifies that the mobile phone takes a central part in an everyday context that raises a wide range of profiles (regardless of age, gender, or cultural differences) and can be doubtless exploited in benefit of these healthcare settings. Therefore, the goal is to implement systems that allow applications to be friendly, reliable, and easily incorporated in targeted infrastructures. In order to fulfill these user-centered challenges, we can take advantage of mobile phone strengths. Its ubiquitous and continuous access, familiarity, and ease of use has the potential to make health-related activities and assistive technologies more convenient, less stigmatic, and more engaging. It is also cost effective to connect the wireless sensor units to the phone and utilize the computing and communication facilities of the already existing device in healthcare applications.

Ethical issues will require specific attention when providing healthcare services on personal mobile phones. Privacy protection is crucial as the applications gather, store, and redirect personal health data. Autonomy gets important as AmI technology facilitates many automatic functions. However, the user needs to be aware what is happening and (s)he has to have ways to control AmI actions. Integrity and dignity are emphasized when discussing about AmI applications targeted to elderly people, children, and disabled people. It should be considered how the actual user feels the system and whether (s)he can accept it. Reliability is an important issue when a consumer device is used as a platform for safety of critical applications. E-inclusion highlights the importance of ensuring that new technical solutions are accessible for people with limited abilities. Finally, the role of the technology in the society raises important issues related to how mobile technology will be used in healthcare and how it will change our lives.

In terms of further work concerns, the identification of new scenarios and the deeper data acquisition has to continue. Our laboratory tests were based on the mobile phone interface and included peripheral devices (the pillbox and the simulated array) in a rather basic way responding just to emulation purposes. In future investigations, it would be necessary to carry on tests implying them in a more central role, so a complete system interaction is acquired and can be evaluated. From this perspective, future investigations must try to analyze the interaction paradigms emerging between the mobile phone and the health environment and assure that ethical and user-centered challenges are properly fulfilled.

References

Abran, A., Khelifik, A., Suryn, W., and Seffah, A. (2003). Consolidating the ISO usability models. In *Proceedings of the 11th International Software Quality Management Conference*, Glasgow, U.K., April 23–25, 2003.

Beaudin, J. S., Intille, S. S., and Morris, M. E. (2006). To track or not to track: User reactions to concepts in longitudinal health monitoring. *Journal of Medical Internet Research*, 8(4): e29, October–December 2006.

Bohn, J., Coroamă, V., Langheinrich, M., Mattern, F., and Rohs, M. (2005). Social, economic, and ethical implications of ambient intelligence and ubiquitous computing. In *Ambient Intelligence, Part I*, pp. 5–29. Springer, Berlin, Germany.

Broens, T. H. F., van Halteren, A. T., van Sinderen, M. J., and Wac, K. E. (2005). Towards an application framework for context-aware m-health applications. In *EUNICE 2005: Networked Applications 11th Open European Summer School*, pp. 1–7, Colmenarejo, Spain, July 6–8, 2005.

European Respiratory Journal (2005). Sleep apnoea more likely to kill younger sufferers. *European Respiratory Journal*, 25: 3.

Friedewald, M., Vildjiounaite, E., Punie, Y., and Wright, D. (2006). The brave new world of ambient intelligence: An analysis of scenarios regarding privacy, identity and security issues. In *Security in Pervasive Computing*, Lecture Notes in Computer Science, pp. 119–133, Springer, Berlin, Germany.

Honkanen, M., Lappeteläinen, A., and Kivekäs, K. (2004). Low end extension for bluetooth. In *IEEE Radio and Wireless Conference 2004*, pp. 199–202, Atlanta, GA, September 19–22, 2004.

IPTS/ESTO (2003). *Science & Technology Roadmapping: Ambient Intelligence in Everyday Life*. http://fiste.jrc.es/download/AmIReportFinal.pdf.

Ikonen, V., Kaasinen, E., Niemelä, M., and Leikas, J. (2008). Ethical guidelines for (mobile-centric) Ambient Intelligence. MINAmI project public deliverable D1.4. Version 1.2. http://www.fp6-minami.org/uploads/media/MINAmI_EthicalGuidelinesforAmI_v12.pdf

ISO 9241-11 (1998). Ergonomic requirements for office work with visual display terminals (VDTs)—Part 11: Guidance on usability.

Iso-Ketola, P., Karinsalo, T., Myry, M., Hahto, L., Karhu, H., Malmivaara, M., and Vanhala, J. (2005). A mobile device as user interface for wearable applications. In *Proceedings of PERMID 2005*, pp. 5–9, Munich, Germany.

Islam, R., Ahamed, S.I., Talukder, N., and Obermiller, I. (2007), Usability of mobile computing technologies to assist cancer patients. In Holzinger, A. (ed.), USAB 2007 *LNCS*, 4799, pp. 227–240, Springer, Heidelberg.

Jantunen, I., Laine, H., Huuskonen, P., Trossen, D., and Ermolov, V. (2008a). Smart sensor architecture for mobile-terminal-centric ambient intelligence. *Sensors and Actuators A: Physical*, 142: 352–360.

Jantunen, J., Lappeteläinen, A., Arponen, J., Pärssinen, A., Pelissier, M., Gomez, B., and Keignart, J. (2008b). A new symmetric transceiver architecture for pulsed short-range communication. In *Proceedings of the Global Telecommunications Conference 2008*, IEEE GlobeCom 2008, pp. 1–5, New Orleans, LO, November 30–December 4, 2008.

Jeong, H. J. and Arriaga, R. I. (2009). Using an ecological framework to design mobile technologies for pediatric asthma management. In *Proceedings of the Mobile HCI'09: The 11th International Conference on Human-Computer Interaction with Mobile Device and Services*, Bonn, Germany, September 15–18th.

Jung, J., Ha, J., Lee, J., Kim, Y., and Kim, D. (2008). Wireless body area network in a ubiquitous healthcare system for physiological signal monitoring and health consulting. *International Journal of Signal and Image Processing*, 1(1): 47–54. http://www.sersc.org/journals/IJSIP/vol1_no1/papers/06.pdf

Kaasinen, E., Ermolov, V., Niemelä, M., Tuomisto, T., and Välkkynen, P. (2006). Identifying user requirements for a mobile terminal centric ubiquitous computing architecture. In *Proceedings of the International Workshop on System Support for Future Mobile Computing Applications* (*FUMCA'06*), Irvine, CA, pp. 9–16, IEEE, Washington, DC.

Kaasinen, E., Tuomisto, T., Välkkynen, P., Jantunen, I., and Sierra, J. (2008). Ubimedia based on memory tags. In *Proceedings of the 12th International Conference on Entertainment and Media in the Ubiquitous Era*, MindTrek 2008, Tampere, Finland, pp. 85–89, ACM, New York.

Karat, J. and Karat, C. (2003). The evolution of user-centered focus in the human computer interaction field. *IBM Systems Journal*, 42: 532–542.

Korhonen, I., Parkka, J., and Van Gils, M. (2003). Health monitoring in the home of the future. *IEEE Engineering in Medicine and Biology Magazine*, 22(3): 66–73.

Kosta, E., Pitkanen, O., Niemela, M., and Kaasinen, E. (2008). Ethical-legal challenges in user-centric AmI services. In *Proceedings of Internet and Web Applications and Services* (*ICIW '08*), pp. 19–24, Athens, Greece, IEEE Computer Society, Washington, DC.

Kummer, T.-F., Bick, M., and Gururajan, R. (2009). Acceptance problems of ambient intelligence and mobile technologies in hospitals in India and in Germany. In *Proceedings of the 17th European Conference of Information Systems*, Verona, Italy.

Lopez, R., Chagpar, A., White, R., Hamill, M. H., Trudel, M., Cafazzo, J., and Logan, A. G. (2009). Usability of a diabetes telemanagement system. *Journal of Clinical Engineering*, 34(3): 147–151, July/September 2009.

Matthews, M., Doherty, G., Sharry, J., and Fitzpatrick, C. (2008). Mobile phone mood charting for adolescents. *British Journal of Guidance and Counselling*, 36(2): 113–129.

Morak, J., Schindler, K., Georzer, E., Kastner, P., Toplak, H., Ludvik, B., and Schreier, G. (2008). A pilot study of mobile phone-based therapy for obese patients. *Journal of Telemedicine and Telecare 2008*, 14: 147–149.

Nielsen, J. (1994b). Enhancing the explanatory power of usability heuristics. In *Proceedings of CHI 94*, pp. 152–158, Boston, MA, ACM, New York.

Nielsen, J. (2005). Medical usability: How to kill patients through design. *Jakob Nielsen's Alertbox*, April 11, 2005. http://www.useit.com/alertbox/20050411.html

Niemelä, M., Rafael Gonzalez Fuentetaja, R. G., Kaasinen, E., and Gallardo, J. L. (2007). Supporting independent living of the elderly with mobile-centric ambient intelligence: User evaluation of three scenarios. In *Proceedings of the 2007 European Conference on Ambient Intelligence*, Lecture Notes in Computer Science, pp. 91–107, Springer, Berlin, Germany.

Osterberg, L. and Blaschke, T. (2005). Adherence to medication. *New England Journal of Medicine*, 353: 487–497.

Quero, J., Tarrida, C., Santana, J., Ermolov, V., Jantunen, I., Laine, H., and Eichholz, J. (2007). Health care applications based on mobile phone centric smart sensor network. In *Proceedings of the 29th Annual International Conference of the IEEE Engineering in Medicine and Biology Society* (*EMBS 2007*), pp. 6298–6301, Lyon, France.

Robbins A. M., Renshaw J. J., and Berry S. W. (1991). Evaluating meaningful auditory integration in profoundly hearing-impaired children. *American Journal of Otology*, 12: 144–150.

Roussos, G., Marsh, A. J., and Maglavera, S. (2005). Enabling pervasive computing with smart phones. *IEEE Pervasive Computing*, 4(2): 20–27, January–March 2005.

Scandurra, I., Hägglund, M., Koch, S., and Lind, M. (2008). Usability laboratory test of a novel mobile homecare application with experienced home help service staff. *Open Medical Informatics Journal*, 2: 117–128. http://www.pubmedcentral.nih.gov/articlerender.fcgi?artid=2669647

Slovis, B., Andreoli T. E., and Brigham, K. (2001). Disordered breathing. *Cecil Essentials of Medicine*, pp. 210–211. Philadelphia, PA: W.B. Saunders.

Wu, J.-H., Wang, S.-C., and Lin, L.-M. (2005). What drives mobile health care? An empirical evaluation of technology acceptance. In: *Proceedings of the 38th Hawaii International Conference on System Sciences*, pp. 150–151, Big Island, HI, IEEE Computer Society, Washington, DC.

Chapter 19

Secured Transmission and Authentication

Fahim Sufi and Ibrahim Khalil

Contents

19.1 Introduction

After the introduction of Health Information Protection and Privacy Act (HIPPA) of 1996 in the United States [14], transmissions of physiological signals are envisaged to be secured. A telemonitoring platform that ignores protection of private health information is a threat to patients' privacy. Unfortunately, none of the existing telemonitoring platforms integrate any encryption, obfuscation, or anonymization techniques for the conformance of HIPPA regulations. However, few researchers [3] argue that if physiological signals (such as electrocardiography (ECG)) are sent without the name of the person (whose physiological signal is being sent), then there can be no way to determine (by a hacker) whose physiological signal is transmitted. Unfortunately, recent studies in ECG-based biometric [5] show that ECG can successfully be used to identify a person. Hence, even though the name of the patient is disassociated from the physiological signal, it is possible to identify that person and retrieve his health information. Therefore, research in encryption, obfuscation, and anonymization is deemed to be crucial for a health-monitoring platform seeking wide acceptance by society.

As we will learn from Chapter 20, adoption of compression technology is often required for wireless cardiovascular monitoring, due to the enormous size of ECG signal and limited bandwidth of Internet. After compressing the ECG signal, the compressed ECG packets must be secured before releasing them into the public Internet. To ensure secured transmission encryption algorithm with higher security strength should be applied on the compressed ECG. Secured ECG data transmission protects patients' private health information as well as biometric data (since ECG can be used to identify a person) [20,22,23,26–28,30]. The protection of private health information with encryption ensures that patients' cardiovascular details, stress levels, and respiratory details remain concealed. On the other hand, the protection of ECG as biometric data ensures protection against replay attacks (or spoof attack) (Figure 19.1).

Once the hospital (or monitoring service) receives the compressed ECG, the packets must be decompressed before performing human identification using present research on ECG-based biometric techniques. This additional step of decompression creates a significant processing delay for identification task. This becomes an obvious burden on a system, if this needs to be done for a trillion of compressed ECG per hour by the hospital. Therefore, in this chapter, we present the

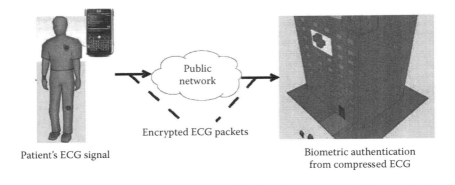

Patient's ECG signal

Public network

Encrypted ECG packets

Biometric authentication from compressed ECG

Figure 19.1 Patient-side ECG encryption on mobile phone and hospital-side ECG-based biometric authentication.

concept of person identification directly from their compressed ECG. This chapter is organized into two sections: transmission security and biometric authentication.

19.2 Transmission Security

The concealing ECG feature waves (P wave, QRS complex, and T wave) can be performed by the following three methodologies.

1. *ECG Encryption*: For ECG encryption, we have used permutation cipher during the encoding phase [20,22]. After this encoding, with character shuffling the ECG becomes secured and without the knowledge of the permutation key, original ECG cannot be retrieved. Moreover, using existing compression techniques (such as AES, DES, Rinjadel) with existing compression algorithms (such as WinZip, bzip, pkzip, etc.), security strength of the encoded ECG can be further raised. This method was also compared with conventional AES and DES encryption techniques. The proposed encryption method provides a substantially higher security strength (approximately $4 \times !255 \times !255 \times 18 \times 4$ times than that of existing encryption mechanisms), which ensures real-time performance on mobile platform (experimented on smart phones and PCs).

2. *ECG Obfuscation*: A new ECG obfuscation method was designed and implemented on different subjects using added noises corresponding to each of the ECG features [23,28]. The features are first detected, followed by the application of feature-specific noise on top of the detected ECG feature. Thus, all the features are smeared with specific noises that act as a key. Only authorized personnel can remove these specific noises from the corrupted (obfuscated) ECG to retrieve the original ECG. This obfuscated ECG can be freely distributed over the Internet without the necessity of encryption, since the original features needed to identify personal information of the patient remain concealed. Only authorized personnel possessing a secret key will be able to reconstruct the original ECG from the obfuscated ECG. Distribution of the key is extremely efficient and fast due to small size (only 0.04%–0.09% of the original ECG file). Moreover, if the obfuscated ECG reaches the wrong hand (hacker), it would appear as regular ECG without encryption. Therefore, traditional decryption techniques including powerful brute force attack are useless against this obfuscation. Figures 19.2 and 19.3 show the whole obfuscation process.

   Even though previous research implemented the ECG obfuscation process on desktop solutions [23,28], the noise-based obfuscation techniques can easily be deployed on the mobile solution. Noises (for obfuscation) can be kept and maintained within the mobile devices as vectors, or it can be generated from equations. One alternative solution for J2ME and .Net implementation on mobile devices is obtaining the random noise from a random noise generating server in a secured fashion. The authorized personnel from the

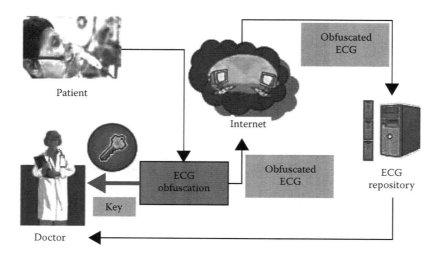

Figure 19.2   Basic concept of ECG obfuscation. (Adapted from Sufi, F. and Khalil, I., *J. Med. Syst.*, 23(2), 121, 2008.)

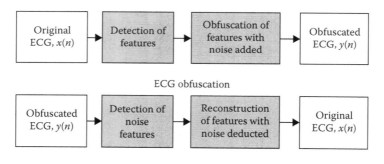

Figure 19.3   Obfuscation and de-obfuscation process. (Adapted from Sufi, F. and Khalil, I., *J. Med. Syst.*, 23(2), 121, 2008; Sufi, F. et al., A new ECG obfuscation method: A joint feature extraction and corruption approach, in *International Conference on Technology and Applications in Biomedicine, 2008 (ITAB 2008)*, Shenzhen, China, pp. 334–337, May 2008.)

hospital (or the cardiologist) can de-obfuscate the obfuscated ECG by downloading the same noise key from that server.

3. *ECG Anonymization*: The anonymization technique, which was designed and developed, segregates the low frequency from the ECG signal and performs partial encryption with DES. The wavelet coefficient from representing the lowest ECG frequency is distorted and the ECG is regenerated using the distorted coefficient (Figure 19.4). We used two novel ECG anonymization techniques based on discrete wavelet transform [30] and wavelet packets [29].

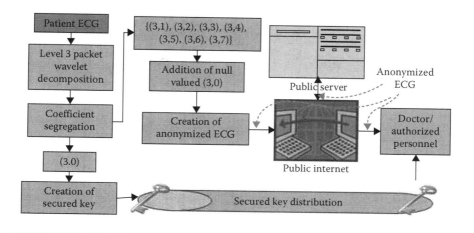

Figure 19.4 Anonymizing ECG with wavelet packet [29] (performed in patient side). (Adapted from Sufi, F. et al., *Int. J. Biomet.*, 1(2), 191, 2008.)

The wavelet packets were proven to provide 100% anonymization, showing robustness against replay attack by the spoofer [29]. Even with the most recent available technology, the anonymized ECG remained totally unidentified. A key, which is only 5.8% of the original ECG, is securely distributed to the authorized personnel for reconstruction of the original ECG. During the packet wavelet implementation, the secret key size was nearly halved, compared to that of DWT. Figures 19.4 and 19.5 show the anonymization and de-anonymization of the ECG with wavelet packet.

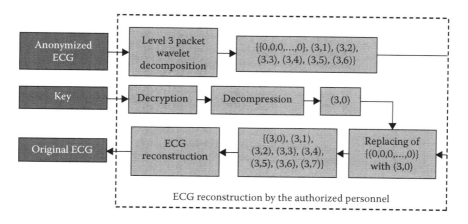

Figure 19.5 De-anonymizing the anonymized ECG with wavelet packet [29] (performed in hospital/cardiologist side). (Adapted from Sufi, F. et al., *Int. J. Biomet.*, 1(2), 191, 2008.)

Even though earlier research demonstrated the wavelet-based obfuscation on desktop-based solutions (with MATLAB®), the same implementation can be deployed on the mobile .Net framework with minor modifications. The wavelet functionalities implemented with MATLAB can be translated using MATLAB Builder NE (http://www.mathworks.com) to .NET and .COM components, which can then be easily integrated into mobile solutions.

ECG obfuscation and anonymization techniques work on plain ECG. To utilize ECG obfuscation and anonymization [23,28–30] the compressed ECG must be decompressed first. As we are using encoded or compressed ECG for our mobile phone–based telecardiology system, application of permutation cipher (ECG encryption) is the most efficient one, as previously shown by [20,22]. This permutation-based ECG encryption scheme can work while encoding (or compressing) the ECG segments. In the next section, we explore ECG encryption in detail.

19.2.1 Overview of ECG Encryption

As depicted in Figure 19.6, the system architecture contains three major building blocks: secured acquisition device, secured mobile phone, and central server (CS) [22]. There are two major communication links to be secured: the link from the acquisition device to the mobile phone and the link from the mobile to the VPN mesh of CSs.

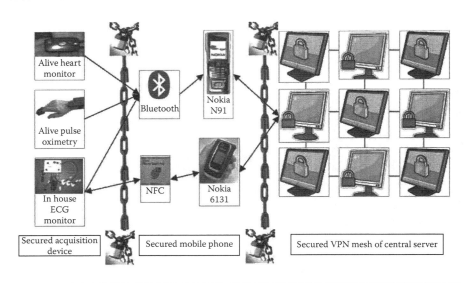

Figure 19.6 System architecture of three-phase encoding–compression–encryption mechanism. (Adapted from Sufi, F. and Khalil, I., *Security Commun. Netw.*, 1(5), 389, 2008.)

19.2.1.1 Secured Acquisition Device

We used Alive Heart Monitor (www.alivetek.com) to collect ECG signal and transmit to the mobile device using well-known Bluetooth protocol. Securing transmission of ECG signal from acquisition device to patients' mobile communication was done with 6 byte long globally unique Bluetooth device address (BDA), authentication, authorization, encryption, and PIN exchange. However, this Bluetooth-based authentication consumes nearly 10 s of handshaking time and has security flaws, which may compromise integrity of the ECG data [22]. To minimize the authentication time while using in-house ECG acquisition device, we proposed a prototype RFID-based touch scheme using near field communication (NFC) (www.nfc-forum.org) technology. The touch scheme allows us to initiate and establish Bluetooth link between the acquisition device and the mobile phone within seconds (nearly 2 s). Nokia 6131 phone was utilized along with NFC starter kit for deployment of NFC-based authentication. We used JSR-257 Contactless Communication API and JSR-82 Bluetooth API within the mobile phone permitting NFC and Bluetooth-based authentication communication with the acquisition device. Almost all of the existing devices (including Alive) transmit the ECG signal without encryption, ignoring privacy. However, our low complexity encoding method can be implemented directly on a microprocessor-based acquisition device to secure transmission of ECG signals. This proposed in-house ECG acquisition device can support compact flash-based memory and has enough internal memory (buffer) to support both real-time and store-now-forward later operations. In the latter mode, large amount of ECG data (several hours) can be stored depending on the size of the compact flash. Two major purposes of NFC module deployed in in-house acquisition device are fast and authorized establishment of Bluetooth link, and receiving of paired permutation key from the mobile phone to be used in our proposed encoding method (Figure 19.7). Since NFC is restricted by transmission range of approximately 10 cm for passive RFID devices, it is considered to be relatively secured than other protocols like Bluetooth.

19.2.1.2 Secured Mobile Phone

Once the mobile phone receives the ECG data, it is expected to be already encoded with the proposed encoding method (e.g., when ECG is acquired from the in-house acquisition device). However, that may not always be the case when the data is coming from commercially available monitoring devices like Alive Heart Monitor. In such a case, data must be encoded on the mobile phone with the proposed method, before performing compression and encryption. As our three-phase ECE mechanism requires existing compression and encryption on the encoded text, we obtained algorithms of ICSharpCode.SharpZipLib (http://www.icsharpcode.net) library to compress the data. This library supports four

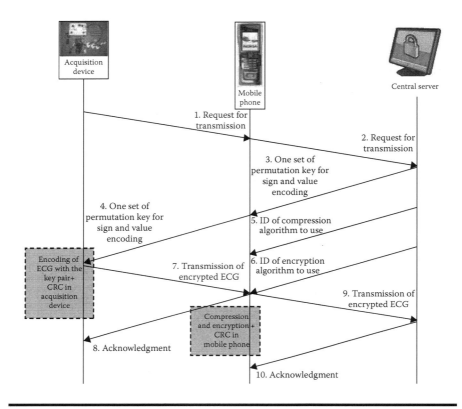

Figure 19.7 **Communication protocol among acquisition device, mobile phone, and CS. (Adapted from Sufi, F. and Khalil, I.,** *Security Commun. Netw.,* **1(5), 389, 2008.)**

types of compression: Zip, GZip, Tar, BZip2. Since this compression library was available in open-source format, we could easily implement the algorithms on J2ME platform. For encryption, System.Security. Cryptography library was used in.Net platform. However on J2ME platform, our implementation of encryption mechanism on regular mobile phone was through the usage of cryptographic algorithm API provided by Legion of the Bouncy Castle, which supported AES, AES Fast, AES Light, Blowfish, CAST5, CAST6, 3DES, DES, IDEA, RC2, RC5 32 bits, RC5 64 bits, RC6, Rijndael, Serpent, Skipjack, and Towfish block cipher as well as RC4 stream cipher [33]. The usage of this API ensures data transfer security from the mobile phone to the medical servers, as shown in Figure 19.8. Although several alternative algorithms are available for both compression and encryption on mobile platform, the choice can be made at the CS, which then notifies the mobile platform about the specific algorithms to be used (Figure 19.7).

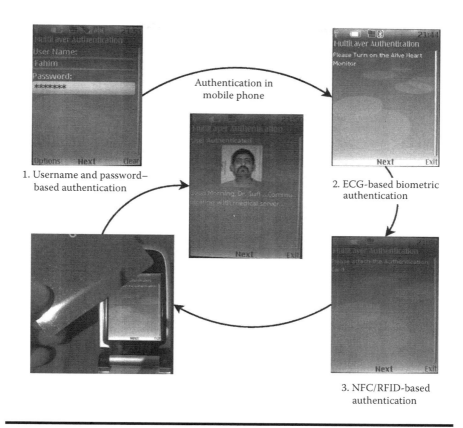

1. Username and password-based authentication

Authentication in mobile phone

2. ECG-based biometric authentication

3. NFC/RFID-based authentication

Figure 19.8   The proposed cardiac authentication system.

19.2.1.3 Central Server

This is often referred as medical server. The predominant purpose of the CS is to act as a patient data repository. It collects ECG and other physiological data from IP-enabled mobile phones carried by a large number of patients that are remotely monitored. Generally, medical servers are distributed in nature. Hence, the doctors and medical practitioners can issue queries to many of these servers. Security among these medical servers can be achieved with modern virtual private networking built using IPsec protocol. Since both acquisition devices and mobile phones (or smart phones) have lower processing power, few of the computationally intensive tasks can be delegated to the CSs. For example, number of permutations for 256 ASCII characters is $256P_{256}$ = 256! For this tremendous value, processing and selection of a particular permutated set of ASCII value is not feasible on a mobile phone, because of immense computational requirements. Therefore, before transmission of the bio-signal, the encoding mechanism may request a pair of permutation keys from the CS for a typical scenario presented in Figure 19.7. Following the request, the CS sends a pair of permutation keys (for sign and value encoding) (refer to [20,22]) in a secured manner. These permutation keys are then used to perform the encoding with p(·) primitive. To overcome the delay in obtaining the permutation keys, the mobile phone should receive them from the server (as encrypted message) on regular intervals before the expiry of the existing keys. Therefore, before transmission of the actual ECG signal begins, the acquisition device can quickly receive the key pair from the mobile phone without invoking the CS and incurring additional delay.

While sending the permutation keys, the CS can also notify about the specific compression and encryption methods to be used, since there are many possible schemes. Therefore, during our experimentation phase, the CS packed permutation keys, compression and encryption identifiers, and encrypted this piece of information with DES symmetric algorithm before sending to the mobile phone. As explained in [22], this encrypted information is only needed to be transmitted before the commencement of actual ECG transmission, which may continue for few hours or even days. The CS randomly selects the existing compression and encryption schemes for the three-phase ECE mechanism. This not only enhances the security of ECG data transfer, but also significantly improves transmission time, as shown in [22].

19.2.2 Discussion: Security Strength of the ECG Encryption Scheme

A brute force attack is an exhaustive procedure that tries all possibilities until the right combination is determined. Therefore, the time required to complete the brute force attack primarily depends on the size of the search space, Δ, which can be defined as

$$\Delta = \prod_{f=1}^{F} \Delta_f \qquad (19.1)$$

where
  Δ_f is the factor search space
  $F = 3$ for the encoding mechanism
  $F = 5$ for three-phase ECG mechanism

For encoding mechanism, the value of Δ_1, Δ_2, and Δ_3 are Δ_4, 256!, and 256! respectively, which are defined in Table 19.1. Apart from the values of Δ_1, Δ_2, and Δ_3, the three-phase ECE mechanism requires $\Delta_4 = 4$ and $\Delta_5 = 18$ to expand its search space. Therefore, it is evident that for the encoding method, the search space for brute force attack to obtain the source of ECG (for determining C) and right permutation keys (value and sign) is an enormous number $4 \times 265! \times 256!$ This enormous number increases further when the number of possible supported devices (Δ_1)

Table 19.1 Notations for Security Strength of the Scheme

Notation	Definition
Δ_1	For ECG acquisition alone, we used three different acquisition devices (GE MAC 5500 [23], Alive Heart Monitor [24], In-house developed ECG monitor) during experimentation. Apart from these, MIT BIH Arrhythmia database entries were also being evaluated. Therefore, the total number of ECG sources was 4.
Δ_2	This is the number of all ASCII permutations for sign encoding, which is 256!
Δ_3	This is the number of all ASCII permutations for value encoding, which is also 256!
Δ_4	This is the number of supported compression algorithm. Eventually, only one compression algorithm is selected from 4.
Δ_5	This is the number of supported encryption algorithms. During our experimentation one encryption is selected from 18 encryption mechanisms.

Source: Sufi, F. and Khalil, I., *Security Commun. Netw.*, 1(5), 389, 2008.

is raised. As it tends to infinity, the probability, P $(1/\Delta)$ of the retrieving the right combination for deciphering tends to zero.

All the combinations of Δ result in possible ECG samples (numerical floating points), only one of which constructs the right ECG segment. The only way to discern the right floating point from the wrong ones is to plot at least one segment of ECG comprising all ECG features (P wave, T wave, and QRS complex). Unlike dictionary-based brute force attack, there is no automated solution to match ECG morphology in order to ascertain right combination of ECG samples from enormous search space. If in near future, a grid of supercomputers can compare a trillion trillion trillion (10^{36}) combinations of one ECG segment (comprising 500 ECG samples) per second for ECG morphology matching, it will take approxi-

mately 9.333×10^{970} years $\left(\dfrac{(4 \times 256! \times 256!)}{(3600 \times 24 \times 365 \times 1036)} \approx 9.333 \times 10^{970} \right)$ to enumer-

ate all the combinations. On average, the correct combination would be found in half of that time. In addition, the three-phase ECE mechanism conceals the statistical model of the encryption by allowing multiple compressions and encryption algorithms giving $(4 \times 18) \times 9.333 \times 10^{970}$ years, even without considering the time required to decipher keys for existing encryption mechanisms. In fact, a device that could check a billion billion (10^{18}) AES keys per second would require about 3×10^{51} years to exhaust the 256 bit key space. Eventually, at this point, one might question about the necessity of this ridiculous strength of security for ECG data transmission.

19.3 Authentication

According to the researchers, a user can be authenticated primarily by three modes: knowledge-based authentication, possession-based authentication, and biometric-based authentication. Knowledge-based authentication that generally uses a user name and password (or PIN) combination is the most common form of authentication. In mobile phone–based telecardiology, it has been used by few researchers [3]. Possession-based authentication can be RFID-based authentication, which is recently being adopted in the form of NFC for mobile phones. There are few mobile phones and solutions available that support NFC-based authentication model (http://www.nexperts.com/). Lastly, ECG-based biometric authentication can also be used to identify a patient subscribed to wireless monitoring facility [12,21,26,27]. Combining all these different authentication mechanisms can provide a substantially higher level of security in patient authentication, as shown in Figure 19.8.

19.3.1 Why Authentication from Compressed ECG?

Person identification from their ECG signal was made possible just 9 years ago [2]. Till then, there have been a surge of research on ECG-based biometric [2,4,5,9,10,12,16,21,26,27,33]. Basically, ECG-based biometric can be achieved by comparing the enrolment ECG template and recognition ECG template. These ECG biometric templates are generated utilizing direct time domain methods (e.g., duration and amplitude of P wave, QRS complex, T wave, etc.), signal processing methods (e.g., wavelet distance measurement [WDM], percentage root-mean-square distance [PRD], etc.) or some other methods (e.g., polynomial distance measurement (PDM), etc.) [26]. According to the literature, techniques depicted in [21,26] were the first usage of ECG biometric on mobile platform. Irrespective of underlying methods used for the creation of the template, all the existing ECG biometric works on plain ECG signal (i.e., uncompressed and not encoded).

However, during a remote telecardiology scenario, ECG packets are often kept dealt in compressed format, where the patient is attached to miniature ECG acquisition devices that transmit the patients ECG signals to remote locations using mobile communication devices [3,8,11,13,17–20,22,25,32]. In such a scenario, ECG signal acquired by the ECG acquisition device is directed to their own mobile phones or personal computers as ECG packets (via Bluetooth, Wi-Fi, Zigbee, or NFC protocol), and from there on it is redirected to the patients' Internet service providers. Through the public Internet infrastructure, the compressed ECGs reach the hospital that is remotely monitoring the cardiac patients. If required, the hospital may use public Internet to send the compressed ECG to an outside cardiologist for expert opinion. Therefore, it is clearly seen from these telecardiology scenarios [20,22,25,32], compressed ECG packets are often preferred for efficient transmission and storage purposes.

Now for these scenarios (remote telecardiology applications), if a patient is to be authenticated either by the hospital or the remote cardiologist, then the compressed ECG packets are required to be decompressed first before applying existing biometric techniques [2,4,5,9,10,12,16,21,26,27,33]. This added step of decompression before biometric authentication generates slight delay in cardiovascular patient authentication. These scenarios basically enable remote monitoring facility for the monitored patients, when they feel uncomfortable (i.e., a possible cardiac anomaly event). According to literature, any delay in treatment and diagnosis can be fatal, after a cardiac abnormality symptom [6]. Apart from posing threat to the patient's health, delay in authentication can be an unnecessary computational burden on the hospital system, as the hospital may have thousands of subscribed patients and decompression of all their ECG segments at a given point of time is an enormous task.

Therefore, like diagnosis directly from compressed ECG [1,20,24], authentication directly from compressed ECG will be faster, since the compressed ECG packets do not need to be decompressed.

19.3.2 Authentication from Compressed ECG

Application of ECG compression is crucial for mobile phone–based cardiac patient monitoring, since enormous amount of ECG data is required to be transmitted over limited bandwidth of mobile networks [20,22,25]. The compression algorithm reported in our earlier work [20,22] performs lossless ECG compression. Therefore, the encoding function $\varepsilon(.)$ transforms the ECG signal, X_n to a compressed ECG, C_r (Equation 19.2). As the ECG features set, F is a subset of ECG signal X_n (Equation 19.3); therefore, feature waves are subset of encoded ECG C_r (Equation 19.4). Innovative algorithm can be designed to reveal these encoded ECG feature set (that represents original ECG feature set) for authentication.

An example will clarify the objectives of the research outlined in this chapter. A CVD patient can be monitored in real time by a hospital (i.e., the patient's mobile phone transmits ECG packets one after another to the hospital) or a patient can subscribe to event-based monitoring, where the patient's mobile phone transmits ECG packets only if an abnormality is detected. Figure 19.9 shows the normal ECG signal for an event-based monitored patient (Entry ID CU1 of CU Ventricular Tachyarrhythmia Database [15]) at instance A. At the next instance, all on a sudden, that patient encounters abnormal heartbeats, as seen in Figure 19.10. The patient's mobile phone detects the abnormality using abnormality detection algorithms [7]. Then the patient's mobile sends both normal and abnormal ECG packets in compressed format using [20,22]. The normal compressed ECG (Figure 19.11) is used in identifying the patient, and the abnormal ECG is used for diagnosis of the cardiac abnormality. After identifying the patient directly from the compressed ECG (Figure 19.11), the patient can then be provided for the services he is entitled to (e.g., ambulance facility, onsite specialist, etc.).

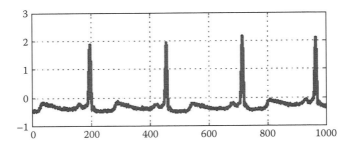

Figure 19.9 A normal ECG segment of a monitored patient at instance A (ECG obtained from CU1 entry MIT BiH CU Ventricular Tachyarrythmia Database). (Adapted from Online, Physiobank: Physiologic signal archives for biomedical research, available at http://www.physionet.org/physiobank/(accessed 2009).)

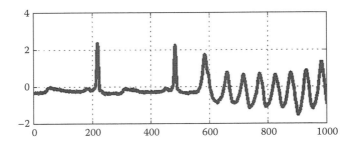

Figure 19.10 An abnormal ECG segment of the monitored patient at instance A + 1 (ECG obtained from CU1 entry MIT BiH CU Ventricular Tachyarrythmia Database). (Adapted from Online, Physiobank: Physiologic signal archives for biomedical research, available at http://www.physionet.org/physiobank/ (accessed 2009).)

Figure 19.11 Compressed ECG for Figure 19.9 (using algorithms presented earlier in [20,22]).

Equations 19.2 and 19.3 basically demonstrates the fact that lossless compression algorithm (presented in our earlier work in [20,22]) preserves the subtle ECG feature waves. Therefore, these features can be directly obtained from the compressed ECG (Equation 19.4). Based on this principle, we have successfully demonstrated the fact that authentication and diagnosis can be performed directly from compressed ECG [1,20,24].

$$\varepsilon(X_n) = C_r \tag{19.2}$$

$$F \subset X_n \tag{19.3}$$

$$F \subset C_r \tag{19.4}$$

Our experiments suggest that this biometric authentication directly from the compressed ECG can be performed in two different approaches:

- *Rule-Based Methods*: Within this approach, the different conditions are established by extensive experimentation, and those conditions are programmed in a rule-based system. This approach can be compared to the methods described in detection of heart rate (HR) from compressed ECG in [20]. For detecting the HR in mobile, frequency count of high-pitched characters were performed within a set window. If that frequency count reaches above a set threshold, then a heartbeat is confirmed.
- Similar mechanisms may be adopted, where the frequency counts for all the encoded characters (Figure 19.12) are performed. Our preliminary experimentation suggests that for a particular person, his or her frequency counts for some selected compressed characters (Figure 19.12) remain within a boundary.
- These observations can be coded within the mobile devices to perform biometric authentications. As seen from Figure 19.13, patient authentication can be performed on the patient side (in the patient's mobile phone) for the authentication between the patient's mobile phone and the patient's ECG acquisition device. Otherwise, the authentication can be performed on a remote cardiologist's mobile phone (hospital service). Moreover, these lightweight authentication mechanisms can be implemented within the routing nodes (as seen in Figure 19.14) to determine the routing information for transmitting right ECG segment to the right hospital (or cardiologist).
- *Intelligent Methods*: The subtle relationship between the ECG feature waves and some encoded characters (compressed characters) can be hard

Characters

@£$¥èéùìòçøøÅå_^{}[~]|€ÆæßÉ
!#¤%&()*+,-./:;<?¡§¿äöñüàÀÁÂ
ÄÅÈÉÊÌÍÎÏÐÑÒÓÔÕÖÙÚÛÜÝPþáãâãçê
ëíîïðóôõúûÿabcdefghijklmnopq
rstuvwxyzABCDEFGHIJKLMNOPQRS
TUVWXYZ\

Numeric sub groups

0–50
50–100
100–150
150–200
200–250
250–300
300–350
350–400
400–500

Figure 19.12   157 Character and numeric subgroups (attributes) used for generating compressed ECG (from plain ECG signal). Details of this character substitution-based compression techniques have been described in [20,22].

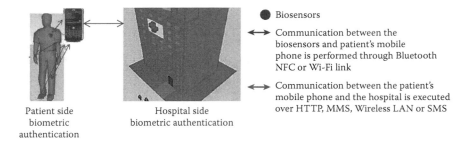

● Biosensors

⟷ Communication between the biosensors and patient's mobile phone is performed through Bluetooth NFC or Wi-Fi link

⟷ Communication between the patient's mobile phone and the hospital is executed over HTTP, MMS, Wireless LAN or SMS

Patient side
biometric
authentication

Hospital side
biometric authentication

Figure 19.13   Patients authentication with compressed ECG biometric can be performed on patient side or hospital side.

to determine by manual inspection (as suggested in our earlier approach). Intelligent mechanisms like subset selection, clustering, principal component analysis, and neural network may play a very important role in compressed ECG biometric. Using these intelligent techniques, we have successfully identified ECG feature waves signaling specific cardiovascular diseases ([1,24]).

■ Our experimentation with different entries of normal sinus rhythms database [15] shows that using attribute selection techniques we can

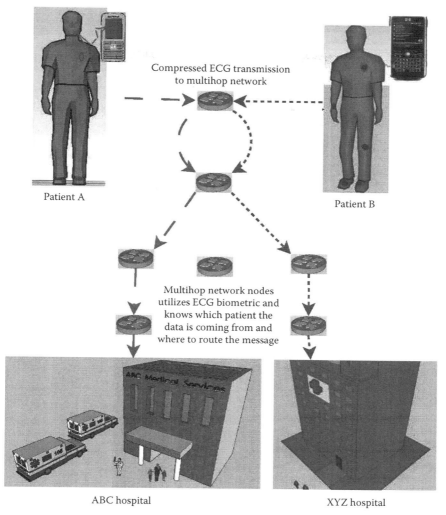

Compressed ECG transmission to multihop network

Patient A

Patient B

Multihop network nodes utilizes ECG biometric and knows which patient the data is coming from and where to route the message

ABC hospital

XYZ hospital

Hospital uses ECG-based biometric from compressed and knows what type of facilities the patient should receive from the hospital

Figure 19.14 ECG-based biometric authentication can be useful for the interme-diate routing nodes to ascertain the routing information.

identify 48 attributes from the list of 157 attributes shown in Figure 19.12. By applying clustering (e.g., EM and K-Means [1,24]), person identifi-cation is possible. However, due to algorithmic complexity, data min-ing or intelligent approaches are only limited to high-end smart phones or mobile phones.

19.4 Conclusion

Within this chapter, techniques that can be used to ensure transmission security while transmitting ECG packets in mobile telecardiology were explored. As we have seen, we can perform this with permutation cipher on mobile phone, noise-based obfuscation, or even wavelet-based anonymization techniques [22,23,29]. On the other hand, to perform patient authentication in mobile telecardiology, we can utilize polynomial-based approach [25,26,31] or some other simpler ECG-based biometric techniques [27]. ECG biometric on mobile phones (to be used by patient's mobile phone Figure 19.13, cardiologist's mobile phone Figure 19.13, or intermediate routing nodes Figure 19.14) is faster when it works directly on compressed ECG. Therefore, both rule-based methods and intelligent methods that perform compressed ECG biometric were also discussed. In a nutshell, this chapter provides overview of encryption and biometric authentication for mobile telecardiology [3,8,11,13,17–20,22,25,32], which can protect patients' privacy (e.g., sensitive health information) and uphold HIPAA ([14]).

References

1. Ayman, A., I. Khalil, and F. Sufi. Cardiac abnormalities detection from compressed ECG in wireless telemonitoring using principal components analysis (pca). *International Conference on Intelligent Sensors, Sensor Networks and Information Processing*, 2009, *ISSNIP*, pp. 207–212, Melbourne, Australia, December 2009.
2. Biel, L., O. Petersson, L. Philipson, and P. Wide. ECG analysis: A new approach in human identification. *IEEE Transaction on Instrumentation and Measurement*, 50(3):808–812, June 2001.
3. Blount, M. et al. Remote health-care monitoring using personal care connect. *IBM Systems Journal*, 46(1):95–113, March 2007.
4. Bui, F. M. and D. Hatzinakos. Biometric methods for secure communications in body sensor networks: Resource-efficient key management and signal-level data scrambling. *EURASIP Journal on Advances in Signal Processing* 2008, Article ID 529879, p. 16, 2008.
5. Chan, A. D. C., M. M. Hamdy, A. Badre, and V. Badee. Wavelet distance measure for person identification using electrocardiograms. *IEEE Transaction on Instrumentation and Measurement*, 57(2):248–253, February 2008.
6. Bradley, E. H. et al. Achieving rapid door-to-balloon times: How top hospitals improve complex clinical systems. *Circulation*, 113(8):1079–1085, March 2006.
7. Friesen, G. M., T. C. Jannett, M. A. Jadallah, S. L. Yates, S. R. Quint, and H. T. Nagle. A comparison of the noise sensitivity of nine QRS detection algorithms. *IEEE Transactions on Biomedical Engineering*, 37(1):85–98, January 1990.
8. Hung, K. and Y. T. Zhang. Implementation of a wap-based telemedicine system for patient monitoring. *IEEE Transactions on Information Technology in Biomedicine*, 7(2):101–107, June 2003.
9. Irvine, J. M. et al. Heart rate variability: A new biometric for human identification. *International Conference on Artificial Intelligence*, pp. 1106–1111, Las Vegas, NV, 2001.

10. Israel, S. A., J. A. Irvine, A. Cheng, and B. K. Wiederhold. ECG to identify individuals. *Pattern Recognition*, 38(1):133–142, 2005.

11. Jasemian, Y. and L. Arendt-Nielsen. Evaluation of a realtime, remote monitoring telemedicine system using the bluetooth protocol and a mobile phone network. *Journal of Telemedicine and Telecare*, 11(5):256–260, 2005.

12. Khalil, I. and F. Sufi. Legendre polynomials based biometric authentication using QRS complex of ECG. *International Conference on Intelligent Sensors, Sensor Networks and Information Processing*, 2008, *ISSNIP 2008*, pp. 297–302, Sydney, Australia, December 2008.

13. Lee, R.-G., K.-C. Chen, C.-C. Hsiao, and C.-L. Tseng. A mobile care system with alert mechanism. *IEEE Transaction on Information Technology in Biomedicine*, 11(5):507–517, September 2007.

14. Online. Health insurance portability accountability act of 1996 (hipaa), centers for medicare and medicaid services (1996). Available at: http://www.cms.hhs.gov/hipaageninfo (accessed 2008).

15. Online. Physiobank: Physiologic signal archives for biomedical research. Available at: http://www.physionet.org/physiobank/ (accessed 2009).

16. Poon, C. C. Y., Y. T. Zhang, and S. D. Bao. A novel biometric method to secure wireless body area sensor networks for telemedicine and m-health. *IEEE Communication Magazine*, pp. 73–81, April 2006.

17. Sufi, F. Mobile phone programming java 2 micro edition. *Proceedings of the 2007 International Workshop on Mobile Computing Technologies for Pervasive Healthcare*, pp. 64–80, Phillip Island, Melbourne, Australia, ISBN: 978-0-646-48230-9, Dec. 2007.

18. Sufi, F., I. Cosic, and Q. Fang. A mobile phone based remote patient assessment system. *International Journal of Cardiovascular Medicine*, 9(Supplement):8–9, 2008.

19. Sufi, F., Q. Fang, and I. Cosic. ECG r-r peak detection on mobile phones. *Proceedings of the 29th Annual International Conference of the IEEE Engineering in Medicine and Biology Society (EMBS 2007)*, pp. 3697–3700, Lyon, France, August 2007.

20. Sufi, F., Q. Fang, I. Khalil, and S. S. Mahmoud. Novel methods of faster cardiovascular diagnosis in wireless telecardiology. *IEEE Journal on Selected Areas in Communications*, 27(4):537–552, May 2009.

21. Sufi, F. and I. Khalil. An automated patient authentication system for remote telecardiology. *International Conference on Intelligent Sensors, Sensor Networks and Information Processing*, 2008, *ISSNIP 2008*, pp. 279–284, Sydney, Australia, December 2008.

22. Sufi, F. and I. Khalil. Enforcing secured ECG transmission for realtime telemonitoring: A joint encoding, compression, encryption mechanism. *Security and Communication Networks*, 1(5):389–405, 2008.

23. Sufi, F. and I. Khalil. A new feature detection mechanism and its application in secured ECG transmission with noise masking. *Journal of Medical Systems*, 23(2):121–132, DOI: 10.1007/s10916-008-9172-6, 2008.

24. Sufi, F. and I. Khalil. Diagnosis of cardiovascular abnormalities from compressed ECG: A data mining based approach. *Proceedings of the 9th International Conference on Information Technology and Applications in Biomedicine*, Larnaca, Cyprus, November 2009.

25. Sufi, F., I. Khalil, Q. Fang, and I. Cosic. A mobile web grid based physiological signal monitoring system. *International Conference on Technology and Applications in Biomedicine*, 2008 *(ITAB 2008)*, pp. 252–255, Shenzhen, China, May 2008.

26. Sufi, F., I. Khalil, and I. Habib. Polynomial distance measurement for ECG based biometric authentication. *Security and Communication Network*, Published Online on December 3, 2008, DOI: 10.1002/sec.76, 2008.

27. Sufi, F., I. Khalil, and J. Hu. *ECG Based Biometric: The Next Generation in Human Identification*, P. Stavroulakis (ed.), Handbook of Communication and Information Security (Springer), February 2010, pp. 309–331.

28. Sufi, F., S. Mahmoud, and I. Khalil. A new ECG obfuscation method: A joint feature extraction and corruption approach. *International Conference on Technology and Applications in Biomedicine, 2008 (ITAB 2008)*, pp. 334–337, Shenzhen, China, May 2008.

29. Sufi, F., S. Mahmoud, and I. Khalil. A novel wavelet packet based anti spoofing technique to secure ECG data. *International Journal of Biometrics*, 1(2):191–208, 2008.

30. Sufi, F., S. Mahmoud, and I. Khalil. A wavelet based secured ECG distribution technique for patient centric approach. *5th International Summer School and Symposium on Medical Devices and Biosensors, 2008, ISSS-MDBS 2008*, pp. 301–304, Hong Kong, China, June 2008.

31. Sufi, F., Q. Fang, and I. Cosic. A mobile phone based intelligent scoring approach for assessment of critical illness. *International Conference on Technology and Applications in Biomedicine, 2008 (ITAB 2008)*, pp. 290–293, Shenzhen, China, May 2008.

32. Sufi, F., Q. Fang, S. S. Mahmoud, and I. Cosic. A mobile phone based intelligent telemonitoring platform. *3rd IEEE/EMBS International Summer School on Medical Devices and Biosensors, 2006, ISSMDBS*, pp. 101–104, Cambridge, MA, September 2006.

33. Wubbeler, G., M. Stavridis, D. Kreiseler, R. D. Bousseljot, and C. Elster. Verification of humans using the electrocardiogram. *Pattern Recognition Letters*, 28:1172–1175, February 2007.

Chapter 20

Efficient Transmission in Telecardiology

Fahim Sufi and Ibrahim Khalil

Contents

20.1 Introduction

Life without mobile phones is becoming increasingly impossible. In our day-to-day life, mobile phones are no longer devices of communication only. Entertainment, personal management, social activities, work, and healthcare are some of the broader categories that the latest mobile phones cater to. In the entertainment sector, mobile phones are used for gaming, music, video, and other interactive (e.g., match calculator)/Web applications for amusement. Personal management services offered by mobile phones varies from organizers, schedulers, task list, notes, reminders, and even pocket finance management softwares. Social activities services may include Facebook, Twitter, and other chat room applications. Many of these social activities can be performed over Wi-Fi/Bluetooth protocols rather that Hyper Text Transfer Protocol (HTTP), offering a cheaper option. Pocket Word, Pocket Excel, Pocket PowerPoint, finance and budgeting software, e-mailing suits, etc., provide greater support for work/business activities. Lastly, healthcare is an area that is currently in focus and where extensive research is being conducted [13,15,17,25–29]. Calorie and activity monitoring softwares running on mobile phones are already available. In this chapter, we focus on cardiovascular monitoring services over mobile phone network.

Electrocardiogram (ECG) signal is primarily used by the cardiologists to diagnose cardiovascular diseases (CVDs), which is the number one killer in modern times. In remote telemonitoring faster treatment mandates faster authentication, communication, and diagnosis between the patient and the hospital. Since the latest mobile phones are capable of performing computations as well as data transmission, we utilize them to efficiently detect a cardiac anomaly first. After detecting a CVD symptom, the mobile phone transmits the ECG signal containing the CVD signature to the hospital or cardiologist in a secured manner. Security is crucial, since ECG contains private health information of the patient. Moreover, recent research shows that ECG can be used as a biometric entity. Therefore, encryption must be applied to protect patient's private health information and identity. If the hospital identifies a life-threatening event, ambulance will be sent directly to that patient. GPS locations (also mobile cell location information) of the CVD-affected patient are used to successfully locate the patient. On the patient's way to hospital, the patient can transmit continuous ECG, which assists the cardiologists within a hospital to prepare treatment facilities for the patient. Therefore, door-to-ballooning time, the time between the hospital being aware of patient's condition and the time to insert mesh inside catheterized theater [27], is minimized.

Mobile phone involvement in the monitoring of CVD patients reduces unnecessary delays in some of the key areas (as seen by the triangle—Figure 20.1)

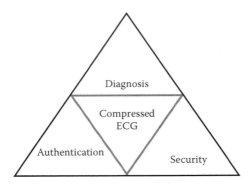

Figure 20.1   Issues in mobile phone–based telecardiology.

- Faster transmission of enormous ECG signals.
- Diagnosis of CVD running efficient algorithm within the mobile phone.
- Authentication mechanism between the patient and hospital.
- Secured transmission of ECG signals from the patient to the hospital.

Finally, the door-to-ballooning time and (symptom-to-ballooning time) time is drastically reduced by advance information sent in advance from patient before the patient is admitted to the hospital [4].

20.2 System, Concept, and Architecture

The first objective of the mobile phone is to ensure efficient transmission of ECG signal. This can be achieved by the usage of compression technologies. Unlike using the off-the-shelf generic compression algorithms (e.g., WinZip, PKZip, Gzip, etc.), we propose the use of a specialized compression algorithm for ECG. The benefits of using this specialized algorithm to compress ECG are:

1. This compression mechanism provides substantially higher compression ratio (up to 95%).
2. This compression algorithm executes faster in mobile phones, compared to other LZW or wavelet-based algorithms [27,29].
3. ECG data can be made secured with a substantially higher security strength by the scrambling of encoded (with this compression algorithm) characters [27,29].
4. Authentication between the hospital and the patient can be performed directly from the compressed ECG.
5. Diagnosis can be performed directly from the compressed ECG [30].

To reduce the delay in treating the CVD-affected patients and save their lives, authentication, diagnosis, and encryption algorithm can be made faster by implementing our compression, encryption, authentication, and diagnosis algorithms (Figure 20.2).

The patient is attached with an ECG acquisition device that collects the ECG signal from the patient's body. The collected ECG is continuously transmitted to the patient's mobile phone through bluetooth protocol. The patient's mobile phone then compresses the ECG signal and generates compressed ECG packets. The ECG packets are then routed through different Internet nodes and finally reach the hospital medical servers [27].

Once the compressed ECG packets are received by the hospital, the hospital may use the compressed ECG packets to authenticate the patient using compressed ECG-based biometric techniques. Once the hospital recognizes the patient as its

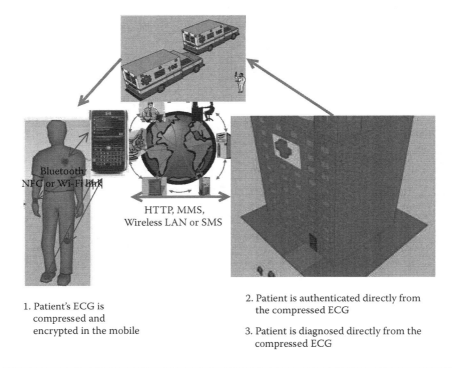

1. Patient's ECG is compressed and encrypted in the mobile

2. Patient is authenticated directly from the compressed ECG

3. Patient is diagnosed directly from the compressed ECG

Figure 20.2 Mobile phone–based ECG compression, encryption, authentication, and diagnosis.

subscribed monitoring customer, then the hospital can perform a preliminary detection of CVD using compressed ECG. This faster preliminary detection of CVD can determine the serious patients requiring cardiologist's attention.

20.3 Background Compression

The existing compression algorithms can be divided into two main categories: lossless [18–35] and lossy [14–33]. Lossless compression algorithms are based on techniques such as null suppression, run-length encoding, diatomic encoding, pattern substitution, inter-beat differencing, intra-beat differencing techniques, and statistical encoding [7]. In null suppression method, repeated null values or zero values are replaced with shorter code word followed by the length of the null occurrence. Run-length encoding is similar to null suppression, except for its application for all the repeated symbols. In diatomic encoding method, occurrence of two consecutive symbols is coded with a single code symbol. Pattern substitution method looks for specific patterns and encodes them with short code symbols. All the above-mentioned lossless compression methods can be implemented in single pass. But in two-pass statistical encoding methods, the complexity level is higher, where symbol probability or frequency is calculated before encoding. The knowledge of symbol probability or frequency is used to encode the information optimally, where more frequent symbols are encoded with the shortest code word and less frequent symbols are encoded with the larger code word. Many forms of entropy coding are used in the literature [6] for ECG compression. Huffman encoding [12] is also used frequently as the final step for many ECG compression algorithms [1,9,32], since it compresses the information further without any loss. In Huffman encoding, most frequently occurring values are represented by simple and shortened binary code. LZW [37]-based algorithms pioneers Huffman and arithmetic encoders by searching for the repeated patterns in the input stream and encoding them with shorter codes. Since all the two-pass encoding schemes scan the whole document in the first pass, these methods cannot be utilized for the purpose of real-time compression as the whole ECG session is required to be completed before applying compression. Transformational techniques like wavelet transform also have been used for ECG compression [6,8,18,19]. After wavelet decomposition, set partitioning in hierarchical trees (SPIHT) encoding is often adopted to exploit the inherent similarities across sub-bands in a wavelet decomposition and perform uniform quantization and bit allocation. SPIHT codes the most important wavelet transform coefficients in priority, and transmits them according to that order. Even though SPIHT encoding is generally used for lossy compressions [8,19], recent research demonstrates its applicability in lossless ECG compressions [18]. Research on lossy ECG compression has outnumbered the research in lossless ECG compression. There are mainly three types of lossy ECG compressions: direct time domain methods [9,14,22], feature extraction methods [1,8–10,16,32,34,36], and transformational methods, which generally exploit wavelet transformation, discrete cosine transformation, KLT, and Fourier transformation [1,8,10,16,19,33]. Direct time domain methods are of lower complexity than the feature extraction methods or transformational methods. Therefore, techniques presented in direct time domain methods can be easily implemented in small devices [22]. Generally, these methods employ the knowledge of previous samples, often referred to as the prediction algorithm or utilize the knowledge of both previous and future samples, referred to as interpolation algorithm [14]. Differential pulse code modulation (DPCM) has been one of the most popular direct transformational ECG compression methods. The simplest form of DPCM, applies zero order prediction (ZOP), where the previous sample is thought to be the predicted sample, and only the difference between the current sample and the previous sample is transmitted [14]. This technique is commonly termed as first difference or intra-beat difference. In feature extraction methods of lossy ECG compression, different sections of ECG curve is recognized by the algorithm. Many of the ECG compression techniques that start with QRS detection, before performing the actual compression, fall into this category [1,8–10,16,32,34,36]. QRS detection is needed for detecting each of the beats. Many of the methods perform inter-beat difference where current beat is subtracted from the previous beat or an average beat [1,9,35].

The residue from inter-beat subtraction is generally less than intra-beat difference. This is mainly because in inter-beat difference, complex areas like QRS and T are also normalized, leaving minimal residue. But, for patients, having irregular beats, inter-beat difference doesn't offer much reduction of information. To deal with this varying period (TT interval) issue, beats are averaged [8,9,16,32,34] before any other operations. Some techniques [1] engage beat template databases, which are modified and updated with every ECG sample. Beat templates from the database are used for inter-beat difference calculations resulting in complex PC-based telecardiology applications [1]. Moreover, since the baseline is not always straight or even, some methods adjust or remove the baseline [1,36]. Beat alignment/period alignment, amplitude normalization [8], or baseline adjustment [1,36] require execution of further operations. All these additional operations make the compression procedure more complex. Recently, transformational methods, especially wavelet transform–based compression algorithms are becoming popular. Almost all of the transformational methods are targeted for PC-based solutions. Unlike the PC-based programming environment, the small device environment, e.g., CLDC of J2ME, has hardly any supporting libraries which make the development extremely difficult. Moreover, the limitations of the hardware such as the lower memory size and the number of I/O ports challenge the implementation of transformational methods on mobile phones.

20.4 Why Compression Is Necessary

First of all, in a wireless telecardiology application, huge ECG data is transmitted from the patient to the medical server or to the doctors. Transmission and storage of enormous amount of physiological signal imposes high level of challenge for the research community. To help us understand precisely about how large data we are dealing with, examples from previous research [5] is necessary. Data volume from a single patient with only few biosignal acquisition devices (ECG, EEG, EMG, SpO2, Accelerometer) can easily reach 13 GB in 24 h [5], in case of continuous monitoring. If a patient with heart abnormality is remotely monitored, then with 12 lead ECG, 10 bit resolution, and 360 Hz sampling frequency, the data can easily reach up to 2.77 GB in 1 day. And, if we intend to transmit this enormous amount of data using available telecommunication technologies (like, PSTN, GSM, GPRS, 3G, etc.), then a dedicated minimum speed of 269 kbps is required (Required transmission speed = Size of data/transmission time = 2.77 GB/24 h = 269 kbps) Unfortunately, this type of transmission speed (upload) is guaranteed by a very limited number of telecommunication service providers. During this transmission of enormous ECG data, compression technology can be applied for faster transmission on limited bandwidth wireless link. At the end, faster transmission means faster treatment for the patient.

Secondly, some of the messaging protocols like SMS can only allow a limited set of characters or message size. For example, SMS can only accommodate a message size of 140 bytes or 160(140 × (8/7)) characters. Since each SMS involves cost for the patient, it is imperative to transmit compressed data for economic reasons.

Lastly, compression algorithm also adds value to a real-time telemonitoring scenario by allowing more storage of physiological signals.

Compression algorithms for ECG can be broadly classified into three major groups, namely, direct domain method [14,22], feature extraction method [16,36] and transformational method [19, [33]. The existing ECG compression techniques are somewhat unsuitable for mobile phone–based wireless telecardiology applications for the following reasons:

- *Computationally expensive*: Most of the existing ECG compression algorithms were designed and tested on PC [14,16,19,22,33,36]. However, a regular mobile phone (not high end) is capable of running only 10,000 operations per second, while executing Java MIDlets on Java kilobyte virtual machine (KVM) [23]. CLDC and MIDP restrict the usage of floating-point operations, which means all the floating points must be removed before performing any operations on the mobile devices. Multidimensional arrays are not supported as well, hence, any algorithm performing matrix-based calculation cannot be implemented on JavaTM-based mobile devices directly. Complex functions comprising a large number of basic operations such as QRS detection, beat/period alignment, and transformations will provide unexpected delay or even deadlock in small devices. Furthermore, due to the memory restriction, algorithms requiring large memory for maintaining lookup table [6], codebook [36], and frequency table [1,9,11,16,32] is not feasible for mobile phone platforms (Table 20.1). Frequency tables were widely used by some two-pass encoding techniques such as Huffman [1,9,32,33], LZW [11], significant bit encoding [16], and SPIHT [8,18,19]. Therefore, algorithms for biosignal compression, analysis, display, secured transmission, etc. must be shared by limited computational power offered by the mobile devices. Under these limitations, existing PC-based biosignal compression algorithms are ineffective for mobile phone–based wireless telecardiology applications.
- *Unsuitable for MMS/SMS transmission*: Mobile phone–based wireless telecardiology application often requires transmission of ECG signals over existing MMS and SMS protocols [17,31]. These MMS and SMS protocols only support limited character sets (e.g., GSM 03.38) during transmission. However, most of the existing ECG compression algorithms use Huffman encoded optimally arranged binary representation of the compressed ECG as the final output. Therefore, if existing ECG compression algorithms are adopted for transmission of compressed ECG over existing text messaging (SMS/MMS) communication, many of the characters will be lost, since they are not supported by the underlying transmission infrastructure. This will result in lossy and mutilated transmission of ECG, which is not at all suitable for diagnostic purposes.

Table 20.1   Comparison of RAM and CPU Speed among Mobile Phone, Implantable Devices, and PC

Device Type	RAM (kb)	CPU Speed (kHz)
Pacemaker	About 10	30–100
Implantable loop recorder	About 100	30–100
Implantable cardioverter defibrillator	About 250	30–100
Mobile phone	About 250	50–100
Personal computer	128,000	1,200,000

Encoded ECG (with algorithms of [27,29]) can be transmitted via MMS on a Java-enabled mobile phone (such as Nokia 6280) with the following code:

```
class mms_Send implements Runnable{
    public void run(){
        try{
            mc = (MessageConnection) Connector.open(connURL);
            mmsMsg = (MultipartMessage)
    mc.newMessage(MessageConnection.MULTIPART_MESSAGE);
            mmsMsg.setSubject("Compressed ECG Data 102 of
                MIT_BIH");
            mmsMsg.addAddress("to", connURL);
            mmsMsg.setHeader("X-Mms-Priority", "normal");
            byte[] bytECG = encSB.toString().getBytes("UTF-8");
            ecgMessagePart = new MessagePart(bytECG, "text/
                plain", "id1", "message text", "UTF-8");
            mmsMsg.addMessagePart(ecgMessagePart);
            mc.send(mmsMsg);
            frm.deleteAll();
            frm.append("Messase Sent");
            mc.close();
        }
        catch(Exception e){
            e.printStackTrace();
        }
        // Sending via MMS Ends
    }
}
```

■ *Need for Real-time Performance*: Real-time processing in telemonitoring simply means the time required to process (compute) the physiological signal must be less than the time required to receive that signal, during a continuous acceptance of physiological signal from the acquisition. If the mobile phone, which receives physiological signal, consumes less than 1 s to process 1 s worth of ECG data, then real-time operation is executed [26]. This real-time factor [26], often determines the efficiency and effectiveness of a remote telemonitoring platform. Therefore, any algorithm pertaining to real-time telemonitoring should be evaluated for real-time performance.

■ *Requirement of decompression for further analysis on compressed EGG:* Finally, the main objective of existing ECG compression techniques is to achieve high compressibility by having redundancy free output, which do not preserve any ECG features within the encoded text. This means that to perform analysis from the compressed ECG, one must decompress the compressed ECG signals, which results in unwanted delays. This is especially true in case of resource-constraint mobile devices, where decompression time could be long.

Moreover, the vast majority of the literature related to existing ECG compression algorithms are lossy, which make them unsuitable for medical diagnosis, as mandated by the law requirements in many countries including the United States. Therefore, lossless compression is mandatory and the main focus of this section. The unsuitability of existing compression algorithms for mobile phone–based telemonitoring is briefed in Table 20.2.

Table 20.2   Unsuitable Criteria for Mobile Phone–Based Real-Time ECG Compression

Unsuitable Criteria for Mobile Phone–Based Real-Time ECG Compression	Listed References (Reviewed Literatures)
Complex calculation involving floating points	[2,3,7,8,10,14,18,20,21,32]
Matrix-based calculation (involves usage of multidimensional array)	[2,7,32]
Reconstruction error	[2,7,8,10,14,18,20,21,32,33]
QRS detection	[2,3,8,10,20,21,32,33]
Beat alignment (period normalization/ base alignment/amplitude normalization)	[2,3,8,10,20,32,33]
Two-pass encoding	[3,7,8,14,20,32,33]
Lookup table–based encoding	[10,18]
Application of transformational methods	[7,8,14,18,21,32,33]

20.5 ECG Compression Method

Our ECG compression algorithm is based on symbol substitution technique [5,24,27–29,31], which has been successfully applied in our earlier research. Previous symbol substitution algorithms [5,24,28,31] achieved the same level of compression by symbol substitution of four consecutive characters from genome sequence with one ASCII character. According to [5,24,28], the basic compression achieved with symbol substitution is faster and computationally inexpensive compared to other types of compression algorithms. Therefore, it is reasonable to adopt a symbol substitution–based ECG compression technique for inexpensive processing on mobile phones. By efficiently substituting ECG samples with a character, which is supported by MMS or SMS transmission, we will be able to transmit compressed ECG without any loss or distortion. However, there are few challenges in performing symbol substitution for ECG:

1. Unlike genome sequences, where there is a single character (any character from A, T, G, C group) representing a single nucleotide, each of the ECG sample has multiple digits (1–6 digits for MIT BIH Database ECG entries [?]). The number of digits depends on the acquisition device resolutions.
2. Unlike genome sequences, where the four characters can be represented with four digits (1, 2, 3, and 4), a single ECG sample is a noninteger floating point value. Therefore, before the start of symbol substitution process, ECG samples should be translated into integer values.
3. Unlike genome sequences, where the sequences do not have any negative values, a single ECG sample can be either positive or negative. Therefore, the signs and the values require separate encoding processes.

Figure 20.3 shows the character sets used for performing the encoding compression. These characters can be transmitted through MMS/SMS as per our experiments.

The proposed symbol substitution–based ECG compression deals with the above challenges while performing compression. Before compression, original ECG is normalized first.

20.6 Implementation in Mobile Platform with J2ME

Figure 20.4 shows the implementation of specialized compression algorithm on mobile platform including the MMS/SMS/HTTM transmission functionalities. This is how it looks like on a patient's mobile phone. Following J2ME code performs the ECG compression on a J2ME platform (runs on Java KVM-supported mobile phones).

Characters

@£$¥èéùìòçøøÅå_^{}[~]|€ÆæßÉ
!#¤%&()*+,-./:;<?¡§¿äöñüàÁÅ
ÄÄÉÉËÌÍÏÐÑÒÓÖÕÖÙÚÛÜÝÞþáâãçê
ëíîïðóôõúûýabcdefghijklmnopq
rstuvwxyzABCDEFGHIJKLMNOPQRS
TUVWXYZ\

Numeric sub groups

0–50

50–100

100–150

150–200

200–250

250–300

300–350

350–400

400–500

Figure 20.3   157 character and numeric subgroups (attributes) used for generating compressed ECG (from plain ECG signal). Details of this character substitution–based compression techniques have been described.

```
public void encode(){
        dte1 = new Date();
        dte2 = new Date();
        tme1 = dte1.getTime();
        noOfSmpl = 1;
        intDiff = new int[8];
        endOfEcg = false;
        int rrComp1, rrComp2, rrCount;
        rrSB = new StringBuffer();
        boolean rrOn;
        rrOn = true;
        rrCount = 0;
        ecgSmpl1 = 200*getEcgSample();
        rrComp1 = 0;
        while (endOfEcg! = true){
            for(int i=0; i<=7; i++){
                ecgSmpl2 = 200*getEcgSample();
                noOfSmpl+ = 1;
                diff = (ecgSmpl2-ecgSmpl1);
                if((Math.ceil(diff)-diff)>0.5){
                    intDiff[i] = (int) Math.floor(diff);
                }
```

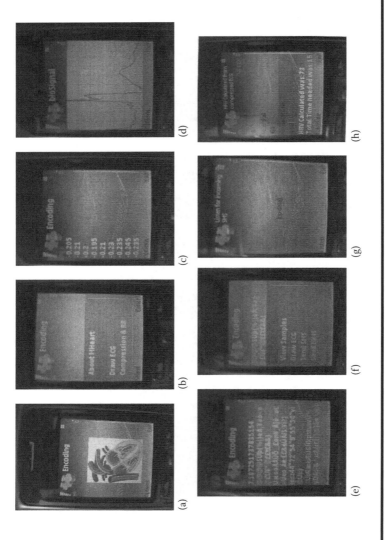

Figure 20.4 Implementation of compression in Java-supported mobile phone. (a) Main screen on patient's mobile. (b) Different menus in patient's mobile. (c) Viewing ECG samples in patient's mobile. (d) Drawing ECG in patient's mobile. (e) Compressed ECG on patient's mobile before transmission. (f) Patient's mobile transmit compressed ECG via SMS/MMS. (g) Doctor's mobile phone listens for messages. (h) Proposed HR detection on doctor's mobile. (From Sufi, F. et al., *IEEE J. Select. Area. Commun.*, 27(4), 537, May 2009. With permission.)

```
            else{
                intDiff[i] = (int) Math.ceil(diff);
            }
            ecgSmpl1 = ecgSmpl2;
    }
    signVal = 0;
    for (int i = 0; i< = 7; i++){
        if(intDiff[i]< = -1){
            switch (i){
                case 0:
                    signVal = signVal+1;
                    break;
                case 1:
                    signVal = signVal+2;
                    break;
                case 2:
                    signVal = signVal+4;
                    break;
                case 3:
                    signVal = signVal+8;
                    break;
                case 4:
                    signVal = signVal+16;
                    break;
                case 5:
                    signVal = signVal+32;
                    break;
                case 6:
                    signVal = signVal+64;
                    break;
                case 7:
                    encSB.append('');
                    break;
            }
            intDiff[i] = Math.abs(intDiff[i]);
        }
    }
}
encSB.append(gsm[signVal]);
for(i=0; i<=6; i+=2){
    if((intDiff[i]<10) && (intDiff[i+1]<10)){
        join = intDiff[i]*10+intDiff[i+1];
        encSB.append(gsm[join]);
    }

    else if ((intDiff[i]< 47) && (intDiff[i+1]< 47)){
        encSB.append(gsm[intDiff[i]+100]);
        encSB.append(gsm[intDiff[i+1]+100]);
    }
```

```
            else {
                encSB.append(String.valueOf((int)intDiff[i]));
                encSB.append('"');
                encSB.append(String.valueOf((int)intDiff[i+1]));
                encSB.append('"');
                // rr detection proceeding
                if (rrOn==true){
                  if((rrComp1<intDiff[i])&&(intDiff[i]<intDiff
                    [i+1])){
                        rrComp1 = intDiff[i+1];
                  }
                  else{
                      if (rrComp1<intDiff[i]){
                          rrCount-;
                      }
                      rrVal = (float) rrCount;
                      rrVal = rrVal/360;
                      rrSB.append(rrVal);
                      rrCount = 0;
                      rrOn = false;
                       intHrv++;
                  }
                }
                // rr detection ending
            }
        }
```

After receiving the compressed ECG, the doctor's mobile phone retrieves the ECG. The doctor's mobile phone has a listener enabled that listens for arrival of compressed ECG (e.g., via MMS, as seen in the following code). After the listener catches a compressed message, it instantly decompresses with the following code:

```
/*
 * MMS_Receive.Java
 *
 * Created on 11 May 2007, 12:23
 */
import java.io.InputStream;
import javax.microedition.midlet.*;
import javax.microedition.lcdui.*;
import java.io.IOException;
import javax.microedition.io.*;
import javax.wireless.messaging.Message;
import javax.wireless.messaging.MessageConnection;
import javax.wireless.messaging.MessageListener;
import javax.wireless.messaging.MessagePart;
import javax.wireless.messaging.MultipartMessage;
import javax.wireless.messaging.TextMessage;
```

```
/**
 *
 * @author Researcher
 * @version
 */
public class mReceive extends MIDlet implements MessageListener,
  CommandListener{
    MessageConnection mc;
    Command Exit, clrScrn, decECG, test;
    Form frmECG;
    Message msg;
    char[] chrCompressed;
    Display disp;
    int msgLen, position, intDiff[], pos, next;
    boolean done, eof, _7thSign;
    short diffNum;
    double prevECG, dblDiff;
    char gsm[];
    StringBuffer dgt, sbECG;
public mReceive(){
        frmECG = new Form("ECG Display");
        Exit = new Command("exit", Command.EXIT, 1);
        clrScrn = new Command("Clear Screen", Command.OK, 2);
        decECG = new Command ("Decompress ECG", Command.OK, 2);
        test = new Command("Debug", Command.OK, 2);
        frmECG.addCommand(decECG);
        frmECG.addCommand(Exit);
        frmECG.addCommand(clrScrn);
        frmECG.addCommand(test);
        frmECG.setCommandListener(this);
        disp = Display.getDisplay(this);
        pos = 0;
        intDiff = new int[8];
        _7thSign = false;
        eof = false;
        msgLen = 0;
        position = 0;
        diffNum = 0;
        sbECG = new StringBuffer();
        String strGSM = "@$_^{}[~]| !#%&()*+,-./:;
        <?ABCDEFGHIJKLMNOPQRSTUVWXYZ
abcdefghijklmnopqrstuvwxyz\\";
        gsm = new char[148];
        strGSM.getChars(0,148, gsm,0);
        dgt = new StringBuffer(3);
        prevECG = -0.2OO;
        disp.setCurrent(frmECG);
    }
```

```
public void startApp() {
    done = false;
    msg = null;
      try{
          mc = (MessageConnection) Connector.open
            ("mms://:mmsReceive");
          mc.setMessageListener(this);
      }catch(IOException ioe){
          ioe.printStackTrace();
      }
    disp.setCurrent(frmECG);
}
public void pauseApp() {
}
public void destroyApp(boolean unconditional) {
}
public void notifyIncomingMessage(MessageConnection mc){
    readMessageThreaded(mc);
}

public void readMessageThreaded(final MessageConnection mc){
    Thread th = new Thread(){
          public void run(){
                int intPos;
                intPos = 0;
                try{
                    Message message = null;
                    Message = mc.receive();
                    if (message! = null){
                        if(message instanceof
                          MultipartMessage){
                        MultipartMessage mpmsg =
                          (MultipartMessage) message;
                        MessagePart[] mps = mpmsg.
                          getMessageParts();
                        for(int i=0; i<mps.length; i++){
                        MessagePart mp = mps[i];
                        byte[] content = mp.getContent();
                        String ecgMsg = new String
                          (content, "UTF-8");
                        done = true;
                        msgLen = ecgMsg.length();
                        intPos = ecgMsg.indexOf("</smil>");
                        chrCompressed = new char[msgLen-
                          intPos-7];
                        ecgMsg.getChars((intPos+9),
                          msgLen, chrCompressed, 0);
```

```
                        String strECG = ecgMsg.
                          substring(intPos+9);
                        msgLen = strECG.length();
                        frmECG.append(strECG);
                      }
                    }
                  }
                }catch(Exception ex) {
                }
        }
    };
    th.start();
}
public void commandAction(Command c, Displayable d){
    if (c==Exit){
        mc = null;
        notifyDestroyed();
    }
    if (c==clrScrn){
        if (done = true){
            frmECG.deleteAll();
            disp.setCurrent(frmECG);
        }
    }
    if (c==decECG){
    frmECG.deleteAll();

    while(!eof){
        diffNum = 0;
        decSign(readChar());
        if (_7thSign){
            decSign(readChar());
        }
        while(diffNum<8){
            decode(readChar());
        }
        if (pos<200){
            for (short n = 0;n<8;n++){
                sbECG.append(precision(prevECG));
                sbECG.append('\n');
                dblDiff = intDiff[n];
                dblDiff = dblDiff/200;
                prevECG = prevECG+(dblDiff);
                pos++;
            }
        }
    }
}
```

```
        frmECG.append(sbECG.toString());
        }
}
char readChar(){
    char c;
    c = '\uffff';
        if(position<msgLen){
            c = chrCompressed[position];
            position++;
    }
    else{
        eof = true;
        //Forcing the algorithm to exit from the main while loop
//(where !eof will terminate the program)
        diffNum = 8;
    }
    return (c);
}
    short findChar1(char c){
    short id;
    id = -1;
    for (short i=0;i<100; i++){
        if(c==gsm[i]){
            id = i;
            break;
        }
    }
    return id;
}
short findChar2(char c){
    short id;
    id = -1;
    for (short i=100;i<148; i++){
        if(c==gsm[i]){
            id = i;
            break;
        }
    }
    return id;
}

void decode(char c){
    short join;
    if (!Character.isDigit(c)){'
        if (c=='"'){
            intDiff[diffNum] = intDiff[diffNum]*Integer.
              parseInt(dgt.toString());
            diffNum++;
            dgt.setLength(0);
```

```
            }
            else {
                join = findChar1(c);
                if (join == -1){
                    //When difference values were encoded
                        separately
                    join = findChar2(c);
                    intDiff[diffNum] = intDiff[diffNum]*
                        (join-100);
                    diffNum++;
                } else
                {
                    // When 2 consecutive (single digit)
                        differences
            //were encoded
                        intDiff[diffNum] = intDiff[diffNum]*
                            ((int) (Math.floor ((float)
                            (join/10)))));
                        diffNum++;
                        intDiff[diffNum] = intDiff[diffNum]*
                            ((int) (join%10));
                        diffNum++;
                }
            }
    }
    else{
        //When encoded characters are digits
        dgt.append(c);
    }
}
//Following function decodes the sign
void decSign(char c){
    if (c==''){
        _7thSign = true;
    }
    else{
        short signNum;
        signNum = -1;
        for (short i=0; i<128; i++){
            if (c==gsm[i]){
                signNum = i;
                break;
            }
        }
        for (short i=0; i<7; i++){
            if ((signNum%2)==1){
                intDiff[i] = -1;
            }
            else{
```

```
                      intDiff[i] = 1;
                }
                signNum = Short.parseShort(String.valueOf
                  (signNum/2));
        }
        if (_7thSign){
            intDiff[7] = -1;
            _7thSign = false;
        }
        else{
            intDiff[7] = 1;
        }
    }
}
public String precision(double d){
    int len, i;
    String strFix = String.valueOf(d);
    StringBuffer sb = new StringBuffer(6);
    len = strFix.length();
    if (len>6){
            sb.append(strFix.substring(0,6));
    }
    else{
        if ((len==1)||(len==6)){
                sb.append(strFix);
        }
        if (len<6){
                sb.append(strFix);
                for (i=len-1; i<5;i++){
                      sb.append('0');
                }
        }
    }
    return sb.toString();
    }
}
```

After decompressing the compressed ECG, the doctor can view the plain ECG either graphically or numerically by using the J2ME canvas class. Following is the code snippet for viewing the ECG signal on mobile phone.

```
/*
 * MIDPCanvas.java
 *
 * Created on 20 December 2006, 16:51
 */

import javax.microedition.lcdui.*;
import javax.microedition.io.*;
```

```java
import java.io.*;

/**
 *
 * @author Researcher
 * @version
 */

public class MIDPCanvas extends Canvas implements
  CommandListener {

    /**
     * constructor
     */

    float[] fltEcgSamples;
    int x1,y1, x2,y2, w, h, mag;
    int middle;
    InputStream dis;
    Form frm;
    int cnt;
    StringBuffer sb = new StringBuffer(6);

    public MIDPCanvas(float[] fltEcg) {
        cnt = 0;
        x1 = 0;
        y1 = getHeight()/2;
        mag = 150;
        fltEcgSamples = fltEcg;
    }

    /**
     * paint
     */

    public void paint(Graphics g) {
        w = g.getClipWidth();
        h = g.getClipHeight();
        g.setColor(220,255,200);
        g.fillRect(0,0,w,h);
        g.setColor(0,100,0);
        g.setStrokeStyle(g.DOTTED);
        for(int i=1; i<=8; i++){
            g.drawLine(0,i*w/8,h,i*w/8);
        }
        for(int i=1; i<=8; i++){
            g.drawLine(i*h/8,0,i*h/8,w);
        }
```

```
            g.setStrokeStyle(g.SOLID);
            middle = h/2;
            float ecg;
            ecg = 0;
            while((cnt<21600) && x1<(w-1)){
                        ecg = fltEcgSamples[cnt++];
                        x2 = x1+1;
                        y2 = (int) (ecg*mag);
                        y2 = middle - y2+ 20;
                        g.drawLine(x1,y1,x2,y2);
                        cnt++;
                        x1 = x2;
                        y1 = y2;
            }
}
/**
 * Called when a key is pressed.
 */

protected void keyPressed(int keyCode) {
    switch (getGameAction(keyCode)){
        case RIGHT:
            x1 = 0;
            repaint();
            break;
        case UP:
            mag = mag+50;
            x1 = 0;
            repaint();
            break;
        case DOWN:
            mag = mag-50;
            x1 = 0;
            repaint();
            break;
    }
}

/**
 * Called when a key is released.
 */

Protected void keyReleased(int keyCode) {
}

/**
 * Called when a key is repeated (held down).
 */
```

```
Protected void keyRepeated(int keyCode) {
}

/**
 * Called when the pointer is dragged.
 */
protected void pointerDragged(int x, int y) {
}

/**
 * Called when the pointer is pressed.
 */

protected void pointerPressed(int x, int y) {
}

/**
 * Called when the pointer is released.
 */

protected void pointerReleased(int x, int y) {
}

/**
 * Called when action should be handled
 */

public void commandAction(Command cmd, Displayable d) {
}
}
```

20.7 Conclusion

Within this chapter, the focus was on how to implement compression technology on mobile platform for wireless telecardiology. Usage of secured compression technology, such as [27,29] benefits faster transmission. As discussed earlier, faster transmission (along with faster authentication and diagnosis) results in faster treatment for the patient.

Once the compressed ECG is received by the doctors/cardiologists, patient authentication is performed directly from the compressed ECG (i.e., decompression is not required). After the patient is identified as a valid subscriber of the cardiovascular monitoring facility, preliminary CVD diagnosis is performed from the compressed ECG. Authentication (and security) and diagnosis from the compressed ECG have been detailed in Chapters 19 and 21.

References

1. A. Alesanco, S. Olmos, R. S. H. Istepanian, and J. Garcia. Enhanced real-time ecg coder for packetized telecardiology applications. *IEEE Transactions on Information Technology in Biomedicine*, 10(2), 229–236, 2006.

2. G. D. Barlas, G. P. Frangakis, and E. S. Skordalakis. Dictionary based coding for ecg data compression. *Proceedings of Computers in Cardiology*, London, U.K., 1993.

3. T. Blanchett and G. C. Kember. Klt-based quality controlled compression of single lead ecg. *IEEE Transactions on Biomedical Engineering*, 45(7), 942–945, 1998.

4. E. H. Bradley et al. Achieving rapid door-to-balloon times: How top hospitals improve complex clinical systems. *Circulation*, 113(8), 1079–1085, March 2006.

5. L. Chen, S. Lu, and J. Ram. Compressed pattern matching in dna sequences. *Proceedings of the 2004 IEEE Computational Systems Bioinformatics Conference, 2004. (CSB 2004)*, Stanford, CA, pp. 62–68, August 2004.

6. K. Duda, P. Turcza, and T. P. Zielinski. Lossless ecg compression with lifting wavelet transform. *Instrumentation and Measurement Technology Conference, Budapest, Hungary*, 2001.

7. H. Gilbert. *Data Compression: Techniques and Applications Hardware and Software Considerations*. John Wiley & Sons Ltd., New York, 1987.

8. M. M. Goudarzi, M. H. Moradi, and A. Taheri. Efficient method for ecg compression using two dimensional multiwavelet transform. *Transactions on Engineering, Computing and Technology*, 2, 2004.

9. P. S. Hamilton and W. J. Tompkins. Compression of the ambulatory ecg by average beat subtraction and residual differencing. *IEEE Transactions on Biomedical Engineering*, 38(3), 253–259, 1991.

10. Y. Hao, P. Marziliano, M. Vetterli, and T. Blu. Compression of ecg as a signal with finite rate of innovation. *Proceedings of the 2005 IEEE Engineering in Medicine and Biology 27th annual conference*, Shanghai, China, 2005.

11. R. N. Horspool and W. J. Windels. An lz approach to ecg compression. *Proceedings of the 1994 IEEE Seventh Symposium on Computer-Based Medical Systems (CBMS'94)*, Winston-Salem, NC, pp. 71–76, 06 1994.

12. D. A. Huffman. A method for the construction of minimum-redundancy codes. *Proceedings of the IRE*, 40, 1098–1101, 1952.

13. K. Hung and Y. T. Zhang. Implementation of a wap-based telemedicine system for patient monitoring. *IEEE Transactions on Information Technology in Biomedicine*, 7(2), 101–107, 2003.

14. S. M. S. Jalaleddine, C. G. Hutchens, R. D. Strattan, and W. A. Coberly. Ecg data compression techniques-a unified approach. *IEEE Transactions on Biomedical Engineering*, 37(4), 329–343, April 1990.

15. Y. Jasemian and L. Arendt-Nielsen. Evaluation of a realtime, remote monitoring telemedicine system using the bluetooth protocol and a mobile phone network. *Journal of Telemedicine and Telecare*, 11(5), 256–260, 2005.

16. B. S. Kim, S. K. Yoo, and M. H. Lee. Wavelet-based low-delay ecg compression algorithm for continuous ecg transmission. *IEEE Transactions on Information Technology in Biomedicine*, 10(1), 77–83, January 2006.

17. R.-G Lee, K.-C. Chen, C.-C. Hsiao, and C.-L. Tseng. A mobile care system with alert mechanism. *IEEE Transaction on Information Technology in Biomedicine*, 11(5), 507–517, September 2007.

18. S. G. Miaou and S. N. Chao. Wavelet-based lossy-to-lossless ecg compression in a unified vector quantization framework. *IEEE Transactions on Biomedical Engineering*, 52(3), 539–543, 2005.

19. S. G. Miaou and C. L. Lin. A quality-on-demand algorithm for wavelet-based compression of electrocardiogram signals. *IEEE Transactions on Biomedical Engineering*, 49(3), 233–239, 2002.

20. G. B. Moody, K. Soroushian, and R. G. Mark. Ecg data compression for tapeless ambulatory monitors. *Computers in Cardioliology*, 467–470, 1998.

21. I. M. Rezazadeh, M. H. Moradi, and A. M. Nasrabadi. Implementing of spiht and sub band energy compression (sec) method on two-dimensional ecg compression: A novel approach. *Proceedings of the 2005 IEEE Engineering in Medicine and Biology, 27th Annual Conference*, Shanghai, China, 2005.

22. P. Rossi, A. Casaleggio, M. Chiappalone, M. Morando, G. Corbucci, M. Reggiani, G. Sartori, and S. Chierchia. Computationally inexpensive methods for intra-cardiacatrial bipolar electrogram compression. *Europace*, 4, 295–302, November 2002.

23. F. Sufi. Mobile phone programming java 2 micro edition. *Proceedings of the 2007 International Workshop on Mobile Computing Technologies for Pervasive Healthcare*, Melbourne, Australia, ISBN: 978-0-646-48230-9, pp. 64–80, December 2007.

24. F. Sufi, I. Cosic, and Q. Fang. Design and implementation of an intelligent agent system for genome data retrieval. *International Conference on Biomedical and Pharmaceutical Engineering, 2006 (ICBPE 2006)*, pp. 475–480, December 2006.

25. F. Sufi, I. Cosic, and Q. Fang. A mobile phone based remote patient assessment system. *International Journal of Cardiovascular Medicine*, 9(Suppl.), 8–9, 2008.

26. F. Sufi, Q. Fang, and I. Cosic. Ecg r-r peak detection on mobile phones. *Engineering in Medicine and Biology Society, 2007 (EMBS 2007), 29th Annual International Conference of the IEEE*, Lyon, France, pp. 3697–3700, August 2007.

27. F. Sufi, Q. Fang, I. Khalil, and S. S. Mahmoud. Novel methods of faster cardiovascular diagnosis in wireless telecardiology. *IEEE Journal on Selected Areas in Communications*, 27(4), 537–552, May 2009.

28. F. Sufi, Q. Fang, I. Cosic, and R. Ferguson. Client side decompression technique provides faster dna sequence data delivery. *27th Annual International Conference of the Engineering in Medicine and Biology Society, 2005. (IEEE-EMBS 2005)*, Shanghai, China, pp. 2817–2820, 2005.

29. F. Sufi and I. Khalil. Enforcing secured ecg transmission for realtime telemonitoring: A joint encoding, compression, encryption mechanism. *Security and Communication Networks*, 1(5), 389–405, 2008.

30. F. Sufi and I. Khalil. Diagnosis of cardiovascular abnormalities from compressed ecg: A data mining based approach. *Ninth International Conference on Information Technology and Applications in Biomedicine*, Larnaca, Cyprus, November 2009.

31. F. Sufi, Q. Fang, S. S. Mahmoud, and I. Cosic. A mobile phone based intelligent telemonitoring platform. *Third IEEE/EMBS International Summer School on Medical Devices and Biosensors, 2006 (ISSMDBS)*, Boston, MA, pp. 101–104, September 2006.

32. K. Urar and Y. Z. Ider. Development of compression algorithm suitable for exercise ecg data. *Proceedings of the 23rd Annual EMBS International Conference*, Istanbul. Turkey, 2001.

33. M. B. Velasco, F. C. Roldan, F. L. Ferreras, A. B. Santos, and D. M. Munoz. A low computational complexity algorithm for ecg signal compression. *Medical Engineering and Physics*, 26, 553–568, 2004.

34. J. J. Wei, C. J. Chang, N. K. Chou, and G. J. Jan. Ecg data compression using truncated singular value decomposition. *IEEE Transactions on Information Technology in Biomedicine*, 5(4), 290–299, 2001.

35. T. Zhang, S. Simske, and D. Blakley. Scalable ecg compression for long-term home health care. *Proceedings of the 2005 IEEE Engineering in Medicine and Biology, 27th Annual Conference*, Shanghai, China, 2005.

36. Y. Zigel, A. Cohen, and A. Katz. Ecg signal compression using analysis by synthesis coding. *IEEE Transactions on Biomedical Engineering*, 47(10), 1308–1316, 2000.

37. J. Ziv and A. Lempel. Compression of individual sequences via variable-rate coding. *IEEE Transactions on Information Theory*, 24(5), 530–536, 1978.

Chapter 21

Efficient Cardiovascular Diagnosis

Fahim Sufi and Ibrahim Khalil

Contents

21.1 Introduction

For wireless telecardiology, limited bandwidth is one of the major bottlenecks of faster diagnosis [1]. Since the cardiac cell damage is an irrecoverable process that may be initiated after a cardiac arrest, faster diagnosis and treatment can be a life saver [2,3]. Therefore, for cardiologists, the term "time is muscle" is of utmost importance. To reduce the delay in treatment and diagnosis, in Chapters 19 and 20

Figure 21.1   Cardiovascular diagnosis on mobile phone.

we focused on faster ECG transmission via mobile telephony network and faster (and secured) authentication mechanisms. In this chapter, the focus will be on faster diagnosis of cardiovascular abnormality. The key to faster diagnosis is to perform diagnosis directly from the compressed ECG as opposed to decompressing the compressed ECG and then perform diagnosis on the plain ECG. Recent research in wireless telecardiology shows that cardiovascular diagnosis from compressed ECG is faster than decompression followed by diagnosis [4–10]. Many of the recent telemonitoring platforms suggest using innovative compression technologies, so that the massive amount of ECG data (e.g., ECG from 12 lead) can be accomodated with today's bandwidth-constrained mobile Internet [9–12]. However, since ECG packets are kept in compressed format by these efficient telecardiology platforms, the packets need to be decompressed before using any existing diagnosis algorithms [13]. This extra step of decompression before performing diagnosis entails delay in diagnosis, which can be a killer. After having a cardiac arrest, the cardiac cell damage may start, which is an irrecoverable process. The dead cardiac cells will never revive. That is why every second counts, when a patient is having the symptom of a cardiac arrest [2,3].

Following this demand of faster cardiovascular diagnosis and treatment, a new breed of ECG diagnosis directly from the compressed ECG is being flourished. Cardiac abnormality detection directly from the compressed ECG is the main focus of this chapter (Figure 21.1).

21.2 ECG Diagnosis from Plain ECG

The diagnosis of cardiovascular diseases with ECG signal has been well researched and well established for the last few decades. However, almost all these algorithms were designed for the plain (uncompressed) ECG. However, within the last 5 years, research on mobile phone–based cardiovascular monitoring is progressing well, since cardiovascular diseases is the number one killer of modern era [1,9,11–14]. Existing ECG diagnosis algorithms can be classified into three different categories as follows:

1. Direct approach
2. Transformational approach
3. Intelligent approach

21.2.1 Direct Approach

Direct methods encompass diagnosis algorithms that are of relatively lower complexity. They tend to apply some simple mathematical operations on the original ECG to extract the feature waves. Amplitude-based, first derivative–based, and second derivative–based methods are some examples of direct methods to locate the QRS complex of ECG [13,15,16]. Because of lower complexity, they are suitable for mobile and embedded devices with low computational capacity. In [13], we have implemented few of the direct methods to locate QRS complex and calculate RR interval, as seen in Figures 21.2 and 21.3.

Following codes shows the .Net implementation (Figure 21.3) of amplitude-based, first derivative–based, and second derivative–based methods.

Figure 21.2 Implementation of direct methods for diagnosing ECG [1313] on J2ME-supported mobile phones. (From Sufi, F. et al., Ecg r-r peak detection on mobile phones, *29th Annual International Conference of the IEEE on Engineering in Medicine and Biology Society, 2007 (EMBS 2007)*, Lyon, France, pp. 3697–3700, August 2007. With permission.)

Figure 21.3   Implementation of direct methods for diagnosing ECG [1313] on .Net supported PDAs/smart phones. (From Sufi, F. et al., Ecg r-r peak detection on mobile phones, *29th Annual International Conference of the IEEE on Engineering in Medicine and Biology Society, 2007* (*EMBS 2007*), Lyon, France, pp. 3697–3700, August 2007. With permission.)

```
/*Amplitude based method*/

detected = False
         peakDetected = 0
         peak = Val(strECG)
         Do Until strECG Is Nothing
            strECG = fsReader.ReadLine
            temp = Val(strECG)
            If temp > 0.4 And detected = False Then
                If temp > peak Then
                    peak = temp
                    peakDetected = peakDetected + 1
                ElseIf peakDetected > 1 Then
                    strRR = strRR & (i/360) & ","
                    i = 0
                    peak = 0.4
                    detected = True
                    peakDetected = 0
                End If
            End If
            If i = 10 Then
                detected = False
            End If
            i = i + 1
     Loop
  ⋮
```

```
/*First Derivative based method*/

GetSlop(3, fsReader)
        Dim m As Int16
        Do Until strECG Is Nothing
            If Val(ar(0)) > 0.1375 And Val(ar(1)) > 0.1375
And Val(ar(2)) > 0.1375 Then
                    GetSlop(12, fsReader)
                    For m = 0 To 11
                        If Val(ar(m)) < -0.2 Then
                            If Val(ar(m + 1)) < -0.2 Or
Val(ar(m + 2)) < -0.2 Then
                                    count = MaxIndex(ecg, m - 1)
                                    strRR = strRR &
      CStr((i + count 13)/360) & ","
                                    i = 12 - count
                                    Exit For
                            End If
                        End If
                    Next
            Else
                GetSlop(3, fsReader)
            End If
        Loop
:
/*Second Derivative based method*/

Dim fd(5) As String
        For i = 1 To 5
            fd(i - 1) = vECG(i + 1) vECG(i - 1)
        Next
        strECG = vECG(6)
        Dim sd As String
        Dim detection As Double
        count = 7
        sd = fd(4) - 2 * fd(2) + fd(0)
        Dim cnt As Int16
        cnt = 0
        Do Until strECG Is Nothing
            count = count + 1
            strECG = fsReader.ReadLine
            For i = 1 To 6
                vECG(i - 1) = vECG(i)
            Next
            vECG(6) = strECG
            For i = 1 To 4
                fd(i - 1) = fd(i)
            Next
```

```
        fd(4) = vECG(6) - vECG(4)
        sd = fd(4) - 2 * fd(2) + fd(0)
        detection = 1.3 * System.Math.Abs(CDbl(fd(0))) +
1.1 * System.Math.Abs(CDbl(sd))
            If (detection >= 0.9) And count > 9 Then
                cnt = count
                strRR = strRR & ((count - 6 + MaxIndex(vECG, 6))
    / 360) & ","
                count = 0
            End If
        Loop
```

⋮

21.2.2 Transformational Approach

These methods include application of transformations such as wavelet transform, Fourier transform, cosine transforms etc. [17,18]. They require higher computational power compared to most of the direct methods. Often, these methods transform the ECG from time domain to some other domains like the frequency domain.

21.2.3 Intelligent Approach

These methods employ various intelligent techniques like neural network (NN), clustering, support vector machine, attribute selection, Hermite polynomial, etc. [19]. They consume the highest level of processing power, and are almost unsuitable for processing by the mobile or embedded devices.

21.3 ECG Diagnosis from Compressed ECG

With the advent of modern mobile phone–based telecardiology applications, the ECG messages are being transmitted in compressed format to suit the limited bandwidth of mobile telephone network [1,9,12,20]. The ability to perform diagnosis straight from the compressed ECG packets has been proven to be faster than applying existing ECG diagnosis on the compressed ECG, after decompression [4,9,10]. Existing research suggests that ECG diagnosis from the compressed ECG can be performed by the following three ways:

1. Instant detection approach
2. Direct approach
3. Intelligent approach

21.3.1 Instant Detection Approach

This technique provides the likelihood of a particular compressed ECG packet containing abnormality. The heart rate estimation method, described in [9], is an example of instant detection technique based on compressed ECG. In [9], through experimentation, a clear relationship between the number of beats and compression ratio was drawn. It was found that an ECG containing more heart beats will have lower compression ratio when compared with an ECG packet with less heart beat, under the following two constraints:

1. If the person is same
2. If the numbers of ECG samples are the same for the packets

The heart rate estimation algorithm does not require reading the compressed ECG. It just looks the payload size within a particular ECG packet (containing compressed ECG). Therefore, these algorithms provides the fastest diagnosis (since there is no decompression or even reading of compressed character involved) decision at the cost of accuracy. Therefore, they are suitable for preliminary diagnosis (not for final diagnosis).

21.3.2 Direct Approach

Direct methods check the frequency of some particular compressed characters (or encoded characters) and apply a threshold for diagnosis decision. As an example, in [9], the beat detection algorithm counts the frequency of high-pitched *V* characters, and when this count is more than a threshold value (e.g., 5 or 6), a peak is determined within a set window. Figure 21.4 shows how the *V* character set is embedded within the compressed ECG. When implemented on regular mobile phone, this algorithm executed efficiently on J2ME platform, as seen in Figure 21.5.

Following is the J2ME code for the algorithm.

```
/*
 * HRV_CmprsECG.java
 *
 * Created on 27 July 2007, 14:53
 */

import java.io.InputStream;
import java.io.IOException;
import java.io.InputStreamReader;
import javax.microedition.midlet.*;
import javax.microedition.lcdui.*;
import java.util.Date;

/**
 *
 * @author Researcher
 * @version
 */
```

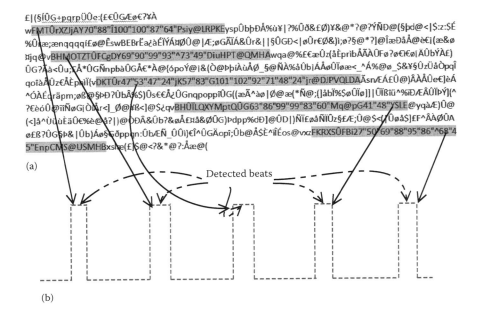

Figure 21.4 The composition of *V* character set (denotes the high-pitched values) and *U* character set (denotes the low-pitched values) within compressed ECG [9]. The *V* characters (that denote the beats) are highlighted in the figure. (a) Compressed ECG stream. (b) Beats detected with Algorithm described in [9]. (Adapted from Sufi, F. et al. *IEEE J. Select. Area. Commun.*, 27(4), 537, 2009.)

```
public class HRV_CmprsECG extends MIDlet
implements CommandListener{
    InputStream isECG;
    StringBuffer sb;
    InputStreamReader isrECG;
    Command calculate, ext;
    Date dt1, dt2;
    long tm1, tm2, tmSpan;

    Form frm;
    Display disp;
    short cntSlope, clrCntSlope, intHRV, intPause;
    boolean pause;
    public HRV_CmprsECG(){
        sb= new StringBuffer();
        frm= new Form("HRV Calculated from compressed ECG");
        calculate= new Command("Calculate", Command.OK, 1);
        ext= new Command("Exit", Command.EXIT, 2);
```

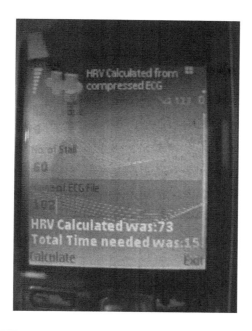

Figure 21.5 Implementation of direct method algorithm to detect heart rate from compressed ECG. (From Sufi, F. et al. *IEEE J. Select. Area. Commun.*, 27(4), 537, May 2009. With permission.)

```
        frm.addCommand(calculate);
        frm.addCommand(ext);
        frm.setCommandListener(this);
        pause=false;
        cntSlope=0;
        clrCntSlope=0;
        intPause=0;
        intHRV= 0;

        isECG=getClass().getResourceAsStream("com105.txt");
        try{
            isrECG=new InputStreamReader(isECG, "UTF-8");
        } catch(IOException ioe){
        }
        //sb=new StringBuffer();

        disp=Display.getDisplay(this);
        disp.setCurrent(frm);
    }
```

```
public void startApp() {
}

public void pauseApp() {
}

public void destroyApp(boolean unconditional) {
}

public void commandAction(Command c, Displayable d){
    if(c==ext){
        notifyDestroyed();
    }
    if (c==calculate){
            dt1= new Date();
    dt2= new Date();
    tm1=dt1.getTime();
    int next;
    try {
        while((next=isrECG.read())!=-1){
                char ch= (char)next;

                if (pause!=true){
                    if (Character.isDigit(ch)||
((ch>='A')&&(ch<='Z')) ||(ch=='"')){
                        cntSlope++;
                    }
                    else{
                        clrCntSlope++;
                        if (clrCntSlope>=3 && cntSlope>5){
                            cntSlope=0;
                            intHRV++;
                            clrCntSlope=0;
                            pause=true;
                        }
                    }
                }
                else {
                    intPause++;
                    if(intPause>51){
                        pause=false;
                    }
                }
            }
        }
    }
```

```
            catch(IOException ioe){

            }

            tm2=dt2.getTime();
            tmSpan=tm2-tm1;

            frm.append("HRV Calculated was: ");
            frm.append(String.valueOf(intHRV));
            frm.append("Total Time needed was: ");
            frm.append(String.valueOf(tmSpan));
            //frm.append(sb.toString());
            }
        }
}
```

Apart from heart rate detection, there were other algorithms in [9] such as wide QRS detection algorithm, RR detection algorithm, falling under direct method approach of compressed ECG.

21.3.3 Intelligent Approach

Intelligent methods involve usage of subset selection, clustering (e.g., K-Means, Expectation Maximization etc.), principal component analysis (PCA), nueral network etc., on compressed ECG [4,10].

In [10], the frequency of the different characters used for encoding (compressing) the ECG was used first. Since there were about 158 characters (including the numbers), using all these characters as attributes for clustering will result in misclassification. Therefore, an attribute selection was performed before clustering. The attribute selection mechanism identifies few key attributes (or encoding character) responsible for distinguishing a normal ECG segment (compressed) from an abnormal ECG segment (compressed). Expectation maximization (EM) is our favorite (over K-means clustering) for identifying abnormal ECG packets in compressed form. However, K-means algorithms are easier to be implemented on mobile platform.

In [4], PCA was used first to identify the principal components from the 158 frequency counts for all the encoded characters (i.e., characters used for compressing/encoding the ECG [9,11]). Then, using the few principal components, a linear NN was trained. When the trained NN was tested with a test set, the trained system could correctly identify abnormality directly from compressed ECG, without decompressing them. Implementing this system on mobile and embedded devices requires proper optimization of coding since both PCA and NN are computational intensive.

Finally, Figure 21.6 shows the suitability of different detection methods on mobile platform, described in this chapter.

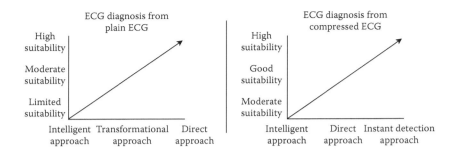

Figure 21.6   Suitability of different algorithms on mobile platform.

21.4 Conclusion

In this chapter, the focus was on mobile phone–based faster cardiovascular diagnosis. To execute the diagnosis algorithms efficiently on mobile platform, the algorithms need to be of lower complexity. Also, faster diagnosis is possible if the ECG file (from where diagnosis is to be performed) lengths are smaller by harnessing specialized ECG compression technology (as compared to larger plain ECG) [9,11]. Since a lossless compression (such as the one in [9,11]) transforms a larger file into a smaller one, if diagnosis is performed from the smaller compressed file (containing same information in different form), there will less file reading I/O operations.

Therefore, there has been recent research performed on cardiovascular diagnosis from compressed ECG ([4,9,10]) for wireless telecardiology. These researches demonstrate accurate diagnosis from ECG (compressed) more efficiently than conventional approaches of mobile phone–based diagnosis (such as the implementation demonstrated in [21,13]).

Appendix 21.A

Complete code for running amplitude-based, first derivative–based, and second derivative–based beat detection algorithms.

```
/*
 * RRCompare.java
 *
 * Created on 4 January 2007, 16:44
 * Completed on 27th February 2007, 1:00 am by Fahim Kamal Sufi
 */

import java.util.Date;
import javax.microedition.midlet.*;
```

```java
import javax.microedition.lcdui.*;
import javax.microedition.io.*;
import java.io.*;

/**
 *
 * @author Researcher
 * @version
 */

public class RRCompare extends MIDlet
implements CommandListener {
    Command m1;
    Command m2;
    Command m3;
    Command ext;
    Command selComm;
    int ecgId;
    List selEcg;
    Date dte1, dte2;
    long tme1, tme2, tmSpan;
    Form frm;
    float ecg[];
    float fd[];
    float sd;
    float tEcg;
    double rrVal;
    int mxID, i, intHR;
    double detection;
    boolean endOfEcg;
    InputStream dis;
    StringBuffer ecgFile;
    public RRCompare(){
        ecgId=0;
        ecgFile=new StringBuffer();
        m1=new Command("Amplitude", Command.SCREEN, 1);
        m2=new Command("First Derivative", Command.SCREEN, 2);
        m3=new Command("Second Derivative", Command.SCREEN, 3);
        ext=new Command("Exit", Command.EXIT, 1);
        String elements[]={"100", "102", "105", "111", "114",
"117", "201", "207", "210", "213", "219", "222", "228", "231",
"234", "901", "902", "903", "904","905","906","907","908","909",
"910"};
        selEcg=new List("Select the ECG File", List.IMPLICIT,
        elements, null);
        selComm= new Command("Select ECG", Command.ITEM, 1);
        selEcg.setSelectCommand(selComm);
        selEcg.setCommandListener(this);
```

```
        frm= new Form("RR Comparison");
        frm.addCommand(m1);
        frm.addCommand(m2);
        frm.addCommand(m3);
        frm.addCommand(ext);
        frm.setCommandListener(this);
        Display.getDisplay(this).setCurrent(selEcg);
        intHR = 0;
}
public void startApp() {
}

public void pauseApp() {
}

public void destroyApp(boolean unconditional) {
        notifyDestroyed();
}

public void commandAction(Command cmd, Displayable dp){
        if (cmd==ext){
            destroyApp(true);
        }

        if (cmd==m3){
            //Second Derivative Based
            StringBuffer sb;
            StringBuffer rr;
            int next, cnt, count;
            sb=new StringBuffer(6);
            rr=new StringBuffer();
            ecg=new float[8];
            fd=new float[6];
            dte1= new Date();
            dte2=new Date();
            tme1=dte1.getTime();
            try{
            dis=getClass().getResourceAsStream(ecgFile.
              toString());
            cnt=0;
            while((next=dis.read())!=-1 && cnt<=6){
                char c= (char)next;
                sb.append(c);
                if(next=='\n'){
                    ecg[cnt]=Float.parseFloat(sb.toString());
                    sb.setLength(0);

                    if (cnt>=2){
                        fd[cnt-2]= ecg[cnt]-ecg[cnt-2];
                    }
```

```
                if (cnt==6){
                    sd=fd[4]-2*fd[2]+fd[0];
                }
                cnt++;
            }
        }

        //Appending the previous character (often '-')
//that was pulled out during checking last while condition
        char c=(char) next;
        sb.append(c);
        count=7;

        while((next=dis.read())!=-1){
            c= (char)next;
            sb.append(c);
            if(next=='\n'){
                tEcg=Float.parseFloat(sb.toString());
                sb.setLength(0);
                count++;
                int i;
                for(i=1; i<=6; i++){
                    ecg[i-1]=ecg[i];
                }
                ecg[6]=tEcg;
                for(i=1;i<=4;i++){
                    fd[i-1]=fd[i];
                }
                fd[4]=ecg[6]-ecg[4];
                sd=fd[4]-2*fd[2]+fd[0];
                detection=1.3*
                java.lang.Math.abs(fd[0])+1.1*java.lang.
                  Math.abs(sd);

                if ((detection>=0.9) && count>9){
                    int m, id;
                    id=0;
                    float maxVal;
                    maxVal=ecg[0];
                    for(m=1;m<7;m++){
                        if(maxVal<ecg[m]){
                            maxVal=ecg[m];
                            id=m;
                        }
                    }
                    intHR++;
                    rrVal=(count-6+id);
                    rrVal=rrVal/360;
                    rr.append(rrVal);
```

```
                        count=0;
                        //rr.append(',');
                    }
                }
            }
        }
        catch(IOException ioe){
        }
        tme2=dte2.getTime();
        tmSpan=tme2-tme1;
        frm.append(Integer.toString(intHR));
        intHR=0;
        frm.append("Time needed for second derivative
          based RR detection");
        frm.append(Long.toString(tmSpan));
        //frm.append(rr.toString());
    }
    if (cmd==selComm){
        switch (selEcg.getSelectedIndex()){
            case 0:
                ecgFile.append("/100fx.txt");
                break;
            case 1:
                ecgFile.append("/102fx.txt");
                break;
            case 2:
                ecgFile.append("/105fx.txt");
                break;
            case 3:
                ecgFile.append("/111fx.txt");
                break;
            case 4:
                ecgFile.append("/114fx.txt");
                break;
            case 5:
                ecgFile.append("/117fx.txt");
                break;
            case 6:
                ecgFile.append("/201fx.txt");
                break;
            case 7:
                ecgFile.append("/207fx.txt");
                break;
            case 8:
                ecgFile.append("/210fx.txt");
                break;
            case 9:
                ecgFile.append("/213fx.txt");
                break;
```

```
      case 10:
         ecgFile.append("/219fx.txt");
         break;
      case 11:
         ecgFile.append("/222fx.txt");
         break;
      case 12:
         ecgFile.append("/228fx.txt");
         break;
      case 13:
         ecgFile.append("/231fx.txt");
         break;
      case 14:
         ecgFile.append("/234fx.txt");
         break;
      case 15:
         ecgFile.append("/901fx.txt");
         break;
      case 16:
         ecgFile.append("/902fx.txt");
         break;
      case 17:
         ecgFile.append("/903fx.txt");
         break;
      case 18:
         ecgFile.append("/904fx.txt");
         break;
      case 19:
         ecgFile.append("/905fx.txt");
         break;
      case 20:
         ecgFile.append("/906fx.txt");
         break;
      case 21:
         ecgFile.append("/907fx.txt");
         break;
      case 22:
         ecgFile.append("/908fx.txt");
         break;
      case 23:
         ecgFile.append("/909fx.txt");
         break;
      case 24:
         ecgFile.append("/910fx.txt");
         break;
   }
      frm.append("RR detection for ");
      frm.append(ecgFile.toString());
```

```
            Display.getDisplay(this).setCurrent(frm);
}
if (cmd==m1){
    //Amplitude based detection
    StringBuffer sb;
    StringBuffer rr;
    float rrVal;
    int next, cnt, count;
    sb=new StringBuffer(6);
    rr=new StringBuffer();
    ecg=new float[8];
    fd=new float[6];
    dte1= new Date();
    dte2=new Date();
    tme1=dte1.getTime();
    double peak,temp;
    boolean detected;
    int peakDetected;

    detected= false;
    peakDetected=0;
    peak=0;

    try{
     dis=getClass().getResourceAsStream(ecgFile.
       toString());
     cnt=0;
     while((next=dis.read())!=-1){
         char c= (char)next;
         sb.append(c);
         if(next=='\n'){
                 if (cnt>=1){
                 temp=Float.parseFloat(sb.toString());
                 sb.setLength(0);
                 //for MIT_BIH it is more that 0.4
                 //for BioPac it is 1.2
                     if ((temp>0.40)&&(detected==false)){
                         if (temp>peak){
                             peak=temp;
                             peakDetected++;
                         }
                         else if (peakDetected>1){
                             intHR++;
                             rrVal=(cnt-1);
                             rrVal=rrVal/360;
                             rr.append(rrVal);
                             //rr.append(',');
                             cnt=1;
                             peak=0.40;
```

```
                              detected=true;
                              peakDetected=0;
                    }
            }
            if(cnt==10){
                detected=false;
            }
        }

        else{

        // For the first ECG value
        peak=Float.parseFloat(sb.toString());
        sb.setLength(0);
        cnt++;
        }
        cnt++;

        }
      }
    }

    catch(IOException ioe){
    }
    tme2=dte2.getTime();
    tmSpan=tme2-tme1;
    frm.append(Integer.toString(intHR));
            intHR=0;
    frm.append("Time needed for amplitude based RR
      detection");
    frm.append(Long.toString(tmSpan));
    //frm.append(rr.toString());
}
// First Derivative based approach
if (cmd==m2){
    StringBuffer sb;
    StringBuffer rr;
    float rrVal,ecgVal;

    int next, cnt, m;
    sb=new StringBuffer(6);
    rr=new StringBuffer();
    endOfEcg=false;
    ecg=new float[13];
    fd=new float[13];
    dte1= new Date();
    dte2=new Date();
    tme1=dte1.getTime();
    i=0;
```

```
            dis=getClass().getResourceAsStream(ecgFile.
              toString());
            ecg[0]=getEcgSample();
            //rr.append(ecg[0]);
            ecg[1]=getEcgSample();
            //rr.append(ecg[1]);
            frm.append(rr.toString());
            getSlop(3);
            while(endOfEcg!=true){
                //for MIT BIH onset fd is 0.1375 and
//offset is -0.2
                    if((fd[0]>0.05)&&(fd[1]>0.05)&&(fd[2]>0.05)){
                        getSlop(12);
                        //m was 11 before which can fire an
//error of array index out of bound, since
                        //m=10
                        for(m=0; m<=11; m++){
                            if(fd[m]<-0.1){
                                if((fd[m+1]<-0.)|| (fd[m+2]
                                    <-0.1)){
                                    cnt= MaxIndex(ecg,(m-1));
                                    rrVal=i+cnt-13;
                                    rrVal=rrVal/360;
                                    intHR++;
                                    System.out.println(intHR);
                                    rr.append(rrVal);
                                    i=12-cnt;
                                    getSlop(3);
                                    break;
                                }
                            }
                        }
                    }
                    else {
                        getSlop(3);
                    }
            }
            tme2=dte2.getTime();
            tmSpan=tme2-tme1;
            frm.append(Integer.toString(intHR));
            intHR=0;
            frm.append("Time needed for first derivative
              based RR detection");
            frm.append(Long.toString(tmSpan));
            //frm.append(rr.toString());
        }
    }
```

```java
public float getEcgSample(){
    int next;
    float ecgVal;
    ecgVal=0;
    StringBuffer sb;
    sb=new StringBuffer();
    try {
        while((next=dis.read())!=-1){
            char c= (char)next;
            sb.append(c);
            if(next=='\n'){
                ecgVal=Float.parseFloat(sb.toString());
                sb.setLength(0);
                break;
            }
        }

        if (next==-1){
            endOfEcg=true;
        }
    }
    catch(IOException ioe){
    }
    ecgId++;
    return ecgVal;
}

public int MaxIndex(float ec[], int idx){
    float mxVal;
    int m, id;
    id=0;
    mxVal = ec[0];
    for (m=1;m<=idx+1; m++){
        if (mxVal<ec[m]){
            mxVal=ec[m];
            id=m;
        }
    }
    return id;
}
// Following method calculates the first derivative
//and the parameter ind
// takes the number of ecg sample from which the first
//derivative (fd) is calculated.
public void getSlop(int ind){
    int m;
    if (ind==12){
```

```
ecg[0]=getEcgSample();
ecg[1]=getEcgSample();
for(m=2; m<=ind; m++){
        ecg[m]=getEcgSample();
        fd[m-2]=ecg[m]-ecg[m-2];
    }
    i+=13;
}
else{
    //for(m=0; m<=(ind-1); m++){
    // Following 2 lines added 29th November, 2007
        fd[0]=fd[1];
        fd[1]=fd[2];
    // Addition ends
        ecg[2]=getEcgSample();
        fd[2]=ecg[2]-ecg[0];
        ecg[0]=ecg[1];
        ecg[1]=ecg[2];
        i++;
    //}
}
}
}
}
```

References

1. F. Sufi, I. Khalil, Q. Fang, and I. Cosic. A mobile web grid based physiological signal monitoring system. *International Conference on Technology and Applications in Biomedicine, 2008 (ITAB 2008)*, Shenzhen, China, pp. 252–255, May 2008.

2. E.H. Bradley et al. Achieving rapid door-to-balloon times: how top hospitals improve complex clinical systems. *Circulation*, 113(8):1079–1085, Mar. 2006.

3. G. De Luca, H. Suryapranata, J.P. Ottervanger, and E.M. Antman. Time delay to treatment and mortality in primary angioplasty for acute myocardial infarction: Every minute of delay counts. *Circulation*, 109:1223–1225, 2004.

4. A. Ayman, I. Khalil, and F. Sufi. Cardiac abnormalities detection from compressed ecg in wireless telemonitoring using principal components analysis (pca). *International Conference on Intelligent Sensors, Sensor Networks and Information Processing, 2009 (ISSNIP 2009)*, Melbourne, Australia, December 2009.

5. K. Hung and Y.T. Zhang. Implementation of a wap-based telemedicine system for patient monitoring. *IEEE Transactions on Information Technology in Biomedicine*, 7(2):101–107, June 2003.

6. Y. Jasemian and L. Arendt-Nielsen. Evaluation of a realtime, remote monitoring telemedicine system using the bluetooth protocol and a mobile phone network. *Journal of Telemedicine and Telecare*, 11(5):256–260, 2005.

7. I. Khalil and F. Sufi. Mobile device assisted remote heart monitoring and tachycardia predictiovn. *International Conference on Technology and Applications in Biomedicine, 2008 (ITAB 2008)*, Shenzhen, China, pp. 484–487, May 2008.

8. R.-G. Lee, K.-C. Chen, C.-C. Hsiao, and C.-L. Tseng. A mobile care system with alert mechanism. *IEEE Transactions on Information Technology in Biomedicine*, 11(5):507–517, September 2007.

9. F. Sufi, Q. Fang, I. Khalil, and S.S. Mahmoud. Novel methods of faster cardiovascular diagnosis in wireless telecardiology. *IEEE Journal on Selected Areas in Communications*, 27(4):537–552, May 2009.

10. F. Sufi and I. Khalil. Diagnosis of cardiovascular abnormalities from compressed ecg: A data mining based approach. *Ninth International Conference on Information Technology and* Applications *in Biomedicine*, Larnaca, Cyprus, November 2009.

11. F. Sufi and I. Khalil. Enforcing secured ecg transmission for realtime telemonitoring: A joint encoding, compression, encryption mechanism. *Security and Communication Networks*, 1(5):389–405, 2008.

12. F. Sufi, Q. Fang, S.S. Mahmoud, and I. Cosic. A mobile phone based intelligent telemonitoring platform. *Medical Devices and Biosensors, 2006. Third IEEE/EMBS International Summer School on ISSMDBS*, Boston, MA, pp. 101–104, September 2006.

13. F. Sufi, Q. Fang, and I. Cosic. Ecg r-r peak detection on mobile phones. *29th Annual International Conference of the IEEE on Engineering in Medicine and Biology Society, 2007 (EMBS 2007)*, Lyon, France, pp. 3697–3700, August 2007.

14. F. Sufi, I. Cosic, and Q. Fang. A mobile phone based remote patient assessment system. *International Journal of Cardiovascular Medicine*, 9(Suppl.):8–9, 2008.

15. G.M. Friesen, T.C. Jannett, M.A. Jadallah, S.L. Yates, S.R. Quint, and H.T. Nagle. A comparison of the noise sensitivity of nine qrs detection algorithms. *IEEE Transactions on Biomedical Engineering*, 37(1):85–98, January 1990.

16. P.S. Hamilton and W.J. Tompkins. Quantitative investigation of qrs detection rules using the mit /bih arrhythmia database. *IEEE Transactions on Biomedical Engineering*, 33(12):1157–1165, December 1986.

17. Z. Dokur, T. Olmez, and E. Yazgan. Comparison of discrete wavelet and fourier transforms for *ecg* beat classification. *Electronics Letters*, 35(18):1502–1504, September 1999.

18. D. Lemire, C. Pharand, J. Rajaonah, B. Dube, and A.R. LeBlanc. Wavelet time entropy, t wave morphology and myocardial ischemia. *IEEE Transactions on Biomedical Engineering*, 47(7):967–970, July 2000.

19. L. Gang, Y. Wenyu, L. Ling, Y. Qilian, and Y. Xuemin. An artificial-intelligence approach to ecg analysis. *Engineering in Medicine and Biology Magazine, IEEE*, 19(2):95–100, March–April 2000.

20. F. Sufi, I. Khalil, and I. Habib. Polynomial distance measurement for ecg based biometric authentication. *Security and Communication Network*, Published Online on December 3, 2008, DOI: 10.1002/sec.76, 2008.

21. F. Sufi. Mobile phone programming java 2 micro edition. *Proceedings of the 2007* International *Workshop on Mobile Computing Technologies for Pervasive Healthcare*, Melbourne, Australia, ISBN: 978-0-646-48230-9, pp. 64–80, December 2007.

Index